Springs and Bottled Waters of the World

Springer

Berlin
Heidelberg
New York
Barcelona
Hong Kong
London
Milan
Paris
Singapore
Tokyo

Philip E. LaMoreaux · Judy T. Tanner (Eds.)

Springs and Bottled Waters of the World

Ancient History, Source, Occurence, Quality and Use

with 209 Figures and 53 Tables

Editors

Dr. Philip E. LaMoreaux

P.E. LaMoreaux & Associates, Inc.
P.O. Box 2310
Tuscaloosa, AL 35403
USA
E-mail: pela@btech.net

Judy T. Tanner

2610 University Boulevard
Tuscaloosa, AL 35401
USA

ISBN 3-540-61841-4 Springer-Verlag Berlin Heidelberg New York

Library of Congress Cataloging-in-Publication Data

Springs and bottled water of the world : ancient history, source, occurence, quality and use / Philip E. LaMoreaux, Judy T. Tanner (eds.).
p. cm.
Includes bibliographical references (p.).
ISBN 3540618414
1. Springs. 2. Health resorts. 3. Bottled water industry. I. LaMoreaux, Philip E. (Philip Elmer), 1920- II. Tanner, Judy T., 1954-

GB1198 .S67 2001
553.7--dc21

2001042003

Springer-Verlag Berlin Heidelberg New York
a member of the BertelsmannSpringer Science+Business Media GmbH

Printed in Germany

Cover Design: *design & production*
Dataconversion: Büro Stasch · Bayreuth

SPIN: 10528238 32/3130 – 5 4 3 2 1 0 – Printed on acid-free paper

Dedication

William James Powell

(June 30, 1921 – September 1, 1995)

"Springs and Bottled Waters" is dedicated to the memory of William J. Powell, who helped make this book possible. Mr. Powell, with his jovial and pleasing personality, was dedicated to the service and enhancement of his professional associates and friends.

Preface

This book, describing select springs and bottled waters around the world, provides a broad spectrum of information on springs, as they have provided man with a source of water and have materially affected the evolution and development of civilization. Springs have played a prominent role in history, agriculture, military campaigns, religion, and science. Dr. O. E. Meinzer, the father of hydrogeology, once made the statement that certain parts of the Bible read like a water supply paper. To prove his point, he cited the stories of Rebecca at the well, and Moses "smiting" the rock and "water gushed forth". Perhaps the most famous religious springs would be Ayun Musa near Suez in northwest Sinai, or the springs at Kadesh Barnea in northeast-central Sinai, both associated with springs of the Exodus story in the Bible and the Koran.

In the preparation of this book data was obtained from major hydrogeologic data bases of the principal water resources investigatory groups around the world, for example, the US Geological Survey, the State Geological Surveys in the USA, and from contacts with leading hydrogeologists with knowledge about springs. References were also obtained from publications of scientific societies such as the American Geological Institute, American Geophysical Union, Geological Society of America, American Water Resources Association, American Institute of Hydrology, International Association of Hydrogeologists, Geological Surveys of Ireland, Sweden, Great Britain, France, China, Japan, and many others. A select group of scientists voluntarily wrote about famous springs that they know, providing a wide variety of site-specific geological, geographical, historical, and regulatory information. Their help is gratefully acknowledged.

Walt Schmidt, *USA*
J. B. W. Day, *UK*
R. T. Sniegocki †, *USA*
Uri Kafri, *Israel*
Arie S. Issar, *Israel*
C. R. Aldwell, *Ireland*
David Drew, *Ireland*
Hussein Idris, *Egypt*
Petar Milanović, *Yugoslavia*
Stefan Wohnlich, *Germany*
John Gunn, *UK*
J. R. Vegter, *South Africa*
Soki Yamamoto, *Japan*
Jaroslav Vrba, *Czech Republic*
Paolo Bono, *Italy*

Carlo Boni, *Italy*
Heinz Hötzl, *Germany*
I. Povara, *Romania*
Kara-kys D. Arakchaa, *Russia*
Árpád Lorberer, *Hungary*
Ren YuanPei, Zhang Yun Zheng, *Tibet*
Ladislav Melioris, *Slovak Republic*
Josef G. Zötl, *Austria*
Bernard Blavoux, Jacques Mudry, Jean-Michel Puig, *France*
M. Maggiore, F. Santaloia, F. Vurro, *Italy*
G. A. Kellaway, *UK*
A. Afrasiabian, *Iran*
Eliyahu Rosenthal, *Israel*
Vello Karise, Peeter Vingisaar, *Estonia*
Yuan Daoxian, *China*
Gunter Dörhöfer, Gottfried Goldberg, *Germany*
Anthony W. Creech, R. D. Dowdy Jr., *USA*
RNDr. Ondrej Franko, *Slovak Republic*
Shane O'Neill, *Ireland*

Contents

Contributors

A. Afrasiabian
(Dr., Director)
National Applied Study and Research Center, Water Resources Research Organization
PO Box 15875-3584, Tehran, I.R. Iran, Phone: 98-21-7520474, Fax: 98-21-7533186

C. R. Aldwell
(Dr., Principal Geologist)
Geological Survey of Ireland, Beggars Bush, Haddington Rd., Dublin 4, Ireland
Phone: +35-31-6715233, Fax: +35-31-6681782

Caryl Alfaro
(Geologist)
Ogden Environmental, 2904 West Corp Boulevard, Suite 204, Huntsville, AL 35805, USA
Phone: 205-539-3016, Fax: 205-539-3074, E-mail: clalfaro@oees.com

Kara-kys D. Arakchaa
(Dr.)
Russia

Bernard Blavoux
Hydrogeology Laboratory, Faculty of Sciences
33 Louis Pasteur Street, F-84000 Avignon, France

Carlo Boni
(Dr.)
Institute de Geologia, Citta Universitaria, Largo Forano 18, 00199 Rome, Italy

Paolo Bono
(Dr.)
Professore Associato Di Idrogeologia, Dipartimento Di Scienze Delia Terra
Universita Degli Studi "La Sapienza", P.le A. Moro 5, 00185 Roma, Italy

Karen L. Bryan
(Vice President)
Oil and Gas and Legal Services, P. E. LaMoreaux & Associates, Inc.
PO Box 2310, Tuscaloosa, AL 35403, USA
Phone: 205-752-5543, Fax: 205-752-4043, E-mail: pela@dbtech.net

Anthony W. Creech
(P.G., Senior Environmental Scientist)
Resource International, Ltd, 9560 Kings Charter Drive, PO Box 6160, Ashland, VA 23005-6160, USA
Phone: 804-550-9200, Fax: 804-550-9259

Yuan Daoxian
(Dr.)
Institute of Karst Geology
40 Qixing Road, Guilin, Guangxi 541004, Peoples Republic of China
Fax: 4-86-733-5813708, E-mail: yuandx@sun.ihep.ac.cn

J. B. W. Day
(Dr.)
"Oakwood", Dippenhall, Farnham, Surrey, GU105EP United Kingdom
Phone: 0252-850360, Fax: 0252-851592

Gunter Dörhöfer
(Dr.)
Geological Survey of Lower Saxony
PO Box 510153, Stilleweg 2, D-30655 Hannover, Germany
E-mail: gunter.doerhoefer@bgr.de

R. D. Dowdy, Jr.
Camp Holly Springs, 4100 Diamond Springs Drive, Richmond, VA 23231, USA
Phone: 804-795-2096

David Drew
(Dr.)
Department of Geography, Trinity College
Dublin 2, Ireland
Phone: 353-1-772941, Fax: 353-1-772694, E-mail: ddrew@sun1.tcd.ie

Ondrej Franko
(RNDr., DrSc.)
Milynske Nivy 42, 821 09 Bratislava, Slovakia, Phone: 07-5211109, Fax: 07-5211109

Lois D. George
(Vice President, Senior Hydrogeologist)
Environment and Ecology, P. E. LaMoreaux and Associates, Inc.,
PO Box 2310, Tuscaloosa, AL 35403, USA
Phone: 205-752-5543, Fax: 205-752-4043, E-mail: pela@dbtech.net

Gottfried Goldberg
Geological Survey of Lower Saxony
PO Box 510153, Stilleweg 2, D-30655 Hannover, Germany

John Gunn
(Prof.)
Geographic & Environmental Scs. Dept., University Queensgate
Huddersfield HD1 3DH, United Kingdom, E-mail: j.gunn@hud.ac.uk

Heinz Hötzl
University of Karlsruhe, Geologisches Institut
Kaiserstraße 12, D-76131 Karlsruhe, Germany
Phone: 0149-721-685654, E-mail: heinz.hoetzl@bio-geo.uni-karlsruhe.de

Hussein Idris
(Col.)
17 Bahgat Aly St. (Apt. 18), Zamalek – Cairo, Egypt, Phone: (202)3403170, Fax: (202)3411752

Arie S. Issar
(Prof., Visiting Professor)
Departemento de Geodinamica, Facultad de Ciencias Geologicas, Universidad Complutense
28040 Madrid, Spain, Fax: 34-1-3944845, E-mail: issar@bgumail.bgu.ac.il

Uri Kafri
(Dr., Director)
Geological Survey of Israel, 30 Malkhei Yisrael St., Jerusalem 95501, Israel

Vello Karise
Geological Survey of Estonia
Kadaka tee 80/82, EE0026 Tallinn, Estonia, Phone: (372)6579661, Fax: (372)6579664

G. A. Kellaway
(Dr.)
14 Cramedown, Lewes, East Sussex, BNF3NA, United Kingdom

Árpád Lorberer
(Dr.)
Institute of Hydrology of the VITUKI Pic., H-1095 Budapest, Kvassay J. út 1. Hungary

M. Maggiore
Department of Geology and Geophysics, University of Bari, Italy

Ladislav Melioris
(Prof., RNDr., DrSc.)
Department of Hydrogeology, Faculty of Natural Sciences, Comenius University, Bratislava
Slovak Republic, Mlynská dolina, Pov. G, 842 15 Bratislava, Czechoslovaia

Bashir A. Memon
(Dr., Executive Vice President, Senior Hydrogeologist)
P. E. LaMoreaux and Associates, Inc., PO Box 2310, Tuscaloosa, AL 35403
Phone: 205-752-5543, Fax: 205-752-4043, E-mail: pela@dbtech.net

Petar Milanović
(Dr.)
Strumicka 19, 11000 Belgrade, Yugoslavia, Phone: (011)444-9109

Jacques Mudry
Hydrogeology Laboratory, Faculty of Sciences
33 Louis Pasteur Street, F-84000 Avignon, France

Shane O'Neill
(Principal Hydrogeologist)
O'Neill Ground Water Engineering, 86, Arconagh, Naas, Co. Kildare, Ireland
Phone: 045-895668 or 087-2300933

Ioan Povara
(Dr.)
Institutul De Speologie, str. Frumoasa 11, R-78114 Bucuresti 12, Romania

Jean-Michel Puig
Hydrogeology Laboratory, Faculty of Sciences, 33 Louis Pasteur Street, F-84000 Avignon, France

QuDengNiMa Mineral Water Company
Tibetan, Gangba, China

Eliyahu Rosenthal
(Dr., Deputy Director, Senior Research Associate)
State of Israel, Deputy Director for Scientific Projects, Senior Research Associate, PO Box 6381,
Jerusalem, 91063 Israel, Phone: 9722-384403, Fax: 9722-388704, E-mail: elirose@vms.huji.ac.il

F. Santaloia
Department of Geology and Geophysics, University of Bari, Italy

Walt Schmidt
(Dr., State Geologist and Chief)
Florida Geological Survey, Gunter Building, 903 W. Tennessee Street
Tallahassee, FL, 32304-7700, USA, Phone: 904-488-4191, Fax: 904-488-8086

R. T. Sniegocki †
c/o Mrs. Irene Sniegocki, 8616 Linda Lane, Little Rock, AR 72207, USA

Johannes R. Vegter
(Dr.)
Private Box 59739, 0118 Karenpark, 0001 Pretoria, South Africa

Peeter Vingisaar
Geological Survey of Estonia, Kadaka tee 80/82, EE0026 Tallinn, Estonia
Phone: (372)6579661, Fax: (372)6579664, E-mail: vingisaar@estgeol.egk.ee

Jaroslav Vrba
(Dr.)
DHV CR, s.r.o. Drahobejlova 48, 190 00 Prague 9, Czech Republic
Phone: 42-2-66037219, Fax: 42-2-66037207

F. Vurro
Department of Geomineralogy, Univeristy of Bari, Italy

Mary Wallace Pitts
(Hydrogeologist)
P. E. LaMoreaux and Associates, Inc., PO Box 2310, Tuscaloosa, AL 35403, USA
Phone: 205-752-5543, Fax: 205-752-4043, E-mail: pela@dbtech.net

Zhou Wanfang
(Dr., Hydrogeologist, Engineering Geologist)
P. E. LaMoreaux and Associates, Inc., PO Box 4578, Oak Ridge, Tennessee 37831-4578, USA
Phone: 423-483-7483, Fax: 423-483-8739, E-mail: pelaor@usit.net

Stefan Wohnlich
(Dr.)
University of Munich, Dept. of Hydro and Eng. Geology
Luisenstraße 37, D-80333 Munich, Germany, Phone: 49(089)5203-422, Fax: 49(089)5203-286

Siki Yamamoto
(Prof. Dr.)
Department of Geography, University of Rissho, 4-2-16, Oosaki, Shinagawa-Ku, Tokyo 141, Japan

Josef G. Zötl
(Dr.)
St. Leonharderstraße 7, A-4293 Gutau, Graz, Austria

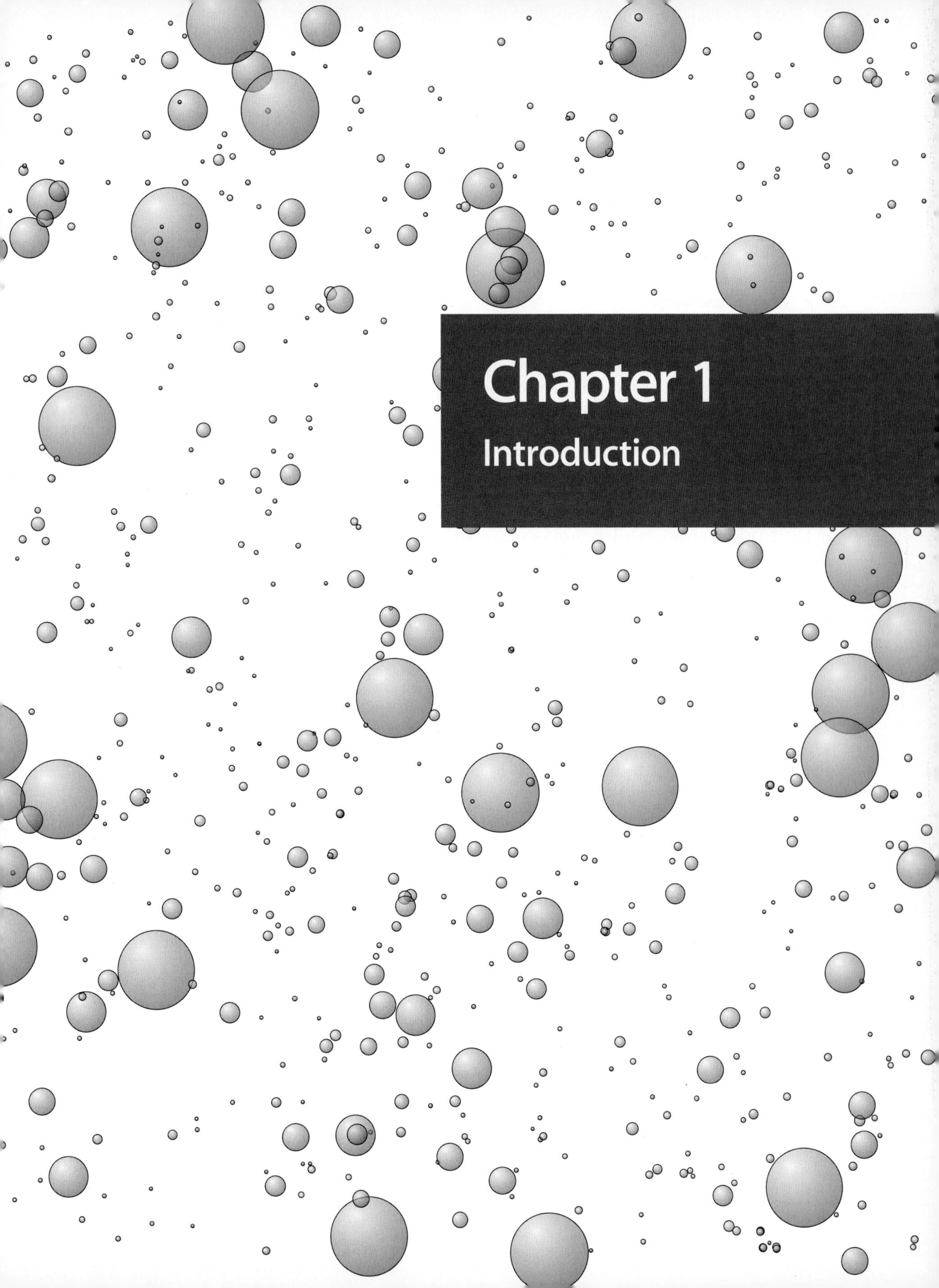

Chapter 1
Introduction

Chapter 1
Introduction

Philip E. LaMoreaux

1.1 Purpose and Scope

Springs and Bottled Waters of the World is a combined text and data reference book, designed to provide information for the general public, politicans considering legislation, legal and scientific professions, and to governmental employees about springs, springs that provide humans with a major source of water around the world. The book has been prompted by the explosive development of water for private, municipal, and industrial use as well as for the relatively new bottled water industry. The role of springs is described for ancient civilizations, military campaigns and in more recent times, for tourism and health spas around the world (see also Chap. 8 for significant historical facts on springs and bottled waters). No effort has been made to identify all bottled waters or describe them. The editor and authors have travelled extensively around the world and a large collection of bottles and labels for bottled waters has been brought together with a significant number of brochures, technical and popular reports, and newspaper articles on the subject. It is from this source of information that a selected set of bottled waters has been described and illustrated and from which examples of development, use, legislation, regulations, and rules governing bottled waters has been taken. Chapter 8, Famous Springs, provides selected historical material on springs and spring studies along with descriptions of famous springs by leading scientists from different countries. These papers follow a general agenda: source, occurrence, history of development and use, and methods for development.

One of the earliest scientific works about springs in the USA was published in 1841, a memoir of the *First Delaware Geological Survey* by James C. Booth. This report describes spring water, its chemical composition, use in geological mapping and for medicinal purposes. A more comprehensive book, *Mineral Springs of North America* by J. J. Moorman, M.D., Physician to White Sulphur Springs and a Professor of Medical Jurisprudence and Hygiene at Washington University, Baltimore, documented his 35 years of experience and investigation of the nature and medicinal applicability of mineral waters. His work, accomplished during his residency at White Sulphur Springs, described his observation of effects of the water on a variety of diseases. This fascinating book was published by J. B. Lippincott & Co. in 1873 and describes mineral waters in general, mineral waters used as medicines, and classified springs as to red sulphur, sweet, sweet chalybeate, hot, warm, alum, and healing springs.

There are many other early publications on springs and mineral waters and spas in Europe. One of the earliest of these is *Les Sources de France* by Me Stanislas Meunier published by the Libraire Hachette et Cie in Paris 1886. Subsequently, there have been hundreds of scientific and popular reports, pamphlets, and books written on the subject. Again, it would be extremely difficult to list all of these references, however, a selection has been made to aid more detailed research as the need arises.

1.2 The History of Mineral Water Exploitation in China

In China, the modern mineral water industry has been described by Dr. Yuan Daoxian, the Director of the Institute of Karst Geology at Guilin, Guangxi, China as follows (Daoxian 1996):

> According to historical documents, Chinese people knew about mineral water, especially hot springs, for a long time. A book, *Annotation on Water Scripture*, by Li Daoyuan who lived in the Beiwei Dynasty (A.D. 386–543), described 41 hot springs, most of them in North China, but also 9 in South China. In Volume 38 of the Geography Dictionary, compiled in the time of Kangxi Emperor (A.D. 1662–1722) of the Qing Dynasty, 78 hot springs were noted. The knowledge on mineral and hot springs grew rapidly in China, and in the book *Study on Hot Springs in China* (Chen Yanbing 1939), 584 hot springs

were described, including 103 from Guangdong Province, and one from Tibet. Another book *Major Hot Springs in China* (Zhang Hongzhao 1956, Geological Publishing House), collected data of 972 hot springs in China. In *Methodology for Geothermal Water Survey and Exploration*, compiled by the Institute of Hydrogeology and Engineering Geology (MGMR, Geological Publishing House 1973), more than 2 000 hot springs and exploration wells were identified and shown on a map.

As the middle reach of the Yellow River is the cradle of Chinese culture, and the political center of China originated in North China, the exploration of mineral and hot waters was also started in the north, and expanded gradually to the south. For example, the Lishan hot spring, 25 km to the east of Xian City, now the provincial capital of Shaanxi province, but which used to be the capital of China for thousands of years (1134 B.C.–A.D. 907), has been used by many monarchies of China for medical treatment since the time of King Guangwu (A.D. 25–56) of the eastern Han Dynasty. When General Su Wenda remained there because of serious pernicious malaria, a kind of vital subtropical disease, and was not able to return to the capital (Luoyang, Henan Province) with his army, Su was soon cured by daily bathing in the hot spring. The Xinzi hot spring at the south of Lushan Mountain, a well-known summer resort in Hiangxi province on the middle reach of the Yangtze River, it was recorded from the time of the eastern Jin Dynasty (A.D. 317–907), that Xue Yong, the governor of Xixian County was cured from an epidemic by bathing in it.

The first scientific summary of mineral water in China was made by Li Shizhen (A.D. 1518–1593), the great ancient Chinese pharmacologist in Ming Dynasty. In his book, *Compendium of Materia Medica*, he classified mineral water in China according to various criteria and gave examples. For instance, according to the chemical contents, he classified mineral waters into sulfur springs, cinnabar springs, vitriol springs, reagar springs, and arsenic springs. On the basis of water taste, he classified mineral waters into sour springs, bitter springs, salty springs, cold springs, and hot springs. He also made remarks on the medical effects of mineral waters for treatment of dermatosis, rheumatism, etc.

1.2.1 The Uses of Mineral Water in Ancient China

The most popular use of mineral water in ancient China was for medical treatment. However, there were also records of employing hot springs for cooking and agriculture.

Medical Use. In the book, *Annotation on water scripture*, published in Beiwei Dynasty (A.D. 386–543), Li Daoyuan wrote, "the Huangnu hot spring on the Shahe River, Lushan, Henan is so hot that rice can be cooked in it. A Taoist Prist drinks it three times a day in addition to bathing in it. All his diseases were cured in 40 days." According to historical records, the King Yingzheng, the first Emperor of Qin Dynasty (221–210 B.C.), cured his dermatosis, King Li Shiming (A.D. 627–649) of Tang Dynasty recovered from rheumatism by taking a bath in the Lishan hot spring at the east of Xian City.

Agricultural Use. In the book, *History of Han Dynasty (continued)*, written by Sima Biao in the period of Jin Dynasty (A.D. 265–420), it is recorded that there is a hot spring 3 km to the south of Cunzhou City, Hunan province. Several hectares of ricefield downstream from the spring, irrigated by warm water, can be sown in December and cropped in March of the next year. Using water from the hot spring, the ricefield has three crops every year. Similar practices are recorded using water from Dongze hot spring, Xingyang county, Hubei province. In the Tang Dynasty (A.D. 618–907), melons are irrigated by water from Lishan hot spring at the east of Xian and can be cropped in February (Daoxian 1996).

The first commercial bottled mineral industry in China was Laoshan Mineral Water near the coastal city of Qingdao, Shandong province. It was started in 1931, but stopped during the Second World War. Since 1962, the production of Laoshan Mineral Water has been restored and sold on the Hong Kong market. Meanwhile, production from several other mineral water enterprises have also been put onto the market since the 1960s, including the Weina Mineral Water of the Inner Mongolia Autonomous Region, Longchuan Mineral Water of Guangdong province, the Pikou and Tanggangzi Mineral Waters of Liaoning province. In 1987, the Standard for Natural Drinking Mineral Water of the People's Republic of China (GB-8537-87) was issued by the National Technology Supervision Bureau. In less than 10 years, the known sites of mineral water in China have increased to more than 2 000, with an exploitable resource of 280 000 000 m^3/yr. Among them, more than 500 sites have been authenticated, and about 250 sites exploited, with an annual production of more than 1 million t. It is expected that the annual production of bottled mineral water in China will be about 5 million t in the year 2000.

In the United States, the US Geological Survey published *Bulletin US 32, 1886 the Analyses of Mineral Springs in the USA*. One of the earliest and most comprehensive references on mineral waters from springs is contained in the book *Mineral Waters of the United States and American Spas* by William Edward Fitch, M.D., a member of the International Society of Medical Hydrology, as well as a surgeon, gynecologist, and educator. This book, published by Lea & Febiger, 1927, contains 799 pages of information on mineral water classification, dosage and physiological impact, application of mineral waters to the treatment of disease, and a comprehensive chapter on the mineral waters of the United States by state.

Perhaps one of the most detailed and descriptive scientific books on springs in the USA was published as *Water Supply Paper 557, Large Springs in the United States* by Oscar Edward Meinzer in 1927. This document originally sold for 30 cents per copy from the Government Printing Office, Washington, DC. It contains 94 pages of text and illustrations and is a relatively rare publication. It described the distribution and character of large springs and their classification with respect to size. This classification is still in use today (Table 1.1; Meinzer 1927). Meinzer also provided information on discharge, quality of water, and the occurrence of springs by geological settings in Florida, Georgia, Alabama, Arkansas, Texas, California, Oregon and elsewhere in the United States.

In the late 1800s and early 1900s there was much popular interest and considerable scientific and economic importance attached to springs. Yet at that time, information concerning springs was widely scattered and difficult to obtain. The report, *Large Springs in the United States* (Meinzer 1927), was an effort on the part of the US Geological Survey, through its Cooperative Program with universities and states to provide these data. Of significance in this report is the attempt to classify and describe source, occurrence, geology, and hydrology of springs. Perhaps one of the most prophetic parts of *Water Supply Paper 557* was its recognition and emphasis of the importance of "springs in Tertiary limestone in Florida" for within this state occurs a very notable group of large springs of first magnitude classification (Meinzer 1927); springs that discharge large quantities of water from solution cavities and openings in the cavernous limestone of the Floridan Aquifer in quantities such that the resulting river can be navigated by sizeable passenger or freight boats. At one time in the late 1800s in the Tallahassee area of northern Florida a number of such springs, sinkholes and rivers were connected to form an intricate inland passageway for freight boats moving supplies and produce from farmlands to cities across the northern part of Florida.

Table 1.1. Classification suggested for practical use in the United States (Meinzer 1927)

Magnitude	Average discharge
First	100 s-ft or more
Second	10–100 s-ft
Third	1–10 s-ft
Fourth	100 gal/min to 1 s-ft (448 gal/min)
Fifth	10–100 gal/min
Sixth	1–10 gal/min
Seventh	1 pint to 1 gal/min. About 200–1 500 gal, or 5–50 barrels/d
Eighth	Less than 1 pint/min. Less than about 180 gal, or about 5 barrels/d

1.3 The Floridan Aquifer – A Type Locality for Springs – The Fountain of Youth

Geologically, much of Florida is underlain by cavernous limestone of Tertiary age. These limestone and dolomite beds of the Oligocene, Miocene, and Eocene contain an intricate network of solution cavities developed along bedding and vertical joints, fracture and fault openings. This interconnected system or primary and secondary openings comprise one of the largest limestone aquifers in the world – the Floridan Aquifer, which has been described in hundreds of publications. Water, before discharging from the large springs has moved in the hydrologic cycle from rainfall percolating down through the soil and then underground into and through large channels in the limestone to discharge points. The discharge from springs in Florida fluctuates with rainfall and near the Atlantic Ocean or the Gulf of Mexico shorelines, fluctuates with tides. Owing to the low relief of the land, dense vegetation, and the mantle of sandy soil through which the water moves before entering the limestone, the spring waters remain clear even during times of greatest discharge. Some of the more famous of these springs are: Silver, Blue, Homasassa, Wakulla, Weeki Wachee, Suwannee, Sulphur, Crystal River, Seminole, Lithia (Fig. 1.1), and Kissengen.

Few states, and perhaps no other part of the world, has publicized or developed its springs as tourist attractions as has Florida. For example, nearly every publication of the Geological Survey of Florida at Tallahassee has some information pertaining to carbonate hydrology of the limestones of that state. A recent issue of the Florida State Parks Guide, developed by the Division of Recreation and Parks and the Department of Natural Resources (DNR), Office of Communications in Tallahassee, is another example. Each year, over 15 million visitors visit state parks for relaxation, sports and beauty.

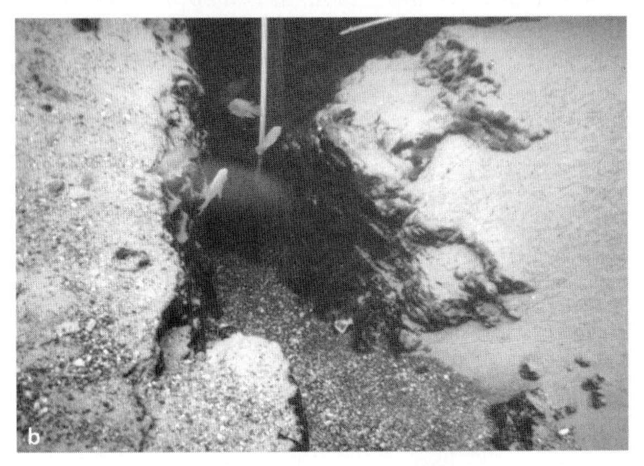

Fig. 1.1. **a** Lithia Springs near Tampa, Florida. The spring is used for recreation and an industrial water supply (7 million gal/d). **b** Underwater photography of primary fracture at Lithia Spring

Most of these parks have crystal clear springs, meandering rivers, lush gardens, and peaceful campgrounds. Some of the most famous of these are: Blue Spring State Park, Homosassa Springs State Wildlife Park, Ichetucknee Springs State Park, Manatee Springs State Park, Peacock Springs State Recreation Area, Ponce de Leon Springs State Recreation Area, Rock Springs Run State Reserve, Wakulla Springs State Park, Weeki Wachee Spring, and Ginnie Springs.

One of the most intriguing aspects of karst springs in Florida is the variety of their settings – geologic, hydrologic, and geographic. Some discharge at the land surface, some flow upward into bodies of fresh or salt water. For example, Homosassa Springs State Wildlife Park is on the Homosassa River just north of Tampa, ebbs and flows with the tide from the Gulf of Mexico. Homosassa discharges into an estuary bay and a river, and is connected at the surface and in the subsurface with salt water from the Gulf. Thus, at different stages of tide, the spring will discharge freshwater or salt water, and fresh and salt water fish exist in the main spring at different times of the day.

The Florida Department of Natural Resources, Division of Recreation and Parks, manages many of these parks, however, in addition, there are many commercially operated springs such as Rainbow Spring, Weeki Wachee Spring, and Silver Springs that attract tourists by the thousands and brings millions of dollars in revenue to Florida each year. These springs are well publicized and have beautiful colored pamphlets exalting their beauty. The Chamber of Commerce of Silver Springs provides the following description:

> The deep, cool water of Silver Spring, clear as air, flows in great volume out of immense basins and caverns in the midst of a subtropical forest. Seen through the glass-bottom boats, with the rocks, under-water vegetation, and fish of many varieties swimming below as if suspended in mid-air, the basins and caverns are unsurpassed in beauty. Bright objects in the water catch the sunlight, and the effects are truly magical. The springs form a natural aquarium, with 32 species of fish. The fish are protected and have become so tame that they feed from one's hand. At the call of the guides, hundreds of them, of various glistening colors, gather beneath the glass-bottom boats.

It was Indian legends about Silver Springs that brought the Spanish Explorer Ponce de Leon to Florida in search of the "Fountain of Youth".

1.4 The Edwards Aquifer, Texas

The Edwards Aquifer in Texas is perhaps one of the most intensely scientifically studied spring systems in the world. The Aquifer is also one of the most productive in the United States and is the sole source water supply for over 2 000 000 people, including the city of San Antonio, Texas. All types of geologic, hydrologic, geochemical and geophysical studies, test drilling, and pumping tests have been applied to the wells and springs producing water from the Edwards. Owing to the expansion of population and development in the area, ground water discharge from the Edwards has increased steadily with growth. Discharge is literally from thousands of artesian and non artesian wells, as well as from large springs of the natural discharge from the system, as at Comal Springs. Natural discharge from springs has decreased as pumpage from wells and natural springs has increased (Crowe and Sharp 1997).

The Edwards Aquifer consists of nearly 200 m or 550 ft of thin to massively bedded limestone-dolomite. The updip limestone and dolomite of the system in its surface outcrop is highly karstified in the recharge area, as well as in the subsurface in the southern extent of the freshwater producing part of the aquifer. A bad water line (highly mineralized) bounds the aquifer to the south and east. The aquifer extends eastwest approximately 360 m and varies in width from about 8 m to 64 km. Natural discharge from the aquifer occurs at Large Spring, San Marcos, Comal, Hueco, San Antonio, San Padro, and Leona.

Owing to extensive withdrawal from the aquifer by springs and wells for competitive uses (municipal, industrial, military, and agriculture), some springs have decreased substantially in flow or stopped flowing all together. Multiple environmental problems have resulted and legislation and regulations have been developed (see Legal Aspects of Springs; Chap. 7.3).

1.5 Springs – Sources of Water, Industry, Art Form, Mythology

Springs in all parts of the world are of a diverse character, from small trickles of water into an old hollowed out log to large pump lift systems into stainless steel tanks for municipal and industrial use and by the bottled water industry. Springs are the lifeblood of small farm houses in the foothills of the Appalachians, Rocky Mountains, Alps or Urals, to large municipal and industrial water supply developments such as Lez Spring – water supply for the city of Montpellier, France, or Big Spring for the city of Huntsville, Alabama, and the karst spring systems that supply water to villages and communities of the Swabian Alb in Germany. There are spectacular springs such as Old Faithful, a world-renown hot water geyser in Wyoming, or the geysers of Iceland, Kamchatka, Russia, or Hot Springs, Arkansas. Most of these springs have a long history of development and use. Many are the source of legends or myths. As an example, Hot Springs, Arkansas, has been developed over the past millennium for its therapeutic value, first by the Indians and then by "white man". It is now famous as Hot Springs, Arkansas, the home of the former President of the United States.

Fig. 1.2. Samples of brochures for touristics

Fig. 1.3. Labels illustrate artistic renditions of springs and legends, plus chemical analyses and nutritional facts

Fig. 1.4. Variety of bottles in colors, sizes, shapes, and materials

Fig. 1.6. Special plastic bottles with mouth pieces designed for hikers, joggers, runners, bikers, and others

Fig. 1.5. Perrier art deco bottles

Springs are harnessed for their energy and generate electrical power as at Trebinje near the Adriatic Ocean in Yugoslavia (Milanović 1979), or for thermal energy as at many places in Russia, Europe and the United States.

One of the most interesting historical uses for spring water in the foothills of the southern Appalachians was its cooling power, when run over coils that were integral parts of a whiskey still, distilling "illicit moonshine" that often sold for $1.00 per gallon. Some products were high quality, some very bad, but at a price much less than today's cost of bottled spring water. Nonetheless, just as publicized by the public relations merchants of today, its important attribute was that "it was made with fresh spring water".

Springs have captured the imagination of authors. For example, the book, *Historic Alabama, Hotels and Resorts* by James F. Sulz, University Alabama Press 1960, describes the use of mineral waters in the early development of tourism in Alabama. An excellent recent scientific report by J. S. McColluch was published by the West Virginia Geological and Economic Survey in Springs of West Virginia in 1986. A book by Maureen and Timothy Green titled *The Best Bottled Waters in the World*, published by Simon & Schuster, Inc., describes a select group of spring waters and spas around the world and bottled water labels, as a form of art. Springs have excited collectors who have made hobbies of collecting spring water, bottles, labels, caps and other paraphernalia.

Early reports, water supply papers of the US Geological Survey (USGS), commercial pamphlets, and hundreds of newspaper articles have now been written on the subject of springs. Present day textbooks on hydrogeology nearly all contain sections describing springs. For example, *Hydrogeology* by Davis and DeWiest, Wiley Publishing, 1966, and similar textbooks, contain basic concepts

Fig. 1.7. Different shapes and sizes of plastic bottles

Fig. 1.8. Special shapes of **a** bottles and **b** cans

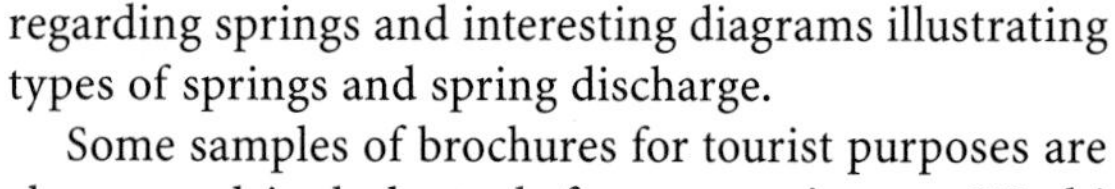

regarding springs and interesting diagrams illustrating types of springs and spring discharge.

Some samples of brochures for tourist purposes are shown and include such famous springs as Weeki Wachee and Silver Springs (Fig. 1.2). Photographs of labels illustrate artistic renditions of springs and legends, plus chemical analyses and nutritional facts (Fig. 1.3). Spring waters are bottled in a great variety of bottles – colors, sizes, shapes, and materials – glass, plastic, and cans (Fig. 1.4 and 1.5). Attractive labels emphasize the product as clean, sparkling spring, artesian, mountain, clear, crystal, springtime, and many others. Cans and glass bottles are used in dispensers like soft drinks. Special plastic bottles with mouth pieces are designed for hikers, joggers, runners, and bikers, and special shapes and materials, metal and plastic, are used for shipping purposes (Fig. 1.6, 1.7, and 1.8).

1.6 Political/Legal Aspects of Springs[a]

During the past 20 years, groundwater from wells and springs for municipal, industrial, and common use has expanded greatly. During the past 15 years, the last half of the 1980s and the 1990s, there has developed a great concern on the part of the public for a safe source of drinking water. In many parts of the world, municipal water supply systems fall under strict regulations of a great variety of regulatory agencies – local, state, and federal in the United States, and within the Economic European Community (EEC), and other parts of the world. Public perception is that bottled water is safer. For example, in 1995, over $2.5 billion of bottled water were purchased in the USA alone.

[a] See Chap. 7 for details.

This rapidly expanding market in Europe and the USA has resulted in some rather imaginative marketing procedures and unique bottles, labels, and brochures have been developed to sell the product. The product comes under such phraseology as "sparkling water", "natural sparkling water", "artesian water", "mountain spring water", "pure spring water", and many other catch phrases. In the United States, the US Food and Drug Administration (USFDA) is carefully analyzing slogans, labels, and advertising for misrepresentation.

Professor Dr. Jose A. Cuchi Oterino provided the following information on Spanish bottled water:

There are over 95 bottling companies in Spain; 90 of which are affiliated to ANEABE (Asociation Nacional De Empresas De Aguas De Bebida Envasadas). These affiliated companies produce 98% of the total bottled water volume. Most of the companies are very small, with only 3 or 4 workers and a daily production of about 10 m^3. Three companies produce more than 50% of the total bottled water. Fontvella, located in the northeast of Spain, is the main producer. The factory is located on the Guillerias massif, a granite outcrop. Several other bottling companies are located in the same area. The old and small factories use springs. Some use horizontal galleries, locally called minas. The new factories use water from drilled wells, which provide greater yield and security against pollution. The total water bottled for 1995 was 2989.730 millions of liters. Average annual consumption per capita is 74. The economic value of production is around 78 000 million Spanish pesetas. That is about 577 million US dollars. Spanish bottled waters must follow the EC directive 80/777/CE. The Spanish law for bottling and trading of this water was held by the Real Decreto 1164/1991. Underground waters are divided into medicinal-mineral waters, industrial-mineral, spring and potabilized water. The first and second classes are forbidden any type of treatment, except physical filtration (Oterino 1996).

Shane O'Neill in Dublin, Ireland, provides another example of the development of the bottled water industry in Ireland as follows:

The production of the first Irish bottled water occurred in 1981. Production has grown such that in 1995, 36.7 million of Irish-produced bottled waters were exported while 33 million were consumed indigenously. The infancy of the Irish bottled water industry means that all sources are less than 15 years old and all plants are less than 10 years old. About 45 Irish companies are in the business of bottling water but only four are recognized as Natural Mineral Water. Examples are Glenpatrick Natural Mineral Water, and Kerry Spring Natural Mineral Water. Consumption has grown from 0.3 per capita in 1985 to 9 per capita in 1995. Four companies, Ballygowan Ltd., Kerry Spring Natural Mineral Water, Deep River Rock Spring Water, and Tipperary Natural Mineral Water command over 80% of the Irish market of 33 million/yr. Though the bottled water market is now worth over $160 million/yr, Ireland still consumes only 12% of the European average per capita of bottled waters (Fig. 1.9).

There are three standards controlling the production of bottled waters in Ireland. There is 80/778/EC, the Drinking Water Directive, I.S. 432, the voluntary national bottled water standard, and 80/777/EC, the Natural Mineral Water Directive. The Irish Bottled Water Standard, I.S. 432 is voluntary and was developed to ensure safety of both the raw water and the finished product. The emphasis is on prevention rather than detection. The standard applies to all packaged water regardless as to whether they are Natural Mineral Water, other bottled waters or water coolers. The Standard provides similar controls for all bottled waters to engender consumer confidence in the products. The Standard has about 70 specifications and over 100 recommendations.

Fig. 1.9. Ballygowan Irish Spring Water

Irish natural mineral waters have a low total dissolved solids content of between 150 and 500 mg/l. The main water types in Ireland are: Ca $(Mg)(HCO_3)_2$ with TDS ranging from 150 to 500 mg/l and pH of between 6 to 8. Calcium bicarbonate waters with TDS of up to 600 mg/l found throughout the country; Calcium sulphate waters found just in the northern portion; and Sodium bicarbonate waters in certain confined aquifers.

There are warm springs in Ireland in the range of 14 to 25 °C, found in discrete parts of the country but their chemistry, with one notable exception, would be classified as low mineral content. Two natural mineral water sources are abstracted from Devonian Sandstones while the other two natural mineral waters are abstracted from Lower Carboniferous Limestones. Natural Mineral Water sources have specific yields of between 6.67 to 48.8 m^2/d. No direct health claims are made for Irish natural mineral waters and no health benefits are attributed to Irish bottled waters (O'Neill 1996).

In a paper presented at a workshop for Irish and Russian Hydrogeologists in St. Petersburg, 24–30 August 1996, O'Neill reported:

Natural mineral waters are defined by EU Directive 80/777/EC (OJ No. L 229, 30.8.80 [a]). All natural mineral waters are, by definition, groundwaters (Robins and Fry 1992, in O'Neill 1996). In the Directive, natural mineral waters are defined as being microbiologically wholesome, emerging from a spring or tapped source which is distinguished from ordinary drinking water by its nature and original state and has been protected from risks of pollution (Robins and Fry 1992; Misund and Banks 1994, in O'Neill 1996). The chemical composition must be stable and it must be regularly tested by official agencies (IBWA 1996, in O'Neill 1996). In Ireland this agency is the National Standards Authority of Ireland (NSAI).

The main components of the Directive are:

1. There is no treatment permitted other than filtration, at or above 1 μ pore size;
2. The NMW should contain naturally occurring minerals and microorganisms;
3. The chemical composition needs to be officially registered;
4. There are specific labelling requirements that permit the use of the words "natural mineral water";
5. The NMW must be bottled on site and it is not allowed to be transported before being bottled;
6. A specific geographical location of source must be given;
7. The mineral analysis must be available to the public by being either printed on the label or by reference to a particular analysis;
8. There must be a tamper proof seal on the bottle (IBWA 1996, in O'Neill 1996).

In Ireland the regulations have been promulgated into Irish law as S.I. 11/86. The EU have taken the view that a natural mineral water by definition does not require any treatment and so the emphasis is on demonstrating and maintaining the integrity of the source (O'Neill 1993 in O'Neill 1996). Therefore, recognition requires a very detailed hydrogeological survey, prolonged physico-chemical and microbiological sampling and a source protection plan. In Ireland, recognition takes between 12 and 24 months for the first source and 6 months for every additional borehole exit tapping the same source (IBWA 1996 in O'Neill 1996).

For a more detailed discussion of legal aspects of bottled water, see Chap. 7.

The bottled water industry has also spawned relatively new hobbies: art, collectibles, mythology, bottle caps, labels, bottles, and brochures. One of the most artistic new concepts in promotion is the art deco bottle of Perrier (see Fig. 1.4). The most recent design development is the bottled water mouth piece that can be used by joggers, bikers, and travellers needing portability. Labels portray beautiful environmental settings as on the labels of Coors Beer, Rolling Rock, and many others.

1.7 Alcoholic Beverage Manufacturing

Some alcoholic beverage manufacturing companies likewise take advantage of spring water as an important ingredient in their products. For example, Coors Beer of Colorado plays up the importance of the use of water from the Rocky Mountains. Rocky Mountain spring water is related to snow melt, rainfall runoff, and ground water. As advertised, it is clear, cool, and pure. The extra pale premium beer, Rolling Rock, advertises "brewed from mountain springs". It is manufactured in LaTrobe, Pennsylvania, and on the bottles and packaging, uses the terms, "from the mountain springs to you" (Fig. 1.10). One of the most famous products of all is the whiskey from the distributor Jack Daniel's which advertises its "sipping whiskey" as made with spring water, and in their advertisement show the discharge point of the spring from a limestone cave system. Jack Daniel's, established in 1866, in Lynchburg, Tennessee, advertises "whiskey made as our fathers made it". They state that Jack Daniel's Tennessee Whiskey is a completely natural product and is made from pure spring water, yeast, and 100% whole natural grains; corn, rye, and barley malt. It is the oldest registered distillery in the United States and is a sour mash whiskey. Jack Daniel's has a distinctive label on each bottle that plays up the importance of pure spring water. In its brochures, it describes the importance of spring water even more, showing a picture of the spring and cave (Fig. 1.11).

Fig. 1.10. Rolling Rock advertises "clear, cool, and pure"

Glenfiddich and Glenmorangie, pure malt scotch whiskies, are examples of alcoholic beverages emphasizing the importance of spring water in the manufacture of its product. From its brochures we learn that the origins of Glenfiddich go back to 1886 when William

Fig. 1.11. Jack Daniel's advertises its whiskey as being made with iron-free water from a Tennessee cave spring

Grant built the Glenfiddich distillery. Today, a century later, his descendants still own and manage the distillery. Glenfiddich is the only highland malt to be bottled at its own distillery, and its pale golden color and smooth taste have made it one of the world's most sought-after pure malt scotch whiskies. Glenfiddich is the only 'château-bottled' malt whiskey made in the Highlands of Scotland. The distillery states that to make Glenfiddich it uses a single source of pure natural water, from a spring on the hillside above the Glenfiddich distillery, and this gives it its unique purity of taste. Glenfiddich has bought 1200 acres of surrounding hillside just to protect that small spring.

1.8 Bottled Water from Wells

In recent years, with the great increase in demand for bottled water, operators have searched for new sources for a supply and artesian well water is being used. "Artesian" refers to groundwater confined under hydrostatic pressure. One of the most popular of these in the United States is Artesia. Artesia, bottled in Austin, Texas, is advertised as "100% pure Texas-sparkling artesian water". It is extracted from a well in San Antonio, Texas from a depth of "53 stories". The water comes from the Edwards aquifer that also supplies San Antonio (Fig. 1.12).

Some large local chain stores have their own bottled water such as Delchamps in the southeastern USA with its own label, "natural artesian water", which is sold in some hotels, the Westin Chain has its own label on bottled water in their hotel rooms. Bottled waters are advertised as filtered and ozonated. For example, Kentwood Spring Water Company's water source is from a Gulf Coastal Plain aquifer beneath Kentwood, Louisiana, which is typical of several 100 water-bearing beds in the Atlantic and Gulf Coastal Plain. These aquifers come to the surface in their northern extremities and dip between alternating layers of clay, sand, gravel, chalk and limestone to the south and southeast in the Gulf Coastal Plain and to the southeast in the Atlantic Coastal Plain. Near their outcrop or exposure "up dip", the water is recharged by rainfall, the water takes the average temperature of the rocks in the area and depth at which it occurs, and generally has a low mineral content. However, as the beds dip deeper underground the water is in longer contact with the rocks and becomes warmer and contains more and more minerals. Interestingly, some of this water from the deeper beds has been migrating downward under the force of gravity for thousands of years; some is 20 000 or 30 000 years old based on ^{14}C dating, a factor apparently unknown to the bottled water industry. This massive sequence of aquifers in the Gulf and Atlantic Coastal Plain can potentially provide water depending on geographic location, selected depth, mineral content, temperature or age of origin.

Fig. 1.12. Artesia, pure, sparkling Texas artesian water, extracted from wells in the Edwards Aquifer, San Antonio, Texas; Mountain Valley Spring Water, Hot Springs National Park, Arkansas; Ballygowan, Irish Spring Water, Limick, Ireland; Crystal Geyser Sparkling Spring Water, Nappa Valley, California

Fig. 1.13. **a** Hartwell Industries, Lahti, Finland, production line for Erikos Beer. **b** Dr. Pertti Lahermo, hydrogeologist, Geological Survey of Finland, and Dr. Jan Dowgiallo, Poland, President of Mineral and Thermal Waters Commission of the International Association of Hydrogeologists

There are many bottled waters that are taken from commercial wells or from city water supplies as a source of water. For example, at Lahti, Finland, north of Helsinki, bottled water is taken from city wells developed in a "glacial esker". An esker is a long, narrow, sinuous, steep-sided ridge composed of irregularly stratified sand and gravel that was deposited by a subglacial or englacial stream flowing between ice walls or in an ice tunnel of a stagnant or retreating glacier, and was left behind when the ice melted. The water in an esker is of excellent quality and the deposits are generally very permeable and rapidly recharged by rainfall or tangential freshwater lakes or streams. The water is low in mineral content. However, at the Holland Industry plant this low mineralized water is fortified by adding minerals to convert it to a mineral water for bottling purposes (Hartwell Industries). The water is also used to manufacture Erikos Beer (Fig. 1.13).

On 4 September 1996, a trip to Estonia provided information on the use of mineral-thermal water for bottling and medicinal use. Dr. Reim Vaikmae, Director, Institute of Geology and Dr. Peeter Vingsaar, hydrogeologist with the Geological Survey of Estonia, provided information about the potential sources of mineral and thermal water in Estonia and supplied samples of Varska Bottled Water, one of the leading brands primarily produced for a domestic demand (Fig. 1.14).

One of the travesties of bottled water from well sources is that often the quality of the bottled water that costs a premium is no different than water taken from a tap from a municipal water supply. Municipal water supplies must meet rigid health standards, just as in recent

Fig. 1.14. a Dr. Peeter Vingsaar, hydrogeologist with the Geological Survey of Estonia, provided information about the potential sources of mineral and thermal water in Estonia. b Mrs. Ura M. LaMoreaux holds samples of Varska bottled water

years, bottled waters in many parts of the world have come under strict regulation. If one is travelling, however, drinking water from the same source may be a consideration, so it pays to read the labels.

In the southern half of Alabama, USA, nearly all municipalities south of the "fall line" in the Gulf Coastal Plain, except Mobile, obtain their water supply from wells. Most of these supplies are from an "artesian" well source. Some of this water was recharged from rainfall or went underground thousands of years ago. Villages and towns like Demopolis, Uniontown, Linden, Eutaw, and Montgomery obtain their water supply from wells drilled into an aquifer that was deposited in an ancient sea 60 million years ago. Some of this water contains about 1 ppm fluoride, and is beneficial to the development of children's teeth by resisting decay. So why not a bottled water, from wells at a depth of about 800 ft, that entered the ground over 10 000 years ago, that is artesian, slightly mineralized, naturally fluoridated, thus when used by children would result in healthy teeth? What a label that would make!

1.9 Sources of Reference

Excellent sources of reference for springs and bottled waters are easily available from the *Guide to Geoscience Departments* in the United States and Canada (1977), from the American Geological Institute (AGI). Of particular help in providing information on springs would be the hydrogeological scientists with the Water Resources Division of the US Geological Survey or the same type of professional with each of the 50 State Geological Surveys in the USA. In other countries, similar geological or water resource agencies can be located.

Scientific societies with geoscientist members are likewise a good source of reference for information about springs. For example, members of the Hydrogeology or Environmental Geoscience Divisions of the Geological Society of America, the American Geophysical Union, the Geological Society of Great Britain, and others. The International Association of Hydrogeologists has several commissions with groundwater programs. Most pertinent would be: the Karst Commission, the Groundwater Protection Commission, and the Commission on Mineral and Thermal Waters.

Evidence of the rapid escalation in sources of reference, research, and publications on springs, plus the broad range of sources of publications concerning spring and bottled waters can be illustrated by the random selection of the references that follows:

1. The origin and occurrences of carbon dioxide and gaseous mineral waters in the area of the Variscian Platform of central Europe, extracted from Memoirs of the International Association of Hydrogeologists, Tome VI, Reunion De Belgrade (1963), Beograd, Bulevar vojvode Misica 17, pp. 327–332, Vrba, J., (Vrba 1966).
2. Simulation of spring discharge from a limestone aquifer in Iowa, USA, appeared 1966 in the Hydrogeology Journal (Zhang et al. 1966).
3. International Symposium on Hydrogeochemistry of Mineralized Waters, Cieplice Spa, Poland, 31 May–3 June 1978, Chairman, Jan Dowgiallo, with reports: (1) mineral and thermal waters of the Sudetes Mountains against the background of the geological structure; (2) the dependence of the chemical composition of mineralized waters upon the rock environment (general report based on 21 papers); (3) the isotopic composition as an indicator of the origin of mineralized waters and of their components (general report based on 9 papers); (4) mineralized waters as a source of mineral raw materials as well as a source of information on the occurrence of mineral deposits (general report based on 10 papers); and (5) depth of the freshwater-mineralized water boundary (general report based on 6 papers) (Dowgiallo 1978).

4. Assessing groundwater flowtype using spring flow characteristics, in: The Professional Geologist vol. 33, No. 8, pp. 4–7, (Werner 1996).
5. Short report on IGCP 379 Meeting "Karst Processes and Carbon Cycle", Lipu, Guangxi, China, 26–30 April 1997. The Deep Source CO_2 by Dr. Do Tuyet (The Institute of Geology and Mineral Resources of Vietnam) reported that there are many geothermal springs in Vietnam along the northwest-oriented plate margin zone. The hottest one is at Bang, Le Chuy district of Guang Pinh province. It is 105 °C. Dr. Ching-Nan Liu (Taroko National Park Headquarter, Taiwan, China, and others reported one of the systems along Liwi Hsi River, north of Hualien County (Daoxian 1997).
6. An example of a current international symposium is Mineral and Thermal Groundwater, organized by the Romanian Association of Hydrogeologists, under the auspices of the International Association of Hydrogeologists, 24–27 June 1998, Miercurea Ciuc, Romania.

From the above, it can be determined that there is a broad base of references that one should avail oneself of, which include individual scientific efforts, plus local, state, and national agencies, and national and international societies that have objective research and data collection on groundwater, springs, and bottled waters.

References

Crowe JC, Sharp JM Jr. (1997) Hydrogeologic delineation of habitats for endangered species: the Comal Springs/River System. Environmental Geology 30(1/2) March 1997, Springer-Verlag, pp 17–28

Daoxian Y (1996) The history of mineral water exploitation in China. The Institute of Karst Geology, Guilin, Guangxi, China 541 004, personal correspondence, 12 December 1996, 3 p

Daoxian Y (1997) Short Report on IGCP 379 meeting in Lipu, personal correspondence dated 14 May 1997, 8 p

Davis SN, DeWiest RJM (1966) Hydrogeology. John Wiley and Sons, New York, 463 p

Dowgiallo J (1978) International Symposium on hydrogeochemistry of mineralized waters. Cieplice Spa, Poland, 31 May–3 June 1978. Summary by Jan Dowgiallo, Chairman, 3 p

Fitch WE (1927) Mineral waters of the United States and American spas. Lea & Febiger, Philadelphia and New York

Green M, Green T (1985) The best bottled waters in the world. Simon & Schuster, Inc., New York, 172 pp

McColluch JS (1986) Springs of West Virginia. West Virginia Geological and Economic Survey

Meinzer OE (1923) Outline of groundwater hydrology: US Geological Survey Water-Supply Paper 494. Government Printing Office, 71 p

Meinzer OE (1927) Large springs in the United States: US Geological Survey Water-Supply Paper 557. Government Printing Office, Washington 94 p

Milanović PT (1979) Hidrogeologija Karsta, I Metode Istrazivanja, Institut za koristenje i zastitu voda na krsu. Trebinje, 302 p

Oterino Jose A. Cuchi (1996), Laboratorio de Hidrologia. Escuela Universitaria Politecnica de Huesca, 22 071, Huesca, Spain, personal communication, 13 December 1996, 2 p

O'Neill S (1996) A Review of the Irish bottled water industry with particular emphasis on natural mineral waters, presented at a workshop of Irish and Russian hydrogeologists in St. Petersburg, 24–30 August 1996. Environmental Geology (submitted), 12 p

Sulz JF (1960) Historic Alabama, hotels and resorts. University of Alabama Press, Tuscaloosa, Alabama

Vrba J (1966) The origin and occurrences of carbon dioxide and gaseous mineral waters in the area of the Variscian Platform of Central Europe, extracted from Memoirs of the International Association of Hydrogeologists, Tome VI: Reunion De Belgrade (1963). Beograd, Bulevar vojvode Misica 17:327–332

Werner RL (1996) Assessing groundwater flow-type using spring flow characteristics. The Professional Geologist 33(8):4–7

Zhang Y-K, Bai E-W, Libra R, Rowden R, Liu H (1966) Simulation of spring discharge from a limestone aquifer in Iowa, USA. Hydrogeology Journal 4(4):41–54

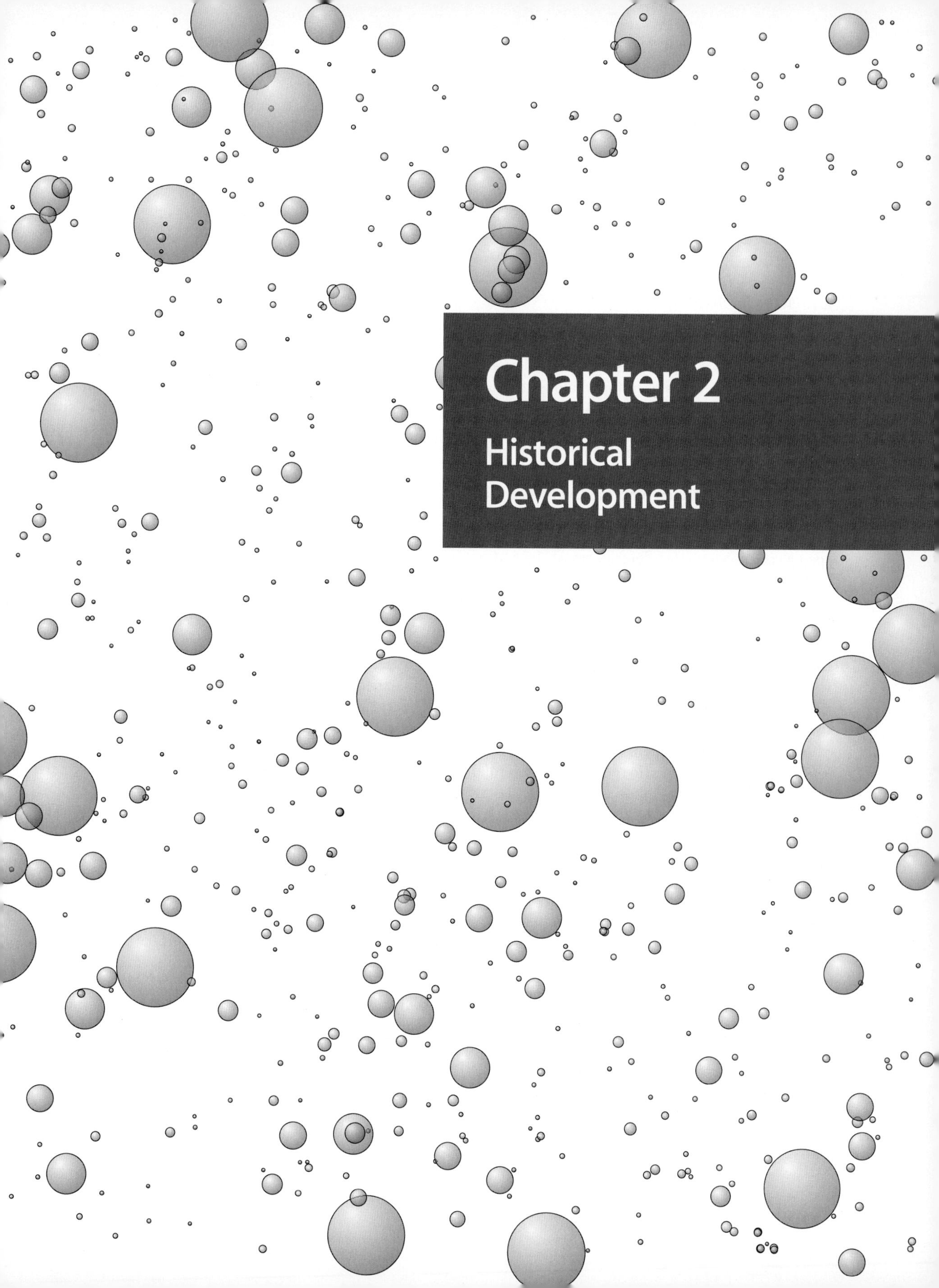

Chapter 2
Historical Development

CHAPTER 2
Historical Development

Philip E. LaMoreaux

2.1 Introduction

From the beginning of time water has been essential for survival. Thus, it is not surprising that evidence of the earliest civilizations has been found along the banks of rivers: the Tigris and Euphrates in Mesopotamia, the Nile in Egypt, the Indus in India, the Huang-He (Yellow River) in China and near large springs. The Chinese classified their emperors as being 'good' or 'bad' depending on whether they maintained their waterworks carefully or whether they allowed them to fall into disrepair. The earliest hydrologic concepts evolved by man concerning springs and the accepted description of the hydrologic cycle began to evolve during the Hellenic Civilization (600 B.C.).

According to Aristotle, "The earth floats on the water", and "Water is the original substance, and hence is the material cause of all things." Aristotle's experience was based on the knowledge of springs issuing from limestone rocks in Greece. The Egyptian priests believed that the earth was created out of the primordial waters of Nān and that such waters were still everywhere below it and that springs discharged these primordial waters. Xenophanes of Colophon lived within the period 570 to 470 B.C. He believed that the "sea is the source of water, and the source of wind. For neither could (the force of the wind blowing outward from within) come into being without the great main (sea), nor the stream of rivers, nor the showery water of the sky; but the mighty main (sea) is the begetter of clouds and winds and rivers." Thus Xenophanes presented an argument to prove his point stating that clouds, rains, springs, and streams all originate from the sea.

2.2 Origin of Rivers and Springs

One of the most complete sets of references to the origin, occurrence, and concepts related to the hydrologic cycle, groundwater and the occurrence of springs is contained in the book *History of Hydrology* (Biswas 1970). It must be recognized that the earliest concepts of water on earth evolved philosophically in early Greek times and the discussions by Plato, Tartarus, Pliny, Critias, Vitrivius, Aristotle, and in early biblical documents, for example, Ecclesiastes. Biswas' research on these early concepts is thorough and commendable.

Plato was born in the month of Thargelion (May–June) of the 1st year of the 88th Olympiad (428–427 B.C.). He accepted the concept of the four basic elements of matter, fire, air, water, and earth. Two possible explanations are available in the dialogues of Plato on the origin of rivers and springs, of which the most quoted hypothesis is the Homeric ocean concept. Plato believed that there were numerous interconnected perforations and passages, broad and narrow, in the interior of the earth. He imagined the existence of a huge subterranean reservoir called Tartarus. This was the largest of all chasms, and it penetrated the entire earth. The watery element had neither a bed or a bottom; it always surged to and fro. Plato wrote, "the water retires with a rush into the inner parts of the earth, it flows through the earth into those regions, and fills them up like water raised by a pump. When it leaves those regions and rushes back hither, it again flows into the nearby hollows, and when these are filled, it flows through subterranean channels and finds its way to several places, forming seas, lakes, rivers, and springs". All waters of rivers and streams flow back to Tartarus directly or through a circuitous route. Like the Egyptians, Plato was aware of one of the fundamental principles of water, that it always flows downhill. Perhaps that was why he stated that the exit of the rivers back into the earth was always lower than the level at which they originated. The flow from Tartarus to the rivers and vice versa was a continuous process.

An alternate explanation of the origin of springs and rivers is in Critias. Referring to conditions at Athens about 9 000 years before his time, Plato said:

Furthermore, it (the land of Attica in ancient times) enjoyed the fructifying rainfall sent year by year from Zeus; and this was not lost to it by flowing off into the sea, as nowadays because of the denuded nature of the land. The land (then) had great depth of soil and gathered the water into itself and stored it up in the soil we now use for pottery clay, as though it were a sort of natural water-jar; it drew down into the natural hollow the water which it had absorbed from the high ground and so afforded in all districts of the country liberal sources of springs and rivers; and surviving evidence of the truth of this statement is afforded by the still extant shrines, built in spaces where springs did formerly exist.

Explanation for the origin of rivers and springs is given by Aristotle:

Just as above the earth, small drops form and these join others, till finally water descends in a body as rain, so too we must suppose that in the earth the water at first trickles together little by little and that the sources of rivers drip, as it were, out of the earth and then unite. This is proved by facts. When men construct an aqueduct they collect water in pipes and trenches, as if the earth in the higher ground was sweating the water out. Hence, too, the headwaters of rivers are found to flow from mountains, and from the greatest mountains there flow the most numerous and greatest rivers. Again, most springs are in the neighborhood of mountains and of high ground, whereas if we except rivers, water rarely appears in the plains. For mountains and high ground, suspended over the country like a saturated sponge, make the water ooze out and trickle together in minute quantities but in many places. They also receive a great deal of water falling as rain.

Vitruvius (A.D. 400), perhaps the first hydrogeologist, described ways of finding water in Chap. 1 of his book. In the absence of surface springs, he stated water has to be sought and collected from underground. He suggested a test for locating underground water was to lie flat on the ground before sunrise (Fig. 2.1) in the area where water is to be sought, and with one's chin on the ground, to take a close look at the countryside, the reason being that the search will then be limited approximately to the same level on the ground. Water can be expected to be found in places where vapors arise from the earth. Water may also be sought in localities where there are plants which generally grow in marshy areas. When a promising location is found, a hole, not less than three feet square and five feet deep, is to be dug, and a bronze or lead vessel with its inside smeared with oil, is to be placed upside down in that hole at about the time of sunset. The hole is then to be filled with rushes or leaves and earth. If drops of water are found within the vessel on the subsequent day, water should be found at that location. Nowhere in his book does Vitruvius advocate using a divining rod.

Vitruvius also related types of soils to groundwater, and hence the nature of the ground should be studied carefully. Details regarding the availability of water in various types of soils (according to Vitruvius) are shown in Table 2.1.

Fig. 2.1. Vitruvius' method for locating water (Biswas 1970)

The German Jesuit, Athanasius Kircher (1602–1680) wrote on the subject of the origin of rivers and springs. He borrowed fundamental concepts from Ecclesiastes, namely that rivers receive their supply of water from the sea. He considered Aristotle's concept of transformation of air into water to be rather ludicrous. He, however, used part of the Stagirite's idea for promoting his own concept. He thought that there were many great hydrophylacia (caverns containing water) within the major mountain ranges of the world, and that they were formed by God in his great wisdom during the creation of the world. Rivers flow out of these caverns in various parts of the world in order that man can use them either for irrigation or navigation. Since the quantity of water in a hydrophylacium is not limitless (as stated in Plato's Tartarus), he had to seek another explanation for

Table 2.1. Details of water available in various types of soils (Biswas 1970)

Type of soil	Depth at which water may be available	Amount	Taste	Remarks
Clay	Near the surface	Scanty	Not good	–
Loose gravel	Lower down in the surface	Scanty	Unpleasant	Muddy
Black earth	–	–	Excellent	Available after winter rains
Gravel	–	Small and uncertain	Unusually sweet	–
Coarse gravel, common sand and red sand	–	More certain	Good	–
Red rock	–	Copious	Good	Difficult to obtain due to percolation
Flinty rock and foot of mountains	–	Copious	Cold and wholesome	–

Fig. 2.2.
Kircher's explanation of the origin of rivers and springs (Biswas 1970)

their sources of supply. He soon found it in a verse from Ecclesiastes. There were two major problems for the Jesuit to surmount, and he must be given credit for identifying them, even though his explanations were wrong. He recognized that difficulties lie in the nature of the connections between the hydrophylacia and the sea, and in the problem of raising seawater to a higher level than it was originally. Kircher believed that such difficulties could be overcome, saying that "there is no man with mind so dull, as not to be ready to follow on hands and feet, as the saying is, my trains of thought". Solving the first problem was comparatively easy, as he visualized seawater passing to a hydrophylacium through openings in the ocean-floor. Figure 2.2 is an idealized sketch of a mountainous area near the sea. The whirlpools in the figure indicate the locations of openings in the sea-bed through which water is carried by subterranean channels to the caverns of mountains from which the rivers originate. The water is returned to the sea by the rivers, thus completing his version of the hydrologic cycle. In the diagram, the subterranean channels appear in a darker shade whereas the rivers are shown in a lighter shade. Figure 2.3 represents a section of a mountain with a cavern being supplied through subterranean channels with water from the sea.

Various theoretical processes for getting water to flow uphill (from the sea to mountain tops) by mechanical methods were explained with elaborate diagrams. The first contemplated the use of a pair of double bellows powered by water-wheels to raise the water, the second, a U-tube filled with water. This provided the justification he needed to claim that seawater is forced through the openings in the bed, and that it flows under pressure to the mountain tops. High winds contributed a share of the pressure on the ocean surface, and they helped to force water through the subterranean channels created by God in his divine wisdom. A second method by which seawater could be raised to higher than its original level was claimed to be produced by the action of fire. Kircher believed that the underground world is:

> A well fram'd house with distinct rooms, cellars, and store-houses, by great art and wisdom fitted together; and not as many think, a confused and jumbled heap or chaos of things, as it were, of stones, bricks, wood, and other materials, as the rubbish of a decayed house, or a house not yet made.

Fig. 2.3.
Kircher's view of a cavern being supplied with seawater (Biswas 1970)

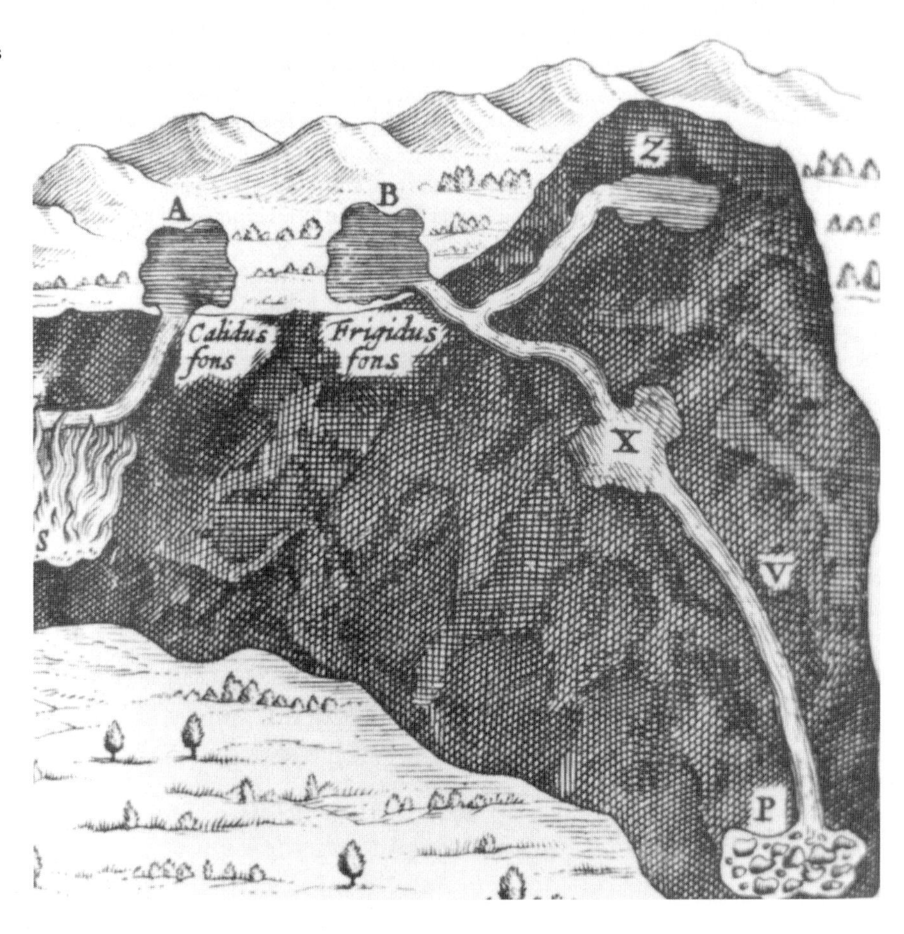

Fig. 2.4.
Origin of hot and cold springs according to Kircher (Biswas 1970)

The caverns are occupied by either fire or water, and all the fire-filled ones have a direct communication with the central fire. If the fire is near the surface, it could break out as a volcano, but if it is located deeply within the earth, it heats up the water in a nearby cavern. The water vaporizes, comes towards the surface, recondenses, and thus generates hot springs. Figure 2.4 shows a hot and a cold spring with origins close to one another. The explanation was very simple: the hot spring *A* is created by the passage of the subterranean channel *L* over the fire-filled cavern *S*, whereas the spring *B* is cold as there is no fire nearby. If, after being heated as in the case of spring *A*, water has to travel a long distance before it comes to the surface, cooling takes place and a cold spring is produced.

Formation of mineral springs is shown in Fig. 2.5. Issuing from a common hydrophylacium *A*, the water qualities of the springs change according to the mineral substances they encounter enroute. For example, spring *H* passes through sulphur-bearing rocks and, hence, it is sulphurous, whereas the water from spring *B* is pure because it does not pass through any soluble materials. It is evident that Kircher was deeply interested in groundwater and origin of springs and rivers but, unfortunately, his basic concepts were not correct.

Johann Herbinius published a book *Dissertationes de admirandis mundi cataractis supra et subterraneis* in Amsterdam in 1678. He listed the various causes for the origin of rivers and springs as follows: "Continuous movement of water in the subterranean abyss which, by virtue of motion drives the water up to the surface of the earth; angels; stars; the spirit of the earth; and air enclosed within the earth." Strangely enough, Herbinius completely disregarded precipitation as even a possible reason! However, he classified his reasons into two types,

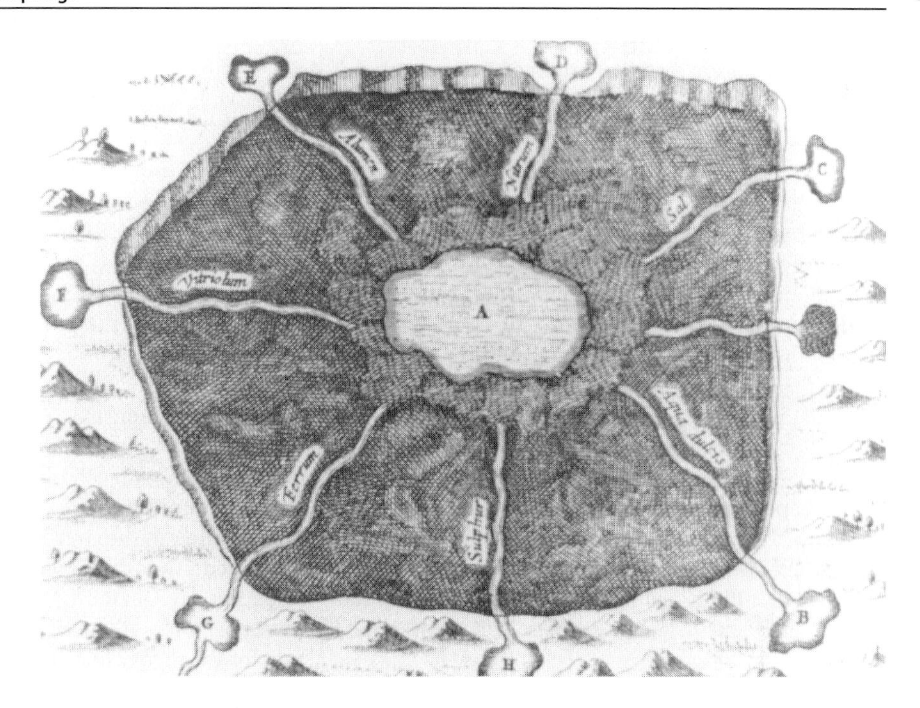

Fig. 2.5. Kircher's explanation of formation of mineral springs (Biswas 1970)

true and false; the first two causes were true and all of the others were false. God is the primary cause; the subterranean abyss is the secondary. Figure 2.6 shows Herbinius' concept of rivers, springs, and artesian conditions originating from water-filled subterranean caverns.

Edmé Mariotte (1620–1684) claimed that rainfall was more than adequate for the creation of springs and rivers. His concept was probably influenced by Perrault's work. The general concept was: rain → infiltration → springs → rivers. Whereas, Pierre Perraults' opinion was: rain → rivers → springs. Percolation takes place after precipitation. When rainfall occurs on hills and mountains, it penetrates the surface of the earth, particularly when the soil is light and mixed with pebbles and roots of trees. Then, if it encounters a layer of clay or a bed of continuous rock, which it is unable to infiltrate, it flows along the surface thereof until it finds a point of egress either at the bottom of the mountain or at a considerable distance below the top. There it breaks out as a spring. Besides proving by experimental investigations that the amount of water that falls as rain is more than enough to supply all the springs and rivers, he demonstrated that the increase or decrease in the flow of springs is directly related to the amount of precipitation. If it does not rain for two months, most streams lose half of their flow, and when a rainless condition continues for a year, most of them become dry. The few which continue to flow do so with a greatly reduced discharge.

Fig. 2.6. A hydrophylacium within a mountain, filled with water, feeds the nearby spring (after Herbinius, Biswas 1970)

2.2.1 Evaporation and the Origin of Springs

There was a fundamental difference between the explanation of the French scientists and the English astronomer on the origin of springs. Perrault and Mariotte concluded that springs originated from intermittent rains, but Halley stated that water is being continually condensed out of vapor on the long mountain ridges, and: "it may almost pass for a rule, that the magnitude of a river, or the quantity of water it evacuates is proportionable to the length and height of the ridges from whence its fountains arise."

2.2.2 Capillary Theory of Springs

Reverend W. Derham (1657–1735), in his book, *Physico-theology*, first published in 1713, put forward the capillary theory of the origin of springs. Switzer described the concept as follows:

> As to the manner how waters are raised up into mountains, and other high lands, and which has all along puzzled so many great men (Mr. Derham says) may be conceived by an easy and natural representation, made by putting a little heap of sand or ashes, or a little loaf of bread, into a basin of water, where the sand will represent the dry land, or an island, and the basin of water the sea about it; and as the water in the basin rises up to or near the tops of the heap in it, so does the water of the sea, lakes, etc. rise in hills: which case he takes to be the same with the rise of liquids in capillary tubes, or between contiguous plains or in a tube fill'd with ashes …

Switzer fully agreed with Derham's concept and considered that the origin of streams and rivers cannot be entirely due to precipitation. He stated that Derham's idea was based on his own meteorological observations, and thus they were very exact, and, hence, beyond dispute.

One of the major chapters of the book *Le spectacle de la nature* by N. A. Pluche (1688–1761), published in 1732, was devoted to the origin of springs. The capillary theory was put forward very clearly and concisely by one of the characters of the book:

> I firmly believe that the seawater deposits its salt on the sands below, and that it rises little and little, distilling through the sands, and the pores of the earth, which have such a power of attraction as is not easily accounted for; and that not only sand, but other earthly bodies have the power of attracting water, I am well assur'd of from an observation which occur'd to me but this very day. When I threw a lump of sugar into a small dish of coffee, I found that the water immediately ascended thro' the sugar, and lay upon the surface of it. Yesterday I observed, likewise, that some water which had been pour'd at the bottom of a heap of sand, ascended to the middle of it. And the case, as I take it, is exactly the same with respect to the sea and the mountains.

This approach, very similar to that of Bernard Pallisey's *Theory and practice*, used a main character, the pundit of the book, who vigorously opposed the theory on three counts. First, water cannot rise more than 32 ft in dry sand, and even then that height is very seldom achieved. Second, the growth of algae will prevent the passage of water after some time, and finally, if it was true, seawater, for the same reason, would saturate all the plains adjoining the coast. Pluche believed in the pluvial origin of springs, and firmly discounted Descartes' concept on the subject. He calculated that if 1 ft^3 of seawater contained only 1 lb of salt instead of the usual 2 lb, the daily flow of the river Seine alone (288 million ft^3, as calculated by Mariotte) would deposit 288 million lb of salt every day. Obviously, the quantity of salt that would be deposited by all the rivers of the world would be too vast for the theory to be true.

2.3 Groundwater Utilization

Undoubtedly, the greatest achievement in the utilization of groundwater in ancient times was the building of qanats (or kanats). A qanat is an artificial underground channel which carries water over long distances, either from a spring or from water-bearing strata, and it solved several problems in water resource engineering. Qanats kept water cool and free of surface pollutants. Figure 2.7 is an aerial photograph of qanat systems originating in the talus deposits at the foot of the mountain near Kashan in Persia. Figure 2.8 shows a typical water supply system by qanats; the cross section shows the sloping main channel and entrance shafts along the tunnel used during excavation to remove the excavated material. The qanats literally brought groundwater along a "spring line" down a slope or gradient to a point of discharge convenient for the use of the water.

A primitive type of water meter was used at the Gadames oasis in North Africa more than 3 000 years ago, and it is still being used without modification. This oasis has a small spring called Ain el Fras ("Spring of the Mare") which, according to legend, was discovered by the horse of an Arabian conqueror. The spring discharges around 180 m^3/h. The water is collected in a basin and distributed through a main canal and two side canals.

Fig. 2.7.
Aerial photograph of qanat systems in Persia (Biswas 1970)

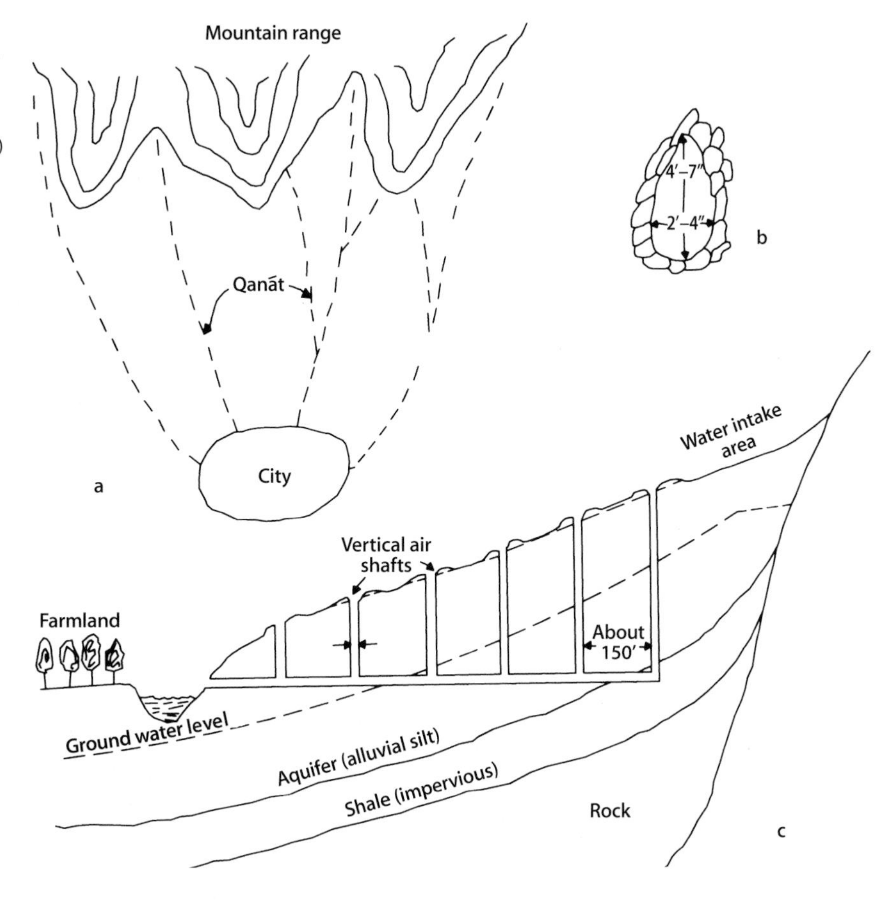

Fig. 2.8.
Details of qanat system (not to scale). **a** Typical qanat layout for water supply; **b** cross-section of a qanat; **c** longitudinal section of a qanat (Biswas 1970)

2.4 Religion, Springs and the Bible

The large springs called Ayun Musa at the northwestern edge of the Sinai Desert seem inexhaustible. They are on the route of Exodus and supplied water at the first stopover of the Israelites after "crossing the sea" on their march toward the promised land. Water from artesian flowing core test wells drilled in recent years in search of coal penetrated the aquifer that supplies Ayun Musa and floods an extensive marshy area partially covered by thick clusters of reeds and tamarisk. These low-lying, salt, marshy areas are called sebkahs. The water from the springs, abandoned core test and wells is slightly salty and sulphurous, however, it is potable. The springs at Ayun Musa are artesian, flowing upward to the land surface along a fault. They are used by the Bedouins today, and were formerly used by the village of Suez. As the spring water surfaces, its carbonates and sulfates precipitate and forms mounds of porous tufa. The mounds have grown to heights of several meters, an indication supported by the ^{14}C dating that the springs have been in existence for many thousands of years.

What is the source of this rich flow of water at the edge of the desert? Israeli hydrogeologists have determined that the Ayun Musa springs provide an outlet for a deep underground aquifer, or reservoir, storing a huge quantity of water (Fig. 2.9). The water has been dated to about 10 000 years and is called paleo or fossil water. Artesian pressure forces this water to the surface to flow from springs or wells (Fig. 2.10). Four distinct lines of evidence support the conclusion that such an aquifer underlies the peninsula. Geological surveys of the peninsula show the existence of the aquifer. Chemical and isotopic analyses indicate that the waters of springs many kilometers apart from recharge of the aquifer to the northeast to discharge at Ayun Musa have similar characteristics and have one common source and suggest that the aquifer holds rainwater that was trapped during the most recent ice age and discharge it down the

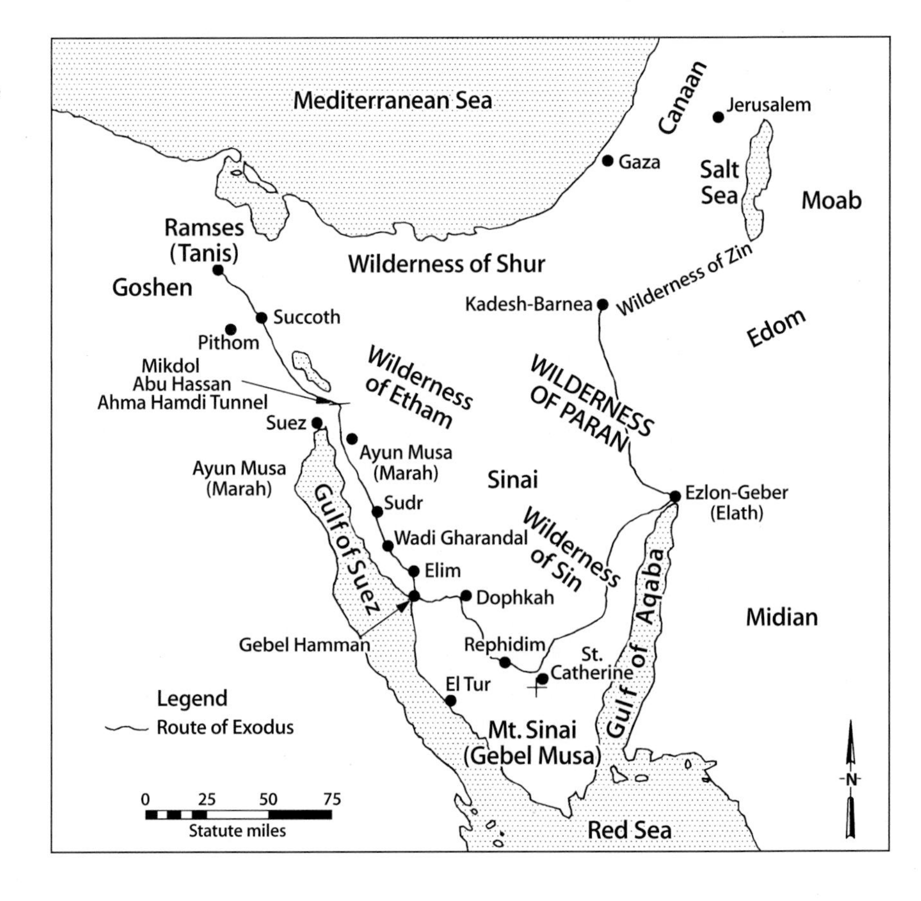

Fig. 2.9.
Generally accepted biblical route of Exodus showing the location of Ayun Musa

Fig. 2.10.
a Spring at Ayun Musa, or known in history as Moses' Spring. There are at least a dozen spring discharge points where water rises along faults from deep-seated beds in the Cretaceous formation (photograph taken by P. E. LaMoreaux in 1953).
b Ayun Musa, Bedouin shacks, and palm trees. In the distance is the Gulf of Suez and Suez. Ayun Musa is about 15 km south of the entrance of the present-day Suez Canal

slope of water-bearing beds toward Ayun Musa where it discharges along a geologic fault.

The Bible describes in many places the importance of springs, wells and water to ancient civilizations. Early mans' activities revolved around these water supplies and they exerted a strong influence over mans' migrations. Some examples from the Exodus are:

1. In the Bible, Genesis 45:4–7 and 50:20, Joseph states that he moved his people into Egypt to save alive "his breathen". They were escaping from the drought. The eastern desert was not heavily populated at that time; the Pelusiac Branch of the Nile still flowed across the area; there was food and water – a veritably large oasis to these nomadic people faced with starvation and meager living in the desert.
2. In the story of Moses, the location of springs controlled the route of Exodus. The story from the King James version is not materially different from the earliest texts in Greek. A few excerpts from the Bible are as follows:

So Moses brought Israel from the Red Sea: and they went out into the Wilderness of Shur: and they went three days in the wilderness and found no water. And when they came to Marah they could not drink of the waters of Marah, for they were bitter (Ex. 15:22–23).

And they came to Elim where there were twelve wells of water and three-score and ten palm trees (Ex. 15:27).

The ancient springs of Ayun Musa are about a 3 days' walk to the south from "Miktol" near the northern tip of the Gulf of Suez. The locality where the Israelites crossed the Red Sea. At Ayun Musa there are dozens of springs, shallow wells, and palm trees. The springs are a unique present-day watering place. To the south, about 50 km (30 miles) at Gebel Hamman, on the Gulf of Suez, the mountains extend to the sea, and where the mountains meet the sea there is an area of artesian-flowing springs (Fig. 2.11). At Gebel Hamman, where the Pharaoh's hot springs have been flowing for thousands of years, Pharoah's soldiers and miners of the Third Dynasty obtained water and took hot baths 1 000 years before Moses' time. The springs at Gebel Hamman are similar to those at Ayun Musa, and yield water from a deep aquifer under artesian pressure. These hot springs, or the "Springs of Pharaoh" are also the result of faulting (Gebel Hamman; Fig. 2.12).

Another famous Biblical story about spring water took place nearby in Wadi Feiran. The Bible tells us that after three month's travel, the Israelites were tired, discouraged, and complained bitterly. They told Moses he had brought them to this remote, desolate place to die. During Moses' earlier travels as a shepherd in exile in this land, he had learned that in Wadi Feiran there were places where water could be found at all times of the year. During rains there were springs and flooding. He knew that in wet periods there were springs that flowed over the rocks, and also in dry periods it was possible to dig shallow wells at these places "marked by a dark reddish rock" exposed in the walls of the valley. This is where Moses "smote the rock" and water gushed forth (Fig. 2.13

Fig. 2.11. Looking south from the mouth of Wadi Gharandal toward the very prominent west-facing scarp of Gebel Hamman. At the base of Gebel Hamman scarp, the present-day springs are still a tourist attraction. A thousand years before Moses' time, these were springs used by the Pharaohs and are still known as Pharaoh's Springs

and 2.14). The Bible (Exodus 17:1–6) thus provides us with an important clue about the unique hydrogeological phenomenon in Wadi Feiran:

And all the congregation of the children of Israel journeyed from the Wilderness of Sin, after their journeys, according to the commandment of the Lord, and pitched in Rephidim: and there was no water for the people to drink. Wherefore, the people did childe with Moses, and said, Give us water that we may drink. And Moses said unto them, Why chide ye with me? Wherefore do ye tempt the Lord? And the people thirsted there for water; and the people murmured against Moses, and said, Wherefore is this that thou has brought us up out of Egypt, to kill us and our children and our cattle with thirst? And Moses cried unto the Lord, saying, What shall I do unto this people? They be almost ready to stone me. And the Lord said unto Moses, Go on before the people, and take with thee of the elders of Israel; and they rod, wherewith thou smotest the river, take in thine hand, and go. Behold, I will stand before thee there upon the rock in Horeb; and thou shalt smite the rock, and there shall come water out of it, that the people may drink. And Moses did so in the sight of the elders of Israel.

Fig. 2.12. Springs at the base of Gebel Hamman are called the "Pharaoh's Springs". Formerly used by the Pharaoh's miners of ancient Egypt, by Moses and the Israelites, and now by tourists. These springs, in former geological times, discharged at a much higher elevation on the scarp of Gebel Hamman, as evidenced by caves and tuffaceous deposits

Fig. 2.13. Steep walls of Wadi Feiran, approximately 90 km east of St. Catherine, showing multiple vertical dark bands or dikes in the mountains

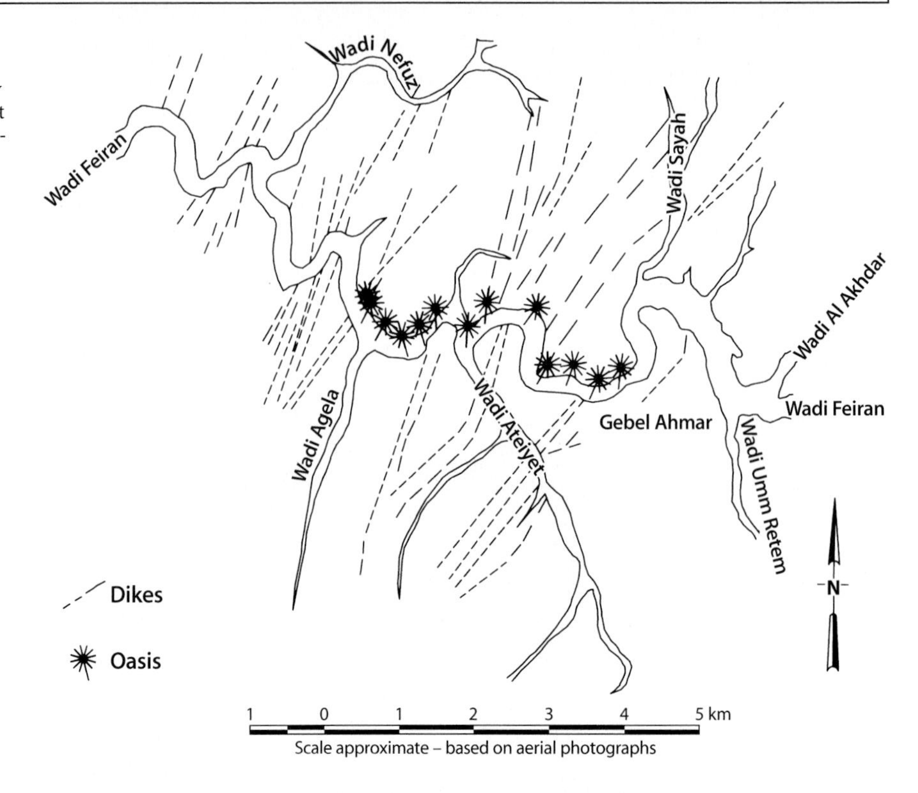

Fig. 2.14.
Map showing numerous dikes along fractures roughly trending northeast/southwest that cross Wadi Feiran (modified from Issar and Eckstein 1969)

2.4.1 Figeh Spring – Damascus, Syria

Figeh Spring is one of the three largest springs in the world. It is the sole source of water supply for the city of Damascus, generally considered to be the oldest continually inhabited city in the world. The Barada River is fed by Figeh Spring and the Barada Valley through the desert is like a green swathe of vegetation across burning sands. Legend is that it was the Garden of Eden of the Bible. Figeh Spring produces a minimum flow of 3 m^3/s into the Barada River. The Romans once used the spring as a health spa, the ruins from which have crumbled down into the spring opening.

Figeh Spring is a coveted source of water, and is guarded 24 h a day to prevent vandalism and contamination. Figeh Spring issues from a huge solution cavity or cavern in limestone rock. The water flow is about 1/3 million or 329 280 gal/min. The water is of exceptional quality – crystal clear and sparkling. It is drinkable as it comes directly from the spring.

Springs have played an extremely important part in history, religion and military expeditions. In Europe and the United States in more recent times, they have become a source of great commercial, industrial and touristic value and as a source of bottled water. In Europe, spas have been developed and around these facilities entire villages and industries have followed. They have been the catalyst for tourism. As we enter the 21st century, springs as a source of bottled water are producing revenues of billions of dollars per year and famous spring waters such as Perrier, Poland, Vichy, Fuggi, Dillons, Anjou, Evian, Excelsior, LaCroix, Apollinaris, Clearly Canadian, Swiss Altima, and many others have become popular names (Fig. 2.15 and 2.16).

The fear of pollution causes people to pay high prices for bottled water. Much pollution of groundwaters in Europe and the United States has resulted from the industrial revolution in the 1940s and 1950s. Municipal, industrial, and agricultural wastes are all primary sources of this pollution. Publicity regarding the environmental movement during the 1960s and 1970s made people aware of harmful chemical constituents in water supplies such as selenium, arsenic, lead, zinc and others. New legislation at national and lower levels in government have attempted to remedy the problem, however

Fig. 2.15. Individual bottled waters

Fig. 2.16. Sample labels of bottled water

people have developed a strong perception that many water supplies are no longer safe to drink, and thus the bottled water industry has exploded in size during the past 10 years. Today while shopping, riding a bus, train, or airplane, at restaurants and fast-food establishments, at athletic events, and around the home, bottled water has become a necessity not a luxury.

This has resulted in "a cost of water", which, when supplied in containers and sold over the counter to the public, exceeds the cost of a barrel of crude oil or a gallon of gasoline. The old adage "water is important to life" has been brought to the public's attention and they are now forced to recognize that water costs money. An interesting article, *Environmental Geology*, compares the value of water purchased from the city of Wichita to Kansas crude and gives a computer value per barrel of some of the more famous bottled waters (Hansen 1992).

Springs, spas, and bottled waters are also preferred for other reasons. On visiting Hot Spring National Park, Arkansas, one of the most famous spas in America, an individual must be impressed with the economic impact of this spring on a small village in the United States. A brochure (Fig. 2.17), beautifully prepared by the National Park Service, US Department of the Interior, explains in greater detail that "water (spring water) attracts people to Hot Springs":

> Water. That's what attracts people to Hot Springs. In fact they have been coming here since the first person stumbled across these hot springs perhaps 10 000 years ago. Stone artifacts found in the park give evidence that Indians knew about and used the hot springs. For them the area was a neutral ground where different tribes came to hunt, trade, and bathe in peace. Surely they drank the spring waters, too, for they found the waters with its minerals and gases to have a pleasant taste and smell. These traces of minerals, combined with a temperature of 143 °F, are credited with giving the waters whatever therapeutic properties they may have. Waters from the cold springs, which have different chemical components and properties, are also used for drinking. Besides determining the chemical composition and origins of the waters, scientists have determined that the waters gushing from hot springs are more than 4 000 years old. And the waters gush at an average rate of 850 000 gallons a day!

The brochure describes what's special about this water, what makes this water hot, provides a history and excellent maps and charts. But what is even more significant is the influence that Hot Springs has on the economy of this area. Merely by observing the variety and number of commercial brochures at the Information Center and studying these brochures about the commercial and public facilities in the area, it is possible to

Hot Springs

Hot Springs National Park
Arkansas

National Park Service
U.S. Department of the Interior

Hot Springs National Park is the unusual blend of a highly developed park in a small city surrounded by low lying mountains abounding in plantlife and wildlife. It is a park with a past, too. The lithograph on the cover (above) shows Bathhouse Row as it looked in 1888. None of these bathhouses exists in this form today, though some of the names live on in present-day structures. They either burned or were pulled down to make way for new, fireproof structures. The Fordyce Bathhouse Visitor Center sits on the site of the Palace in the above drawing. The water boy with his pot and glasses (far left) was a familiar sight at the turn of the century. A similar pot can be seen in one of the visitor center exhibits. From left to right the photographs reveal the variety of scenes of Hot Springs National Park today from the solitude of a forest trail to the wide vistas along the ridgetops of the mountain range. The mule-drawn car echoes the transportation of a century ago in the cover lithograph.

Water. That's what attracts people to Hot Springs. In fact they have been coming here since the first person stumbled across these hot springs perhaps 10,000 years ago. Stone artifacts found in the park give evidence that Indians knew about and used the hot springs. For them the area was a neutral ground where different tribes came to hunt, trade, and bathe in peace. Surely they drank the springwaters, too, for they found the waters with its minerals and gases to have a pleasant taste and smell. These traces of minerals, combined with a temperature of 143°F, are credited with giving the waters whatever therapeutic properties they may have. Waters from the cold springs, which have different chemical components and properties, are also used for drinking. Besides determining the chemical composition and origins of the waters, scientists have determined that the waters gushing from hot springs are more than 4,000 years old. And the waters gush at an average rate of 850,000 gallons a day!

Tradition has it that the first Europeans to see the springs were the Spanish explorer Hernando deSoto and his troops in 1541. French trappers, hunters, and traders became familiar with the area in the late-17th century. In 1803 the United States acquired the area when it purchased the Louisiana Territory from France, and the very next year President Thomas Jefferson dispatched an expedition led by William Dunbar and George Hunter to explore the newly acquired springs. Their report to the President was widely publicized and stirred up interest in the "Hot Springs of the Washita." In the years that followed, more and more people came here to soak in the waters. Soon the idea of "reserving" the springs for the Nation took root, and a proposal was submitted to the Congress by the territorial representative, Ambrose H. Sevier. Then, in 1832, the Federal Government took the unprecedented step of setting aside four sections of land here, the first U. S. reservation made simply to protect a natural resource. Little effort was made to mark the boundaries adequately, and by the mid-1800s, claims and counterclaims were filed on the springs and the land surrounding them.

The earliest bathhouses were crude structures of canvas and lumber, little more than tents perched over individual springs or reservoirs carved out of the rock. Later, wooden structures were built, but they frequently burned, collapsed because of shoddy construction, or rotted due to continued exposure to water and steam. Hot Springs Creek, which ran right through the middle of all this activity, drained its own watershed and collected the runoff of the springs. Generally it was an eyesore; dangerous at times of high water, and mere collections of stagnant pools at dry times. In 1884 the creek was put into a channel, roofed over, and a road laid down above it. Today this is Central Avenue.

In the 1870s the government continued to control the springs and to reserve certain areas as federal property. Private bathhouses, under the supervision of the Federal Government, were allowed to be built. These establishments ranged from the simple to the luxurious. The government even operated a U.S. Free Bathhouse and a Public Health facility. Gradually Hot Springs came to be called "The National Spa," and such slogans as "Uncle Sam Bathes the World" and "The Nation's Health Sanitarium" were used to promote the city. By 1921, the Hot Springs Reservation was such a popular destination for vacationers and seekers of health remedies that the new National Park Service's first director, Stephen Mather, convinced Congress to declare the reservation the 18th national park. Monumental bathhouses built along Bathhouse Row about that time catered to crowds of health seekers. These new establishments, full of the latest equipment, pampered the bather in artful surroundings. Marble and tile decorated walls, floors, and partitions. Some rooms sported polished brass, murals, fountains, statues, and even stained glass. Gymnasiums and beauty shops helped cure seekers in their efforts to feel and look better.

The Army/Navy Hospital (now the Hot Springs Rehabilitation Center) just above the south end of Bathhouse Row contributed to a continued high level of activity during World War II and immediately afterward. Shortly thereafter, however, changes in medical technology and in the use of leisure time resulted in a rapid decline in water therapies. People also started to prefer taking to the open roads in their own cars rather than traveling by train to a specified destination and staying in a hotel a week or two. One by one the bathhouses began to close down as business declined. Today only one of the buildings on Bathhouse Row operates as a traditional bathhouse.

Despite the decline, bathing continues to be a popular pastime. A full range of options is available today: tub and pool baths, shower, steam cabinet, hot and cold packs, whirlpool, and massage. The bathhouses are operated by private concessioners or special use permit holders who provide services in accordance with regulations and inspections by the National Park Service. Information about rates and services can be obtained at the bathhouses of the park's Fordyce Bathhouse Visitor Center.

Do not pass up the opportunity to take advantage of the experience of bathing in the hot spring waters. In a couple of hours you may find more relaxation and pleasure than you had ever imagined. And you will join a long line of people who have bathed in the Hot Springs of Arkansas—a line that goes back 10,000 years.

What's Special About This Water?

The most important thing about Hot Springs' thermal water is that it is naturally sterile. For this reason the National Aeronautics and Space Administration chose this water, among others, in which to hold moon rocks while looking for signs of life. Even during the many early years that the springs were uncovered, the absence of bacteria in the water helped prevent the spread of disease. Today most of the springs have been covered to prevent contamination. Various open springs and a spring cascade on the Arlington Lawn give an idea of what the area would be like if all the springs were open with the water pouring down the hillside into the creek.

What Makes This Water Hot?

In an arc from the northwest around to the east, outcroppings of Bigfork Chert and Arkansas Novaculite absorb rainfall. The pores and fractures in the rock conduct the water deep into the Earth. As the water percolates downward, the increasingly warmer rock heats it, and filters out the impurities. In the process the water dissolves minerals in the rocks. Eventually the water meets the faults and joints in the Hot Springs Sandstone leading up to the lower west side of Hot Springs Mountain where it flows to the surface.

Restoration Efforts

With the decline of bathing in the 1950s, the bathhouses themselves began to close their doors and to fall into disrepair. On Bathhouse Row only the Buckstaff remains open at the present time. In the 1980s local citizens and the National Park Service began exploring ways to return the bathhouses and Bathhouse Row to the splendor, if not the function, of Hot Springs in its heyday. This has resulted in the fortuitous union of private money and public guidance to return the exteriors of the buildings to their original grandeur and the interiors restored and adapted for a multitude of new uses under the provisions of a historic property leasing program. This is an example of the merger of the needs of the future with the preservation of the past and is an essential element in the revitalization of Bathhouse Row and downtown Hot Springs.

The Fordyce

In May 1989 the Fordyce Bathhouse reopened its doors after having undergone extensive restoration. To someone who visited this bathhouse between 1915 and 1920, little is amiss, the restoration has been that thorough. All of the women's side and some of the men's side of the building have been equipped with the furniture, steam cabinets, mechano-therapy machinery, tubs, massage tables, sitz tubs, Hubbard tub, billard table, beauty parlor, and hydrotherapy equipment prevalent in those days.

Fig. 2.17. Brochure for Hot Springs, Arkansas

determine the extent and breadth of this economic impact. A few examples are: Guide to a Dream Vacation, Magic Springs Theme Park, Hot Springs Plaza, Mid America Museum, The Museum of Automobiles, Panther Valley Ranch, Inc., Crafton's Crafts & Stuff, Mountain Brook Stables, Inc., Oaklawn Racing, Was Museum, White and Yellow Ducks Land/Water Sightseeing Tours, The Bath House Show, The Old Country Store, Hot Springs Diamonds, Canoe Rental Service, Country Cupboard Antiques, Wright's Rock Shop, Arkansas Alligator Farm & Petting Zoo, Inc., Belle of Hot Springs-Riverboat Cruising, Hot Springs Mountain Tower-Beyond the Splendor, Gator Golf, Music Mountain Jamboree, The Old Mill Gift Shop, The Basket House, Mountain Valley Spring Company, The Arlington Resort Hotel & Spa, and many others (Fig. 2.18).

Facts are easily obtained that would identify the same type of commercial and industrial opportunities asso-

Fig. 2.18. Types of brochures advertising springs and facilities

ciated with other springs and spas; for example, in Florida: Silver Springs, Homosassa Springs, Weeki Wachee Springs, Ginnie Springs, Rainbow Springs, and many others. These have become veritable tourist gold mines for Florida.

Michael Fricke, in an article, *Bottled-Water, Made in USA*, published by Verlag W. Sachon, 1992, presented an excellent summary on the marketing of bottled waters in America. Pertinent excerpts from this article are as follows:

The United States of America – or USA for short – occupies half of a continent and has a population of over 250 million scattered over this large land mass. The vastly differing population densities create vastly differing demands from the inhabitants for a high standard of drinking water, because water from public drinking-water suppliers, in states with a high to very high population density, could not be said to be of a high standard. Every European tourist who visits the USA for the first time – especially Washington DC –, cannot fail to notice the high chlorine content, as well as the frequently bad taste of the drinking water.

However, the awareness of large parts of the population in respect to drinking water quality must have increased over the last 5 years because it was declared recently at the annual convention of the IBWA (International Bottled Water Association) in Dallas, Texas, that: 'In spite of the general recession there was an increase of 6% in sales of bottled waters in 1990.'

It is true that for some time the IBWA has been trying to push a single Bottled Water Standard for all the states in USA, but to date (June 1991) only 14 states – from Arizona to Wyoming – have taken the essential parts of the IBWA Standards and passed them as legislation.

It's not just the geological conditions in the USA that are varied. Just as varied is the natural offering, the quality and quantity of the groundwater resources, and also the covering laws and legislature, differing from state to state, that provide the regulations for production, extraction, manufacture, labelling, and organisation.

The American fundamental ideas concerning all products for human consumption (food and product safety), becomes especially clear where individual definitions are most identical with those in Europe: artesian water, mineral water, natural water, spring water, and well water.

Due to the populations' fear of bacterial contamination of a daily food product, and the resulting precautions of the legislators, it is almost impossible to purchase untreated water – whether it is bottled, or otherwise packaged.

Because of the hygienic Product Safety, and the high damage claims if it is proved that the manufacturer has violated these rules, every type of water that is filled into containers and then in some way brought into circulation is either ozonised, chlorinated, or otherwise disinfected – a procedure that, for high quality natural mineral water, would be totally unthinkable in Europe. Our French neighbours would have especial difficulty in understanding the process, as they consider water inherent micro fauna as a specific and occasionally characteristic element of their 'natural mineral water' – and not just for balneological baths.

The US market for mineral water is frequently underestimated, as is the capacity and the capability of the American contractor to adjust quickly to changing market situations.

Generally the US waters are treated disdainfully in Europe, and are filed as no comparison, or rather no competition, for the local products.

Nearly all bottled waters in the USA have been without doubt subjected to disinfectant, and without doubt the packaging in PVC or PET milk jugs is not suitable for Europe.

... The US market for bottled water has taken an upward swing in the last 10 years, that is remarkable.

For 11 l consumed per head of the US population in 1980, it was 36 l per head in 1990; bottled water consumption had, in the space of 10 years, tripled.

The value of the total bottled waters sold had, in 6 years, more than doubled.

The volume of business of all bottled water manufactured in 1984 was estimated to be approximately $1.2 billion US. In 1990, the value was approximately $2.6 billion US.

There are great differences between individual sales groups and types:

Although non-carbonated bottled water was the most sold, with nearly 7 billion l in 1990, it had an unproportionally low share of the total 4 billion DM because of its below average wholesale price of approximately 0.35 DM/l.

Water which had been carbonated – sparkling water – reached almost triple the price with 1.00 DM/l, and imported water, lead by Perrier and Evian, reached an average price of over 1.50 DM/l.

In Chap. 7, the legal ramifications of bottled waters are discussed in greater detail.

References

Biswas AK (1970) History of hydrology. Resources Research Centre, Department of Energy, Mines and Resources, Ottaway, North-Holland Publishing Company. American Elsevier Publishing Company, Inc., New York, 336 p

Fricke MJ (1992) Bottled-Water. Made in USA, Verlag W. Sachon

Hansen TJ (1992) Environmental Geology. American Institute of Professional Geologists, 7828 Vance Drive, Arvada, Colorado. The Professional Geologist 29(7):4–5

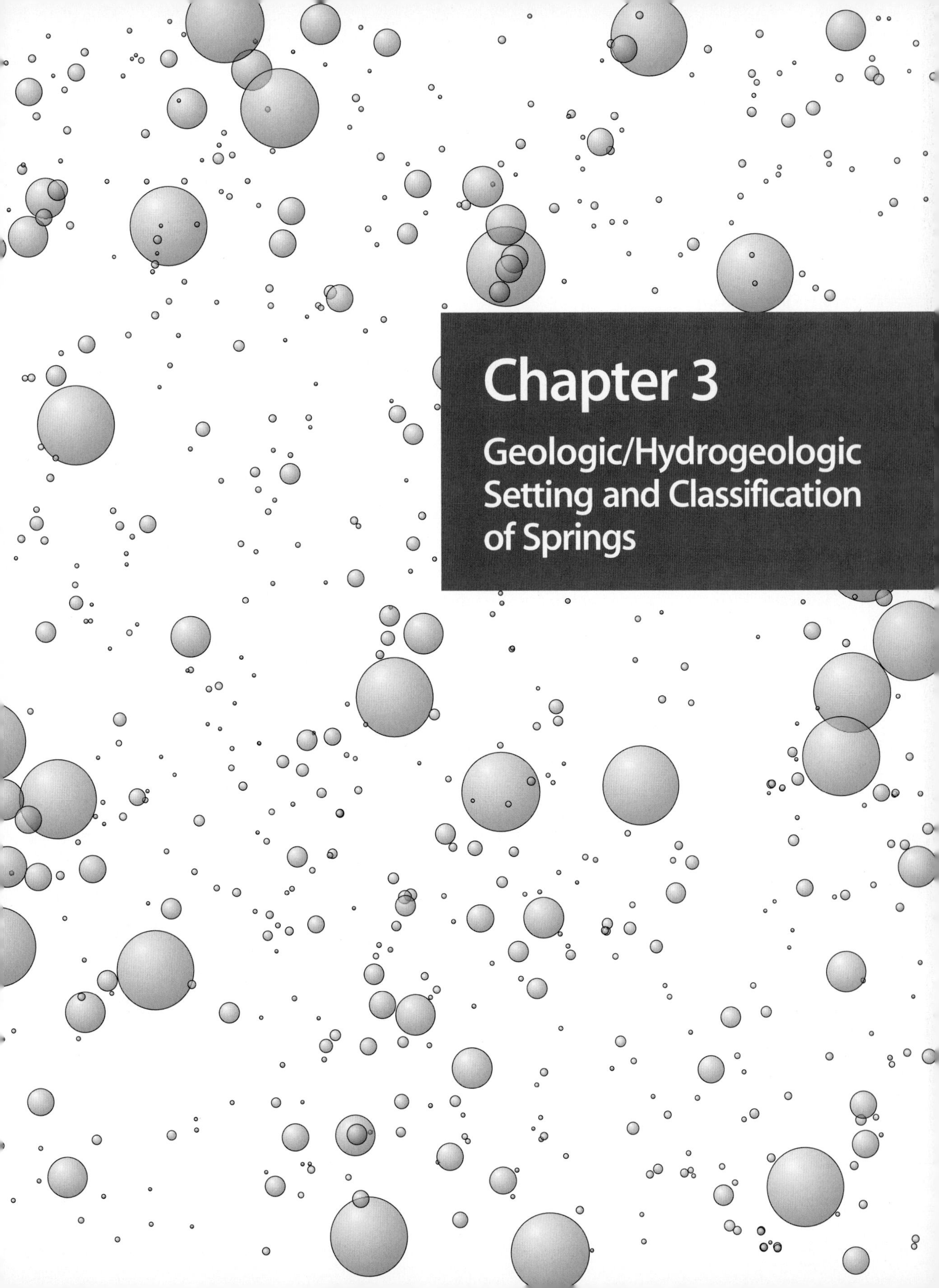

Chapter 3

Geologic/Hydrogeologic Setting and Classification of Springs

CHAPTER 3
Geologic/Hydrogeologic Setting and Classification of Springs

Mary Wallace Pitts · Caryl Alfaro

3.1 Introduction

Springs are defined as places where groundwater flows naturally from a rock sediment or soil onto the land surface or into a body of surface water. Their occurrence is not limited to rocks of a particular age or type or to any specific geological or topographic setting. The diversity of springs is indicative of the wide array of geologic and hydrologic conditions which lead to their occurrence. Springs are dynamic and evolve ebb and flow in response to changes in climatological, topographical, geological, and geomorphological conditions.

It is difficult to adopt an approach for their origin or classification because of such diversity of form and characteristics. Therefore, it is important to recognize that certain conditions must be fulfilled before a spring can occur. From the most insignificant trickle on a hillside to the spectacular rush of water from geysers at Yellowstone Park, all springs share some common criteria.

First, a water-bearing rock unit, or aquifer, must be present. The capacity of this unit to act as an aquifer, plus receiving recharge, generally from precipitation as a surface water source, determines the magnitude and permanency of the discharge from a spring. Crucial also, is the aquifer's capacity to store, transmit and discharge water relative to the geologic units adjacent to the aquifer. Groundwater moves down gradient and along the line of least resistance. A final essential element to spring flow is the size and character of the recharge area (Fig. 3.1 and 3.2).

Perhaps the most interesting criterion to the layman is the character of the point of discharge or its surface expression. The point of discharge can range from an imperceptible seep or swampy area to an opening in a large cave. Scientifically, the most fascinating aspect related to spring flow is the mechanism or group of mechanisms that create the spring: type of source rock, the character of adjacent beds, dynamic face, gravity, permeability, geomorphologic processes, and geologic structure are all important in determining the type and magnitude of spring discharge. Chemical processes and geothermal activity are responsible for some of the most spectacular and diverse forms of springs. If one process

Fig. 3.1. Thunder Spring discharges from above Cambrian Muav Limestone, Grand Canyon of Colorado, USA, known as Vassey's Paradise (photograph by P. E. LaMoreaux, May 1996)

Fig. 3.2.
a TW-2 Spring, Jonah Creek, Childress County, Texas, USA. Salt-water discharges from Blair Formation to Creek. Control pumpage stops flow. **b** Spring discharge point (photographs by P. E. LaMoreaux 10 April 1996)

dominates, it is possible to identify a "typical" form of spring; an intermittent eruption, tufa, chalybeate. However, where a number of processes act together to create conditions for spring flow, the identification of a spring type becomes more difficult.

A spring is a concentrated discharge of groundwater issuing from a defined opening as distinguished from a seep, or seep spring, in which the groundwater appears at the land surface along a zone or area of wetness or oozing type of flow. Meinzer (1942) designates springs which ooze or percolate out of many small openings as "seepage springs". He indicates that at all true springs the water issues from the ground due to hydrostatic pressure. Such areas are to be distinguished from areas of moist soil owing to capillarity. Throughout much of the Piedmont and Appalachian areas of the USA, springs issue from fractures and fissures and/or along contact zones between two rock types of different permeability.

Springs and seeps have a role in altering the landscape simply because they act as an erosion force and the moving water carries with it dissolved minerals which have been assimilated in the passage of water through the rocks and soil. Groundwater discharge by means of both springs and seeps, including groundwater discharge below normal stream level, may be responsible for as much as half of the land erosion taking place in humid areas where subsurface solution of limestone or dolomite can result in many tens or hundreds of feet in lowering of the land surface. Good examples are the karst subsidence areas of Croatia and in parts of the Red River Valley of Texas.

3.2 Hydrologic Cycle

Springs are an important link in the hydrologic cycle. Precipitation is the starting point of the vast unending circulation of the water on the Earth, known as the hydrologic cycle (Fig. 3.3). This system operates chiefly from the energy of the sun. Water is evaporated from the oceans, the land, vegetation, and from smaller bodies of water such as streams, lakes, and ponds. The water moves as vapor through the atmosphere where it accumulates as clouds and eventually returns as precipitation both on land and sea. Springs are a natural part of the hydrologic cycle defined in the *Glossary of Geology* (AGI 1997) as "The constant circulation of water from the sea, through the atmosphere, to the land, and its eventual return to the atmosphere by way of transpiration and evaporation from the sea and the land surfaces". The parameters that comprise the hydrologic cycle include: precipitation, evaporation, transpiration, runoff, recharge, storage, and discharge. Therefore, to understand springs thoroughly, one must understand the hydrologic cycle in which a spring occurs. Each of the parameters in the cycle is defined as follows (AGI 1997):

Precipitation. Water that falls to the surface from the atmosphere as rain, snow, hail, or sleet. It is measured as a liquidwater equivalent regardless of the form in which it fell.

Evaporation. The process, also called *vaporization*, by which a substance passes from the liquid or solid state to the vapor state. Limited by some to vaporization of a liquid, in contrast to sublimation, the direct vaporization of a solid. Also limited by some (e.g. hydrologists) to vaporization that takes place below the boiling point of the liquid. The opposite of condensation (Langbein and Iseri 1960).

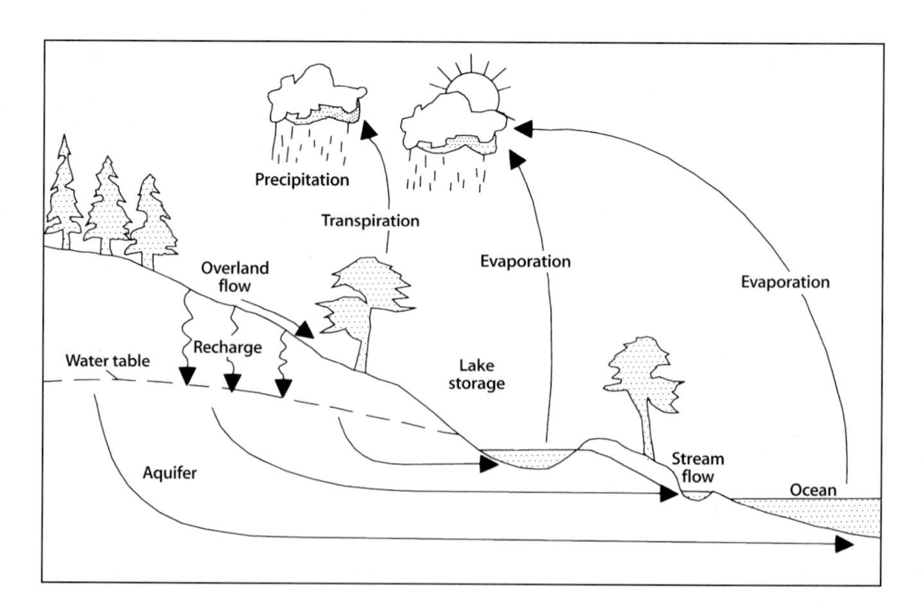

Fig. 3.3. The hydrologic cycle

Evapotranspiration. Loss of water from a land area through transpiration of plants and evaporation from the soil. Also, the volume of water lost through evapotranspiration.

Transpiration. The process by which water absorbed by plants, usually through the roots, is evaporated into the atmosphere from the plant surface.

Runoff (Water). That part of precipitation appearing in surface streams. It is more restricted than *stream flow*, as it does not include stream channels affected by artificial diversions, storage, or other human works. With respect to promptness of appearance after precipitation, it is divided into *direct runoff* and *base runoff;* with respect to source, into *surface runoff, storm seepage,* and *groundwater runoff.* It is the same as *total runoff* used by some workers (Langbein and Iseri 1960). May be expressed as average depth over watershed.

Recharge. The processes involved in the addition of water to the saturated zone, naturally by precipitation or runoff, or artificially by spreading or injection; also, the amount of water added.

Storage. (1) Artificially impounded water, in surface or subsurface reservoirs, for future use. Also, the amount of water so impounded. (2) Water naturally detained in a drainage basin, e.g. groundwater in an aquifer, depression storage, and channel storage.

Discharge. The rate of flow of surface water expressed as volume per unit of time.

As part of the hydrologic cycle, some of the water falling on the land returns to the sea as overland runoff, or enters the soil zone and percolates downward to saturate the underlying rocks. The water level in the saturated rock rises to a position sufficiently high to cause water to discharge into streams or ponds. A lagtime is introduced into the hydrologic system in that some water is stored in the rocks for variable periods of time. Ultimately, however, the system is balanced and, over a long period of time, the input into the rocks (or aquifers) must equal the output: groundwater discharge must equal groundwater recharge. Thus, the water in storage in the rocks must equal a constant in the hydrological equation. This equation in ordinary terms (assuming no groundwater pumping) can be expressed as follows:

$$P = ET + R \pm \Delta S \tag{3.1}$$

with:

- P = precipitation
- ET = evapotranspiration
- R = runoff (or streamflow) and
- ΔS = change in groundwater storage

Precipitation and runoff are relatively easy to measure. Over a long period of time (usually a few years), change in groundwater storage (ΔS) must equal zero but, for short periods (weeks or months), water may be added to or withdrawn from the aquifers. Where a sufficient number of observation wells are available, changes in the quantity of water in storage in the aquifers are shown by changes in the position of the water levels.

An exception to this could occur where geohydrologic conditions provide a means for water to move out of a stream basin and discharge into another basin, or into an estuary bay, or the ocean. Generally, as an example, in the Piedmont and Appalachian regions of the USA, the surfacedrainage basin coincides with the groundwater basin. An exception to this also may occur with surfacedrainage basins in limestone terrains where some of the undergrounddrainage systems may not coincide with the surface drainage. Another exception may occur in mining areas where water is pumped from an operating mine and transferred to another drainage basin.

3.2.1 Precipitation

The source of all water issuing from springs is precipitation. For example, in a humid continental climate where the average annual precipitation is about 50 inches as in the Appalachian Ridge and Valley province in the eastern USA, the rainfall can be highly variable, ranging from 30–60 in. The temperature is also highly variable. Thus, spring discharge would vary with temperature and rainfall. Barometric pressure can also affect spring flow.

East of the mountains in the Piedmont province or in the Gulf and Atlantic Coast Plains, the average annual precipitation increases as the climate is affected by the moisture-laden winds moving northward into the area and dropping their moisture as rain or snowfall. Thus, the flow in these areas may be affected substantially by a factor of 10 or more times during a hydrologic cycle.

3.2.2 Evapotranspiration

Evapotranspiration, in the hydrologic budget equation, is the water returned to the atmosphere by evaporation from vegetation, soil, and water bodies. It includes the water transpired by plants as part of their life processes. Because it is measured directly only with difficulty, it is commonly determined in the hydrologic budget as the remainder after subtracting stream flow from precipitation. As the evaporation of water requires energy, a method has been devised to compute the potential evapotranspiration of water in an area or drainage basin. This method, developed by Thornthwaite and Mather (1955, pp. 348–354), is commonly used to show the theoretical limits of evapotranspiration, based on thermal energy received.

3.2.3 Runoff

Runoff is synonymous with stream flow and is the sum of surface and groundwater flow that discharges into the streams. Surface runoff is that water which moves over the soil surface to the nearest stream channel. It is also defined as that part of the runoff of a drainage basin that has not passed beneath the surface since precipitation; it may, however, include storm seepage, which is that part of the precipitation that infiltrates the surface soil and moves toward the streams as ephemeral, shallow, perched groundwater above the main groundwater level. Groundwater runoff is that water which has passed into the ground, entered the zone of saturation, and discharged into a stream channel from springs or seeps.

Water discharging from transient storage in the aquifers by means of springs and seeps maintains the flow of streams during dry periods. The major factors determining spring discharge are aquifer permeability, hydraulic gradient, areal extent of the aquifer system, and the magnitude and distribution of recharge to the system. In some types of terrain (for example, limestone terrain), moderate to large springs are common and the flow of many of the streams is due largely to discharge from identifiable springs in the upper reaches of tributary streams, but, in shale terrain, fewer identifiable springs exist and the discharge of groundwater occurs mainly as seeps. Because of the poor storage capability of the shale aquifers, the flow from springs in these areas may cease entirely even during minor droughts. This results in a decrease in streamflow and, thus, streams draining shale terrain are characterized by rapidly generated peak flow and by a relatively rapid recession of stream flow during droughts.

3.2.4 Recharge

Aquifers are recharged by precipitation or another surface water source by percolation through the soil and rock. In general, before recharge can occur, any soil moisture deficiency must be satisfied and the demands of the plants must also be met. An example from spring studies in Maryland by Otton and Hilliary (1985) indicates that most groundwater recharge in Maryland occurs during winter and early spring (February, March, and April), except during periods when the ground is frozen. Another period of recharge occurs during the fall (November and December) when the plants have become dormant (Fig. 3.4). However, water level records from observation wells show that recharge to an aquifer can occur at any time of the year if the precipitation is high enough and ground conditions are suitable.

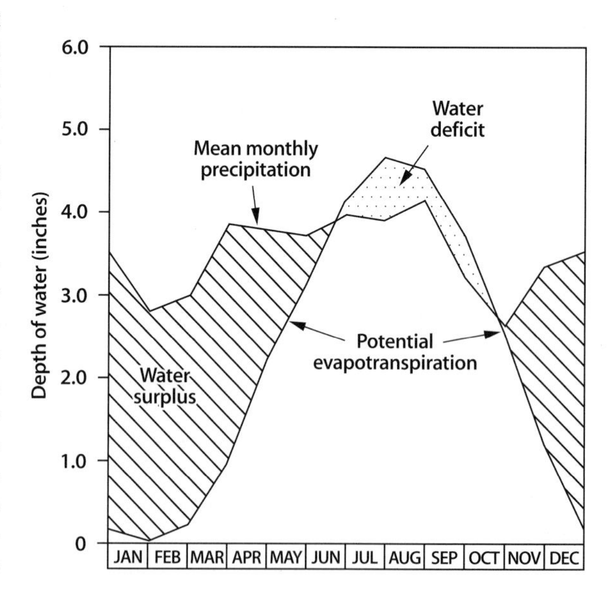

Fig. 3.4. Mean monthly precipitation and potential evapotranspiration at Westminster, Maryland (after Otton and Hilliary 1985)

3.2.5 Storage

All rocks have at least some capacity for storing water. In the Piedmont and Paleozoic areas of the eastern USA, the water is stored in voids in the residuum (weathered soil and bedrock) and in the saprolite and soil above the hard, unweathered rock. In the rocks below, the water is stored in fractures, joints, solution openings, and along bedding-plane partings. Storage coefficients are significant with regard to the flow of springs, inasmuch as springs issuing from aquifers of poor storativity may show wide variability of flow. During prolonged droughts, such springs may cease flowing, which happens in shale-siltstone terrain. In limestone areas with extensive solutional development, the water may be stored in caverns or tubular channels, ranging up to several feet in diameter and with a large storage capacity. Generally, the storage potential of all rocks decreases with depth and, at depths of 1 000 to 2 000 ft, it is very low – generally less than 1%. The storage coefficient (*S*) of an aquifer is defined as the volume of water an aquifer releases from or takes into storage per unit of surface area per unit of change in head. The storage coefficient for an unconfined aquifer is approximately equal to gravity drainage, provided gravity drainage is complete.

Commonly, storage coefficients range between values of 0.30 and 0.00001 (Ferris et al. 1962). The three different conditions shown in Table 3.1 may exist within the common range of coefficients.

3.2.6 Spring Discharge

Physical forces responsible for bringing groundwater to the surface were described by early investigators, Bryan (1919) and Meinzer (1923a), who considered that springs resulted from (1) gravity flow, and (2) artesian flow. More modern concepts do not attempt to make this distinction, recognizing that springs and seeps result from differences in potentiometric head, involving gravity, pressure, and other forces such as differential thermal properties. Also, some springs flow due to the buoyant effect of gases dissolved in the water. Because of the complex reasons causing springs to flow, it is difficult to assign the flow of a spring to any one physical force, although gravity is a major factor.

The use of springs as a source of water depends in large part on their flow characteristics, both the quantity and constancy of flow. To illustrate this factor, the measured flow of springs ranges from less than 1 gal/min to in excess of nearly 10 000 gal/min in spring (Fig. 3.5).

Geysers are a natural form of hot springs. Some of the most reknown are on the North Island of New Zea-

Table 3.1. Storage coefficients

Storage coefficient (dimensionless)	Condition
0.30 – 0.05	Water table
0.049 – 0.001	Intermediate (modified artesian)
0.0009 – 0.00001	Artesian

Fig. 3.5. Harouch-Figeh System. Water supply for city of Damascus. Albion Dolomitic limestone (photograph by P. E. LaMoreaux 1980)

Fig. 3.6. Geyser Springs, Iceland, near Gullfoss Falls, about 20 geysers in volcanic area (photograph by P. E. LaMoreaux 25 November 1995)

land, Old Faithful in Wyoming, USA, and the geysers of Iceland (Fig. 3.6).

Flows of large and medium-sized springs can be measured by conventional stream flow-measuring techniques using a current meter and a stopwatch to measure water velocities across a measured section of the discharge channel. Weirs or flumes are used where their installation is practicable. Figure 3.14 shows an automated weir at Buchbrunnen Spring, Germany. It is electronically/mechanically controlled for diversion of spring water. The system is fully instrumented to control discharge and continuous measurement with recording instruments. Simple flow measurements can be made using a bucket and a stopwatch for most smaller springs. This method is feasible where the spring flows from an open discharge pipe or restricted opening. The flow of some springs can not be measured accurately because of leakage around and under the walls of spring pits or collection boxes.

Spring flows may vary through at least several orders of magnitude during a year. Some springs flow at rather steady rates, while others yield several gallons per minute and dry up completely during droughts. Generally, the temperature of spring water is relatively consistent year-round, not varying more than a degree in temperature. The quality of water from a spring may vary greatly over the year with rainfall causing increases in flow, erosion, sediment load, and chemical character. All of these factors must be determined in the development of a spring as a source of water (see Chap. 5).

The measurements of spring flow compiled during this study as well as flow data reported by other investigators show that most springs, especially those of shallow origin, vary widely in their flow. Meinzer (1923a) recognized this characteristic of springs many years ago and proposed a variability index of spring flow based on the following formula:

$$v = \frac{(m - l)}{a} 100 , \qquad (3.2)$$

with:

- v = variability index, in percent
- m = maximum flow
- l = minimum flow
- a = average flow of a series of measurements

A variable spring is one in which the index number is 100% or more. A subvariable spring is one in which the index number is between 25 and 100%, and a constant spring is one in which the index number is not more than 25% (Meinzer 1923). All springs which fail

Table 3.2. Yield classes of springs, based on their annual mean flow (Otton and Hilliary 1985)

Gallons per minute (gal/min)	Could be considered for
0.7 – 4.9	Household use
5 – 9.9	Household and farm
10 – 19.9	Limited institutional or commercial
20 – 34	Various uses, possibly limited public supply
35 – 70+	Small public supplies

(have zero flow) during part of the year are variable springs. If a specific number of measurements for computing the average flow is not available, it would seem prudent to obtain a series of measurements through at least one annual hydrologic cycle.

In a Maryland study (Otton and Hilliary 1985), the annual mean flow was based on monthly measurements and was determined for 11 springs in the Appalachian province and 23 springs in the Piedmont province. The springs were placed in five yield classes, based on their annual mean flow (Table 3.2).

3.3 Classification and Geologic Setting

Springs may be classified or grouped in several ways according to their (1) mean flow or discharge, (2) geologic setting (rock structure), (3) mean temperature, and (4) chemistry of their water. Classification may also be made by a combination of one or more of the above elements. However, no system of classification is perfect. For example, to classify a spring by its flow, a series of flow measurements must be made over a period of time, preferably at least through one annual seasonal cycle. In his study of large springs in the United States, Meinzer (1927) proposed the classification shown in Table 3.3, based on magnitude of mean flow.

Since Meinzer's classification (1927), a number of different classifications have been developed with our greater understanding of spring origin, discharge and chemical-thermal characteristics. Several classifications have evolved which concentrate on one or more of these specific characteristics, such as type of spring opening, amount of discharge, variability, mineral content, or temperature. These classification systems have evolved slowly and have incorporated numerous refinements. A comprehensive classification of springs however has still not been universally accepted. Reviewing the history of classification and the current classification system, the difficulties and problems in spring classification are apparent. Technological advancements and informational tools (i.e. Internet) will affect future trends in the classification and eventually a definitive classification of springs will evolve.

3.4 Past Civilizations

Springs have provided a source of drinking water for peoples and civilizations all over the world. Towns would develop near water sources such as rivers, lakes, and springs. In ancient times, some rivers like the Nile and Euphrates became the birthplaces of a civilization. However, even these classic rivers were intermittent and either had very low flows or ceased to flow during certain times of the year or droughts. For this reason, springs have remained important for all civilizations.

Over 30 000 years ago in America, archaeological remains near Lewisville, Texas show signs of humans living by local springs (Brune 1975). Indians later came to use springs for irrigation, grinding foods, bathing, and for drinking water. When the white man came to the New World, battles and wars were waged over the possession of springs. Forts and cattle trails in the Old West were routed by springs and "watering-holes". In the

Table 3.3. Classification of springs, based on magnitude of mean flow (Meinzer 1927)

Magnitude	ft^3/s	gal/min	gal/d
First	100	44 900	6.46×10^7
Second	10 – 100	4 490 – 44 900	6.46×10^6 – 6.46×10^7
Third	1 – 10	449 – 4 490	6.46×10^7 – 6.46×10^6
Fourth	0.22 – 1	100 – 449	1.44×10^5 – 6.46×10^5
Fifth	–	10 – 100	14 400 – 144 000
Sixth	–	1 – 10	1 440 – 14 400
Seventh	–	0.12 – 1	173 – 1 440
Eighth	–	0.12[a]	180

[a] 0.12 gal/min = 1 pint/min.

1800s early settlers used the flow of spring water for powering gristmills, flour mills, sawmills, and cotton gins (Brune 1975, p.9) Some settlements, like Huntsville, Alabama, depended on spring water for their water supply. By the late 1800s in the USA, medicinal or health_spas began to flourish, touting the virtues of the therapeutic or regenerative value the waters had on the body. Springs, like Silver, Rainbow, Glenwood, Saratoga, and Hot Springs, Arkansas became popular for recreational purposes.

Water was a primary resource on which people's lives depended. Like wild foods and herbs, ancient people whose daily survival depended on the knowledge of the natural world, would depend on handed-down wisdom and folklore of the area. Springs were a popular topic of discussion of ancient people and the earliest written records of the Greeks, Muslims, and Christians related stories about springs as sources of water.

3.4.1 Spring Identification by Human Senses

Springs were sources of water easily identifiable by our five senses. With the eyes, springs could readily be spotted as flowing waters not necessarily near a river, creek, lake or ocean. The size of the spring was often noted by the size of the opening to the spring. The flow of a spring could visually be observed to be fast, moderate, slow, or not flowing. Objects like bamboo, pumice, or sticks could be thrown into the water to see where the object traveled as well as how rapidly it traveled, and, for example, the springs of Sidon of the Bible were first studied by throwing a shaft into the water and watching it move. The clarity of water could also be visually assessed. People believed that the clearer the water, the more pure it was. Thus, iron was distasteful and caused "red waters". Touch immediately indicated the temperature, texture, and sensation of the water. Smell helped develop a trust in drinking the water. Sulfur in spring water causes a rotten egg odor, an example of how scent creates an aversion to water taste. Some springs are naturally bubbly or carbonated due to the presence of gases which had a tingling effect on the nostrils. Many of the mineral waters of Europe are highly carbonated and are sold as gaseous or non-gaseous. Hearing helped detect the flow of water, therefore, using the senses and observation skills, a first-time visitor to a spring evaluated a source of water before attempting to drink. If the water tasted good or helped a person to feel better, they would most likely note the spring and come again and bring others.

3.5 Classification Prior to the 17th Century

Classification of springs prior to the 17th century was a verbal recitation of the location, size, temperature, and drinkability of water from the spring.

In Ancient Greece, the numerous caves, springs, and karst features influenced Greek philosophers who attempted to explain the origin of water and springs. One school of thought was that water was forced from the sea and driven into rocks and upwards whereby salt would be extracted once it hit the ground surface. Another school of thought, championed by Aristotle, believed springs originated from condensation of moisture in caves.

The Arabs and Persians were knowledgeable in the construction of water supply systems and aware of the importance of springs. The Turks transferred this technology to the Yugoslavian area to build the first water supply system in Sarajevo around 1461 using a gravitational system from a karst spring.

In the American West where surface water was scarce, springs had a strong significance and folklore. Indian trails, then wagon trails and cattle trails, predictably intercepted the locations of springs. The trails of Coronado in the 1500s, La Salle in the 1600s, the Chisom Trail, etc. all were famous historically and passed by many springs.

The first classification of springs was based on whether the water was drinkable. The next most important aspect was whether it was seasonal or perennial. Lastly, the size of the spring determined its adequacy as a water supply. Was the flow large enough to provide groups of people and livestock with plenty of water? Was the water supply enough to sustain the growth of a tribe, town or community? Or, was it just large enough to provide a temporary or sporadic source of supply? Thermal springs or mineral springs were noted and used for medicinal or recreational purposes.

3.5.1 Scientific Awakening

At the close of the 17th century, the ancient greek theories about the origin of springs began to be questioned and the scientific community began to investigate anew the origin of springs. Edme Mariotte and Perrault (1674) began to consider the source and movement of rainfall, stream flow, and groundwater and from their works, a new classification of would evolve. Perrault, in his book Origins of

Springs, refuted the long-standing theories of groundwater movement and origin. Based on his experiments and observations, he argued that rainwater was a catalyst for the movement of groundwater, but was not the origin of springs. He likened springs to daughters of rivers.

By the 19th century, Paramelle (1856) also wrote a book about the origin of springs and credited Perrault with discovering how to measure the flow of a spring and rainfall as a sufficient amount of water to account for the water levels in rivers. But, it was not until the 20th century that Keilhack (1912) attempted a complete classification of springs. He recognized that the origin of springs needed to be understood and universally acceptable before a classification could be developed. In Keilhack's classification, types of springs are not mutually exclusive but are primarily separated by whether a spring is ascending (aufsteigende) or descending (absteigende).

3.5.2 Instrumentation and Quantification

In the late 18th Century, instruments to measure the natural events of rainfall and discharge were developed and with the advent of scientific tools there existed extensions to the five senses. The microscope and binocular would allow us to enlarge our vision. The clock and stream gage would allow us to quantify flow rates of rivers and springs. Pipes and pumps would extend our "touch" and "taste" by bringing water to the surface. Thus, measuring and quantifying of observed relationships and theories could be documented. Earliest data records of a scientific type-water levels, flow rates, temperature, chemical character of water, was begun and the segments of the hydrologic cycle placed on record. There were earlier records obtained, for example the nilometers that measured heights of the Nile in uneven units of pics and carrots, and some rates of flow of springs and streams prior to the 18th Century. However, the first systematic measurements with instruments did not begin until the 1700s.

3.5.3 Scientific Specialization

Philosophers attempted classifications and elaborate systems to explain the origins of the universe, natural world and human existence. Following the philosophical approach, mathematicians and physicist incorporated quantification and classification into their writings. Through this process and the growth of civilization, science began to branch from mathematics, physics and astronomy into chemistry, geology, and biology, beginning a trend of specialization. The sciences, like tools, also became an extension of our senses allowing us to quantify and classify those processes which were observable and quantifiable. Geology and chemistry would become the main sciences to quantify and explain the occurrence of springs and groundwater. Geology would ultimately become the branch of science used to classify springs on the basis of rock types, stratigraphy, lithology, and geological structure or deformation of rocks by stresses in the earth's crust creating joints, fractures, faults, folds, and solutioning or forces creating secondary porosity.

3.5.4 Exploration

Most of the major springs were discovered by people exploring new territories. As populations grew, exploration and mapping of new areas became a driving force for compiling detailed topographic maps on which previously known springs and new springs could be marked. Maps became a universal graphical representation of the world which were found in some form in most cultures. With the exchange of maps, an increasing knowledge of springs and their locations developed. Exploration led to curiosity about the different types of springs, as scientists began to determine the origins of existing springs and the mechanisms which caused different types of springs.

3.6 Documentation and Classification of Springs – Early 1900s

In the early 1900s, the documentation and classification of springs by scientists worldwide began in earnest. Much of this early work was done by government agencies such as geological surveys. In 1902, M. E. Fournier with the Services de la Carte Geologic de la France published two bulletins concerning the origins of groundwater and springs: Etude sur les sources , les resurgences et les nappes aquifers du Jura Franc Comtois, and Etudes sur les projets d'alimentation, le captage, la recherche et la protection des eaux potables. In Germany, K. Keilhack published in 1912 Lehrbuch der Grundwasser- und Quellenkunde which was one of the first attempts to classify all springs by pathways to the surface. M. L. Fuller, with the United States Geological Survey (USGS), published in 1905, Underground Waters of the Eastern United States, which included a classification system. Kirk Bryan, also with the USGS, published Classification of Springs in 1919, and O. E. Meinzer, with the USGS, published Outline of Groundwater Hydrology in 1923, which included a different classification of springs.

In 1933, Dr. Josef Stiny published a text in Austria titled Springs: The Geological Foundations of Springs for Engineers of All Disciplines as well as Students of Natural Sciences, with a chapter devoted to the origin and classification of springs. His classification was divided into three major categories based on flow as: (1) free flowing springs (freifliessende Quellen), (2) overflowing springs (Überfliessquellen), and (3) artesian springs (aufwallende Quellen). A final, fourth category, was reserved for miscellaneous springs which could not be categorized into the other three categories. Examples of these types springs were: geysers (Stossquellen), underwater or cavern springs (Untertagquellen), gas springs (Gasquellen), limestone terrains (Herberquellen), and thermal or mineral springs (Gesundbrunnen und Heilquellen).

From exploration, maps, and local knowledge, springs were known to occur in many parts of the world and in different forms. Springs occurred as mineral springs, thermal springs, or cold springs. They came in all sizes and shapes and in differing topographic and geologic settings.

So, how did these early classifications categorize the various types of springs? The main criteria were:

1. Cause or origin of the spring
2. Rock structure or geologic setting of spring
3. Size or discharge rate of the spring
4. Temperature of the spring
5. Variability of the spring

3.7 Current Classification

Some classification systems borrow heavily from the works of Bryan (1919) and Meinzer (1923a). The current classification systems incorporate parts of these systems and refine or expand portions based on modern quantification and knowledge of springs. The works of Bryan and Meinzer will be elaborated upon to understand the link of their classifications to current classification systems.

Bryan (1919), of the USGS, divided springs into two primary classes: (1) springs resulting from non-gravitational forces, and (2) springs resulting from gravitational forces. Bryan, working within the USGS, may have adopted the then accepted classification for groundwater into his classification for springs. Groundwater was classified as phreatic water and split into two categories: (1) gravity groundwater or (2) groundwater not under the control of gravity.

Bryan used for his classification the character and origin of the water. The origin of springs based on geologic structure was too complex to ascertain in Bryan's opinion, and he gave examples of the vegetative growth around some springs, the accessibility, and the erosional forces operating on the springs as reasons that true structural origin could not be classified. He stated that more than one structural feature or process was operating, making an absolute classification system even more difficult.

The major categories which Bryan established, Key To The Classification of Springs from his original article is as follows:

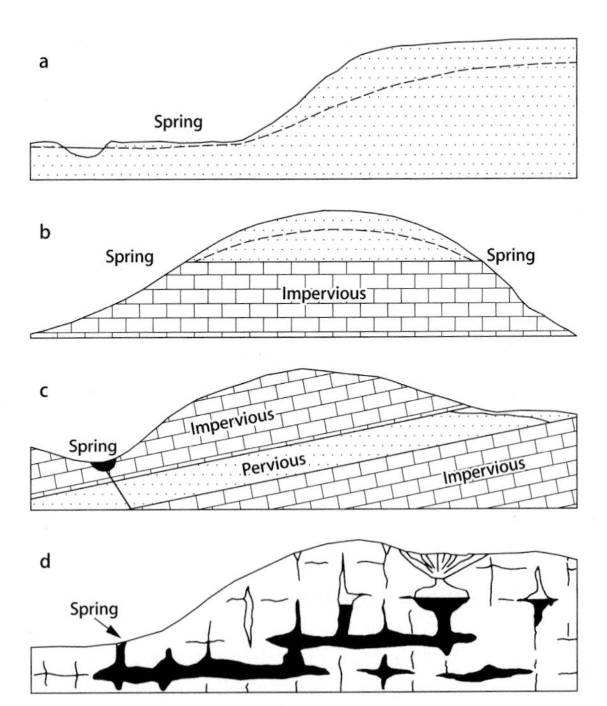

Fig. 3.7. Diagrams illustrating types of gravity springs. **a** Depression spring; **b** contact springs; **c** fracture artesian spring; **d** solution tubular spring (Todd 1980)

1. For springs due to non-gravitational forces or deep-seated waters and not subject to seasonal fluctuations he had two main categories (1) volcanic springs and (2) fissure springs.
2. For springs resulting from gravitational forces or springs due to shallow water and subject to seasonal fluctuations he categorized them as shown in Fig. 3.7 (from Todd 1980, Groundwater Hydrology, p.48).

Key To The Classification of Springs by Kirk Bryan (1919)

I. Springs due to deep-seated waters, juvenile and connate, admixed with deeper meteoric water; do not flow under hydrostatic head and are usually not subject to seasonal fluctuation.
  A. **Volcanic Springs.** Associated with volcanism or volcanic rocks; water commonly hot, highly mineralized and containing gases. Grade from gas vents into springs of normal temperature indistinguishable from those due to other causes.
  B. **Fissure Springs.** Due to fractures extending into deeper parts of the crust; water usually highly mineralized and commonly warm or hot.
    1. *Fault Springs.* Associated with recent faults of great magnitude.
    2. *Fissure Springs.* No direct structural evidence as to origin, but because of temperature and steady flow believed to have deep origin.

II. Springs due to meteoric and occasionally other waters moving as groundwater under hydrostatic head; many fluctuate in flow with the rainfall.
  A. **Depression Springs.** Due to land surface cutting water table in porous rocks.
    1. *Dimple Springs.* Due to depressions in hillsides.
    2. *Valley Springs.* Due to abrupt change in slope at edge of flood plain.
    3. *Channel Springs.* Due to depressions in flood plains or alluvial plains caused by channel cutting of stream.
    4. *Border Springs.* Due to change in slope at border between alluvial plains and playas, lake beds, or river bottoms; relative imperviousness of central clay deposits assists flow.
  B. **Contact Springs.** Due to porous rock overlying impervious rock.
    1. Impervious rock has a horizontal and regular surface.
      a) Underlying bed is of large extent; common in consolidated sedimentary rock.
        (1) Gravity Springs. Overlying material is soft.
        (2) Mesa Springs. The overlying material is hard, usually sandstone or lava flow; water contained in pores and joints of the rock.
      b) Underlying bed is of small extent; common in unconsolidated alluvium; impervious bed is usually clay, cemented gravel, "mortar bed," caliche, or hardpan.
        (1) Hardpan Springs.
    2. Impervious bed has an inclined and regular surface; all springs on the low side unless the overlying bed is very thick and the dip low.
      a) Underlying bed is of large extent.
        (1) Inclined Gravity Springs. The overlying material is soft.
        (2) Cuesta Springs. The overlying material is hard; of same character as mesa springs.
      b) Underlying bed is of small extent; as in hardpan springs.
        (1) Impervious layer dips away from hill; spring possible.
        (2) Impervious layer dips into hill; spring possible only in ravines.
    3. Impervious bed has irregular surface.
      a) Overlying porous material is thick and of wide extent; contact is unconformity. Gravity, inclined gravity, mesa, and cuesta springs may occur, but springs will be sharply localized at lowest parts of contact.
      b) Pocket Springs. Overlying porous material is unconsolidated and more or less discontinuous, residual soil, talus, landslide debris, alluvium, till, stratified drift, wind-blown sand, or volcanic ash.
      c) Overflow Springs. Irregular floor is not continuous, but porous bed is saturated and overflows at lateral contacts; common at receiving end of artesian systems.
      d) Rock Dam Springs. Irregularities of the rock floor under an alluvial plain force water to surface; these may be projections of floor of basin, projections of partly consolidated older alluvium, igneous dikes, or volcanic plugs.
      e) Fault Dam Springs. Dam caused by faulting.
  C. **Artesian Springs.** Due to pervious bed between impervious materials.
    1. *Dip Artesian Springs.* More or less regularly bedded rocks; tilted porous bed crops out in valley; usually sedimentary, also alternations of lava flows, flow breccias, tuffs, gravels.
    2. *Siphon Artesian Springs.* Similar rocks; folded and with outcrops in valley.
    3. *Unbedded Artesian Springs.* Rocks not regularly bedded, but mass of porous material is exposed so as to receive water and crops out in valley; occur in till and perhaps in other rocks.
    4. *Fracture Artesian Springs.* All the conditions above, except that lower end of porous bed does not crop out but an opening allows water to escape. Opening due to fracturing with or without faulting.
  D. **Springs in Impervious Rock.**
    1. *Tubular Springs.* Due to more or less rounded channels in impervious rocks.
      a) Solution Tubular or Cavern Springs. Due to solution channels in limestones, calcareous sandstones, gypsum, salt.
      b) Lava Tubular Springs. Due to caverns and tunnel in lava flows.
      c) Minor Tubular Springs. Due to channels made by movement of water, decay of tree roots, sand streaks, or shrinkage cracks, usually in unconsolidated sediments.
    2. *Fracture Springs.* Due to fractures consisting of joints, bedding planes, columnar joints, openings due to cleavage, fissility, schistosity, cross-bedding planes, and faults in impervious sedimentary, igneous, and metamorphic rocks.
      a) Quadrille Fracture Springs. Due to more or less rectangular system of fractures, one of which is parallel to the horizon.
      b) Crosshatch Fracture Springs. Due to more or less rectangular system of fractures, inclined toward the horizon.
      c) Inclined Fracture Springs. Due to inclined fractures, not necessarily systematic.

Table 3.4. Classification of springs by discharge

Magnitude	Mean discharge (gal/min)	Mean discharge
First	>44 880	>10 m^3/s
Second	4 488 - 44 880	1 - 10 m^3/s
Third	448.8 - 4 488	0.1 - 1 m^3/s
Fourth	100 - 448.8	10 - 100 l/s
Fifth	10 - 100	1 - 10 l/s
Sixth	1 - 10	0.1 - 1 l/s
Seventh	0.12 - 1	10 - 100 ml/s
Eighth	<0.12	<10 ml/s

O. E. Meinzer, in 1923, proposed a classification of springs based on their discharge. Discharge of springs, defined in terms of magnitude, is shown in Table 3.4. The discharge rate of a spring is dependent on the area contributing recharge to the aquifer and the rate of recharge. Seasonal rainfall will determine fluctuations in the rate of discharge of springs. Fluctuations can vary dependent on the rate of recharge and hydrogeologic setting. These fluctuations as a function of the response to variations in the rate of recharge can range from minutes to years. Therefore, a perennial spring will discharge throughout the year as it drains an extensive permeable aquifer. Intermittent springs will only discharge during certain times of the year when there is enough groundwater to maintain the flow.

A variation of Meinzer's classification uses statistics to determine the "characteristic discharge" of a spring. Based on the variability of discharge, R. Netopil (1971) modified Meinzer's classification to consider discharge values that exceeded 10 and 90% instead of using extreme spring discharges. His classification of groups of springs was based on the percentage of discharges exceeded by 10 and 90% as shown in Table 3.5.

Hubert Kriz (1973) used this classification as an aid in his study of the Sulkovy Prameny (Springs) in Czechoslovakia where discharge measurements on the individual hydrological years of 1901–1970 had been made. From these measurements, he calculated the degree of variation of discharge and evaluated different classifications of springs.

Even though elaborate classification systems for mineral springs had already been published, Meinzer and

Table 3.5. Classification of springs

$Q_{10\%}/Q_{90\%}$	Classification of springs
1.0 - 2.5	Extraordinarily balanced
2.6 - 5.0	Well-balanced
5.1 - 7.5	Balanced
7.6 - 10.0	Unbalanced
>10.0	Extraordinarily unbalanced

Bryan did not incorporate those categories into their classification systems. However, more detailed and accurate data on chemistry and geochemistry became more available and the classification systems for mineral springs grew in usage and in accuracy.

One example of a complete classification system for mineral springs was provided by Frank W. Clarke in 1924. Clarke (1924) acknowledged three important criteria in the classification of springs: (1) geologic origin, (2) physical properties and (3) chemical properties. The importance of the geologic origin of the springs was established by the origin of water and by the correlation of springs based on their geologic origin. Physical properties of water, such as temperature, indicated if it was a thermal spring or cold spring. Lastly, the chemical composition of water would indicate the geologic origin of the spring.

Clarke decided that the chemical composition of the water allowed for a more definite classification system and a better comparison of springs. He did not discount the importance of the geologic origin or physical properties of the springs. He simply used a classification of waters based on their water chemistry and applied it to springs. He used literature and recorded analyses and gave an average of ten springs in the US and abroad as an example for each class. The main chemical constituents used for classification of springs were: chloride, sulphate, carbonate, acid, and various waters with different combinations like nitrates, borates, sulphides or silicates.

Clarke's (1924) classification of waters with some selected examples he used for each class are as follows:

I. Chloride waters. Principal negative ion Cl. (Spring at Pahua, New Zealand)
   A. Principal positive ion sodium.
   B. Principal positive ion calcium.
   C. Waters rich in magnesium.

II. Sulphate waters. Principal negative ion SO_4.(St. Lorenzquelle, Leuk, Switzerland)
  A. Principal positive ion sodium.
  B. Principal positive ion calcium.
  C. Principal positive ion magnesium.
  D. Waters rich in iron or aluminum.
  E. Waters containing heavy metals, such as zinc.
III. Sulphate-chloride waters, with SO_4 and Cl both abundant. (King's mineral spring near Dallas, Indiana)
IV. Carbonate waters. Principal negative ion CO_3 or HCO_3. (Spring in Pine Creek Valley, near Atlin British Columbia)
  A. Principal positive ion sodium.
  B. Principal positive ion calcium.
  C. Chalybeate waters.
V. Sulphato-carbonate waters. SO_4 and CO_3 both abundant.
VI. Chloro-carbonate waters. Cl and CO_3 both abundant.
VII. Triple waters, containing chlorides, sulphates, and carbonates in equinotable amounts. (Chaybeale water, Mittagong, New South Wales)
VIII. Siliceous waters. Rich in SiO_2.(Old Faithful Geyser, Yellowstone National Park, Wyoming, USA)
IX. Borate waters. Principal negative radicle B_4O_7.
X. Nitrate waters. Principal negative ion NO_3.
XI. Phosphate waters. Principal negative ion PO_4.
XII. Acid waters. Contain free acids. (Rio Vinagre, Colombia)
  A. Acid chiefly sulphuric.
  B. Acid chiefly hydrochloric.

Chemical analyses of spring water are useful in classifying the water by type and to determine potential uses where water quality is important. For example, high total hardness requires excessive soap or detergent for laundering, high iron causes staining of sanitary fixtures and clothing, high fluoride causes dental fluorosis or staining of teeth. For classification, the so-called trilinear diagram scheme of Hill and Piper can be used (Otton and Hilliary 1985). The chemical analyses of the major cations and anions were used to compute the percentage of total equivalents per liter, and each analysis was then plotted on two triangular diagrams where the percentage values ranged from 0 to 100. Associated with the triangles is a diamond-shaped diagram where plots of values of related cations and anions are used to produce a vector plot. Thus, each analysis is represented by a point on both triangles and on the diamond.

As an example (Otton and Hilliary 1985), the two lower triangles of Fig. 3.8 show the types of water that can occur. All but two of the samples analyzed are either calcium-magnesium bicarbonate types or water of a mixed type.

Another classification of water types according to H. Futak and H. R. Langguth (1986) were used to help classify springs in a hydrogeologic study of central and eastern Pelopommesus, Greece by Steirische Beiträge zur Hydrogeologie (Institute of Geology and Mineral Exploration). In this classification, waters are grouped as follows:

Group I belong to normal earth alkaline, predominately hydrogencarbonatic waters
Group II belong to normal earth alkaline, hydrogencarbonatic sulphatic waters
Group III belong to earth alkaline, predominately hydrogen carbonatic waters with increased alkali content

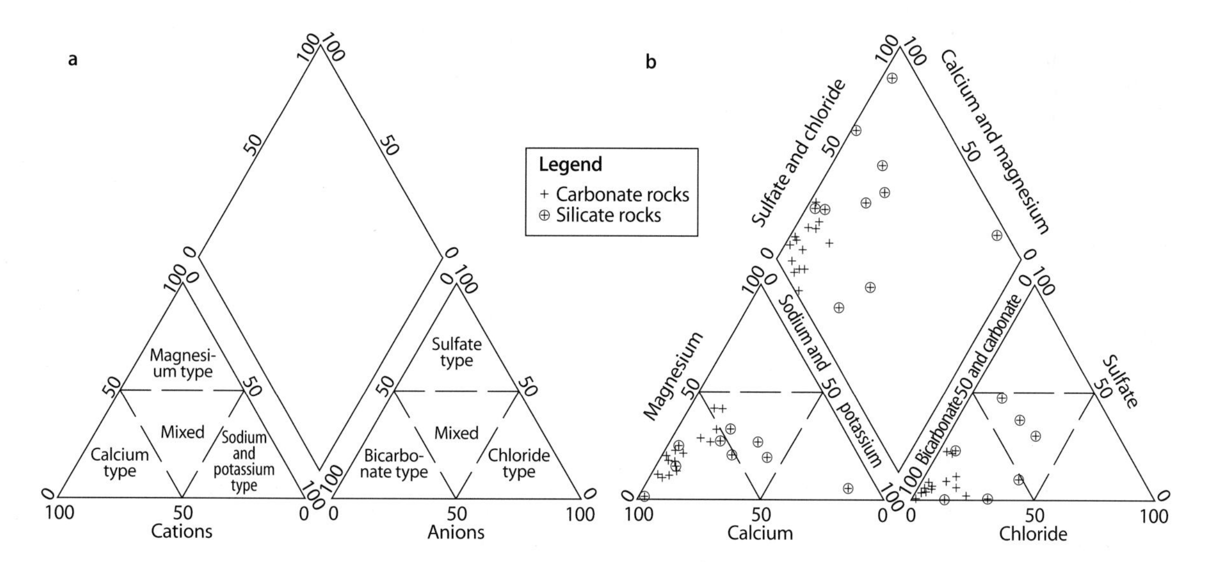

Fig. 3.8. Plot of classification of 24 analyses of Maryland spring water according to the method of Hill and Piper, values in percent of total equivalents per liter. **a** Type of water; **b** spring water analyses (Otton and Hilliary 1985)

Shuster and White (1971) did a hydrogeologic study of the Nittany Valley in Pennsylvania, underlain by a carbonate aquifer. they established a relationship between physical and chemical properties of springs which lead them to classify springs according to the type of flow they exhibited. Diffuse flow springs are caused by water flowing, often laminarily, through primary pore spaces like fractures, joints and bedding planes or secondary small scale pore spaces (in centimeters or less) largely unmodified by solution. Conduit springs are caused by water flowing, often turbulently, through integrated systems of solution passages measured in centimeters or meters.

Scanlon and Thrailkill (1987) studied the Inner Bluegrass Region of Kentucky using dye tracers to determine if there was a relationship between physical and chemical properties of springs in that region. They determined that there was no variation in the chemistry between the local high-level springs discharging from shallow flow paths and major low-level springs discharging from a deep integrated conduit system. Since this type of classification was determined to be dependent on the hydrogeologic setting, it would be considered an incomplete classification.

3.8 Present Documentation and Classification

From the 1920s through World War II, the growth of the scientific community continued with more specialization. For example within geology, specialization into the fields of hydrogeology, geochemistry, meteorology, oceanography, and hydrology. Books, journals, and the national scientific organizations and societies continued using the classifications established by Bryan, Meinzer, and others. The classification of springs employed by a scientist may have more to do with their experience and use of a classification system rather than on the merits of the classification system itself. For example, a geochemist would be more comfortable using a chemical classification system rather than a discharge classification system.

For example, plant taxonomy, the classification of plants, became a subfield specialty of the botanist, although there was already an established classification for plants. The natural or phlyogenetic classification of plants was based on the genetic character relationship between plants. The overall classification of plants traces the evolution of plants and groups them into units based on their structural complexity. Thus, the classification was hierarchically arranged as:

- Division
- Class
- Order
- Family
- Genera
- Species
- Varieties
- Forms

A scientific name was given to each plant and the scientific name constitutes an internationally accepted name that eliminates ambiguity and misunderstandings. So even though the lily, *Erythronium grandiflorum*, has many common names such as Dogtooth Violet, Fawnlily, Glacier Lily, Snow Lily, and Adder' Tongue, it still has only one unique scientific name.

The divergence here is important to illustrate a possible internationally acceptable classification of springs. Herein lies the difficulty however, because who is to say whether springs are similar? As already shown, dependent on the scientist' persuasion similar may mean geologic origin, discharge, or chemical composition, etc. With plants, there were other classification systems referred to as "artificial classifications" which attempted to group plants based on their habitats or colors or number of stamens, but these classifications were considered incomplete because they could not classify completely or handle exceptions well.

Classifications traditionally have aided in research and in the universal sharing of information. Scientific associations have helped serve as watchdogs on professionalism and played a big role in the dissemination of scientific information and discoveries. Scientific journals, conferences, and memberships all share the common goal of furthering the body of scientific knowledge by peer review, providing scientific accuracy and accountability. Government agencies worldwide have promoted scientific research and donated monies to worthy scientific endeavors. Changes in naming a plant, rock type or geologic name, have been traditionally accepted as being strictly monitored by scientific committees and organizations. For example, changing the name of a geologic formation would go through a formal review by national and international committees. Perhaps this process should be followed in developing a final classification of springs.

3.8.1 Instrumentation and Quantification

After World War II, technological advances in instrumentation and the advent of computers began to change the way the scientific community operated. Surveying tools evolved from chains and rods, to transits, to total systems surveying and satellite navigation. Maps were made more accurate and more detailed. Aerial photography and satellite images changed the way maps were compiled and interpreted. Plus, the growth in the geosciences, engineering, and mechanics of water instrumentation and measurement helped make water data more complete and meaningful. Technological advances in general made acquiring and processing data more complete, usually faster, and more precise.

Finally, with the advent of computers, telecommunications, and videos, the "information age" is driven by the fast global exchange of data. For the scientific community this has allowed virtually instantaneous knowledge exchange and awareness of scientific events and discoveries worldwide.

3.9 Classification of Springs Documented by Computer Databases

Computers and the growth of database design to hold large volummes of data allows greater variability for developing a classification system. The relational database design and computer hardware combined create vast storage capability and allow countless searches by sorting on common fields and codes. Computers allow greater and more rapid sharing of data, relative to utilizing a particular database. For example, an international database of springs that could provide all the first order magnitude springs in the world that had carbonate waters. Codes and search criteria must be developed to allow for the correct entry of each spring, but once the database is created, a powerful tool emerges that could automatically classify a spring based on its geologic origin, physical properties, chemical composition, discharge rates, etc. Through global networks and the use of geographic software packages, locations of springs as well as the description of springs could be put in and updated rapidly. This information then could be imported into other programs or used to create hydrogeologic computer models etc.

There are already existing databases which hold information on springs. For example, one database has the following fields and information on springs:

1. Potable/non-potable
2. Water quality parameters
3. Perennial/intermittent
4. Discharge rate
5. Primary water type (geochemical)
6. Usage
7. Location
8. Geological type/occurrence

A publication by the West Virginia Geological Survey used a computer database for classification of *Springs of west Virginia* (McColloch 1986). As with earlier classifications, a primary criterion is necessary to initially divide up springs into groups. In West Virginia, the size of the spring based on discharge was the main criterion, however, the size of the spring could either be a major or minor criterion. In the west Virginia classification, a major spring had a discharge equal to or greater than 2 gal/min or was historically important. A minor spring had all discharge measurements less than 2 gal/min. As a catchall, a third field of "unknown" was added for springs with no available discharge information. The classification of the spring then became more of an inventory. This database included:

1. Location
   - USGS longitude and latitude designation
   - elevation (feet above land surface)
   - 7.5' quad series name
   - hydrologic basin (8-digit code by Office of Water Data Coordination)
2. Topographic (according to WATSTORE User's Guide)
   - depression
   - flat
   - floodplain
   - hillside
   - hilltop
   - sinkhole
   - stream channel
   - terrace
   - undulating
   - valley
3. Ownership
   - private
   - company

- water district
- municipal
- county
- state/federal

4. Water use (defined by WATSTORE User's Guide)
   - abandoned
   - air conditioning
   - bottling
   - commercial
   - domestic
   - other
   - recreation
   - unused
   - industrial
   - institutional
   - irrigation
   - mining
   - observation
   - public supply
   - stock supply
5. Geology (where spring emerges – bedrock formation)
6. Rock type
7. Physical and chemical analyses
8. Comments
9. References

The Illinois Geological Survey has developed a classification system for their spring inventory that generally follows Fetter (1980), Freeze and Cherry (1997), Knighton (1984), and Meinzer (1923a). These include physical, chemical and biological parameters at the spring head and within the spring outflow channel.

Thermal springs, like mineral springs, have been more extensively classified based on geochemical parameters. The studies of oceanography, plate tectonics, and geochemistry have all led to a greater understanding of the mechanisms and occurrence of thermal springs worldwide. Bliss (1983) created a classification of thermal springs based on geochemical parameters and used it to design the GEOTHERM database for the United States.

One of the major problems with geographic and geologic computer databases is coordinating efforts among national organizations, governmental agencies, and international groups to agree on the database design and data entry. Centralization of database design would help eliminate duplication of data, incomplete information or erroneous data. For classifications, it could support some form of national or international standardization. In the United States, the USGS has developed WATSTORE, the Water Information and Retrieval System. On a national scale, as an agency, it has tried to evaluate ways to enhance the utility of this database, however, it is not an easy task for an agency or corporation. And, tracking information on a global scale among different languages, scientific groups, and government agencies becomes a political problem as well as a scientific endeavor. Funding, personnel, computer expertise, etc. are all real factors which can hinder efforts to develop computer databases with complete water information including a classification of springs.

3.10 Future Trends

Classification of springs is an academic pursuit worthy of scientific scholars' attention. In the 1990s, and as the 21st century approaches, water resources and water issues are becoming increasingly important. Natural resource management and planning call for tools which will create a better understanding of our natural world and allow better utilization of finite resources such as water. Looking at water usage will be a key criterion in deciding how water is allocated. Many springs are sources of potable water, so the importance of classifying and locating these springs worldwide will become more commonplace as "pure" water becomes an increasingly valuable commodity. The market for "bottled waters" is one recent trend where the public is buying "pure" water coming from natural springs as an alternative to traditional municipal water supply systems.

3.11 Process

Three different types of processes should be considered: chemical, thermal, and geo/structural. In most cases, no single process is responsible for the resultant landform associated with springs.

3.12 Karst Springs

The formation of a karst landscape is generally related to the occurrence of carbonate rocks that are soluble and leave little or no residue, e.g. evaporites and car-

Fig. 3.9.
a Karst spring at Aak, Germany, near Singen, Germany, Swabian Alb. High flow 24 m^3/s; low flow 7 m^3/s. **b** Donaues Chingen Spring in Germany. Source of the River Danube (photographs by P. E. LaMoreaux September 1992)

bonates (Fig. 3.9a, b). Of these two, carbonates are by far the most significant, limestone $CaCO_3$ and dolomite $CaMg(CO_3)_2$. Typically, karst landscapes develop on pure and compact limestones and dolomites. The higher the percentage of impurities, the more substantially retarded the process of solutioning becomes. The permeability of carbonate rocks depends on porosity, jointing and bedding. Joints have a significant effect on undergroundwater systems and generally connect bedding planes and provide passageways for the flow of water above and below the water table. The dissolution process requires adequate precipitation and may be summarized in Fig. 3.10, the mutual interdependencies in the system are shown by:

$$CaCO_3 + CO_2 + H_2O \longleftrightarrow Ca^{2+} + 2HCO_3^-$$

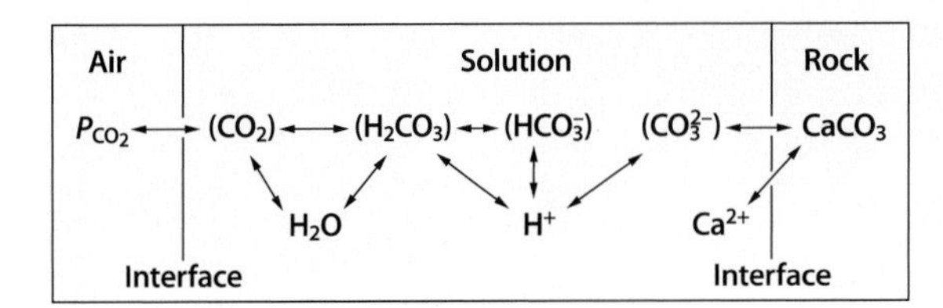

Fig. 3.10. The mutual dependencies in the system CO_2–H_2O–$CaCO_3$; the dissociation of water is not taken into consideration

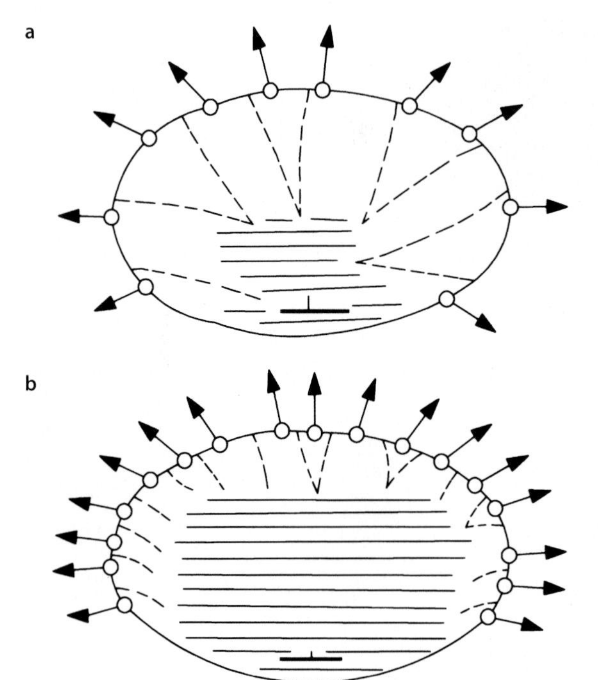

Fig. 3.11. Catchment areas of karst springs **a** in an early and **b** in a late stage of development (*hatched* undifferentiated karst water-body; *dashed lines* underground watershed) (Bögli 1980)

Karst springs represent a natural exit for groundwater to the surface through the hydrologically connected fissures of the karst mass, and appear most commonly at the contact of carbonate mass and an impermeable boundary (Lehman 1932). Lehman introduced the concept of "the karst-hydrological contract" as he noted the numerous seepage points into the karst mass for precipitation and the relatively few karst springs as shown in Fig. 3.11 (Bögli). This can easily be explained

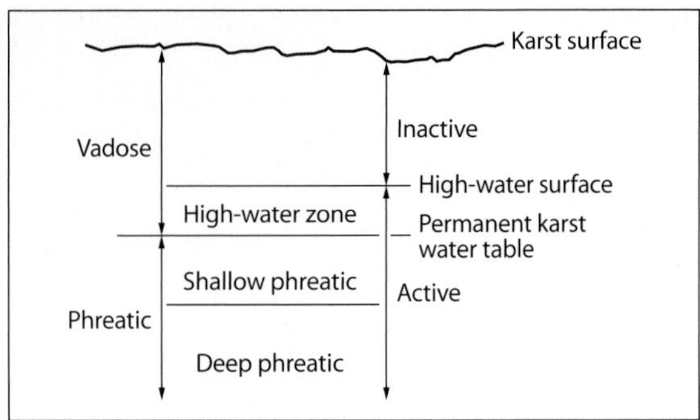

Fig. 3.12. Karst hydrological zones (Bögli 1980)

in relation to the degree of maturity of the system. In the early stages of karstification, only the marginal parts of the water body are oriented towards a spring. With increasing karst-hydrological activity (Fig. 3.12), the catchments of individual springs reach deeper and the more efficient ones capture the drainage from others. In time, the smaller springs are gradually eliminated. If springs are fed from the phreatic zone, the discharge from the lower springs increases at the cost of the higher ones. The faster outflow causes the karst water level to decline and the flow from higher springs ceases. Another interesting aspect of the karst-hydrological contract is that one stream entering the underground system frequently flows out at very different spring outlets at considerable distances from each other.

Most of the larger springs in the world discharge from calcareous rocks because they are relatively soluble to percolating groundwater and therefore susceptible to the solutionally developed channels along joints, fractures, and bedding planes. These channels are capable of transmitting and storing large quantities of water (Fig. 3.13 and 3.14). Good examples include Comal Spring from the Edwards Aquifer in Texas, Silver Springs from the Floridan Aquifer in Florida, and Figeh Spring near Damascus, Syria.

There are many types of karst springs and general principles of classification may be applied to them. Bögli lists three types of classifications: (1) according to outflow, (2) according to geologic and tectonic conditions, and (3) according to origin of the water. Outflow characteristics shall be considered in more detail as they are unique to karst springs.

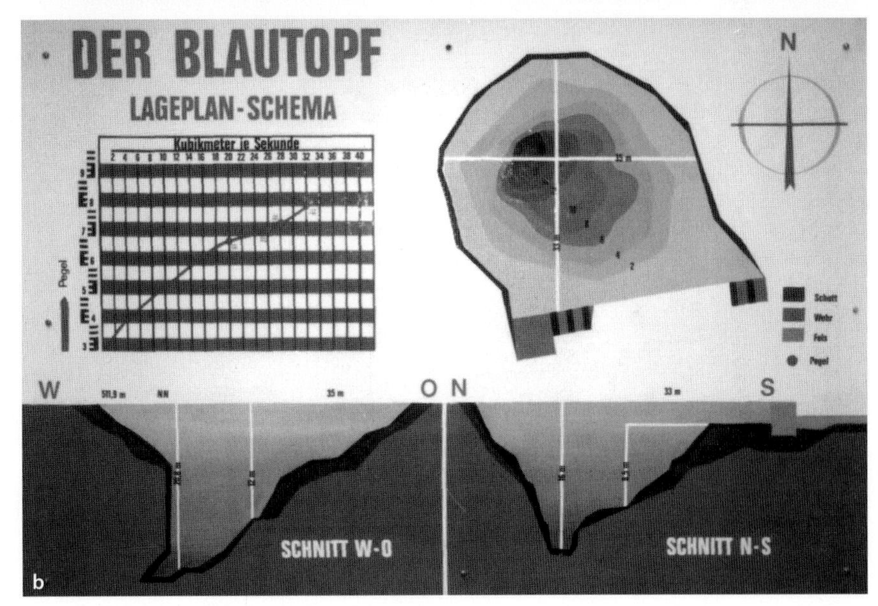

Fig. 3.13.
a Blautopf Spring. Karst spring-water wheel discharge to sink-hole cavity, Blautopf, Germany. **b** Diagram showing profile spring discharge to sinkhole (photographs by P. E. LaMoreaux 20 September 1992)

Periodic. Springs which show in their average discharge a lengthy marked discharge that is *climate* controlled are periodic. Variations in spring flow are normal and occur as a result of changes in precipitation. Many parts of central Europe have a fairly balanced precipitation regime year-round, but evapotranspiration creates an excess of water in winter and a deficiency in summer. This in turn leads to a periodic increase and decrease in recharge and discharge from springs in the area. In Alpine regions, due to snowmelt, the difference between dry winters and early summers is striking, and is reflected in spring discharges. Many regions show a strong annual rhythm. Single precipitation events that cause change in discharge are considered aperiodic.

Fig. 3.14. Buchbrunnen Spring, Germany. Swabian Alb. Karstified limestone. Strict rules protect groundwater in area. **a** Pumping station. **b** Weir discharge (photographs by P. E. LaMoreaux 22 September 1992)

Intermittent. Most springs of this type exhibit great variability in flow. True intermittent springs are rare. Katzer (1909) was the first to recognize the processes behind this rare phenomenon. Figure 3.15 illustrates this process. Water collects in a basin inside a rock mass from which a tube with an elbow first rises and then declines. If the water rises above point k in the elbow, the water flows over and fills the descending branch. If this branch is narrow enough to allow slow flow, a column of water builds up and siphons such water out of basin B until the water level has fallen below A. For the siphon to function correctly, the tube at k must be narrow enough so that when the water runs over, it fills the whole tube. In addition, there cannot be any joints in the elbow which would allow air to enter the system and in so doing break the column of water.

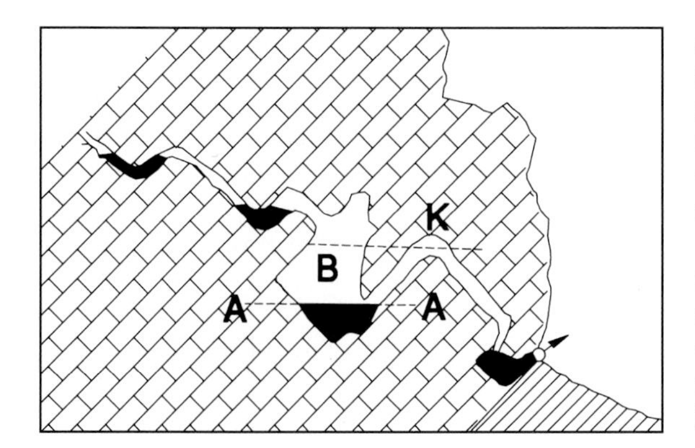

Fig. 3.15. Cross section of an intermittent spring (Bögli 1980)

The siphon phenomenon does not need to be in the immediate vicinity of the spring. If it is close enough, the intermittent water stream may join with another steady stream and the rhythm becomes superimposed on the constant discharge. The drainage of Lilbum Cave in Sequoja National Park has its basic discharge superimposed by a rhythmic swelling and subsidence over a period of several hours. Most intermittent springs do not show the siphon effect at all groundwater levels. If at high water the siphon becomes flooded, it will stop functioning due to lack of air, thus the intermittent effect only occurs at intermediate and low water levels. If the basin is not watertight and the water influx is low, it will not be able to fill the basin to level k, thus the intermittent effect will only occur at higher water levels.

Periodic Spring, Wyoming, discharges from a solution cave in the Madison limestone along the axis of an anticline, as shown in Fig. 3.16. It discharges on and off with a variable cyclicity during low discharges due to the siphon effect. It discharges a continuous but irregularly oscillating stream during high flow periods (June–August). As the discharge diminishes, the amplitude of the oscillations increase erratically. Then the frequency of flow oscillations becomes more regular and the amplitude increases as the yield declines. Ultimately the discharge starts and ceases, initially for a very short period (see Fig. 3.16). At critical transitional discharge rates, the flow pattern sporadically attenuates between continuous but cyclic yields.

Episodic. These show no periodicity and flow is completely irregular, occurring only when there are very high water levels in the karst mass. Binder (1957) noted the great variability in flow for Hungerbrunnen between

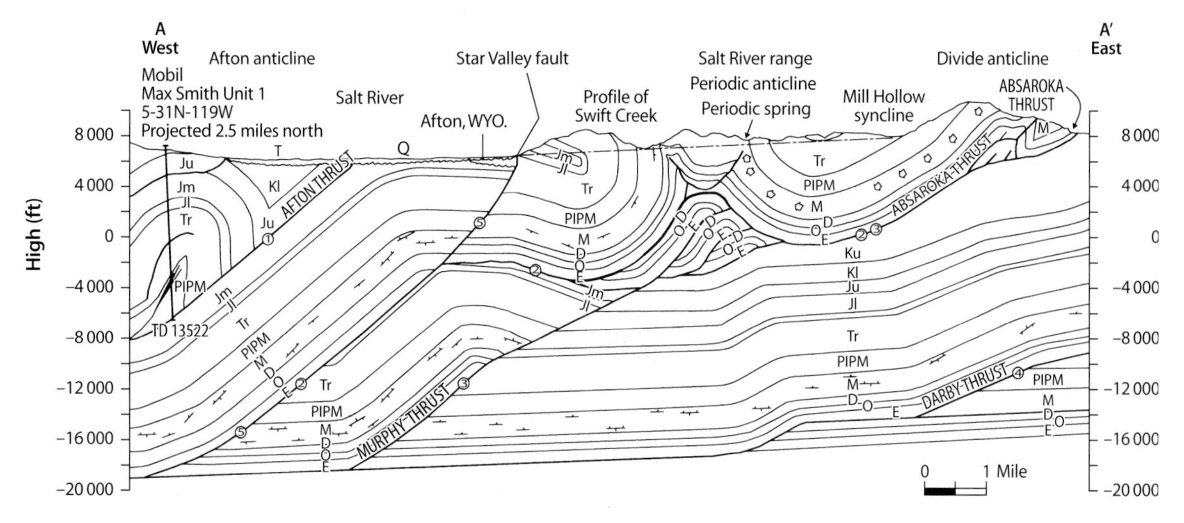

Fig. 3.16. West-east structural cross section through Periodic Spring in the central Salt River Range, Wyoming. Numbers on thrust faults show sequence of faulting with 1 being the oldest. *Q* = Quaternary alluvium; *T* = Tertiary clastic deposits; *Ku* = upper Cretaceous Hilliard, Aspen, Bear River formations; *Kl* = lower Cretaceous Gannett Group; *Ju* = upper Jurassic Stump, Preuss formations; *Jm* = middle Jurassic Twin Creek Limestone; *Jl* = lower Jurassic Nuggett Sandstone; *Tr* = Triassic Ankareh, Thaynes, Woodside, Dinwoody formations; *PIPM* = Permian – Pennsylvanian – Mississippian Phosphoria, Wells, Amsden formations; *M* = Mississippian Madison Limestone; *D* = Devonian Darby Formation; *O* = Ordovician Bighorn Dolomite; *C* = Cambrian Gallatin, Gros Ventre formations (Huntoon and Coogan 1987)

1867 and 1957, ranging from only a few days in 1947 to a continuous period of 18 months around 1939. The Hungerbrunnen (or hunger springs) are so called because they only flow in very wet years which traditionally are years of bad harvest in the eastern Swabian Alb.

Perennial. Perennial springs are the most reliable, having a guaranteed source year-round. However, it should be noted that perennial is a relative term and the karst landscape is dynamic and therefore there is no definite lifespan for a spring, rather it is guaranteed only to be more dependable than the other types. In general, perennial springs are large, thus even in low flow periods continue to discharge.

The most important karst springs are those present at the beginning of strong open streamflows, at river heads. Petrovic (1983) classified them as cave springs, hidden springs, fissure springs, spring eyes and spring systems. For a spring to be economically viable as a water source, there must be a guaranteed year-round supply, or at least a predictable regime. Cave springs, such as Dumanli Spring, Turkey, issue from a visible opening in a cave. This has a mean discharge of 50 m^3/s and a total annual outflow of 1.6×10^9 m^3 (Karanjac and Günay 1980). Analysis of the recession hydrograph allows us to quantify the spring in detail. The Dumanli and several smaller springs in the vicinity discharge 630×10^6 m^3 during the 5 dry months of the year. At the end of the recession, 1370×10^6 m^3 still remain in storage. The coefficient of recession is 0.0026 per day, indicating a large drainage system, a high storage volume and slow drainage. The contributing drainage area is of the order of 1200–1500 km^2. Hidden springs occur when the cave opening has been blocked by landslides or erosion. The Crnojevica River emerges from a hidden spring, but only in dry periods, in wet periods, vast quantities of water flow from the higher Obod Cave. It is often difficult to pinpoint the exact location of hidden springs, as the water may percolate through tens of meters of alluvial or river deposits before discharging at the land surface.

Fissure springs emerge from well-developed, large, linked vertical and sloped fissures and may be explained by the divergence of groundwater flow before its appearance at the surface. In many areas, these are covered by fluvial deposits and so are in effect hidden springs. Spring eyes occur at an advanced stage of kartification and represent the outflow from the open caves of siphon channels. Buläz Spring in Istia, Yugoslavia is a good example and is a typical ascending karst spring (Fig. 3.17). Its capacity ranges from $0.1 \rightarrow 30$ m^3/s (Bonacci 1987). Spring systems consist of a great number of springs belonging to different types. Systems are common in very karstified areas with well-developed watercirculation paths. The Zetna River Basin in Croatia, represents a typical spring system.

The Studenci Spring zone in Croatia consists of four permanent and numerous small permanent and intermittent springs at different levels, 10–20 m above sea level. The levels at which water appears at the surface vary according to the water level in the surrounding karst mass.

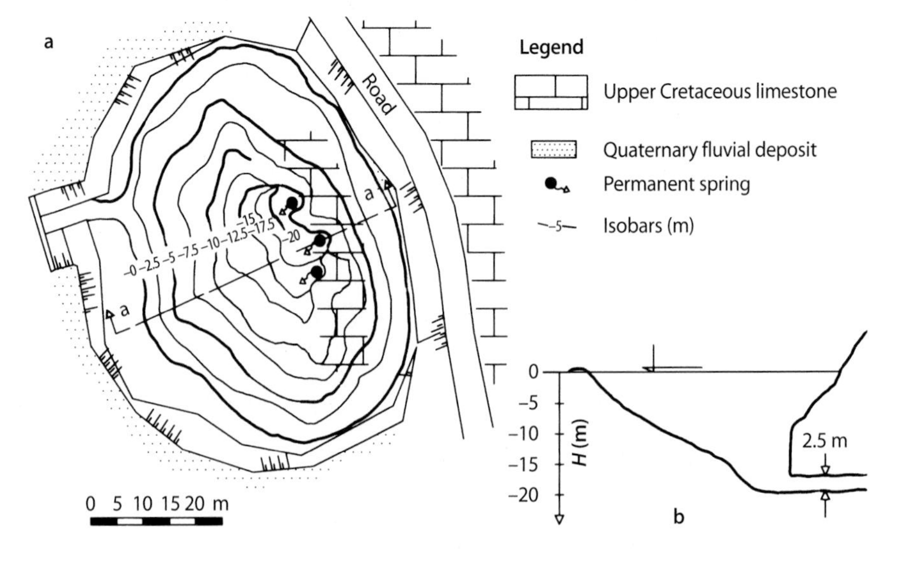

Fig. 3.17. **a** Plan and **b** cross section through the karst spring Bulaž, Yugoslavia (Bonacci 1987)

Riverbed springs can occur in the open riverbed in streams and in karst areas. They are generally numerous and represent spring zones distributed along reaches of a river. Habic (1982) reports the character of Muzleh Spring on the Soca River, Croatia, as being a 400-m-long zone with nine springs on each bank. The existence of a spring zone in riverbeds clearly illustrates the complexity of water circulation in a karst area and the problems inherent in defining catchments and discharge zones. There are many examples of streams flowing over limestone or dolomite bedrock that has become karstified and where, at high water levels, large volummes of water flow upward to the land surface and discharge from buried sinkholes or chimneys. At low groundwater levels in a karst area, surface water can discharge downward through sinkholes or chimneys below the land surface. Reaches of the Danube River in Germany and the Peace River in central Florida are two excellent examples of this phenomenon.

Vauclusian. Trombe (1952) defines a Vauclusian spring as an outflow from an ascending branch of a siphon in a karst area. In a true Vauclusian spring, the water runs under pressure up through the rock. In general, the Vauclusian springs are very large. As with other karst springs, Vauclusian springs are subject to change, and almost all Vauclusian springs show higher dry spring openings which are partly or completely out of use, due to lowering of the local base level of the water table and/or adjacent rivers or streams. Under this condition, the ascending branch now behaves as a piezometric tube, if the waterlevel reaches the rim of the earlier spring, a high water relief spring is formed. In Switzerland, Holloch Spring, 96 m above the present base level at Schleichenden Brunnen, behaves as a high relief spring. The Ruda River in the Cetina River Catchment of Croatia is another typical ascending permanent spring. In the 10-year period 1974–1984, the mean discharge of the spring was 14 m^3/s with a range of 4.42–43 m^3/s. Thus the ratio is 1 : 3.2 : 9.7, indicating that the inflow to the spring has been uniform and thus it has a favorable regime from a hydrological point of view.

Another large group of karst springs is termed subaqueous, and this group is composed of submarine and subclustrine types.

Estavelles. These serve a double hydrological function, as ponors in dry seasons and as springs in wet ones, as illustrated in Fig. 3.18. Estavelles most frequently appear in the middle of large poljes. When the polje is flooded, the estavelle functions temporarily as a submerged spring, and when the water level is lowered in the karst mass, they function as ponors. These are common in Croatian karst and numerous examples may be found in the Cetina River. Dubic (1984) worked on a special type of estavelle, the Gornje-poljski vir in the Niksicho Polje. Throughout most of the year it is a lake with a diameter of 85 m and an average depth of 30 m. In the warm dry period of the year it functions as a ponor. After the first heavy rains in Autumn, the water in the lake

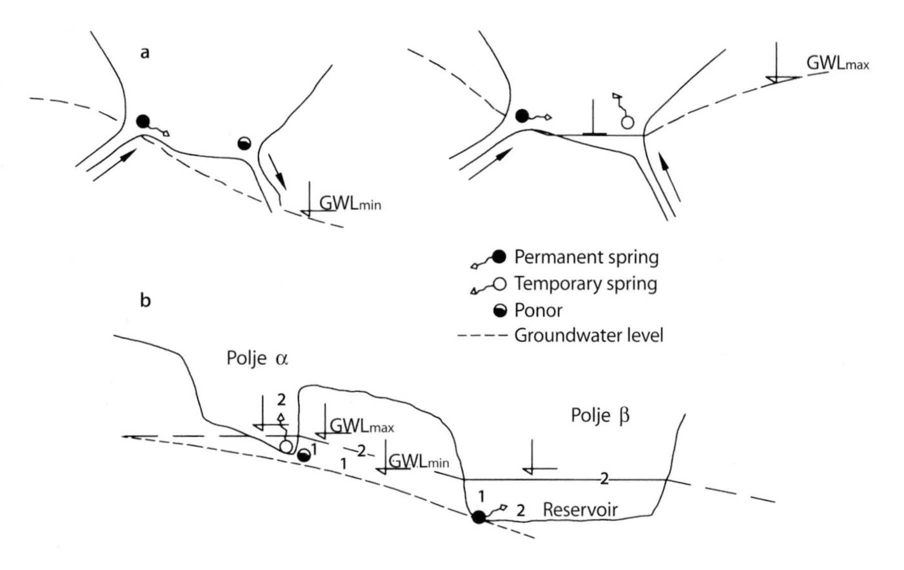

Fig. 3.18. **a** Explanation of the estavelle functioning as ponor and temporary spring. **b** Formation of estavelle in situation when the lower horizons are flooded (Bonacci 1987)

becomes turbulent and disappears into the ponor. After only 50 min the muddy water returns, fills the lake and flows down to the Susica riverbed. The existence and nature of estavelles functioning in this manner provides ample proof that catchment areas in karst change in time, depending on water levels in the karst mass.

Sub-lacustrine. These springs can form in one of three ways. First, the spring may form after the lake due to water pressure in the rock exceeding the hydrostatic pressure of the lake water. Second, division of an existing water course due to erosion, may create such as spring, as seen in glacially over-deepened Alpine valleys. Third, submergence of a karst spring at a valley margin by morphological or tectonic processes may form a sub-lacustrine spring. Water emerging subaqueously into a lake can be recognized by bubbling at the surface, temperature differences, lime content and by cloudiness or color. The Lac d'Annecy, Switzerland, is a good example of a lake containing karst springs due to glacial erosion into karstified limestones.

Vruljas. These emerge from small and large karst joints. They frequently emerge from submarine caves and their openings occur at the bottom of inundated sinkholes. Their existence is due to the lowering of the karst erosion base during the Pleistocene. The change in the position of the springs along coasts depends on the local geologic structure of the coastal belt. The presence of a lot of channels at different levels through which water circulates below and above sea-level results in mixing of fresh and saline waters. Generally, water in vruljes is brackish, salinity increases in dry periods and decreases in wet periods.

Submarine springs, or vruljas (Croatia), are frequently only one element of a larger spring system. Alfirevic (1966) concluded an extensive study of the vruljas of Kastela Bay. He postulates that morphologically, these vruljas correspond to inundated fossil dolines from the pre-alluvial age formed in the continental phase of Dinaric coastal karst. The salinity and temperature of their waters are directly influenced by precipitation in the catchment area. Vruljas occur worldwide, although they are most frequent in the area of the Mediterranean basin, and those with the greatest capacities occur along the coasts of Libya, Israel, Syria, Greece, France, Spain, Italy, and Croatia. About 80% are in water up to 10 m in depth and only a few are located at depths up to 50 m. Discharge capacity varies throughout the year and only a few function permanently as springs. Air photography and heat-sensing imagery are helpful in locating Vruljas.

Brackish karst springs occur along the coast and in submarine environments. According to Kohout (1966) and Breznik (1973, 1978), they may be classified as: (1) springs contaminated due to the hydrodynamic effect, (2) springs contaminated due to the convectional circulation of saline groundwater induced by geothermal leaking, and (3) springs contaminated due to the high density of seawater. Figure 3.19, depicts the functioning of brackish vrulje in fresh, brackish, and coastal springs. In recent years, more attention has been paid to the potential development of sub-marine springs. Potie and Tardieu (date not available) have been working at Port Miou, Marseille over a 10-year period in an attempt to gain a better understanding of submarine springs. An extensive diving program has allowed them to chart the exact topography of the submarine channels, collect samples and make physical and chemical tests in situ. They have noted the difference between the fissure and the channel systems and the independence of the two and their different hydraulic laws at any given time. New methods of investigation are being explored, namely the use of microgravimetric and microseismic logs to locate the interface and fractured zone.

Karst springs in regions of sub-arctic karst are insignificant quantitatively but do exist and so merit a mention. Nahanni karst occurring at 61° N and 124° W is a complex subarctic karst system extending for 28 miles in an 11-mile-wide bank. All water falling on this area (yearly precipitation of 23") eventually discharges to one of two springs, both having catchment areas of approximately 100 mile^2. White Spring issues from dolomites and in July 1973 had an average discharge of 190 ft^3/s. Bubbling Spring resurges from the Nahanni Formation limestones, the water forcing its way through the unconsolidated sands. In August 1973 its average discharge was 260 ft^3/s.

Studies performed in the north-eastern United States have indicated that glacial deposition has had a greater effect on present karst groundwater systems than glacial erosion (Palmer 1977). Many pre-glacial springs were blocked by glacial till causing diversions of groundwater flow to new, generally higher levels, increasing the local storage capacity. In addition, the hydraulic inefficiency of partially blocked springs has been known to cause severe periodic flooding of solutional conduits in the vacluse zone resulting in wide-scale fissure enlargements.

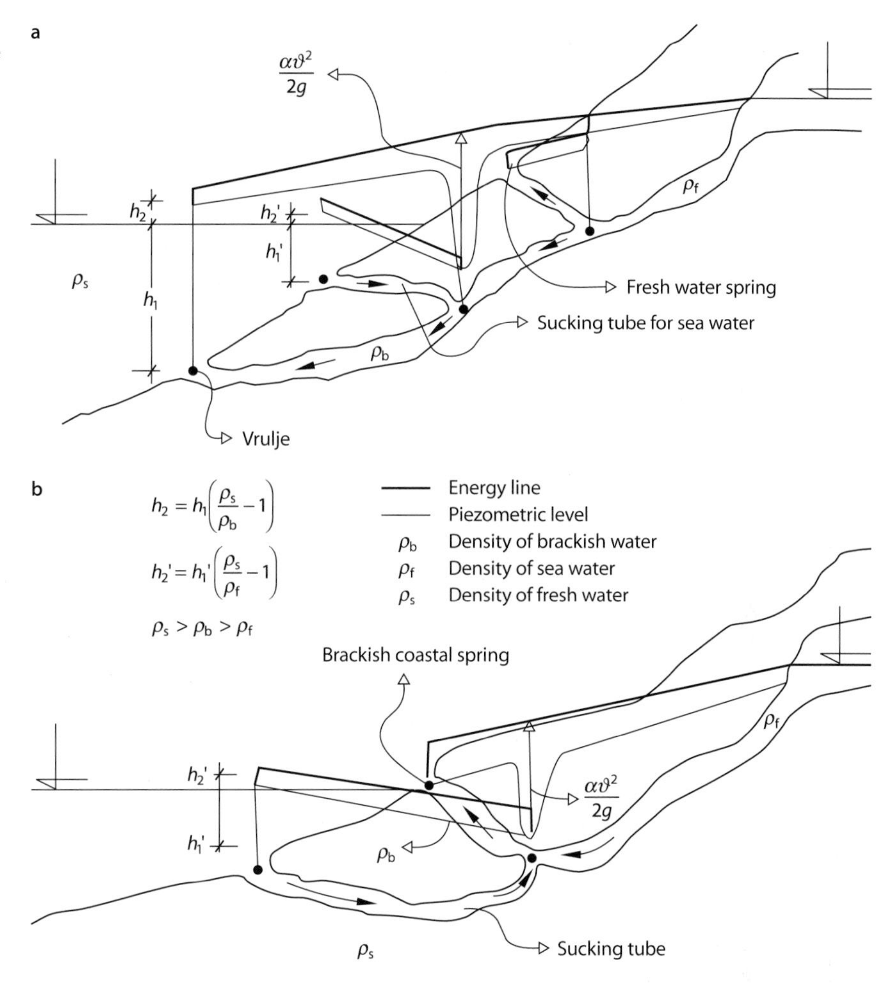

Fig. 3.19. Functioning of a brackish vrulje and fresh (**a**) and brackish (**b**) coastal springs (Bonacci 1987)

One of the largest karst springs in the world has an average discharge of 36.3 m^3/s, water issues from this spring and thirteen others that combine to form the headwaters of the Khabour River. The springs issue from a local anticlinal saddle in a general synclinal structure, tertiary in age. Recharge to the spring appears to be a direct function of surface runoff, the major infiltration area occurring over Eocene limestones. The aquifer discharges into lakes filling depressions formed by karst dolines. The average yield of 36 m^3/s occurs in June and a maximum flow of 41.8 m^3/s in November. The decrease in yield from December to May is associated with flooding and the high stage of the river. Increases in spring discharge occur when river levels decline. This suggests that the yield of the spring is affected by the hydrostatic level of the discharge area.

3.13 Thermal Springs

Thermal springs discharge water having a temperature in excess of the normal local groundwater. They can be defined as springs whose temperature is significantly higher (10 °F) than the mean annual air temperature of the surrounding area. Hot springs can be defined as those whose water has temperatures higher than 100 °F. Usually their discharging water is highly mineralized. In recent years, more attention is being directed toward thermal springs as indicators of areas in which exploitation of geothermal energy might be economically feasible.

The source of thermal energy in groundwater is twofold: the energy in the shallow groundwater is largely from the sun; the energy in the deeper water is derived from upward flow from the Earth's interior. White et al. (1963) indicate that in tropical areas the depth of nearly uniform temperature occurs at about 33 ft (10 m), and in the polar regions this depth increases to 66 ft (20 m). However, this range is imprecise as the ground temperature profile is influenced by several factors, among which are (1) mean annual air temperature, (2) character of the soil and groundwater, (3) thermal conductivity of the zone of aeration, (4) thermal conductivity of the zone of saturation, and (5) presence or absence of circulating groundwater at depth.

Annual air temperature cycles affect groundwater temperatures to a depth of between 40 and 60 ft, except where unusual groundwater flow conditions prevail. Records of subsoil temperature at various depths throughout an annual cycle are given by Smith et al. (1964, from a study at Belgrade, Yugoslavia). The study site is at an elevation of 456 ft and is in the temperature zone at lat. 40°48' N. The temperature difference at various depths is shown in Table 3.6.

Thus, the Belgrade measurements indicate no significant annual temperature change below about a 40-ft depth.

Although no precise depth range can be given below 40 to 60 ft, seasonal temperature effects are generally negligible and the ground temperature is affected chiefly by the heat loss from the interior of the Earth. The increase in Earth temperature with increasing depth is known as the geothermal gradient. It is not a constant value. In 1848, during his second trip to the USA, Sir Charles Lyell noted the geothermal gradient associated with groundwater from artesian wells in Alabama in the Gulf Coastal Plain (Lyell 1849).

Table 3.6. Temperature difference at various depths (Smith et al. 1964, from a study at Belgrade, Yugoslavia)

Depth (ft)	Annual temperature difference (°F)	(°C)
4	25.6	14.2
6.6	18.7	10.4
13	8.5	4.7
26	0.54	0.3
33	0.36	0.2
39	0.18	0.1
46	0.0	0.0

It is known that groundwater provides a viable means of storing thermal energy in the subsurface. As the water discharges from springs, it carries this energy with it. For some of the larger springs, a surprisingly large quantity of energy is available in the water. The temperature of most springs, especially the smaller ones, fluctuates with the annual seasonal climatic cycles.

The normal thermal equilibrium of an area may be affected by: (1) the presence of an intrusive magma rising sufficiently close to the surface to heat the surrounding rock and the interstitial water; (2) the extrusion of igneous rocks (volcanism); (3) heat from radioactive elements; and (4) heat that might be generated by friction along faults. Hypothermal conditions having economic value are usually related to intrusive or extrusive masses of magma. Fault friction may be a minor source of heat for all springs in the vicinity of faults. A *geyser* is a periodic thermal spring resulting from the expansive force of superheated stream within restricted subsurface channels. Figure 3.20 is a schematic diagram showing how water from surface sources and shallow aquifers drains downwards into a deep vertical tube where it is heated to above boiling point (Todd 1980). With increasing pressure, the steam pushes upward, this releases some water at the surface, which reduces the hydrostatic pressure and causes the deeper superheated water to accelerate upwards and flush into steam. The geyser then surges into full eruption for a short interval until the pressure is dissipated. Then, the filling begins again and the cycle is repeated. A mudpot results when only a limited supply of water is available. Water mixes with clay and undissolved particles brought to the surface forming a muddy suspension by the small amount of water and stream continuing to bubble to the surface. The production of steam is dependent on the existence of a source of heat, the presence of a permeable bed and generally on the presence of an impermeable cap rock.

More than half of the 65 first order springs (>2.8 m^3/s) in the US issue from volcanic rocks, fractures being a major control on groundwater movement within the rock mass.

It has been proven that most geothermal water is of meteoric origin. Heat is usually acquired in one of the two ways: (1) water coming in contact with or in close proximity to recently emplaced igneous bodies, e.g., Yellowstone National Park. This area possesses the greatest concentration of thermal springs in the world. Temperatures of 240 °C exist only 300 m below the land

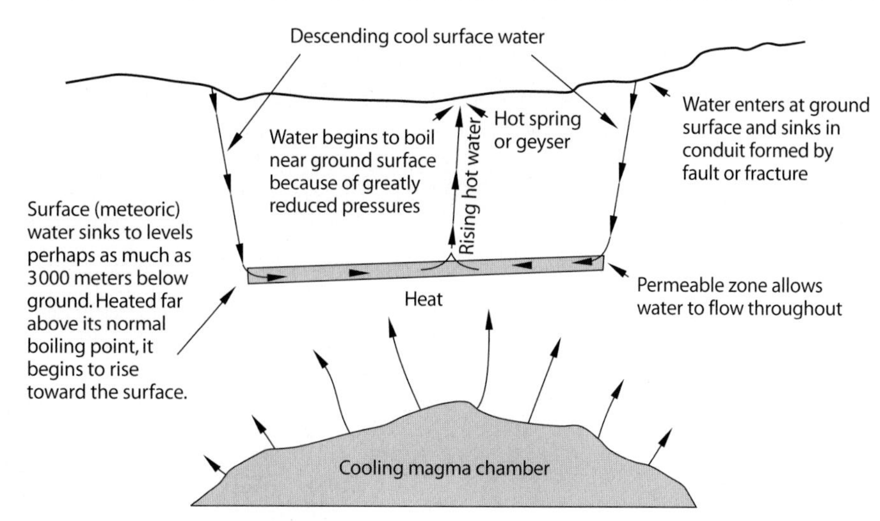

Fig. 3.20. Schematic diagram of a hydrothermal system (Todd 1980)

surface, due to a volcanic eruption 600 000 years ago; and (2) deep circulation, e.g. Hot Springs, Arkansas (Fig. 3.21).

In general, all thermal springs have a deep component of flow which is developed preferentially to the shallow component. In order for this to occur, a favorable structural setting must be present. The normal geothermal gradient provides the source of the heat, the topographically high recharge areas provide the energy to drive the water through the system and the structural setting orients or provides enhanced vertical permeability and porosity. Meteoric water is recharged and moves deeply, gaining heat from the aquifer under a normal geothermal gradient. Fractured or folded high permeability zones allow the heated water to escape without thermal re-equilibration. All thermal spring systems are primarily controlled by structure. Considerable folding and tilting are essential to orientate existing water paths so that hydraulic continuity is achieved between the land surface and great enough depths below to allow the water to become heated (Fig. 3.22). Faulting or fracturing is necessary to allow the heated water to return rapidly to the surface. Springs commonly occur near the crest or on the flanks of an anticline.

The occurrence of thermal springs in Switzerland is limited to the Jura and Alps regions; these areas having the necessary vertical permeabilities. A study of geometric tectonic features, such as axial culminations and depressions, and dislocation and fault planes, has allowed a correlation to be made between their presence and the location of thermal springs. Tectonic features of the dynamic kind, such as recent crustal movements and seismicity also correlate well with thermal spring occurrence. Twenty thermal springs have been identified in the Jura and west Alps areas with temperatures of discharging water ranging from 24–62 °C. Most rocks in the area were fractured, faulted, and sheared during the Alpine orogeny leading to extensive deep circulation. The direct and immediate effect of seismic phenomena or spring occurrence is illustrated by an event which occurred in January 1946. A major earthquake occurred with the epicenter located between Sion and Leuherbud. Immediately following the quake, discharge from a nearby spring increased to five times its normal rate and the temperature rose from 25–28.5 °C. Six weeks later the discharge was still twice as high as that measured before the quake.

The thermal spring at Lavey is probably the most famous in Switzerland. It is located on the Lower Rhone Valley in the most active seismic zone in the country. About 10–20 m of quaternary fill overlays fractured gneiss of the Aiquilles Rouges Massif. North of the spring, the gneiss is overlain by Mesozoic sedimentary rocks. The Mesozoic rocks are thought to be the recharged area at depth. Water then enters the fractured gneiss; and emerges at the spring mouth. Considerable work has been carried out at Lavey to raise the average temperature to 62 °C and the discharge from 1 l/s to 6.6 l/s. Obviously these increases are economically beneficial and suggest that it may be possible to harness the discharge for heating.

Fig. 3.21.
Hot Springs, Arkansas, USA. Hot water discharges along faults to surface, and more recently through drilled wells. A National Monument maintained by the US Park Service. **a** Sign at entrance; **b** a foot bath, one of many discharge points for Hot Springs in area

Thermal springs in eastern Macedonia, Greece occur mainly where two systems of faulting lines meet, one in a north-south direction, the other east-west. Spring location is therefore controlled by tectonics and the continuous crustal movements along the initial fault lines. As was noted in Switzerland, an earthquake in 1932 caused an increase in temperature and discharge of the spring waters, in addition to the appearance of new springs.

In Cyprus, most thermal springs occur in the Trodos ophiolite and for the most part are associated with faults. Some do occur in sedimentary rocks, mainly gypsum and pakhna deposits. The springs at Ayii Anargyri occur at the contact between gypsum and the underlying Pakhna marls. Thus, as in Greece and Switzerland, the thermal springs of Cyprus are associated with areas of past or present tectonism.

Fig. 3.22.
Roman Springs, Ville du Dax, France. Hot water discharges at the surface and to the river at Dax. Developed for Roman Baths, pipe conduits transport hot water to hotel spas around the city. **a** Old Roman structure housing baths during Roman times. **b** Thermal springs discharge points for public use (photographs by P. E. LaMoreaux November 1995)

In Iraq, most thermal springs discharge from the Miocene middle and lower Fars Formation. Groundwater occurrence is controlled by karstic features. The crests of most of the anticlines are highly fractured, jointed and faulted creating an easy pathway for water movement. The deepest known source for groundwater is 250–300 m. Discharge from the springs is directly related to rainfall. Temperature of the discharging water normally falls in the 26–50 °C range. Numerous thermal springs occur in the Suez Gulf in Egypt. The Gulf is a morphotectonic depression running NW-SE between two series of elevated blocks which were formed by complex faulting. The I. Hamman Faraun Spring system is located on a straight line along a fault plane at the foot of the Hamman Farain Mountain. The main group consists of 12 small springs. These are periodic in nature, having a maximum discharge in the morning and a minimum in the afternoon; when some springs may completely stop flowing. This strange behavior has been attributed to tidal influences. Temperature of the discharging water ranges from 60–70 °C and the water is mainly salty. It is thought that the water was originally meteoric and was contaminated by bromine rich brines. The water is too hot to be explained by normal geothermal gradients. It may have come in contact with rocks of extremely high temperature in magmatrically active horizons which are interrupted by basaltic intrusions.

In Romania, the Hercule thermo-mineral spring acts as the main outlet point for the Cerna Syncline which is

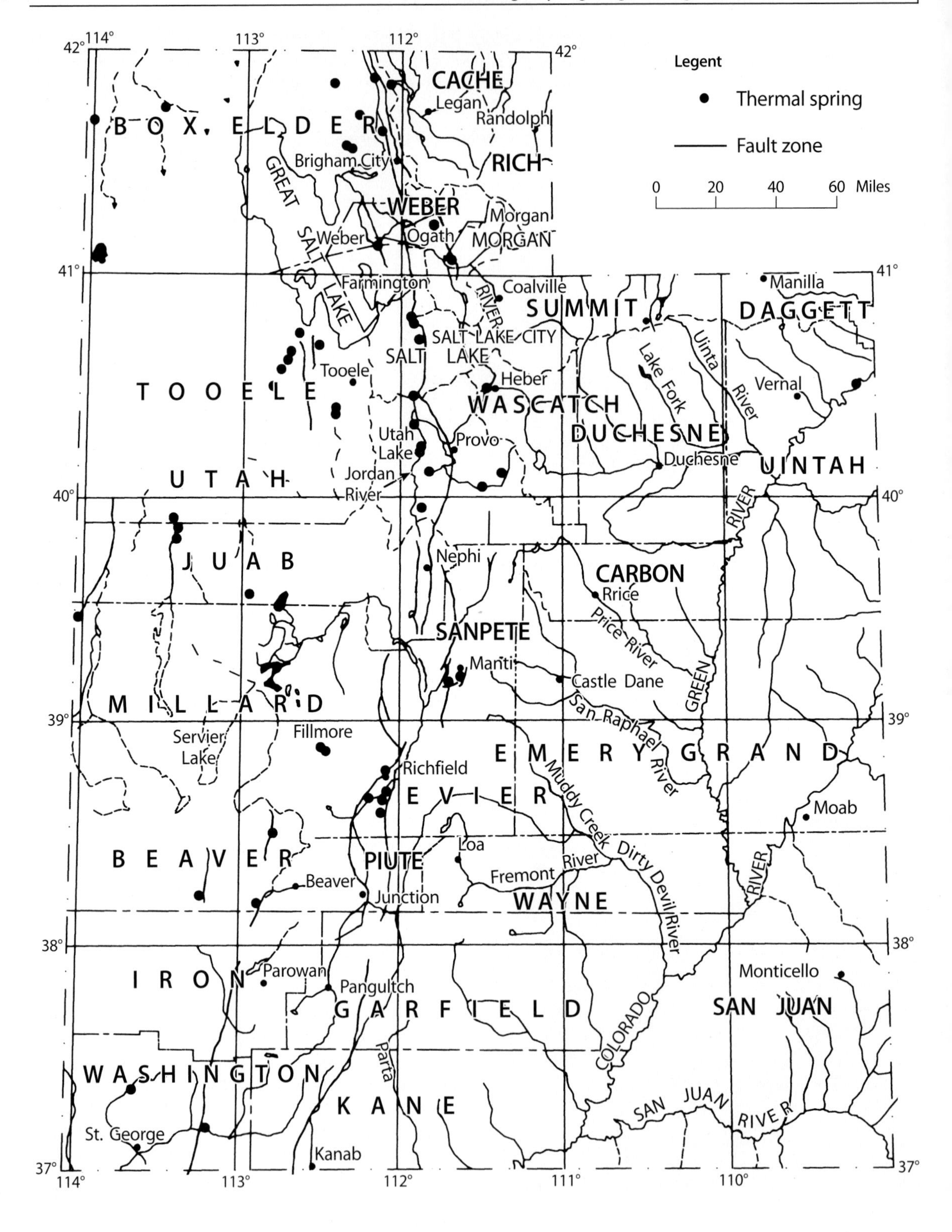
Legent
Thermal spring
Fault zone
0 20 40 60 Miles
42° 114° 113° 112° 42°
41° 41°
40° 40°
39° 39°
38° 38°
37° 37°
114° 113° 112° 111° 110°
CACHE
Legan
Randolph
BOX ELDER
Brigham City
RICH
GREAT SALT LAKE
WEBER
Weber
Ogath
Morgan
MORGAN
Farmington
Coalville
RIVER
SUMMIT
DAGGETT
Manilla
SALT LAKE CITY
SALT LAKE
Tooele
TOOELE
Heber
WASCATCH
Lake Fork
Uinta River
Vernal
DUCHESNE
Utah Lake
Provo
Duchesne
UINTAH
UTAH
Jordan River
RIVER
Nephi
JUAB
CARBON
Rrice
Price River
SANPETE
Manti
Castle Dane
GREEN
MILLARD
San Raphael River
Servier Lake
Fillmore
EMERY
GRAND
Richfield
Muddy Creek
EVIER
Moab
Loa
Dirty Devil River
BEAVER
PIUTE
Fremont River
Beaver
Junction
WAYNE
RIVER
Parowan
Monticello
IRON
Pangultch
GARFIELD
COLORADO
SAN JUAN
Parta
WASHINGTON
KANE
SAN JUAN RIVER
St. George
Kanab

composed of Jurassic and Cretaceous limestones. Within the limestone body mixing occurs between karstic waters of surface origin and thermomineral ascending waters resulting in considerable instability in yields and chemistry.

Most of the warm springs in the eastern United States are in the folded and faulted Valley and Ridge Province of the Appalachian Mountains. Most discharge from sandstone or limestone and are located in a valley on the crest or flank of an anticline. Topography, lithology and structure are the contributing factors in their occurrence. Topography controls the potentiometric head relations which allow the deep circulation patterns to develop. The fact that the water remains hot until its return to the surface indicates the presence of effective flow channels; which are dependent on lithology and structure. Structural features favoring vertical permeability include: openings along bedding planes, open tension fractures, and open faults. Satellite imagery has been used to identify linear features, some of which represent faults or fracture zones. More detailed geological interpretation by Kulander and Dean (1972) was based on the conjunctive use of seismic, well, gravity and magnetic data. Weinman (1976) working in Perry County, Pennsylvania, determined that warm springs frequently occurred at the intersection of lineaments identified from color-composite ERTS imagery. He also determined that the discharge of thermal water was readily identifiable on low altitude thermal imagery.

Many lineaments have been mapped in the warm springs area of Virginia. Denison and Johnson (1971) suggested that these thermal springs are a result of Eocene volcanism or alternatively may represent a younger source of heat (magmatic). However, regardless of whether intrusion has occurred, the combination of deep fracturing, local geology, high relief and structure favors deep groundwater circulation.

Lineaments mapped from a satellite image near Lebannon Springs, New York, agree with evidence that thermal springs occur in valleys where lineaments or faults cut steeply dipping sandstone or limestone beds. Although thermal springs commonly do occur on the crests or flanks of anticlines, the anticline itself is not the sole factor contributing to their occurrence. Tension fractures that run parallel to the anticlinal axis and the dipping rocks are conducive to deep circulation parallel to the structure. However, vertical faults or lineaments, rock type, dip and subsurface thrust faults may be far more important.

There are a number of probable scenarios regarding flow systems in thermal springs. Three situations and probable flow systems which would develop in each are: (1) in a valley near the crest of an anticline; (2) on the flank of a ridge formed by an anticline; and (3) on the flank of a ridge formed by a limb of an anticline.

The three hottest thermal springs of Utah occur in areas of late Tertiary Quaternary vulcanism and range in temperature from 185–189 °F. Nearly all occur on fault zones and although few actually issue from volcanic rocks, they are all close to faulting. Figure 3.23 shows the location of the major spring and fault zones. Becks Hot Springs occur on the line of contact between valley fill of Quaternary age and limestone of Paleozoic Age. Thus, the issuing point is due to lithology differences, but the conduit for the thermal water is the Warm Springs Fault. Volcanic rocks of tertiary age are exposed about 2.5 miles southeast of the springs, but are not believed to be the heat source. Circulation of meteoric water to depths of several thousand feet and contact with saline sediments probably result in the temperatures of 130 °F and the dissolved solids content of 13 000 ppm.

◀ Fig. 3.23. Map showing locations of major thermal springs and major fault zones in Utah (Mundorff 1970)

3.14 Structurally Controlled Springs

Tectonic events and the resultant structures can effect springs in a number of ways, mainly due to differences in permeability. In general, faults and fault zones are often zones of greater permeability than joints or bedding planes. In some cases this does not hold true and faults have been known to retard groundwater flow. An impermeable fault plane will divert flow to other horizons, direct flow along the fault plane, or force water to the surface as springs. Obviously the effects of faults varies with structure and texture of the fault surface. Dip-slip faulting may juxtapose soluble beds against insoluble beds thereby creating a boundary to groundwater flow. Springs, when they occur along the trend of a fault zone, commonly do so on the upthrown side where conduits are near to the surface. Balcones Fault zone in Texas has six springs located on the down-faulted blocks of the Cretaceous Edwards Formation. Fault controlled conduits may carry water directly to springs where fault planes intersect the surface.

Interpretation of Landsat imagery from Madison County, Alabama has shown a concentration of high yield springs along lineaments. Most of the area is underlain by a thick sequence of carbonate rocks of Mississippian Age with clastics capping the carbonate units in outlier hills.

A series of large springs occur in the area and discharge outlets are controlled by geologic structure. Huntsville Spring is an example and was originally developed as the city's water supply. There are many other large springs in Madison County, Alabama (LaMoreaux 1975) that are structurally controlled. A well-developed series of parallel lineaments constituting part of the Huntsville lineament complex in northeast Alabama traverse the area and intersect at a group of high yield springs. The springs fall in a 28×3 mile band passing south of Huntsville, the axis strikes N 43° W along one of the strongly developed lineament trends. Fourteen large springs in the 50–1 000 gal/min range have been identified as either springs issuing from bedrock or resurges from soil. The evidence indicates that the springs are joint and fracture controlled.

A series of springs occurring on both sides of the Colorado River in Arizona discharge from solution widened fractures in the Mississippian Redwall limestone. Vassey Springs, at River Mile 32, is an example of these springs. The water discharges from the Fence Fault or nearby fractures which trend across Marble Canyon. Temperature and water quality data reveal that the spring water is a mixture of distinct waters derived from opposite sides along the fault zone. The groundwater circulates along the fractures. The faults provide laterally and vertically continuous planes of greater permeability through an otherwise massive rock. A series of horizontal conduits and vertical hydraulically continuous channels develop through a series of Paleozoic confining beds. This allows recharge to the groundwater storage. Gradients within the fractures are governed by the fact that the Colorado River serves as a regional sink. The heads at the springs along the river are equal to or slightly greater than the elevation of the river. The Fence Fault is different from other faults traversing the Colorado River in that it has large thicknesses of carbonate rocks that occur below river level allowing the development of a sub-river circulation system.

In the Sierra Madre Oriental hydrogeological region of the United States, most of the large springs (≥1 000 l/s) occur along fractures or valley floors. On the Ozark Plateaus, an area characterized by limestones and dolomites drained by dissecting streams, numerous springs occur. Over 165 have been mapped with minimum discharges of 28 l/s. The springs typically emerge from large solutional openings that follow fractures down to the water table. These are a good example of springs formed due to karst processes and structural effects. Many of these springs have developed solutional openings below the water table and circulation paths up to 240 m deep have been found.

The Mesozoic Tripolitza carbonates of the Molai area of Greece have two distinct flow systems; the western one of which discharges to the shore and the sub-marine springs at Glyfada. The catchment area is connected to the springs by a narrow conduit related to a fracture zone in marbles that accompany the Molai Fault. The unusual nature of karst in this area has been attributed to continuous tectonic activity which has prevented the completion of various stages of karst development, thus precluding the development of the usual type of karst springs.

Some thermal springs are fault controlled, e.g. Ayun Mousa Springs of the Suez Gulf and just east of the City of Suez on the south end of the Suez Canal in Sinai. Twelve springs occur along two parallel lines relating to two distinct faults. The springs are fed from the underlying Miocene confined sandy aquifer at a depth of 300 m. In recent years, exploration for coal, during which test holes were left open, has allowed the discharge of brine to the land surface and large salt water marshes have resulted.

The occurrence in parts of the Niger Delta of seepages of potable water to the sea has been studied in relation to the economic viability of tapping this source as a public supply. It is thought that due to changes in lithology, groundwater has been forced to the surface at the coast. The surface hydrologic system of the area is characterized by streams flowing down the structural basin from the deeply fractured basement complex to younger sedimentary formations at the coast. It is thought that widespread conduits due to fracturing are present in the igneous and metamorphic rock. These conduits intercept some of the surface water and convey it along diffuse channels to emerge at the coast. These seeps are due differential permeability, as a result of fracturing of the basement complex.

In Texas, the Marble Falls and the Ellerburger-San Saba limestones are hydrologically interconnected in

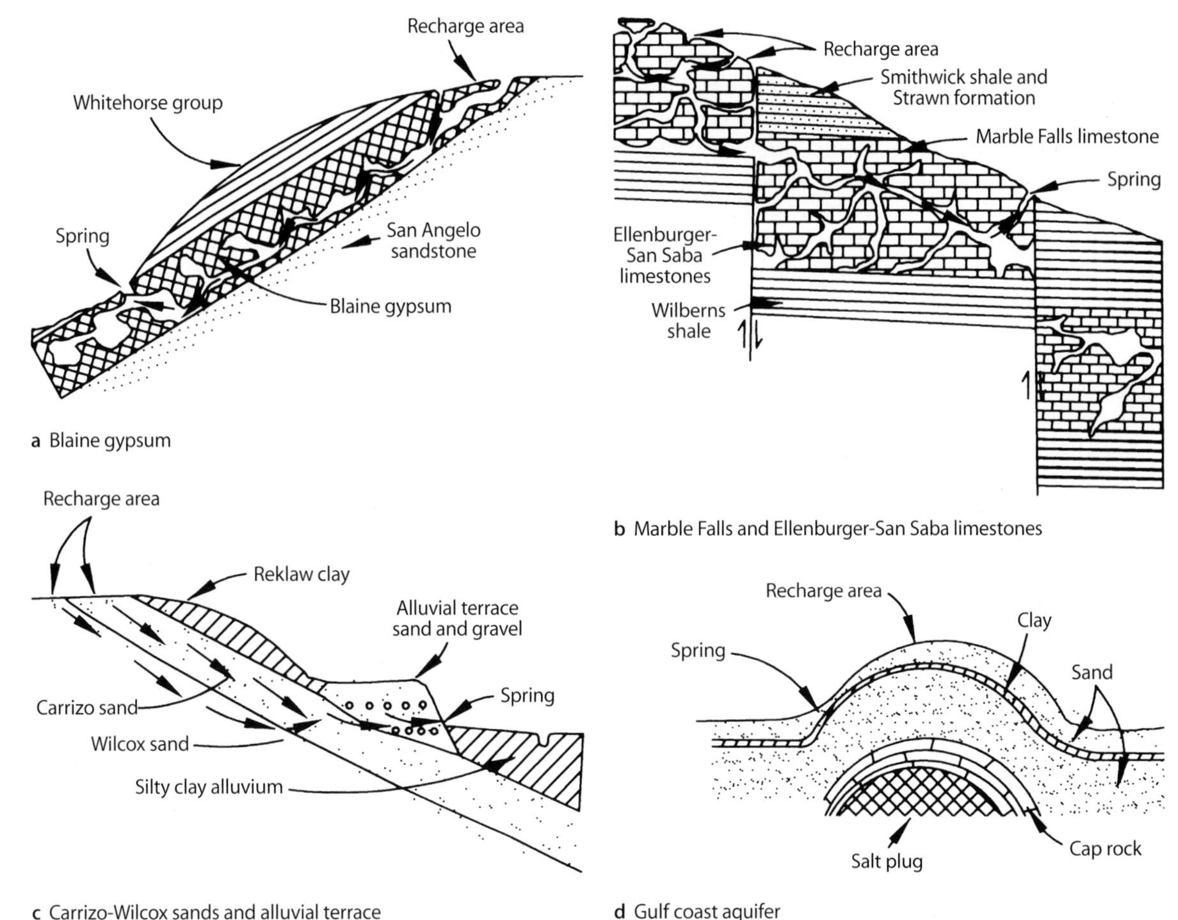

Fig. 3.24. Geologic settings of Texas Springs (Brune 1975)

many areas and function as a common aquifer due mainly to faulting as shown in Fig. 3.24. Also through faulting, the Swithwich Shale and Shawn Formation are so situated as to form a dam against the limestone reservoir. The groundwater escapes as artesian springs through faults. Barnett Springs in San Saba County are a typical example of springs formed in this way.

Another interesting structural relationship which exists along the gulf coast of Texas is that of a salt intrusion raising the overlying gulf coast aquifer into a dome. A layer of confining clay occurs near the surface and prevents infiltrating water from moving to the deeper sands. The water therefore takes the easiest exit route and emerges as a spring at the edge of the dome.

3.15 Miscellaneous

The concentration of 27 first-order springs (discharge exceeds 2.8 m^3/s) within the northern half of Florida is unique. Nearly all occur in the unconfined or thinly confined parts of the Floridan aquifer and are primarily artesian. Vertical zonation of permeability and the southwest dip of the geologic formations tend to control the location of these springs and indeed also their distribution. South of Port Richey, most of the springs issue from the Tampa Formation. North of the Port Richey to Weeki Wachee Springs, most issue from a permeable zone at the base of the Suwanne Limestone. Springs north of Weeki Wachee flow from either the Suwanne or older formations such as the Avon Park Limestone. Lake Tarpon, southeast of Tarpon Springs presents a interesting but complex hydrologic situation. The lake drains intermittently through a sink in the lake bottom that connects with Tarpon Springs in Spring Bayou about 2 miles to the northwest. The conditions leading to the beginning and end of drainage are variable and are a combination of water levels in the lake, in Spring Bayou and in the aquifer in addition to the relative densities of water in all the various parts of a tidal system.

The effect of tidal fluctuations on spring water levels is illustrated in Fig. 3.25. Obviously, with increasing distance from the coast, the effect diminishes and the time lag increases.

The discharge of springs and seeps in permafrost areas provide very useful information regarding the availability of groundwater in a generally deficient groundwater zone. Temperature and composition of the water helps identify the zone as either supra-, intra-, or subpermafrost. Perennial springs commonly discharge water from the subpermafrost zones and derive their water from subpermafrost aquifers via open faults, as shown in Fig. 3.26. Seasonal discharge does occur from supra-permafrost zones.

Springs in areas of glacial deposits in southern Sweden exhibit a marked contrast in terms of discharge depending on whether they issue from "hummocky moraine" which is supraglacial or drumlins. In general, springs issuing from drumlins have yields which are 1–2 orders of magnitude greater than the others. This has been attributed to contrasts between stratigraphic continuity and hydraulic continuity.

In Ohio near Ashland (Norris 1961), a spring on the side of a small tributary valley issues from a glacial outwash deposit of silt and sand with some coarse gravel interbedded with thick till. The spring has a median discharge of about 165 gal/min. This spring illustrates the effects on the regional hydrology of a permeable deposit of bedded-in till. This deposit acts primarily as a conduit through which water from the more extensive bedrock part of the aquifer is discharged to the surface. The relatively large discharge indicates the potential importance of such interbedded deposits as sources of groundwater in a glacial terrain. Deposits of sand, silt and gravel commonly occur in till, many at disconformities between different age till sheets. Buried sand and gravel that originated as outwash plains may be several tens of feet thick.

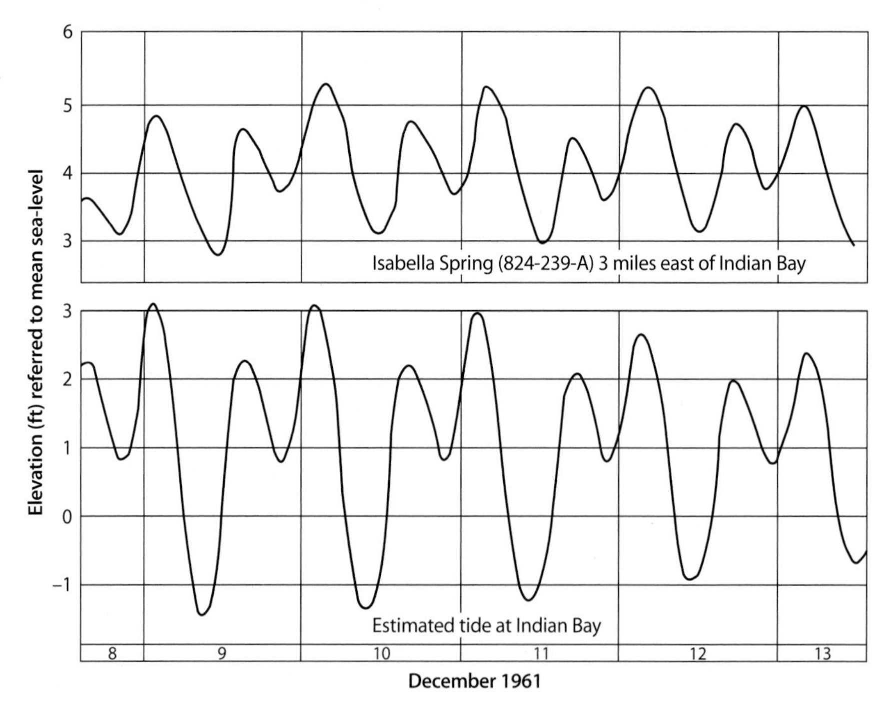

Fig. 3.25. Water level in Isabella Spring (824-239-A) and estimated tide at Indian Bay (Wetterhall 1965)

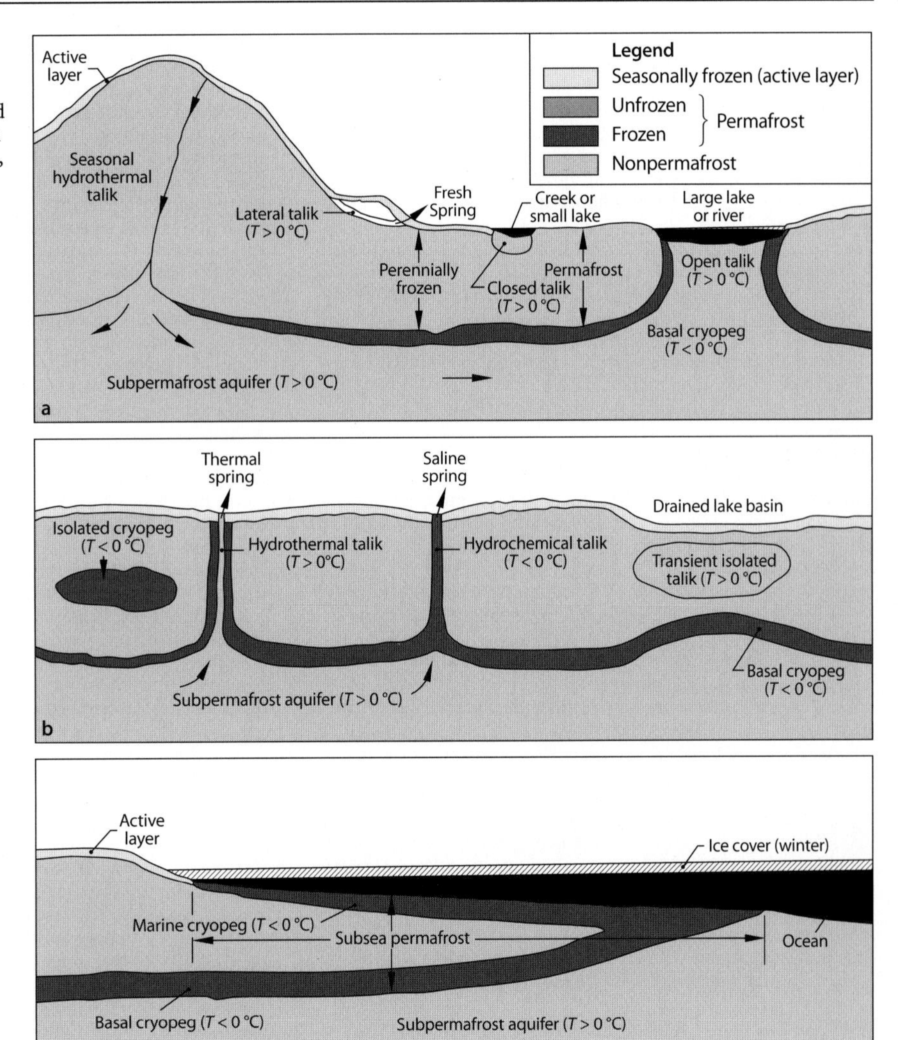

Fig. 3.26. Aquifers and water conditions in permafrost areas. **a** Suprapermafrost: active layer; closed taliks. **b** Intrapermafrost: open taliks (lake, river, hydrothermal, hydrochemical); lateral taliks; transient isolated taliks; isolated, marine and basal cryopegs. **c** Subpermafrost (Sloan and Everdingen 1988).

After reviewing the historical classification of springs and some of the better known current classification systems, the hydrogeologic processes which determine the origin of a spring have delayed the comprehensive and all-inclusive classification for springs simply because of our evolution of knowledge about springs. The sheer diversity of form and the different hydrogeologic settings worldwide have created new classification systems or refined old classification systems broadening our scientific understanding of springs. These hydrogeologic processes create a diversity of springs, a wealth of beauty, economic importance, and scholarly work worldwide.

References

Alfirevic S (1966) Les sources sous-marine de la baie de Kastel. Acta Adriat 6(12):1–38

Amadi P. Abi Bezam (1989) The occurrence of perennial seaward seepages of potable water in parts of Niger Delta, Nigeria. Groundwater 27(4)

Back W, Rosenshein JS, Seaber PR (eds; 1988) Hydrogeology. The geology of North America Volume O-2. The Geological Society of America

Bliss JD (1983) Idaho. Basic data for thermal springs and wells as recorded in geothermal-part A, Idaho, U.S.G.S. Open File Report 83-0431-A, p. 64

Bögli A (1980) Karst hydrology and physical speleology. Springer-Verlag

Bonacci O (1987) Karst hydrology with special reference to the Dinaric Karst. Springer Series in Physical Environment. Springer-Verlag

Breznik M (1973) Nastanke zaslanjenih kraskih izvirov in njihova sanacija (The origin of brackish karstic springs and their development). Geol-Razprave in Porocila 16:83–186

Breznik M (1978) Mechanism and development of the brackish karstic spring Almyros Iraklion. Ann Geol Pays Helleniques, pp. 29–46

Brune G (1975) Major and historical springs of Texas. Texas Water Development Board Report 189, 95 p

Bryan K (1919) Classification of springs. Journal of Geology 27: 552–561

Burdon DJ, Safadi C (1962) Ras-El-Ain: The great karst spring of Mesopotamia, an hydrogeological study. Journal of Hydrology 1(1) Amsterdam, The Netherlands

Clark FW (1924) The data of geochemistry, Chapter VI: Mineral wells and springs. p. 181–217

Davis SN, DeWiest RJM (1966) Hydrogeology. John Wiley, New York, N.Y., 463 p

Dilamarter RR, Csallany SC (eds; 1977) Hydrologic problems in karst regions. Bowling Green, Western Kentucky University, Kentucky

Dubic D (1984) Hidrologija kopna (Land hydrology). Naucna Knjiga, Beograd

Ferris JG, Knowles DB, Brown RH, Stallman RW (1962) Theory of aquifer tests. US Geological Survey Water-Supply Paper 1536-E, 174 p

Fetter CW Jr (1980) Applied hydrology. Charles L. Merrill Publishing Company, 488 p

Fitch WE (1927) Mineral waters of the United States and American spas. Lea & Febiger, New York

Freeze RA, Cherry JA (1979) Groundwater. Prentice-Hall Inc., 604 p

Fuller ML (1904) Underground waters of eastern United States. Government Printing Office, Washington, DC. US Geology Survey Water-Supply Paper 114, 285 pp

Futak H, Langguth HR (1986) Karst hydrogeology of the central and eastern Peloponnesus (Greece). In: Morfis A, Zojer H (eds) Proceedings of the 5th International Symposium on Underground Water Tracing, Athens, Greece. Vereinigung für Hydrogeologie. Forschungen, Graz, Austria

Ghikas IC, Kruseman GP, Leontiadis IL, Wozab DH (1983) Flow pattern of a karst aquifer in the Molai area, Greece. Groundwater 21(4)

Habic P (1982) Kraski izvir Mrzlek, njegovo zaledje in varovalno obmocje (Mrzlek karst spring, its catchment and protection area). Acta Carsol 10:45–73

Hobba WA Jr, Fisher DW, Pearson FJ Jr, Chemerys JC (1979) Hydrology and geochemistry of thermal springs of the Appalachians. USGS Professional Paper 1044-E, 36 p

Huntoon PW (1981) Fault controlled groundwater circulation under the Colorado River, Marble Canyon, Arizona. Groundwater 19(1)

Huntoon PW (1985) Fault severed aquifers along the perimeters of Wyoming artesian basins. V Groundwater 23(2)

Huntoon PW, Coogan JC (1987) The strange hydrodynamics of periodic spring, Salt River Range, Wyoming. Thirty-eighth field conference 1987, Wyoming Geological Association Guidebook

Ingersoll LR, Zobel OJ, Ingersoll AC (1954) Heat conduction with engineering, geological and other applications. Madison, Wisconsin, University of Wisconsin Press, 325 p

Institutul De Speologie (1984) Theoretical and applied karstology. Vol. 1: Proceedings of the First Symposium on Theoretical and Applied Karstology, 22–24 April 1983, Bucharest, Romania

International Congress on Thermal Waters, (1976) Proceedings – Geothermal energy. Vol. 1: Geothermal energy and vulcanism of the Mediterranean area, October 1976, Athen

International Congress on Thermal Waters (1976) Proceedings – Thermal waters. Vol. 2: Geothermal energy and vulcanism of the mediterranean area, October 1976, Athen

Irish National Committee (1979) Hydrogeology in Ireland. Papers and proceedings of a hydrogeological meeting and associated field trips held in the Republic of Ireland 22–27 May 1979. Published by the Irish National Committee of the International Hydrological Programme, Dublin, Ireland

Karanjac J, Günay G (1980) Dumanli Spring, Turkey – the largest karstic spring in the world? J Hydrology 45: 219–231, Elsevier Scientific Publishing Company, Amsterdam

Keilhack K (1912) Lehrbuch der Grundwasser and Quellenkunde, 3rd edn. Geb. Borntraeger, Berlin

Knighton D (1984) Fluvial forms and processes, 218 p. Edward Arnold (Publishers), Ltc., 41 Bedford Square, London SCIB 3DQ

Kohout FA (1966) Submarine springs. A neglected phenomenon of coastal hydrology. Proc Symp on hydrol and water resources, Ankara

Kriz H (1973) Processing of results of observations of spring discharge. Groundwater 11(5):3–14

Kulander BR, Dean S (1972) Gravity and structures across Browns Mountain, Wills Mountain, and Warm Springs anticlines – gravity study of the folded plateau, West Virginia, Virginia, and Maryland. In: Lessing P (ed) Appalachian structures: origin, evolution, and possible potential for new exploration frontiers – a seminar. West Virginia University and West Virginia Geol and Econ Survey, p. 141–180

LaMoreaux PE (1975) Environmental geology and hydrology, Huntsville and Madison County, Alabama. Geological Survey of Alabama, Atlas Series 8. Geological Survey of Alabama, University, AL 35486, 118 p

Lyell Sir C (1849) A second visit to the United States. Harper & Brothers, Publishers. John Murray, London, 287 p

McColloch JS (1986) Springs of West Virginia. West Virginia geological and economic survey, Vol. V-6A, 493 p

Meinzer OE (1923a) Outline of Groundwater hydrology, with definitions. US Geological Survey Water-Supply Paper 494, 69 p

Meinzer OE (1923b) The occurrence of groundwater in the United States. US Geological Survey Water-Supply Paper 489

Meinzer OE (1927) Large springs in the United States: US Geological Survey Water-Supply Paper 557, 94 p

Mundorff JC (1970) Major thermal springs of Utah, Water-Resources Bulletin 13, September 1970: Utah Geological and Mineralogical Survey affiliated with The College of Mines and Mineral Industries, University of Utah, Salt Lake City, Utah

Netopil R (1971) Ke Klasifikaci pramenu podle variability vydatnasti (The classification of water springs on the basis of the variability of yields). Sbornik-Hydrological Conference, Papers. Stud Geogr 22:145–150

Norris SE (1961) Hydrogeology of a spring in a glacial terrane near ashland, Ohio. Geological Water-Supply Paper 1619-A, prepared in cooperation with the Ohio Department of Natural Resources, Division of Water

Palmer AN (1977) Effect of continental glaciation on karst hydrology, northeastern, USA. In: International Association of Hydrogeologists. Memoirs Vol. XII: Proceedings of the Twelfth International Congress, Karst Hydrogeology, Huntsville, Alabama, USA. The University of Alabama in Huntsville Press, Huntsville, AL, USA, p. 109

Paramelle Abbé (1856) L'art de decovrir les sources. 4th edn. Paris 1896

Peale AC (1886) List and analyses of the mineral springs of the US, US Geological Survey Bulletin 32, 235 p

Perrault P (1674) On the origin of springs. Hafner Publishing Co., New York, 209 p

Petrović J (1983) Kraške vode Crne Gore (Montenegro karst water). Posebna izdanja Universiteta u Novom Sadu, p. 111
Potie L, Tardieu B (Date not Available) Development of submarine springs in limestone formations
Rosenau JC et al. (1977) Springs of Florida. Florida Geological Survey Bulletin 31
Sauro U, Bondesan A, Meneghel M (eds; 1991) Proceedings of the international conference on environmental changes in karst areas, I.C.E.C.K.A., Italy, 15–27 September 1991. International Geographical Union
Scanlon BR, Thrailkill J (1987) Chemical similarities among physically distinct spring types in a karst terrain. J Hydrology 89:259–279
Shuster ET, White WB (1971) Seasonal fluctuations in the chemistry of limestone springs: a possible means for characterizing carbonate aquifers. J Hydrology 14(2):93–128
Smith, EJ (1979) Spring discharge in relation to rapid fissure flow. Groundwater 17(4)
Stearns ND, et al. (1937) Thermal springs in the United States. US Geological Survey Water-Supply Bulletin 679-B, 206 p
Stiny J (1933) Springs. The geological foundations of springs for engineers of all disciplines as well as students of natural science, with 154 diagrams in text. J. Springer Publisher, Vienna
Thornwaite CW, Mather JR (1955) The water budget and its use in irrigation. US Department of Agriculture, Yearbook of Agriculture 1955:346–358
Todd DK (1980) Groundwater hydrology. 2nd edn. John Wiley & Sons
Tolson JS, Doyle FL (eds; 1977) Karst hydrogeology
US Geological Survey and Utah Geological and Mineralogical Survey (1971) Nonthermal springs of Utah. Water Resources Bulletin 16
Waring GA (1965) Thermal springs of the united states and other countries of the world, 383 p
Weinman B (1976) Geophysical, geochemical, and remote sensing studies of Pennsylvania's thermal springs. Pennsylvania State University, unpublished MS Thesis.
Wetterhall WS (1965) Reconnaissance of springs and sinks in west-central Florida. Report of Investigations 39, prepared by the US Geological Survey in cooperation with the Florida Geological Survey and the Southwest Florida Water Management District, Tallahassee, Florida
Yevjevich V (ed; 1976) Karst hydrology and water resources. Proceedings of the US-Yugoslavian Symposium, Dubrovnik, 2–7 June 1975. Vol. 1: Karst Hydrology. Water Resources Publications, Fort Collins, Colorado 80522, USA, 1976

Chapter 4

Extracting Methods for Spring Waters

CHAPTER 4
Extracting Methods for Spring Waters

Zhou Wanfang

4.1 Introduction

Public awareness regarding the importance of the quality and reliability of drinking water has increased the popularity of spring waters. The increasing consumption of bottled waters is just one trend where the public is buying "pure" water coming from natural springs as an alternative to traditional municipal water supply systems. Many bottled water products are labeled with "pure spring water", "mountain spring water", or "natural spring water", suggesting that the term "spring water" carries a special connotation of consistently high water quality and good taste.

As defined by Meinzer (1923a,b) and used by the United States Geological Survey and the American Geological Institute (1960 and 1976), a spring is a place where, without the agency of man, water flows from a rock or soil upon the land or into a body of surface water. Thus, water appropriately termed "spring water" must have a clear hydrological association with a spring. In its *model bottled water regulation*, the International Bottled Water Association (IBWA) defined spring water as "... water derived from an underground formation from which water flows naturally to the surface of the earth ... and is not derived from a municipal system or public water supply" (IBWA 1991). According to these definitions, water appropriately termed "spring water" must meet the basic requirement that springs be a natural discharge.

Many of the largest selling and most recognizable brand-name bottled spring waters are collected using boreholes that tap the source of spring waters. Use of the borehole method to extract spring water is acceptable and widely practiced. If boreholes or other engineered interception systems are used, controversy may arise concerning definition and proper labeling of the bottled products. However, to utilize a source of spring water while preserving its integrity and protecting the purity of the water from contamination, protective extractive technologies have to be employed.

Because of the prominence that bottled groundwater has acquired, many countries and states have promulgated regulations or standards which control how spring water may be collected, bottled, labeled, and distributed. Some organizations and agencies on national and international scales have issued guidance and regulations pertaining to commercially bottled spring water. The Food and Drug Administration (FDA) has proposed to amend its regulations to, among other things, define spring water. The FDA definition for spring water provides for the tapping of the source in a way that will ensure sanitation, and therefore, FDA proposed collection of spring water "... at the spring or through a borehole adjacent to the point of emergence..." Also, the "... water shall be from the same underground stratum as the spring and shall retain all the physical properties and be of the same composition and quality as the water that flows naturally or would flow naturally to the surface of the earth if not for its collection below the earth's surface." A more complete description of the legal aspects of spring waters is given in Chap. 7.

4.2 Collection Systems for Spring Waters

Most groundwater that serves as a source of a spring or of a mineral water is meteoric water, i.e. groundwater derived from rainfall and infiltration within the normal hydrological cycle. Although rain is nature's form of distilled water, it typically contains between 10 and 20 mg/l of dissolved materials (Price 1985). Near coastlines, the concentration of sodium chloride may increase, and downwind of industrial areas, sulphur and nitrogen compounds can be more in evidence. The chemistry of meteoric groundwater changes during its passage through rocks, the changes depending on such factors as the minerals with which it comes into contact, the temperature and pressure conditions, and the time available for water and minerals to react. The pathway can range in length from a few tens of feet to hundreds of miles and in depth from a few feet to thousands of feet. The modification of meteoric groundwater in its passage into and through the ground is one of the evolutionary sequence of groundwater chemistry, and it occurs in nearly every aquifer. The larger the period of contact between water and rock, the greater the opportunity that the water exiting the spring will have of acquiring unique physical and chemical characteristics. Therefore, the water exiting the spring is of a different character than the water that originally percolated into the ground at the upgradient area. For this reason alone, it would be inappropriate to define as spring water, groundwater that is near the upgradient end of the migration path way, even though its ultimate destination is a spring system. Similarly, it would be inappropriate to classify as spring water any water that is not clearly hydraulically connected to a spring or that is of a different chemical characteristic than the spring water to which it is identifiable as being derived from the aquifer that has historically fed the natural spring.

It is possible, under some circumstances, for human beings to induce groundwater discharge like a spring to the land surface or into a source body of water where no springs occur. Many bottled waters have as a source "artesian well water". These wells may not be near any given "source spring". They may, however, come from the same aquifer, and even have similar chemical characteristics. However, the producer may wish to extract artesian water from a well. A good example of this type of development are Artesia from Texas and Artesia sold by grocery store chains in Alabama. Such induced discharges, not associated with natural spring discharge, should not qualify as springs. Examples of such non-spring discharge include:

1. Flowing artesian wells in an area not associated with any springs (Fig. 4.1) – although technically similarly to springs, artesian wells fail to meet the definition criteria requirement that springs discharge from a natural opening.
2. Discharges from excavated tunnels, shafts, channels or cuts that tap an aquifer in a location remote from or unassociated with any springs (Fig. 4.2).

4.3 Locating Springs for Spring Water Development

Spring water can be used for many purposes. If the spring water is to be developed for a commercial bottled water, a particular concern must be contamination by the presence of microbes or other organic/inorganic contaminants. When rain water enters soil, bacteria and other microbes attached to very small soil particles can become suspended in the percolating water. As the water-soil-microbes suspension moves through the unsaturated zone, the soil (especially clay) acts as a filtering agent. Wet clay particles have electrically charged surfaces that attract bacteria. Porous and clay-like materials have been used to purify water by filtration. Natural soil bacteria have a competitive advantage over pathogenic bacteria in obtaining nutrients, because the soil bacteria have evolved to be most efficient under the soil conditions and pathogenic bacteria have evolved to be most common in a human host. Virtually all bacteria tend to become attached to the surface of the soil-clay particles that halt and retard their movement. Therefore, the thicker the unsaturated zone or overlying confining layer above the aquifer, the more bacteria will be filtered out of the water. Even when the recharged water contains some other pol-

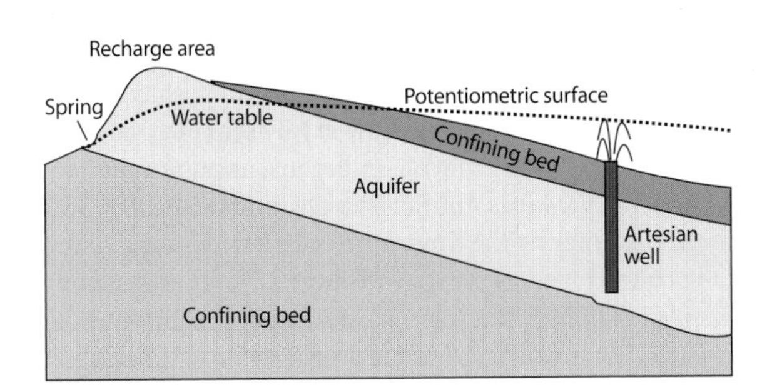

Fig. 4.1. Schematic illustration of an artesian well

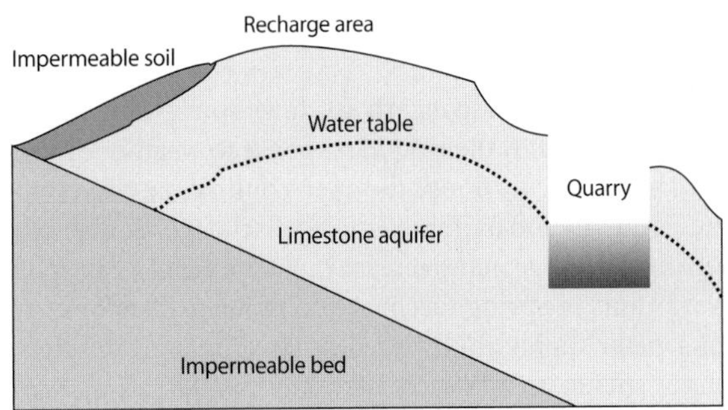

Fig. 4.2. Groundwater interception by a quarry

lutants like heavy metals and/or organic hydrocarbons, the soil and clay in the unsaturated zone will aid in their removal (Fetter 1993). When pollutants move through the saturated zone of the aquifer, many processes such as dilution, dispersion, adsorption, degradation, and diffusion will attenuate pollutant concentrations. The longer the residence time of groundwater in the aquifer systems, the less will be the pollutant concentrations, the more stable will be the chemical composition of the groundwater, and the more suitable will be the groundwater for use as commercially bottled spring water. The aquifer's self-filtering capacity is a very important factor in the use of a spring for spring water development.

It should be noted that some springs, particularly in karst carbonate systems, can have a wide variance in chemical composition because of the fast groundwater flow and the absence of filtering processes in the flow path (refer to Chap. 3 for a detailed description of karst springs). Sinkholes and sinking streams in karst areas allow surface water to recharge to the aquifer rapidly and with very little filtration, degradation or attenuation of particles (Fig. 4.3). The lack of physical filtering of particles is further compounded by the rapid transit of bacteria and viruses before their death or deactivation. Contamination of karst springs has been well documented. Contaminants in spring water can limit its use as a supply for public consumption. Generally, spring water contamination may occur when:

1. The spring water issues from well-developed karst systems or glacial drifts.
2. The recharge area for a spring intercepts contaminated runoff sources.
3. The spring source is close to other near surface sources of contamination such as sewers, livestock, septic tanks, cesspools, and agricultural fields. A karst spring is especially susceptible to those sources of contamination that are located on higher adjacent land.

Another major concern in developing water from springs in karst is their rapid response to weather events, as rainfall quickly floods the open conduits or large fracture systems and increases spring discharge. Conversely, spring discharge may decrease rapidly during a dry spell. Variations in spring discharge complicate development of a water supply, therefore, care must be used in selecting a spring as a source for bottled water, along with monitoring its subsequent use. Proper location, development, and management of a spring water source of

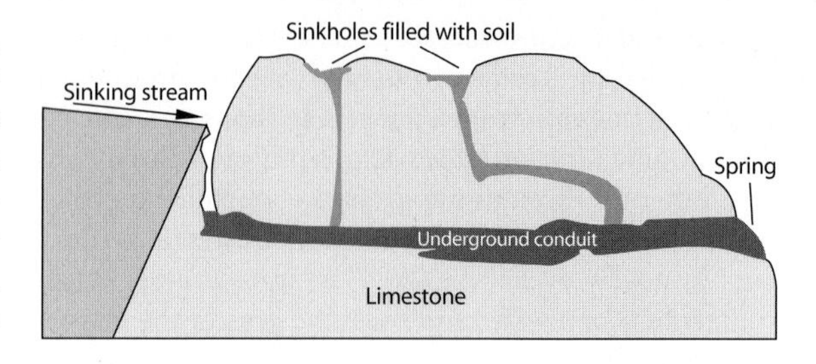

Fig. 4.3. Vulnerability of karst springs to contamination

supply requires a thorough evaluation of the geology, hydrogeology of the area, plus a careful study of activities in the recharge, storage and discharge areas of the spring. Commonly used techniques include:

1. Geologic mapping
2. Geophysical prospecting and remote sensing
3. Well and spring inventory
4. Test drilling
5. Pumping test
6. Groundwater tracing test
7. Water quality analysis
8. Water budget analysis
9. Groundwater modeling
10. Groundwater, spring water and surface water monitoring

Two of the most famous karst aquifer systems in the United States are the Floridan and Edwards aquifers. Probably no two aquifers have been studied in as much detail. The Edwards Aquifer extends 176 miles from Brackettville in Kinney County to Kyle in Hays County (Sharp and Clement 1988). It is one of the nation's most productive sources of groundwater, and in many aspects, the Edwards Aquifer has controlled the development of this area of south central Texas. The largest flowing well in the world was recently drilled into the Edwards limestone near San Antonio, Texas. Referred to as the "Fish Pond well", it had an initial flow in July 1991 of 25 000 gallons per minute (gal/min) (LaMoreaux, unpubl. material). In 1977, the Edwards Aquifer was designated the first Sole Source Aquifer by the Environmental Protection Agency. One of the largest springs in Edwards Aquifer is Comal Springs. Though the springs in Edwards Aquifer are at present not a source of bottled water, one of the largest bottling water companies in the USA is Artesia, which obtains groundwater from artesian wells in the Edwards.

Nearly the entire State of Florida, southern Georgia, and southeastern Alabama, are underlain by the Floridan Aquifer. The State of Florida passed legislation in the early 1970s and developed regulations applied to large withdrawals of groundwater. These regulations require strict adherence to defining the impact on surface water, shallow surficial aquifers, and the deeper aquifers within the Floridan Aquifer System. The regulations require the development of a Regional Impact statement and a Consumptive Use Permit. To meet the requirements of these permits, it is necessary to perform surface- and groundwater studies, extensive geologic and hydrologic studies, pumping tests and to collect detailed water level and stream flow monitoring and water quality data. These permits are under the jurisdiction of the Florida Department of Environmental Regulations and are administered under five Regional Water management districts, such as the Southwest Florida management district. These regional district offices have a regulatory hearing board, hold public hearings that are properly advertised, and have a support staff of geologists, engineers, chemists, and biologists (LaMoreaux 1989).

Discharge from the Floridan Aquifer occurs through springs and by seepage to streams. Florida has 27 springs of the first magnitude at which the average discharge exceeds 2.83 m^3/s (100 ft^3/s). The largest, Silver Springs, has an average discharge of 23.2 m^3/s (530 million gal/d) and reached a maximum discharge of 36.5 m^3/s on 28 Sep. 1960. Water supplies are obtained from the Floridan Aquifer by installing casing through the overlying formations and drilling an open hole in the limestone and dolomite comprising the aquifer. Large withdrawals of groundwater have significantly lowered groundwater levels in some areas, and have affected springs. Thus, in locating a spring for commercial or industrial use, consideration of existing groundwater supply development must be included in the evaluation. There are many examples of adverse affect on either springs or wells near large groundwater development facilities.

A comprehensive study of groundwater and karst springs in the United States at Huntsville, Alabama, was carried out by the United States Geological Survey in cooperation with the Alabama Geological Survey and published as a series of environmental atlases (LaMoreaux 1975). The study was prompted by the contamination of the Huntsville Springs that was the source of water for municipal use in the area. A spring inventory, test drilling, pumping test, groundwater trace test, water quality analysis, and land use practices were investigated in great detail along with the complex recharge, storage, and discharge system in the area. The results of these studies have been used as a guide to the development of water for municipal and industrial use and are the basis for municipal regulations applied to construction and development and for coordination of withdrawals of water from springs, wells, and surface water.

Understanding the aquifer system from which water flows, flow direction, and the rate of water withdrawn allows determination of the area around the associated spring in which environmental restrains should be imposed on development. Some of the spring sources are covered by legislation and regulations. For example, the development of the Figeh spring area near Damascus, Syria is controlled by environmental guidelines (LaMoreaux et al. 1989). Nearly every state in the United States has regulations under the US Drinking Water Act, or the Well/Spring Head Protection Act, or companion state regulations (LaMoreaux et al. 1996). One of the most comprehensive documents controlling industrial development in a karst area is the *Final environmental impact statement for regulatory action associated with the Olin Corporation, remedial Action Plan to isolate DDT from the people and the environment in the Huntsville Spring Branch-Indian Creek System, Wheeler Reservoir, Alabama* (US Army Corps of Engineers 1986).

In southern France, one of the most detailed series of studies of spring flow, water management for municipal use has been carried out by Professor Jacques V. Avias and his students in their studies of the karst spring Source Du Lez. This spring supplies the city of Montpelier, France (Avias 1972, 1977).

4.4 Advantages of Proper Extraction of Spring Water

The following objectives help achieve quality, sustainable use of spring water:

1. Protect and ensure the quality and integrity of the spring water and the spring water source by reducing their vulnerability to alteration or contamination as the result of accidental or intentional introduction of foreign substances such as chemicals or microorganisms.
2. Facilitate the control and conveyance of the water from springs; and
3. Minimize waste of the water resource during capture and management of the water.

Measures to achieve these objectives:

1. Minimizing opportunities for the source water to become contaminated or otherwise altered physically, chemically, or biologically from exposure of the source water to human, other biota, silt, chemical, and other foreign substances from the atmosphere, adjacent land surface, and adjacent surface water body.
2. Delineating protection zones in which environmental restraints can be imposed and eliminate development that may cause physical or chemical change, or give pollution access near surface-solution cavities and rock fractures that are connected to the water in the spring.
3. Develop engineered collection and conveyance systems which facilitate the collection, control, and conservation of the water and provide protective isolation.

Because of the need to protect and ensure the quality and consistency of the intercepted spring water, exposed, unprotected surface withdrawal/diversion storage systems are inappropriate for collecting spring water intended for public consumption. Some examples of unsuitable collection methods include:

1. Open collection boxes and natural, open, unprotected springs accessible to human beings and other animals.
2. Collection systems constructed from unsanitary, unstable, or potentially toxic materials.
3. Collection wells or shafts constructed without proper casing seals and other protective measures.
4. Collection wells or shafts in contact with potentially toxic or unsanitary materials such as adjacent or overlying contaminated areas (waste disposal sites, chemical spills and so on).

4.5 Extracting Methods for Spring Waters

The type of interception method appropriate for a particular spring depends on several site-specific factors, including the quantity of flow to be intercepted, location, recognition and illumination of potential sources of adverse impacts to the spring, type of use intended for the water, and location of processing facilities such as a bottling plant or a treatment facility.

The principle types of interception methods to develop spring water are summarized below. Interception methods can be part of an integrated sanitary system aimed at collecting spring water for human consumption or for a rural domestic water supply. Also, one method can be combined with another to effectively intercept spring waters.

Methods of interception can be classified as direct or indirect. Direct methods are those which collect the water after it has issued from the natural surface discharge point. Indirect methods are those which intercept the water before it emerges from its discharge points. Each of these two types of interception methods can be suitable for a broad range of spring conditions, and each has particular advantages and disadvantages.

4.6 Collecting Spring Water at Original Point of Spring Discharge

This involves constructing a surface structure using appropriate materials to provide a conveyance system to obtain and divert water for delivery to the point of use. The structure could be masonry or concrete with connecting pipes. If water issues from rock fractures, the individual openings should be cleaned and enlarged, as needed, to provide an increase in flow. A fine example of this type of spring development is at Huntsville, Alabama. Big Spring has a maximum discharge of over 30 million gal/d. A sump and pumping system at the spring, coordinated with pumping wells strategically located in a Mississippian rock karst aquifer system provides the city with water. Studies leading up to the karst spring development are described in *Atlas 8, Environmental Geology and Hydrology of Madison County, Alabama* (LaMoreaux 1975). The water from these individual openings can be collected and conveyed to a central sump or spring box by means of a tile or perforated pipeline or by a gravel-filled ditch. The collection works should be an adequate distance below the elevation of the opening to permit free discharge.

If water issues from a single opening, such as from a karst spring or a tunnel in lava, the opening should be cleaned or enlarged as needed. A spring box or sump should be installed at an elevation that will not allow the water to pond over the spring opening at a depth that will not reduce the yield.

Perched or contact springs occur where an impermeable layer outcrops beneath the base of a water-bearing layer. These springs should be developed by intercepting and collecting the flow from the water-bearing formation. French drains or collection trenches extend-

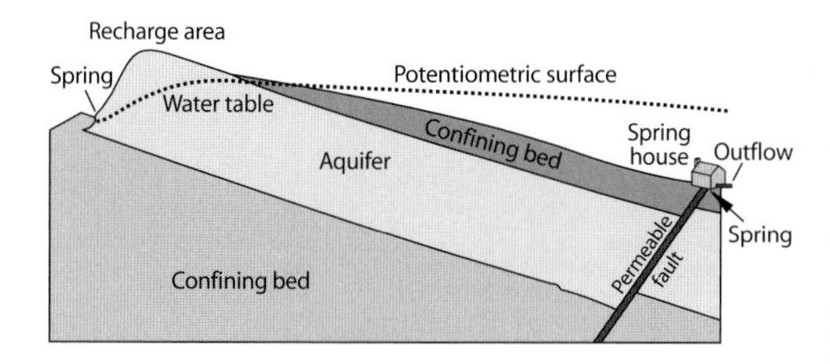

Fig. 4.4. Extracting spring water by building a spring house

ing into the impermeable layer should be used for developing these types of springs. An impervious cutoff wall of compact clay, masonry, concrete, or other suitable materials should be constructed along the downstream side of the trench to insure that the flow enters the system. Spring boxes for perched springs should be floored with concrete unless the underlying materials is solid rock or another stable impervious material.

Artesian springs can be developed by removing obstructions, cleaning or enlarging joints or fractures, or by lowering the outlet elevation. However, sumps and spring boxes should not impound water over the spring outlet.

Collecting spring water at its original discharge point is a traditional extraction method. An ancient structure for collecting and transporting spring water to Rome is the aqueduct (Bono and Boni 1993). These aqueducts are considered to be "monumental works". Samples of existing aqueducts are found in Rome, Greece, France, Germany, and Egypt. This method for development of spring water for public consumption requires a spring house. The house must be secure from insects, other animals, or pollutant sources. After interception, the remaining surplus spring flow can be discharged naturally from the house to the original downstream course. Some spring sites require excavation or other modification around the spring to enhance the physical conditions for constructing the house or other structures and protect the facility from sources of pollution. A classic spring house is shown in Fig. 4.4, which is built to collect the spring water from an artesian spring.

4.7 Diverting and Collecting Spring Water by Blocking Original Discharge Point

For some springs, in particular large karst or fracture springs, it is difficult to collect spring water at its discharge point owing to the wide variation in discharge or to irregular surface conditions. It may be more practical to install a grout curtain to help collect spring water. This practice is similar to building an underground spring house to collect the spring water. The preferential directions of groundwater flow through fractures gradually develop towards the lowest points of flow in the system. The concentration of flow towards one or a number of fractures is a characteristic feature of fracture and karst aquifers. Construction of a barrier across a main fracture can result in increasing storage of water in the subsurface and change the water available, especially during dry periods. This has a notable effect on the increase of dynamic reserves. Under some circumstances, the grout curtain is able to divert the natural discharge to more favorable locations where a spring house or intake can be built. An outstanding example of this type of development is at Source Du Lez, water supply system for the city of Montpelier, France (Avias 1977). If necessary, pipes and pumping wells can be used to get the water from the subsurface. the location and construction of a grout curtain, however, must be preceded by comprehensive investigations, as the correct application of the grout curtain to intercept spring water depends on a detailed knowledge of the geological, geomorphological, hydrogeological and hydrological factors (Milanović 1988), including:

1. Characteristics of geological structure
2. Existing storage of spring-feeding aquifer
3. Hydrodynamic regime of aquifer base flow
4. Characteristics of watershed boundaries
5. Type of spring
6. Position of underground impervious formation
7. Maximum possible backwater level; and
8. Prediction of water quality change

This extraction method can be used when spring water discharges through one single point at the lithological boundary between the aquifer and the impervious formation. It is also a useful method for intercepting spring water in a coastal setting. Some good examples are described in detail by Milanović in the text *Karst Hydrogeology* (1981). The examples provided by Milanović are for the Dinaric region of Yugoslavia. For example, where spring water of good quality could not be obtained under natural conditions because of

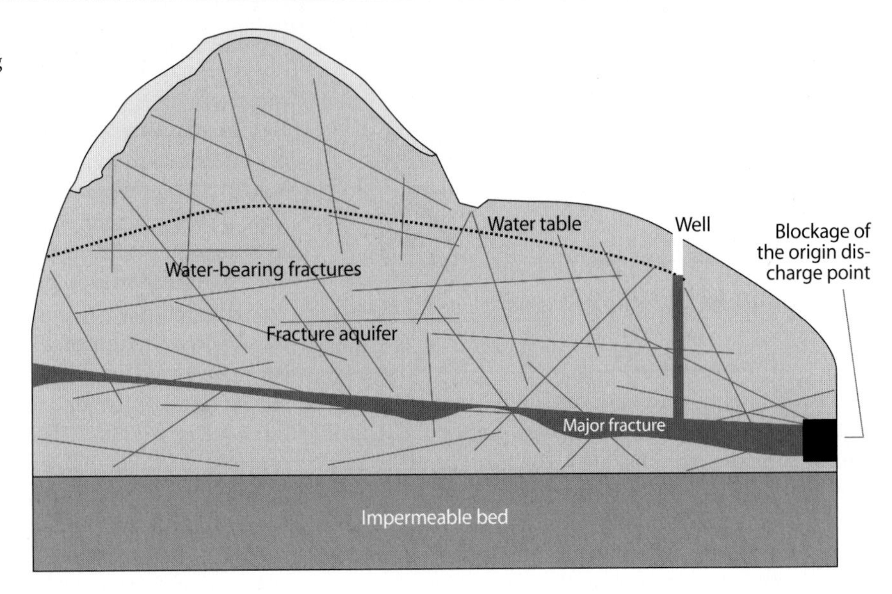

Fig. 4.5. Diverting and collecting spring water by blocking original discharge point

seawater intrusion. In this area, an enormous quantity of water discharged through springs is lost to the sea unused (Komatina 1977). Extensive hydrogeological studies suggested the use of grout and sealing the groundwater discharge points to isolate freshwater from seawater. The water feeding the springs could then be collected at a carefully chosen location and at an appropriate pumping rate. Figure 4.5 shows an example of intercepting spring water by blocking the spring discharge point. A well is used to facilitate the collection of the spring water.

4.8 Extracting Spring Water by Subsurface Channels or Pipes

A covered channel or nearly horizontal pipe can be used to connect directly to a spring flow. The channel or pipe can then divert the flow to a point of use and/or treatment. As shown in Fig. 4.6a, this method is most applicable to contact springs, and the discharge is diffuse. In some circumstances, a drain pipe can be inserted horizontally into a water-bearing zone, as shown in Fig. 4.6b, to collect the spring water. Many examples of this method are found in the uplands of the Appalachian area of the USA. Again, however, the same criteria must be used to protect the spring source, distribution system, and discharge pipe from contamination.

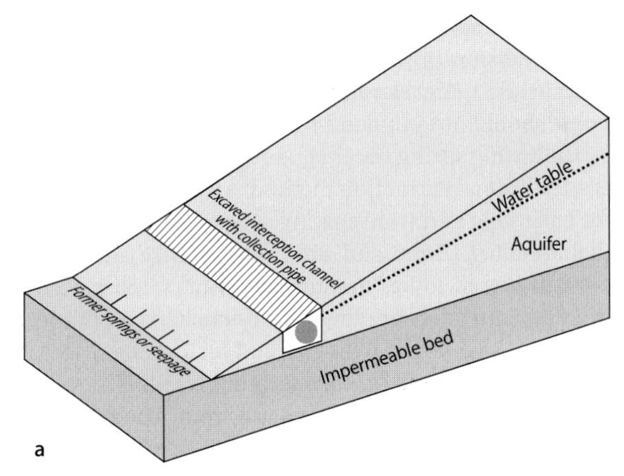

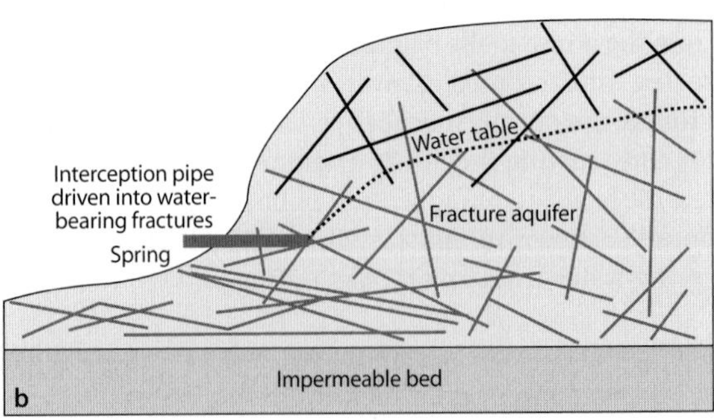

Fig. 4.6. Intercepting spring water by subsurface channels and pipes. **a** Horizontal interception/diversion channel and convergance pipe. **b** Interception pipe inserted into water-bearing fractures

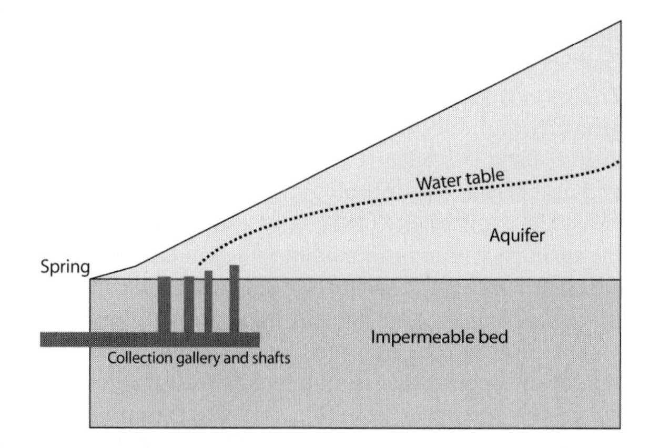

Fig. 4.7. Intercepting spring water by a gallery system with shafts

4.9 Extracting Spring Water by Shaft and/or Collection Gallery System

This indirect method is closely related to the horizontal channel interception method. The basic structure is shown in Fig. 4.7, and it generally consists of one or more of the following features:

1. A vertical shaft or an excavation from which one or more horizontal conduits are bored into the base of the aquifer, intercepting the spring water and allowing the spring water to drain into the conduits to a central collection and routing structure.
2. A horizontal shaft excavated from the hillside into the base of the principal aquifer that feeds an adjacent spring. Additional vertical or lateral galleries connect into drains.

4.10 Extracting Spring Water by Boreholes

Using a properly constructed borehole to intercept spring discharge can prevent the introduction of contaminants from surface sources. Boreholes are typically installed below ground surface with the placement of a sanitary seal that generally extends from the surface to a minimum depth of 20 ft. However, California and New York require that sanitary seals in wells extend to a depth of 50 ft below the ground surface (Department of Water Resources, State of California 1981). These requirements ensure protection of spring water as bacteria often thrive in water at or near the ground surface, where oxygen and nutrients are more plentiful and temperatures are high. Deeper portions of an aquifer are less likely to have high bacterial populations.

Water sources most likely to be contaminated by surface water include inadequately or poorly developed springs, infiltration galleries and shallow wells less than 50 ft deep (Sheahan and Zukin 1993). Deeper wells are less likely to be contaminated by surface water than are springs. The State of Tennessee requires that, if a spring is directly influenced by contaminated surface water, modifications must be made to the system or a properly constructed borehole must be installed that is either "... deeper or at a different location" (State of Tennessee 1991). The borehole method can provide total isolation of water from the surrounding ground surface, atmosphere and biota. It also allows for easy control and conveyance using standard well construction, pumping and piping technology. Another potential advantage of this method is that it can preserve the original pressure and dissolved gas in the water from an aquifer before it discharges to the land surface. An example is shown in Fig. 4.8. Borehole extraction method is applicable to both confined and unconfined aquifers.

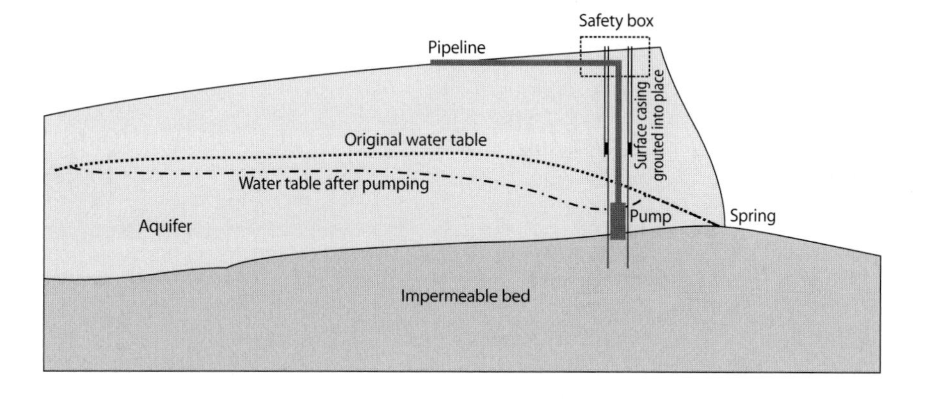

Fig. 4.8. Extracting spring water by a borehole

A primary concern with the borehole method of producing water from a spring source is providing assurance that the extracted water is the same water that feeds a specific spring. Criteria for testing this requirement include:

1. Equivalent chemical and physical characteristics of the waters
2. The same source aquifer between the borehole and the spring
3. Hydraulic connection between the borehole and the spring

4.10.1 Hydrogeologic Methods

If a borehole close to a spring is in the same aquifer that provides the spring with groundwater, the borehole and the spring will produce the same water. To ensure that waters extracted from a borehole and a spring are the same, the borehole should be completed adjacent to the spring. The State of California informally defines "adjacent" as a distance of approximately 250 ft (Sheahan and Zukin 1993). Although the State's definition is somewhat arbitrary and does not consider existing geologic conditions for a site adequately, it is an easily-interpreted and usable rule. Hydrogeologic methods can be used to verify the applicability of the rules and show that either the spring is within or immediately downgradient of the adjacent borehole's capture zone or hydraulic continuity exists between a spring and the adjacent borehole.

The capture zone of a spring or borehole is the portion of the aquifer that is influenced by discharge from either the spring or borehole, and that groundwater within the aquifer in this zone of influence could eventually enter the spring. If a spring is within or immediately downgradient of a borehole's capture zone (Fig. 4.9), then the borehole is likely to intercept the same water that would have naturally discharged at the spring. If an adjacent borehole is within a "capture zone" (Fig. 4.10), then that borehole is also likely to be intercepting the water that would naturally enter the spring.

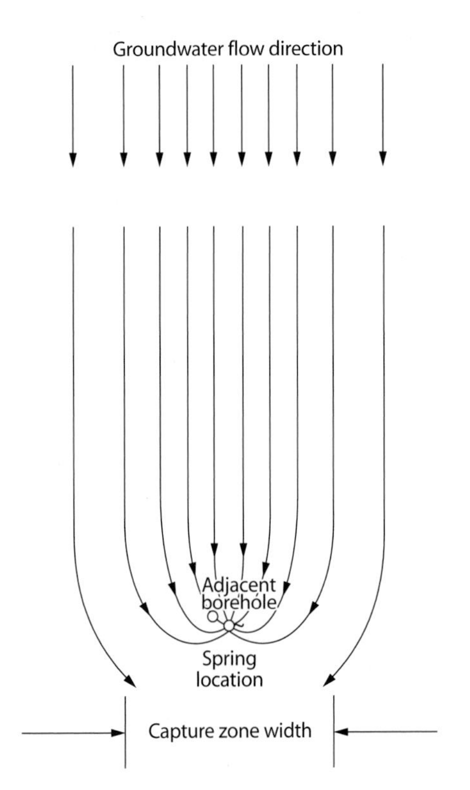

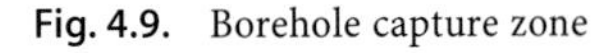
Fig. 4.9. Borehole capture zone

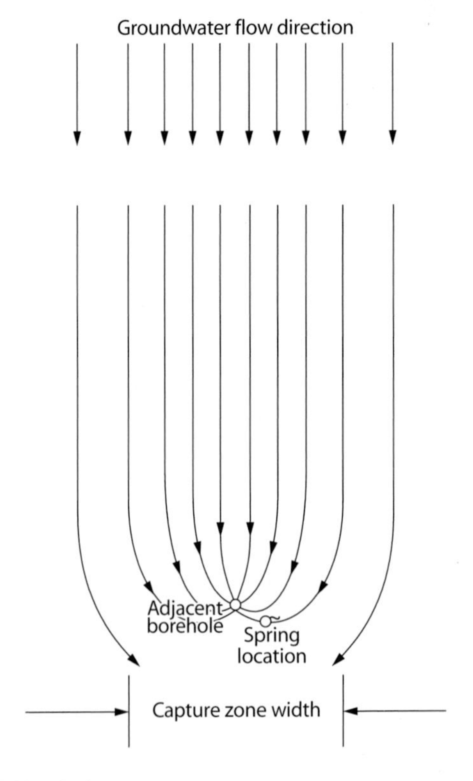

Fig. 4.10. Spring capture zone

Hydraulic continuity between a spring and an adjacent borehole should be proven using at least two or more of the accepted hydrogeologic methods:

1. Comparison of water level and spring flow measurements by monitoring
2. Aquifer-test analyses in the borehole and/or spring
3. Flow net analyses
4. Tracer tests; and
5. Geophysical methods including conductivity test, geothermal test and Geo-bomb test

4.10.2 Geochemical Methods

The chemical quality of the groundwater extracted from a spring and from an adjacent borehole may be chemically compared to determine that major dissolved mineral contents are the same. Several scientific graphical methods are commonly used by hydrogeologists for water quality comparison. A graphic technique commonly used is the Piper trilinear diagram. The Piper diagram, shown in Fig. 4.11, uses percentages of common cations and anions to identify the chemical facies of a water. The concentration of total dissolved solids in the water is not a factor in the Piper diagram methodology, thus the method eliminates extraneous variability from such factors as airborne contamination and evaporation in spring areas.

In using the Piper trilinear diagram for comparing spring water and borehole water, the chemical variability due to natural variations can be accounted for by plotting water analyses of the spring water on the trilinear diagram, and defining the field of variation by encompassing the spring analyses with a tight circle. A percentage adjustment to the field variation, to account for other factors such as laboratory variability, can be made by drawing a second circle around the field of variation. A comparison of water extracted from a borehole and the spring can then be made by plotting analyses of water collected from the borehole on the trilinear diagram and showing whether or not the water falls within the adjusted field of variation. An example of this method is given by Sheahan and Zukin (1993), as shown in Fig. 4.11.

Spring waters are often exploited as they have special chemical compositions or unique physical characteristics. Under these circumstances, those special chemical compositions and physical characteristics should be addressed and monitored regularly.

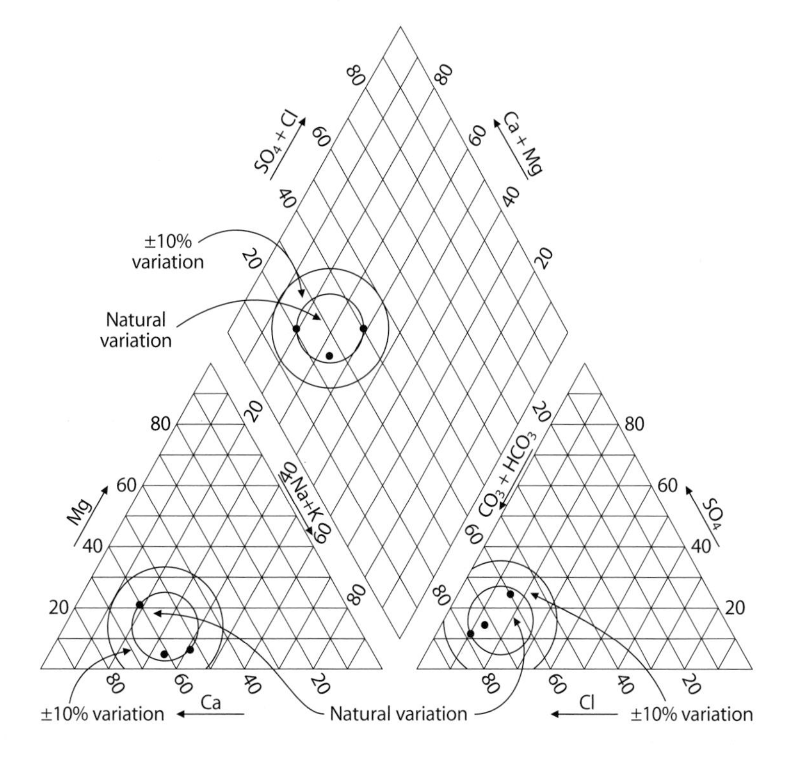

Fig. 4.11. Piper trilinear diagram method (after Sheahan and Zukin 1993)

4.11 Conclusions

Large-scale development of spring water must conform to existing regulatory practices in an area. A universally accepted definition of spring water is a difficult issue that must be addressed on a site-specific basis. The Association of Food and Drug Officials (AFDO) and the IBWA have been active in trying to help regulatory bodies implement technically sound, practical and consistent regulations governing collection methods and sanitary protection of bottled spring water. Both IBWA and AFDO have adopted the *model bottled water regulation* (IBWA 1991; AFDO 1986), under which the practice of extracting water from a spring water source aquifer is explicitly permitted. This aspect of spring development is covered in more detail in Chap. 7.

Following the IBWA Model laws, many different methods can be used effectively for collecting spring water. It is recognized that a wide range of collection methods have been developed at various springs through the world, serving as sources for public water supplies as well as bottled waters. Adjacency to the spring and quality of water provisions are two important factors to select the interception methods. Physical modifications to areas of spring discharge and/or other engineered measures such as covered collection boxes, sealed interception wells upgradient of the discharge area, or encased subsurface collection drains are preferable. Acceptable extraction methods include: surface collection structures at the discharge point of the spring, interception channels, shaft and gallery systems, and boreholes. The surface collecting method is the oldest form of collecting spring waters and can be used in most hydrogeological conditions, provided that the site around the spring is suitable. Diverting and collecting spring water by blocking the original discharge point is mostly applicable to fracture and karst springs. Subsurface channels may be suitable for contact springs and fractures springs. Subsurface extraction methods by boreholes are developed and widely used and are less likely to be contaminated by surface sources of pollutants. Several hydrogeological and geochemical methods are described that can be used to verify that extracted water from the borehole is the same as the water that feeds an adjacent spring.

Two relatively simple criteria may be used to assess the appropriateness of a spring water interception method:

1. Principal chemical and physical characteristics of the intercepted water should correspond to those of the associated spring water at the point of natural discharge.
2. The interception system must be located correctly so that it can be demonstrated through standard hydrogeological principles and investigative practices that the intercepted water is, in fact, the water that would emerge as spring discharge if not intercepted by the collection system (especially important for the borehole interception method).

There are many circumstances in which water collection or extraction systems might not meet the basic criteria described above. In such cases, the system would either be unsuitable as a reliable, sanitary sources or the water provided should not qualify technically as spring water. If an interception system cannot be demonstrated with a reasonable degree of scientific certainty to satisfy the conditions outlined above, then the water produced by that system should not qualify as spring water. Some examples of unsuitable interception systems include:

1. Wells that tap groundwater that would not normally discharge to a nearby natural spring or springs associated with the interception system.
2. Surface sources that do not completely consist of spring water.
3. Systems that capture water that differs consistently and substantially in chemical or physical characteristics from the natural spring water supposedly associated with the interception system.
4. Interception systems that materially alter the chemical or physical characteristics of the water from those characteristics of water normally discharged by the associated spring.

The use of any of the extraction methods requires spring source protection as the major requirement for an acceptable quality program in the bottled water industry. To ensure high quality of water at the spring or from boreholes or other subsurface extraction methods, bacterial, parasitic (*Giardia*, *Cryptosporidium*, etc.) and viral contamination must not occur. In the United States, industry practice typically dictates that bottled water shall be subjected to ozonation, filtration, or other processes, which remove, destroy, or prevent contamination by *Giardia* or lamblia cysts. In Europe, ozonation and some other treatment processes can affect allowable labeling such as prohibition of the term "mineral water". Protection of the source also implies that its recharge mechanism must be protected by natural filtration so that any particles including microbes entering the upgradient end of the pathway must be filtered out long before they enter the aquifer.

References

American Geological Institute (1960) Glossary of geology and related sciences. 2nd edn. National Academy of Sciences, National Research Council, Washington DC

American Geological Institute (1976) Dictionary of geological terms. Anchor Press/Double Day, Garden City, New York

Association of Food and Drug Officials (1986) Model Bottled Water Regulations, Washington D.C.

Avias JV (1972) Karst of France. In: Herak M, Stringfield VT (eds) Karst, important karst regions of the northern hemisphere. Elsevier, Holand, 129–188

Avias JV (1977) Globotectonic control of perimediterranean karstic terranes main aquifers. In: Tolson JS, Doyle FL (eds) Karst hydrogeology. University of Alabama, Huntsville, USA, 57–72

Back W, Landa ER, Meeks L (1995) Bottled water, spas, and early years of water chemistry. Groundwater 33(4):605–614

Biondic B (1988) Tapping and protection of undergroundwater in the ADRIATIC karst region related to the new conception of structure of Dinarides. In: Daoxian Y (ed) Proceedings of the IAH 21st Congress, Guilin, China, 187–193

Bono P, Boni C (1993) Water supply of Rome in antiquity and today. Unpublished Manuscript

Department of Water Resources, State of California (1981) Water well standard. State of California Bulletin 74–81

Fetter CW (1993) Contaminant hydrogeology. Macmilla Publishing Company, New York, 458 p

Ford D, Williams P (1989) Karst geomorphology and hydrology, Academic Division of Unwin Hyman Ltd.

Hanchar DW (1992) Effects of septic-tank effluent on groundwater quality in Northern Williamson County and Southern Davidson County, Tennessee. USGS Water-Resources Investigation Report 91-4011

Hughes TH, Memon BA, LaMoreaux PE (1994) Landfills in karst terrains. Bulletin of the Association of Engineering Geologists XXXI(2):203–208

International Bottled Water Association IBWA (1991) International bottled water association model bottled water regulations, Alexandria, Virginia

Komatina M (1977) Artificial works and efficient interception of groundwater in karst. In: Dilamarter RR, Csallany SC (eds) Hydrologic problems in karst regions, Western Kentucky University, 287–296

LaMoreaux PE (1975) Environmental geology and hydrology, Huntsville and Madison County, Alabama. Geological Survey of Alabama, Atlas Series 8, University, Alabama, 119 p

LaMoreaux PE (1989) Environmental aspects of the development of Figeh Spring, Damascus, Syria. In: Beck BF (ed) Proceedings of the 3rd multi-disciplinary Conference on Sinkholes, 17–23

LaMoreaux PE (1993) Hot springs and bottled water, Geotimes 7

LaMoreaux PE, Wilson BM, Memon BA (eds; 1984) Guide to the hydrology of carbonate rocks. UNESCO, Paris, 345 p

LaMoreaux PE, Hughes TH, Memon BA, Lineback N (1989) Hydrogeological assessment – Figeh Spring, Damascus, Syria. Environ geol Water Sci 13(2):73–127

LaMoreaux PE, LeGrand HE, Powell WJ (1996) Legal aspects in karst areas. The Professional Geologists 5–9

Meinzer OE (1923a) Outline of ground water hydrology, with Definitions. US Geological Survey Water Supply Paper 494, Washington Government Printing Office, 71 p

Meinzer OE (1923b) The occurrence of ground water in the United States. With a discussion of principles. US geological Survey Water Supply Paper 489, Washington Government Printing Office, 321 p

Milanović P (1981) Karst hydrogeology. Water Resources Publications, Littleton, Co., USA, 434 p

Milanović P (1988) Artificial underground reservoirs in the karst experimental and projected examples. In: Daoxian Y (ed) Proceedings of the IAH 21st Congress, Guilin, China, 76–87

Price M (1985) Introducing groundwater. George Allen & Unwin, London, 195 p

Robertson JB, Edberg SC (1993) Technical considerations in extracting and regulating spring water for public consumption. Environmental Geology 22:52–59

Sharp JM Jr, Clement TJ (1988) Hydrochemical facies as hydraulic boundaries in karstic aquifers – the Edwards Aquifer, Central Texas, USA. In: Daoxian Y (ed) Proceedings of the IAH 21st Congress, Guilin, China, 841–845

Sheahan NT, Zukin JG (1993) Developing spring water under the proposed FDA rules. The Professional Geologists 9–11

Song LH (1989) Subsurface reservoirs and karst geomorphology. In: Beck BF (ed) Proceedings of the 3rd multi-discipline Conference on Sinkholes, 369–376

State of Tennessee (1991) Guidance for determining if a groundwater source is under the direct influence of surface water

Stephenson JB, Beck BF, Zhou WF, Smoot JL, Turpin M (1996) Management of highway stormwater runoff impacts to groundwater in karst areas, a status report. Sixth Tennessee Water Resources Symposium, Nashville, Tennessee

Studlick JR, Bain RC (1980) Bottled water – expensive groundwater. Groundwater 18(4):340–345

US Army Corps of Engineers (1986) Final environmental impact statement for regulatory actions associated with the Olin Corporation, remedial action plan to isolate DDT from the people and the environment in the Huntsville Spring Branch-Indian Creek System, Wheeler Reservoir, Alabama. US Army Corps of Engineers, Nashville District

Wermund EG, Cepeda JC (1977) Relation of fracture zones to the Edwards limestone aquifer, Texas. In: Tolson JS, Doyle FL (eds) Karst Hydrogeology. University of Alabama, Huntsville, USA, 239–253

Chapter 5

Quantitative Analysis of Springs

CHAPTER 5
Quantitative Analysis of Springs

Bashir A. Memon

5.1 Introduction

Concern for groundwater resources has increased in the last decade owing to its greater development and use. Strict environmental regulations and growing competition for a limited resource have led to many groundwater investigations. These investigations generally include the determination of aquifer properties, hydraulic conductivity, transmissivity, storativity and leakage, determined primarily by pumping tests. There follows a summarization of quantitative methods that are available to analyze aquifers or aquifer systems hydraulically connected to springs.

A pumping test with observation wells and springs is one of the most reliable means of quantifying hydraulic characteristics and the response of natural springs to discharge and pumping because it yields results that are representative of a large area of the aquifer system in which the spring occurs, rather than results from single points. A pumping test is defined as an in situ field study, directed at obtaining controlled aquifer system response data. Usually, a spring and/or production well is pumped at a constant, predetermined rate, and water levels are measured at frequent intervals in the spring or well and nearby observation wells and springs. Time-drawdown and distance-drawdown data are analyzed with conceptual analytical models such as type-curve matching, straight-line methods, or inflection-point selection techniques (Theis 1935).

Important conditions which may influence pumping test field data are: (1) nearby geologic or hydrologic boundaries or discontinuities; (2) geologic structure and lithology of the aquifer; (3) continuity and character of contiguous beds; (4) aquitard storativity and drawdown in source bed; (5) decreased transmissivity with water table decline; (6) interference from nearby production wells, streams, or lakes; (7) changes in barometric pressure; (8) physical changes to the aquifer system (e.g., catastrophic collapse); (9) tidal fluctuations and/or stream stages; and (10) climatic changes – rain, freezing, etc.

Water level records must be analyzed before, during, and after the tests for climatic changes, barometric pressure, stream stage, and tide changes. Adjusted water level data are plotted and matched to the selected appropriate well function type curve for the conceptual analytical model and equations that best suit the site-specific aquifer system and conditions for the interpretation of test data. Match-point coordinates are substituted into equations to determine hydraulic characteristics. There are numerous conceptual analytical models and corresponding equations available for analysis and interpretation of pumping test data (see the references for examples).

5.2 Test Design Characteristics

The pumping test design must be based on a pretest conceptual understanding of geologic structure, hydrogeologic setting, initial aquifer system, hydraulic characteristics, boundaries, discontinuities, and well construction and operational features based on available lithologic and hydraulic data.

The purpose of a pumping test design is to ensure that a proposed test site and associated equipment will yield accurate results, and to minimize uncertainties and errors in data collection and analysis. Some important design criteria that must be determined prior to performing a test are:

1. The diameter, depth, areal, vertical limits, and position of all intervals open to the aquifer system in the spring and observation well(s) should be determined.
2. The spring/well should be equipped with a reliable pump, and discharge-control equipment.
3. The wellhead and discharge lines should be accessible for installing, regulating, and monitoring equipment.
4. It should be possible to measure water levels in the spring/well before, during, and after pumping.
5. The water discharged should be conducted away from the spring/well to minimize or eliminate recirculation.

6. All production wells within the pumping test area of influence should be capable of being controlled, and their discharges should be known.
7. The rate of discharge of natural springs within the pumping test area of influence should be monitored.
8. Radial distance and direction from any known boundaries including natural springs to each observation well should be determined.
9. Nearby aquifer system discontinuities should be mapped.
10. Fluctuations in nearby surface water stages should be monitored.
11. Adequate monitoring facilities must be provided to produce reliable data on meteorological factors – rainfall, snow, temperature, and barometric pressure; and pertinent surface water features – rivers, lakes, streams, ponds.

Pumping tests are commonly 24 to 72 h in duration. The duration of the test must be determined based on the geologic, hydrogeologic, and climatic setting, discontinuities, boundary conditions, and existence of interfering factors, i.e. well or spring discharge in area, construction, etc.

5.3 Analysis of Pumping Test Data

The development of quantitative methods for interpreting aquifer test data has paralleled the development of mathematical methods for describing the flow of groundwater to wells (Theis 1935). Thiem is generally credited with the development in 1906 of the first equations to describe the flow of groundwater to a well during a pumping test (Table 5.1). The Thiem equation, developed from

Table 5.1. Some of the conceptual models for analysis and interpretation of pumping test data

Model number	Flow condition	Aquifer type	Aquitard leakage	Aquitard storage	Well storage	Partial well penetration	Anisotropic properties	Reference
1	Equilibrium	Confined	No	No	No	No	No	Thiem (1906)
2	Equilibrium	Unconfined	–	No	No	No	No	Thiem (1906)
3	Transient	Confined	No	No	No	No	No	Theis (1935)
4	Transient	Confined	Yes	No	No	No	No	Hantush and Jacob (1955)
5	Transient	Confined	Yes	Yes	No	No	No	Hantush (1964)
6	Transient	Confined	No	No	No	Yes	Yes	Hantush (1964)
7	Transient	Confined	Yes	No	No	Yes	Yes	Hantush (1964)
8	Transient	Confined	No	No	Yes	No	No	Papadopulos and Cooper (1967)
9	Transient	Confined	Yes	No	Yes	No	No	Lai and Su (1974)
10	Transient	Confined (fissure-block system)	Yes	Yes	No	No	No	Boulton and Streltsova (1977)
11	Transient	Confined	No	No	No	No	Yes	Papadopulos (1965)
12	Transient	Confined to unconfined aquifer conversion	No	No	No	No	No	Moench and Prickett (1972)
13	Transient	Unconfined	No	No	No	No	Yes	Neuman (1972)
14	Transient	Unconfined	No	No	No	Yes	Yes	Neuman (1974)
15	Transient	Unconfined	No	No	Yes	Yes	Yes	Boulton and Streltsova (1976)
16	Transient	Unconfined ("aquitard-aquifer")	Yes	Yes	No	Yes	Yes	Boulton and Streltsova (1975)

Darcy's Law, can be used to compute transmissivity from measurements of drawdown in two observation wells. In the development of this equation, Thiem made a number of simplifying assumptions about the geometry and hydraulic behavior of the aquifer, including: (1) the aquifer is isotropic and homogeneous and of infinite lateral extent, (2) the pumping well is screened over the entire saturated thickness of the aquifer (the pumping well is said to fully penetrate the aquifer), (3) the aquifer is bounded above by an aquiclude that is horizontal and of infinite lateral extent, (4) the pumping rate is constant, and (5) pumping has continued for a sufficient length of time to achieve equilibrium conditions.

Theis (1935) was the first to introduce the concept of time to the mathematics of groundwater hydraulics and derived the nonequilibrium formula to solve problems of groundwater flow. The Theis formula is based on the following assumptions: (1) the aquifer is homogeneous and isotropic, (2) it has infinite areal extent, (3) the discharge well penetrates and receives water from the entire thickness of the aquifer, (4) the transmissivity is constant at all times and at all places, (5) the well has an infinitesimal (reasonably small) diameter, and (6) water removed from storage is discharged instantaneously with decline in head.

Collectively, these assumptions constitute a conceptual model for the aquifer system. A large number of alternate conceptual models have been proposed by various authors such as Wenzle, Jacob, Cooper, Lohman, Hantush, Walton, etc. (Table 5.1) for use in the analysis of pumping tests data, and each of these is based on a different set of simplifying assumptions. In practice, the analyst selects a conceptual model based on a variety of data including: geologic descriptions of the aquifer materials obtained from drilling logs, geological maps, etc., the construction records for the production well and one or more observation wells, the drawdown data collected during the test, and a set of assumptions about the geometry and hydraulic behavior of the aquifer system. Analytical solutions to the equations describing groundwater flow during pumping tests have been developed for each of these conceptual models to determine hydraulic properties of the aquifer system feeding the springs. Because the computed values of aquifer properties depend on the choice of conceptual model used to analyze the test data, the selection of an appropriate conceptual model is the single most important step in the analysis of aquifer test data. It must be noted that a thorough study that characterizes the geology of the spring must be available to properly interpret quantitative results.

5.4 Analysis of Aquifer Systems Hydraulically Connected to Springs

Determination of the structure and physical properties of an anisotropic and heterogeneous aquifer poses practical problems because their characteristics can be poorly defined, and water flow into them is of a very particular type. Yet it is essential for water resources estimation, planning and management to be able to answer questions as how much water can be used, where is it coming from, and what are the physical parameters characterizing the aquifer.

With direct observations limited to springs, wells and/or boreholes, inputs and outputs, the rest of the aquifer response can be deduced. A realistic approach is one that uses available information on subterranean conditions to clarify the structure of the aquifer system and to help explain its observed response to recharge-discharge phenomena. Although each aquifer is unique in its individual characteristics, some structural components are widely found, although they vary in relative significance in different systems. A good example was the study of a karst aquifer system involving pumped wells and spring measurements for the Corps of Engineers, Red River project, Childress County, Texas (Fig. 5.1 and 5.2).

Atkinson (1985) suggests that a three end-member spectrum may be a more appropriate way of visualizing the concept of granular, fracture, and conduit aquifers. This conceptual classification of flow media is then related to phreatic flow regimes (Fig. 5.3).

An alternative method of qualitative and quantitative classification is to use the system approach, starting with precipitation leading to discharge.

Comprehensive analysis of aquifer systems involves determining the following:

1. Areal and vertical extent of the system
2. Its boundary conditions
3. Input and output (discharge) sites
4. Interior structure of linkages
5. Capacities and physical characteristics of the storage
6. Response of storage and output to recharge
7. System's response under different flow conditions
8. Potential for physical changes within the aquifer system before, during, and after the test

The information required can be obtained by taking four complementary approaches: water balance estimation, borehole analysis, spring hydrograph analysis, and water tracing.

Fig. 5.1. **a** Childress County, Texas, Jonah Creek, a tributary of Red River. Instrumentation for measurement of water levels and monitoring discharge during pumping test. **b** Pumping test in progress

Fig. 5.2. **a** Childress County, Texas, Jonah Creek, a tributary of Red River. Spring discharge prior to initiation of pumping test. **b** Cessation of discharge from spring during pumping test

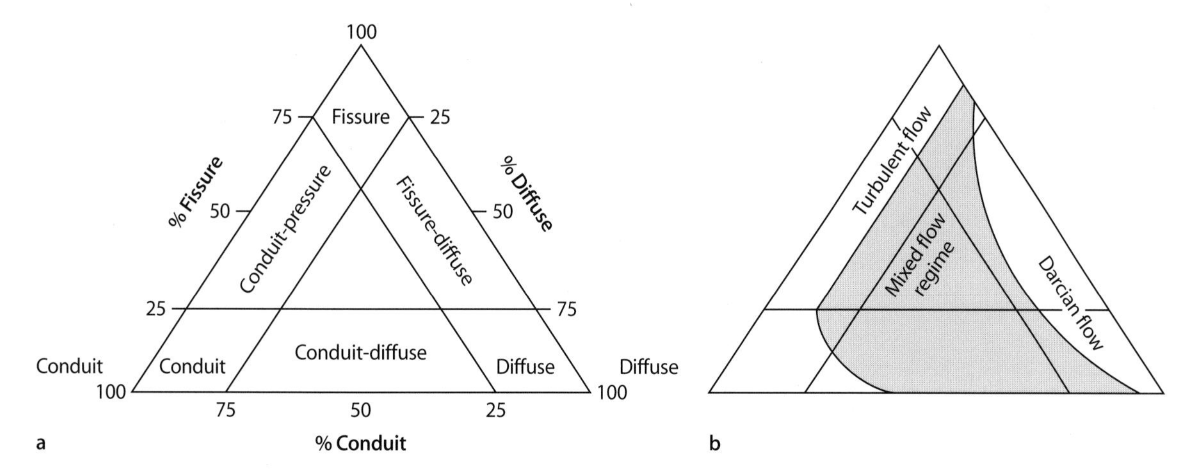

Fig. 5.3. **a** Conceptual classification of aquifers. **b** Aquifer relationship to predominant flow regimes (modified from Atkinson 1985)

5.5 Groundwater Recharge

Recharge rates for aquifers sustaining discharge from springs must be estimated before groundwater resources can be evaluated and the consequences of the utilization of springs predicted. The major sources of recharge to aquifers are direct precipitation on intake (recharge) areas and/or downward percolation of stream runoff, possibly by an artificial recharge method. Under unique conditions, artificial recharge can take place through basins, stream channel, ditches, furrows, pits, flooding, and irrigation practices. Recharge from precipitation on intake areas is often irregularly distributed in time and place.

A small fraction of the annual precipitation percolates from the surface into the underlying aquifer, and moves vertically downward under the influence of gravity until it reaches the level at which all the pores are water-filled. The surface that separates the water-saturated zone from the zone with air-containing pores is the water table (Fig. 5.4). However, a large portion of precipitation runs overland to streams or undergoes evaporation or transpiration before it recharges the aquifer. The amount of precipitation that reaches the zone of saturation depends upon the following factors: (1) the occurrence of precipitation (rain and/or snow), (2) air temperature, humidity, wind velocity, (3) intensity, duration, and seasonal distribution of rainfall, (4) the character and thickness of the soil and other deposits above the water table, (5) vegetative cover, (6) soil moisture content, (7) depth to the water table, (8) topography, and (9) land use.

The quantity of vertical percolation (leakage) varies from place to place and is controlled by the permeability and thickness of deposits through which leakage occurs, the head differential between sources of water and the aquifer, and the area through which leakage occurs.

The water is ultimately lost from the groundwater system by surface discharge through springs and seeps into surface streams, lakes, or the sea.

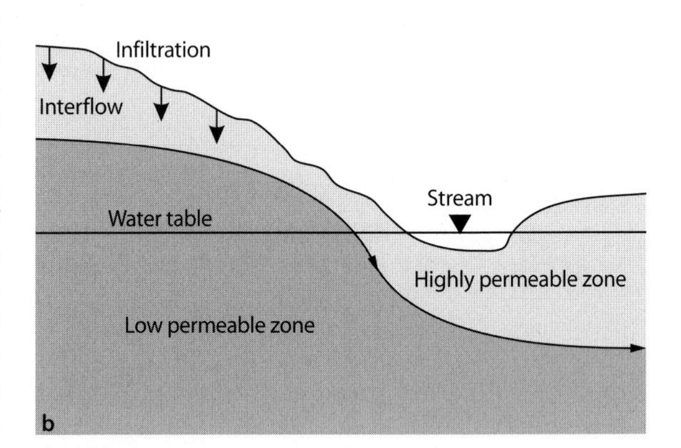

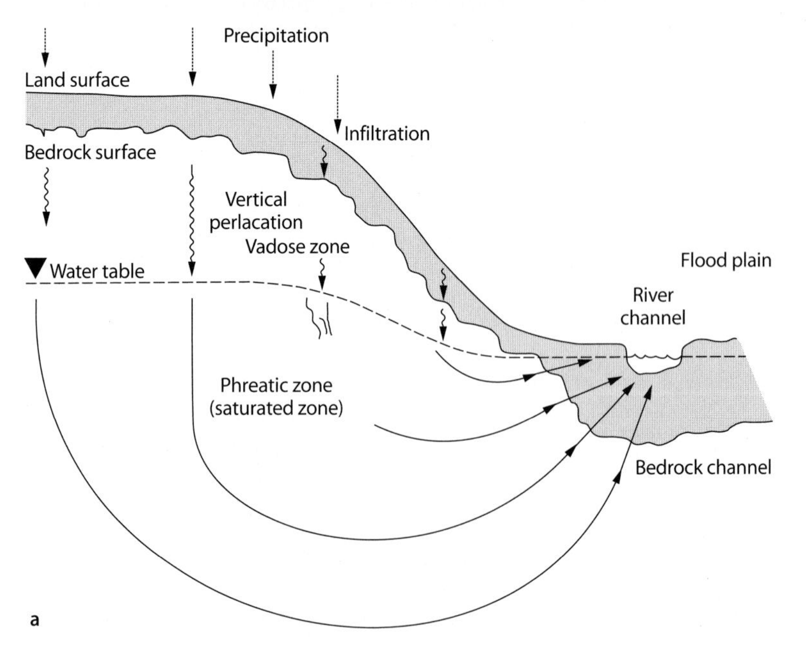

Fig. 5.4.
a Water table and vadose and phreatic zones of an unconfined aquifer (modified from White 1988). **b** Interflow developing where a highly permeable but thin layer of weathered rock overlies a bedrock unit of lower permeability (modified from Fetter 1980)

5.5.1 Precipitation

Groundwater recharge is determined by appraising the various components of the water balance. Precipitation information for a given period for a basin can be obtained either from National Oceanographic Atmospheric Administration (NOAA) or by installing a rain gage and recording the amount of rain on charts. Point rainfall data as measured by gage provide depth of precipitation over a specific area either for an individual storm or on a seasonal or annual basis. Precipitation data over a long period of years must be analyzed for consistency before the data are used in water balance studies.

5.5.2 Evapotranspiration

A large amount of water as precipitation is returned to the atmosphere as vapor, through the combined action of evaporation and transpiration. Evapotranspiration depends on many factors such as temperature, relative humidity of the atmosphere, wind velocity, barometric pressure, and solar radiation. Evaporation is the process by which molecules of water at the surface of water or moist soil acquire enough energy through sun radiation to escape the liquid and pass into the gaseous state. Evaporation data for a given period for a basin are available from NOAA.

Transpiration is the process by which water from plants (weeds, herbs, trees, etc.) is discharged as vapor into the atmosphere. Transpiration varies with the species and density of plants and to a certain extent, with the moisture content of the soil.

5.5.3 Runoff

The term runoff is usually considered synonymous with stream flow and is the sum of surface runoff and groundwater runoff that reaches the stream. Surface runoff is the portion of precipitation that finds its way into the stream channel without infiltrating into the soil. Groundwater runoff is the portion of precipitation that infiltrates into the soil or to the water table and then discharges as spring flow and/or discharges into the stream channel.

Precipitation is partly depleted by interception and by the small amount of water that infiltrates through the soil surface and fills puddles and surface depressions. When the available interception and depression storage are completely exhausted and when the rainfall intensity at the soil surface exceeds the infiltration capacity of the soil, overland flow begins (Fig. 5.5). The soil surface is then covered with a thin sheet of water called detention; once the overland flow reaches a stream channel, it is called surface runoff. Part of the water that infiltrates into the soil will continue to flow laterally at shallow

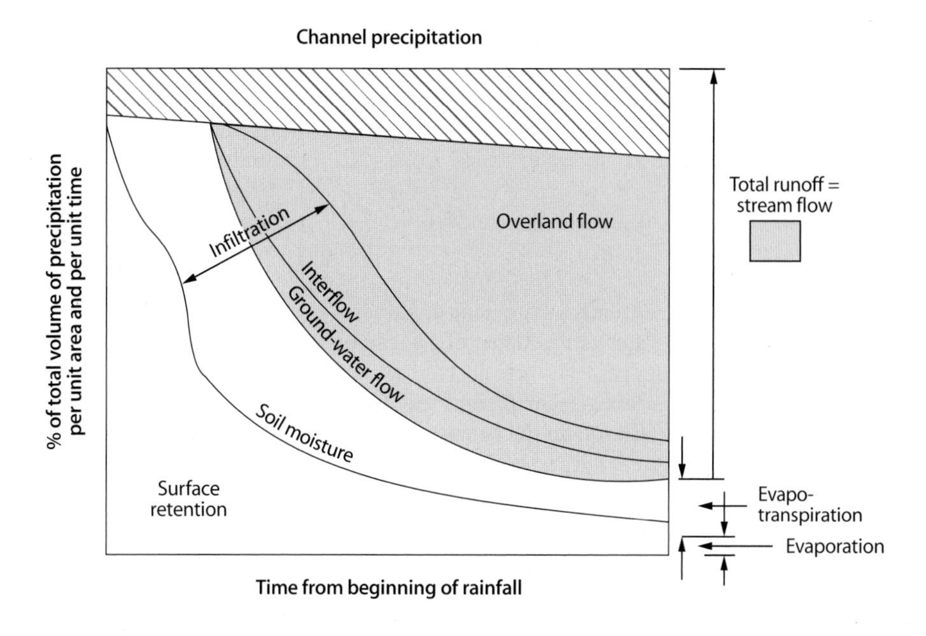

Fig. 5.5. Contribution to stream from storm (modified from DeWiest 1965)

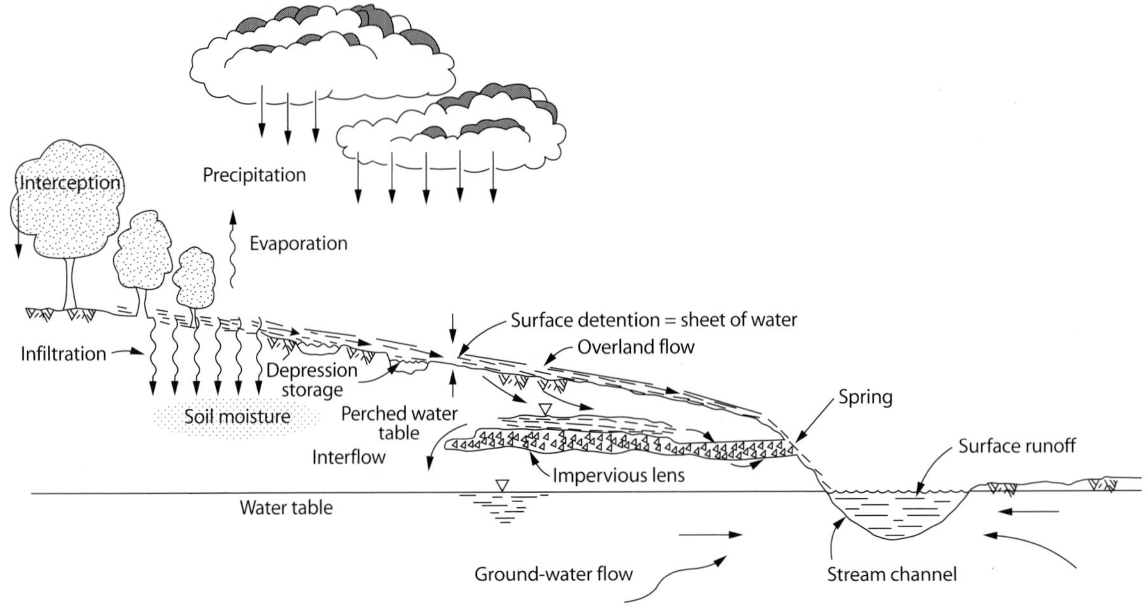

Fig. 5.6. Runoff cycle (modified from DeWiest 1965)

depths as interflow, owing to the presence of relatively impervious lenses just below the soil surface and will emerge as a spring or reach the stream channel in this capacity (Fig. 5.6). Another part will percolate to the groundwater table and eventually will discharge as spring flow or reach the stream channel to become the base flow of the stream. A third part will remain above the water table in the zone of unsaturated flow. Recharge, discharge, and change in storage are graphically illustrated in Fig. 5.7.

The change in groundwater storage can be calculated by substituting the values of precipitation, evapotranspiration, surface runoff, and groundwater runoff in Eq. 5.1.

5.5.4 Water Budget

A water budget is a quantitative statement of the balance between the total water gain and loss of a basin for a period of time. The budget takes into account all water – surface and subsurface – entering and leaving or stored within a basin. Recharge to the basin is equal to discharge plus or minus changes in basin storage.

Drainage basins, in general, are contiguous to headwater reaches of streams, and the boundaries are reasonably congruous with groundwater and topographic divides. There is no surface or subsurface flow into or out of the basins except subsurface underflow from the basins in the vicinity of springs and streams.

Precipitation (rain and snow) entering the basins is considered the only water gain for the water budget. Water leaving basins includes surface runoff, groundwater runoff (spring flow), evapotranspiration, subsurface underflow. Water is stored beneath the surface in soils, and in the groundwater reservoir. Changes in storage of water in the soil are reflected in changes in soil moisture. Changes in spring flow and water levels in wells indicate changes in storage of water in the groundwater reservoir.

Under these conditions, the water budget for a given period is expressed as water gain minus water loss, plus or minus change in storage:

$$P = R_s + R_g + E_T + U_f \pm \Delta S_m \pm \Delta S_g \tag{5.1}$$

where

P = precipitation, including rain and snow
R_s = surface runoff
R_g = groundwater runoff and spring flow
E_T = evapotranspiration
U_f = subsurface underflow
ΔS_m = change in soil moisture
ΔS_g = change in groundwater storage

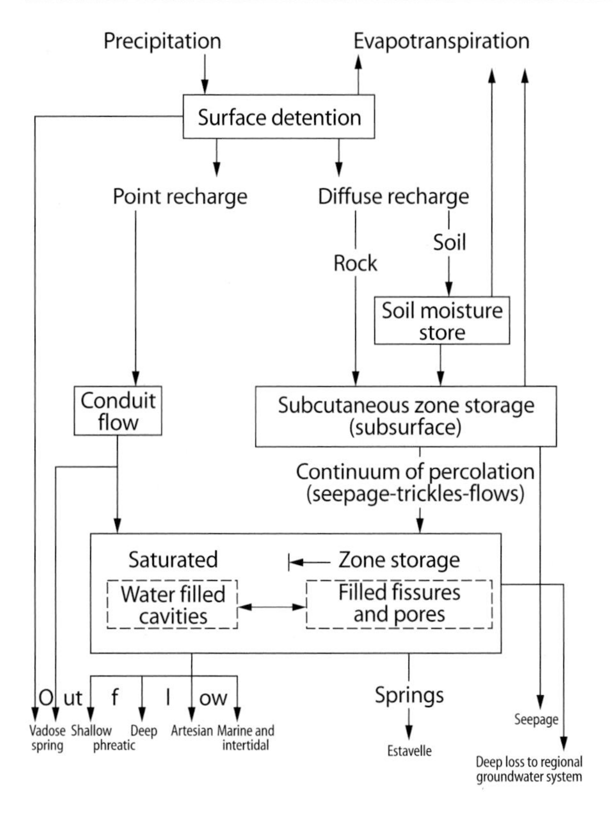

Fig. 5.7. Recharge, discharge, and change in storage (modified from Ford and Williams 1992)

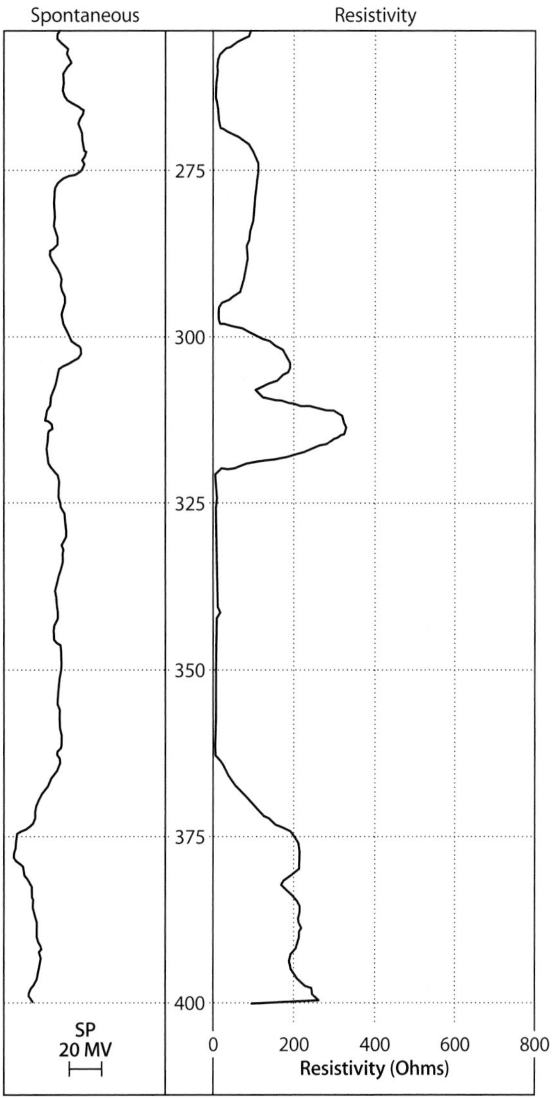

Fig. 5.8. Electric log

5.6 Geophysical Logging of Boreholes Associated with Spring Studies

Borehole logging provides in situ measurements of parameters related to physical characteristics of the rock formation, the fluid in it and the borehole. Geophysical logging is most productive when other methods are employed to characterize geologic and hydrologic parameters of the aquifer under investigation. Electrical, gamma, neutron, caliper, geothermal, downhole camera, and stereophoto techniques are available for logging and have been used with varying success.

Two types of electrical logging, spontaneous potential (or self potential) and resistivity, are obtained by lowering an electrode down a borehole. Spontaneous potential records the naturally occurring potential difference at various depths between a surface electrode and the borehole electrode. Resistivity is obtained by a source of current induced by electrodes to measure the potential difference at different depths. The resistivity log provides an apparent resistivity versus depth. Both properties must be measured by lowering electrodes, or sonde, in uncased wells, and can be interpreted to distinguish rock unit thickness and stratigraphic sequence, i.e. formation characteristics (Fig. 5.8).

Radioactive logging methods vary according to their sources and detectors. One type of radioactive logging, referred to as gamma logging, measures gamma radiation at a penetration distance of about 30 cm. The radiation flux is measured by a scintillation counter lowered down the borehole. Variations in flux can be correlated with rock unit boundaries, with clays and shales being commonly several times more radioactive than sandstones, limestones or dolostones. These measurements can be used in the geologic classification of the spring site.

Gamma-gamma and neutron-gamma logging techniques require artificial radiation sources and detectors and, hence, are less frequently used. Gamma-gamma logging has been successfully applied in the determination of rock density variations down boreholes. Neutron-gamma logging is a standard technique for soil moisture measurement, but it is less frequently used in boreholes. It provides a measure of the hydrogen abundance per unit volume of rock, and so is related to the abundance of water and to porosity (Fig. 5.9).

Caliper logging, using mechanically extended caliper arms on an instrument lowered into a borehole is standard procedure used on many drilled wells and helps define and evaluate variations in drillhole diameter. Caliper logs are mechanical devices used to determine the diameter of an uncased wellbore and are most often run in conjunction with other well-log types (acoustic, density, dipmeter, etc.). Caliper logging may also be used in uncased boreholes. This technique measures the variation of well diameter with depth, which helps to identify features such as solution openings, cavities, fractures, bedding planes and zones of weakness (Fig. 5.10). This can be a particularly helpful tool in areas where springs discharge from cavities in limestone.

Television downhole cameras of 84 mm have been lowered successfully down boreholes to view the fracture and cavities, irregularities in the diameter of boreholes, and, to a certain degree, the lithologic nature of the section drilled. Lighting problems provide limitations when observation distances exceed 30 cm.

Temperature logs measuring fluid properties are often run in association with electrical logs. They use a sonde with a thermocouple and record temperature variations with depth. This is useful because of the known

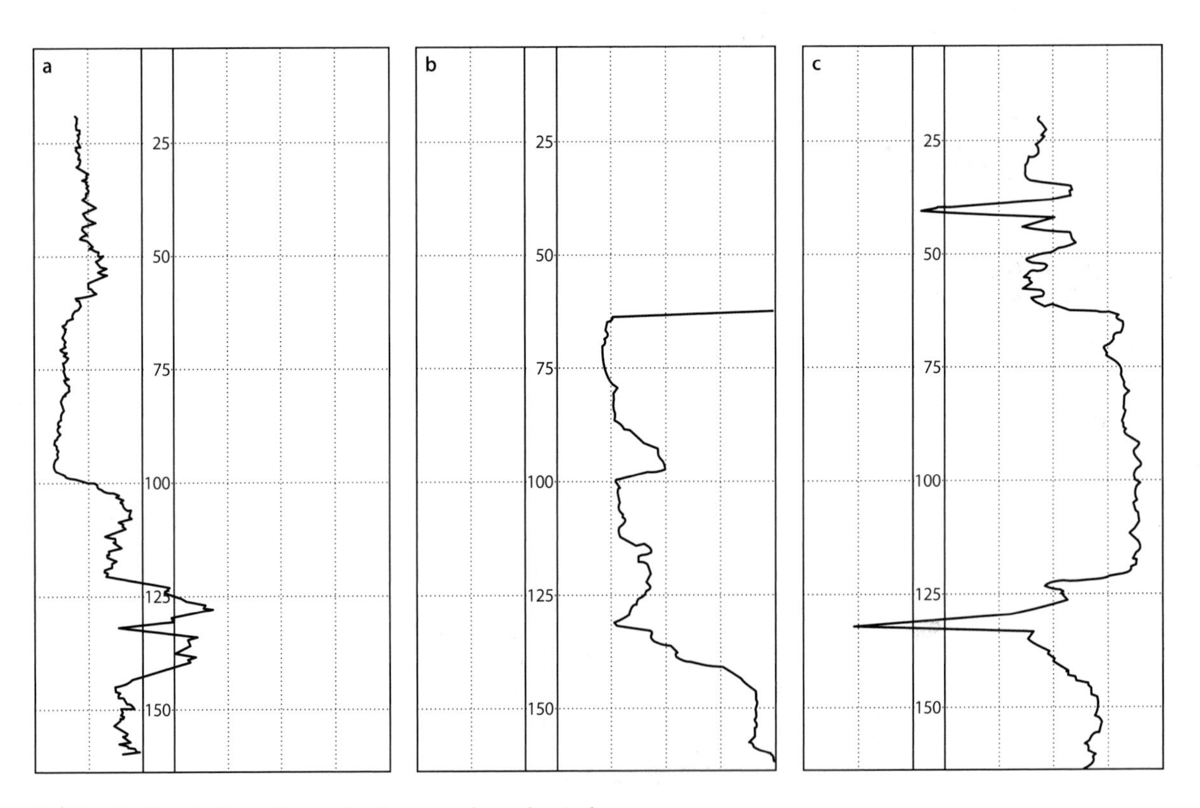

Fig. 5.9. Radioactive log. **a** Gamma log; **b** neutron log; **c** density log

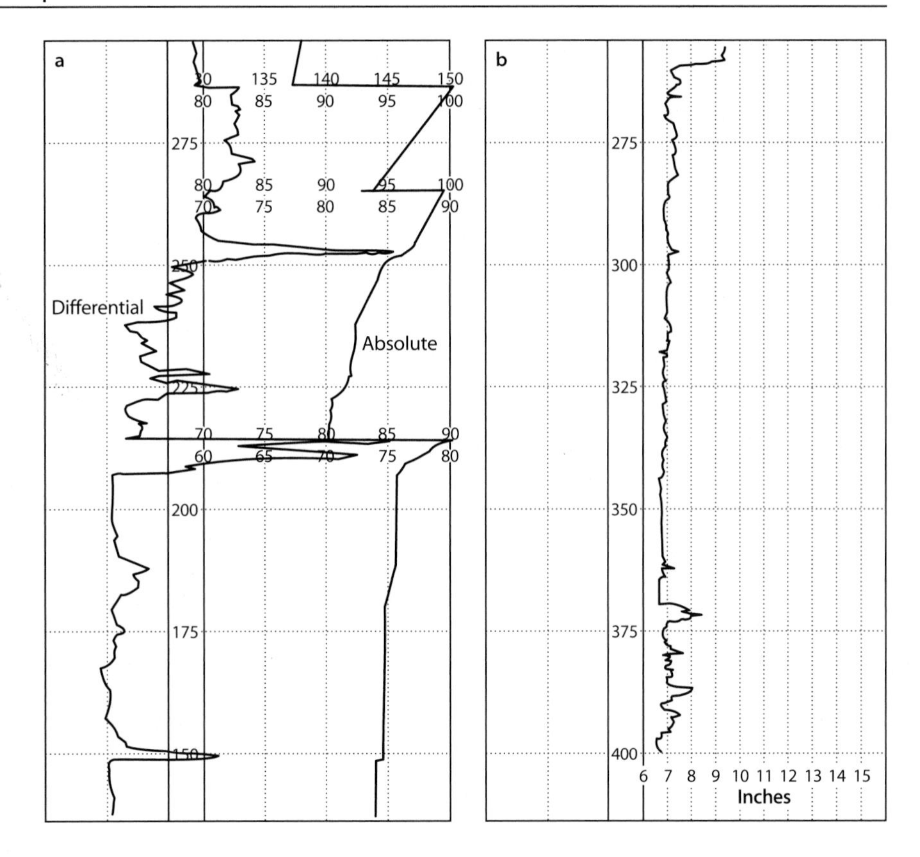

Fig. 5.10.
a Temperature log and
b caliper log

relationship between temperature and electrical conductivity and also because temperature variations can indicate discrete groundwater bodies sometimes related to water movement from different source areas (see Fig. 5.10).

5.7 Remote Sensing Techniques

Remote sensing is the use of reflected and emitted electromagnetic energy to measure the physical properties of distant objects and their surroundings. Thus, remote sensing methods are tools to inventory resources, monitor the environment, solve problems, and plan future investigations. Remote sensing data are recorded on a strip chart, magnetic tape or film. These data are processed, analyzed to obtain information, and interpreted to draw conclusions.

The origin of nonphotographic remote sensing can be traced way back to World War II with the development of radar and side-looking airborne radar (SLAR) and thermal infrared detection systems. Scientists at that time recognized that an exceptional amount of resource information could be obtained from high altitude aircraft and satellite photographs (Hemphill 1958; Morrison and Chow 1964).

Remote sensing data has been recorded by the world's first operational satellite system, ESSA 1, launched in February 1966, and Earth Resources Technology Satellite (ERTS 1 – called Landsat 1) in July 1972. The ERTS 1 is the first satellite designated specifically for study of earth resources. Earth and atmospheric data are acquired by Landsat 2 and 3, SMS/GOES, TIROS N, Nimbus 7, HCMM and NOAA 5. These data extend from the visible through the near infrared and thermal infrared to the microwave part of the electromagnetic spectrum.

A landsat scene enables geologists/hydrologists to look at features of regional size – river basins and systems, geologic structures, snow cover – and to see things that escaped their notice when they were standing right on top of them. Space-acquired data offer repetitive monitoring of large areas to complement conventional data which are both temporally and spatially limited.

Timely, synoptic data in conjunction with point measurements can be used to expand the information base to an entire watershed. However, application of remote sensing data requires an understanding of data needs, hydrogeologic phenomena and models, and capabilities afforded by remote sensing technologies.

Development of new methods for processing, analysis and interpretation of remote sensing data are widely used for a better understanding of diagentic and tectonic processes, as well as the effects of these processes on groundwater occurrence and movement. Landsat images (RBV and bond 4 MSS images) and digital data are used to detect large underwater springs, seeps, and shallow geothermal groundwater by anomalous snow melt patterns. Potential applications of remote sensing technology are as diverse as the scenes that might be photographically recorded while overflying the entire world.

Earth Resources Technology Satellite (ERTS) imagery, high- and low-level aerial photography, and side-looking airborne radar (SLAR) have been used to map surface manifestations of joints, fractures, and faulting. For geological and hydrogeological investigations, comprehensive overlaps are prepared to assist field geologists and hydrogeologists in detailed investigations of the features and to facilitate the planning of geophysical field exploration and drilling programs.

Interpretation of aerial photographs using stereophoto techniques provides information that could be used to delineate lithologic units, geomorphological features, historical land use, structural features (faults, fracture, joints), lineaments (alignments of drainage, topography or vegetation) and preferential flow zones.

National Aeronautics and Space Administration (NASA) flown aerial photography, ERTS data, and other aerial photography can be acquired from the Earth Resources Observations Systems (EROS) Center, which has been established by the US Geological Survey of the Department of the Interior at Sioux Falls, South Dakota.

5.8 Spring Hydrograph Analysis

The hydrograph is a graphic presentation of a plot of discharge from a hydraulic or hydrologic unit or system such as river, drainage basin, or spring versus time. The duration for which distribution of flow of the spring is sought varies from a few days to a year. A sufficient number of values should be plotted to indicate adequately all significant changes in the slope of the hydrograph. The study of the hydrograph of a spring before, during, and after its recharge area is affected by a storm is particularly helpful in determining the amount of natural replenishment that recharges the aquifer-sustaining spring; the shape of the outflow hydrograph recorded for a spring is a unique reflection of the response of the aquifer to recharge.

The form and rate of recession provide significant information on the storage, lithological composition, and structural characteristics (fissures, fractures, joints) of the aquifer system sustaining the spring. Therefore, analysis of spring hydrographs contributes to the understanding of the nature of the aquifer system and the operation of the drainage system. The understanding of the system is further enhanced by simultaneous analysis of both hydrographs (quantity variation) and chemographs (quality variations).

The duration and intensity of precipitation strongly influences the form of flood hydrographs for springs, but it is well known that basin characteristics such as shape, size, slope drainage density, lithology and vegetation modify the spring response. In addition, antecedent conditions of storage strongly influence the proportion of the rainfall input that discharges as spring flow and the lag between the input event and the output response. However, for a particular climatic regime, lithology emerges as one of the dominant controls on hydrograph form. Impermeable rocks yield strongly peaked hydrographs because of little storage and rapid discharge, whereas basins composed of highly permeable formations such as glacial outwash, limestone, and basalts tend to have flatter, broader, and more delayed responses. There is a continuum of rainfall-response functions in nature, with karst terrains with diffuse flow characteristics lying near the permeable end of the spectrum.

In karst systems there is considerable variety in the extent and degree of groundwater reservoir development as well as various mixes of antigenic and allogenic inputs; thus, there are many variations in the outflow responses of karst springs. Some outflow hydrographs are highly peaked, others are oscillatory, and many are broad and relatively flat. According to Smart (1983a,b), a flat-topped apparently truncated hydrograph is an indication that it is an underflow spring, the "emitting" peaked remainder of the hydrograph is emerging at an intermittent high water overflow spring draining the same aquifer. In interpreting hydrograph form, it is important to have a knowledge of the spring output, and to monitor the entire output of the drainage system.

5.9 Hydrograph Recessions

The output for a spring can show certain discharge responses to a given precipitation leading to a discrete recharge event over a basin. Those responses are characterized by: (1) a lag time before response occurs, (2) a rate of rise to peak output (the 'rising limb'), (3) a rate of recession as spring discharge returns towards its pre-storm outflow (the 'falling limb'), and (4) small bumps on either limb, although these as best seen on the recession (falling limb). These bumps are caused by interference and anomalies related to discharge from nearby wells and/or springs tapping the same aquifer and external stresses on artesian aquifers such as changes in barometric pressure.

When the hydrograph is at its peak, storage in the system is at its maximum. The rate of withdrawal of water from that storage is indicated by the slope of the subsequent curve.

Quantitative analyses of hydrograph recession is based on Eq. 5.2 given by Maillet (1905), who proposed that the discharge of a spring is a function of the volume of water held in storage and is described by an exponential equation. If this curve is plotted on semi-logarithmic graph paper, it is represented as a straight line with slope β:

$$O_t = Q_o \bar{e}^{\beta t} \tag{5.2}$$

or

$$\log Q_t = \log Q_o - 0.4343\, t^{\beta} \tag{5.3}$$

and

$$\beta = \frac{\log Q_1 - \log Q_2}{0.4343\,(t_2 - t_1)} \tag{5.4}$$

where
Q_t = discharge (m^3/s) at time t
Q_o = discharge (m^3/s) at time zero
t = time elapsed (in days) between Q_t and Q_o
e = base of the napierian logarithmic
β = recession coefficient (T^{-1})

The value of the recession coefficient β in Eq. 5.4 derives from the hydrogeological characteristics of the aquifer, especially effective porosity and transmissivity. When β is large and the half-flow period $t_{0.5}$ (defined as the time required for the baseflow of the spring to halve, hence $2Q_{t0.5} = Q_o$), is small, the recession is steep, indicating rapid drainage of conduit or porous media and underground storage. If no recharge is occurring but β is small and $t_{0.5}$ is large, then very slow drainage of the aquifer probably occurs from an extensive fissure or porous network with a large storage capacity and high resistance to recharge throughout.

It is important to perform recession analysis of the spring hydrograph to determine volume of water held in storage, lag time, and duration of meteorologic events which influence the discharge at the spring.

Recession analysis of Figeh Spring, Syria, which is one of the largest springs in the world, discharging in the range of 12–15 m^3/s was performed to evaluate the impact of regional and local variations in climatic conditions, lag time, and geophysical compartmentalization within the aquifer system on the storage and discharge spring. The study indicated that geologic setting and the variations in the intensity of rainfall and snowmelt across the recharge area affect the intensity of the impulse at discharge.

5.10 Water Tracing Techniques

The exceptional and complicated natural water phenomena in karst have excited the fancy of the average man as well as the educated specialists. Several connections between the ponors (sinkholes) and springs have been made by accidents and only seldom by conscious tracing by tracers, such as cottonhulls, sawdust, pine needles, water turbidity, charcoal from burnt-out places, etc. Gradually, these tracing methods have been improved, and more accurate observations and scientific evaluations have resulted.

Water tracing over a period of a century is one tool for the hydrogeologist to determine catchment boundaries, groundwater flow velocities, areas of recharge, and sources of pollution. Chemicals such as common salts, ammonium sulfate, and fluorescein have been used for water tracing for more than a century.

The first documented tracer experiment was performed by Tom Brink in 1869 by introducing aniline red dye to determine the hydraulic connection of the Danube River to Aach Spring in Hegane, Germany. Brink and Knop repeated the experiment in 1877 by simultaneously injecting three tracers; shale oil, fluorescein (this was the first hydrogeological use of this very well known tracer) and sodium salt into the river. All three tracers appeared in Aach Spring, confirming the hydraulic connectivity of river water with Aach Spring.

Several studies were performed by injecting various tracers such as fluorescein dyes, salt, colored spores, and nonpathogenic germs to determine underground flow velocity and the catchment area of the Aach Spring.

Extensive research is being carried out to understand groundwater flow in karstified carbonate rock. Methods used include tracing techniques and isotopic studies to provide definitive knowledge of the geology, stratigraphy and structure, degree of solutioning and interconnectively of fracture systems, which control groundwater occurrence and flow in karstified carbonate rocks.

These studies identify in detail the relationship of bedding, fissure, fracture and joint systems, folding and dissolutioning processes, and provide relevant information to support more precise interpretation of quantitative results.

Back and Zötl (1975) reviewed the full range of modern techniques from naturally occurring tracers to artificially introduced radioisotopes. Two classes of water tracing agents are available: natural tracers, which include flora and fauna (principally microorganisms), ions in solution, and environmental isotopes; and artificial tracers, which include radiometrically detectable substances, dyes, salts, and spores.

5.11 Use of Naturally Occurring Tracers

5.11.1 Naturally Occurring Microorganisms

Bacteriological and virological examination of a spring can be undertaken to establish the hygienic quality of water and, if contaminated, to help trace possible sources of pollution. Karst aquifers are notoriously bad water filters; thus, transmission of microorganisms is expected. There is extensive literature on the movement of bacteria through porous media (Romero 1970; Gerba et al. 1975).

5.11.2 Ions in Solution

Chloride has long been used as a natural tracer to determine the freshwater-saltwater interface in coastal aquifers and to detect possible intrusion of seawater into wells used for water supplies in coastal areas. The location of water emerging at springs can be determined by concentration and changes in the concentration of different water quality characteristics. This is the basis of spring chemograph separation. The nature of the flow system sustaining the spring is also interpretable from such data (Shuster and White 1971; Bakalowicz 1977).

5.11.3 Environmental Isotopes

The use of naturally occurring environmental stable and radioactive isotopes in groundwater hydrology has made enormous advances in recent years (International Atomic Energy Agency 1981, 1983, 1984; Fritz and Fontes 1980; Lloyd 1981). Hydrogen possess ^{1}H, ^{2}H (deuterium, D) and ^{3}H (tritium, T), the last being radioactive. Oxygen isotopes include two of interest in hydrology: ^{16}O and ^{18}O, both being stable. Because hydrogen and oxygen constitute the water molecule, they are the best conceivable tracers of water. They occur in various combinations, such as $[^{16}O]$, H_2O, $[^{18}O]H_2O$, HDO, and HTO. Other important environmental isotopes commonly used in groundwater investigations are: ^{12}C, ^{13}C, and ^{14}C. Less frequently used are isotopes of argon, chlorine, helium, krypton, nitrogen, radium, radon, silicon, sulphur, thorium and uranium (International Atomic Energy Agency 1983, 1984).

The main applications of environmental isotopes to groundwater investigations are to provide a signature of a particular groundwater type that can be related to its area of origin, to identify mixing of waters of different provenance, to provide velocities and directions, and to provide data on underground residence time of the water.

5.11.4 Radioactive Isotopes

Radioactive isotopes are unstable and undergo nuclear transformation, emitting radioactivity. Their decay is spontaneous and unaffected by external influence. It occurs at a unique rate for each radioisotope, defined by its half-life, $T_{1/2}$, which is the time required for half of the radioactive atoms to decay, according to an exponential law:

$$N = N_0 e^{-\lambda t} \tag{5.5}$$

where

N = number of radioactive atoms present at time t

N_0 = number of radioactive items present at the commencement of decay

λ = half-life or decay constant

The half-life for ^3H is 12.43 years and for ^{14}C is 5730 years. Consequently, ^3H is useful for dating waters up to about 50 years in age, whereas ^{14}C has an upper limit of approximately 35000 years (Fontes 1983).

Interpretations of radioisotope data at a spring depends very much on the model of flow and mixing adopted (Yurtsever 1983). Several alternatives have been suggested with the piston flow model at one extreme and the completely mixed reservoir model at the other. Piston flow assumes that recharge occurs as a point injection, and that a discrete slug of tracer moves through the groundwater system. In nature, this never occurs perfectly because of mixing and dispersion effects; nevertheless, the flow behavior of well-developed karst is sufficiently similar to it. By contrast, the completely mixed reservoir model assumes uniformity at all times, each episode of recharge mixing instantaneously. It also assumes stationary conditions with respect to reservoir volume, discharge and infiltration rate.

Actual tracer behavior in most groundwater systems lies between these two extreme cases. Dispersive models have also been proposed to describe intermediate situations, taking into account mixing and dispersion within the system but assuming that pulse variations in tracer output can be related back to concentration variations in recharge events.

In many aquifers, including karst, the water may have components of different residence time, and to assign one mean age may be misleading. This point was well made by Siegenthaler et al. (1984) in regard to a spring in the Swiss Jura. Strong evidence showed that in periods of base flow a relatively homogeneous old water sustains the spring, whereas after a storm or meltwater event rapid runoff of new water is also important. It is, therefore, important to determine the average residence time of the older reservoir water.

Tritium concentration has been largely thought of as a means of dating water but, in view of the currently decreasing ^3H values in reservoirs, it is becoming less important for that purpose. Rozanski and Florkowski (1979) and Salvamoser (1984) suggest that as a consequence, the use of environmental krypton-85 ($T_{1/2} = 10.8$ yr) should be considered for dating. In several respects, it is an ideal tracer: its amount is still increasing, it shows no marked seasonal variations, it has virtually no sinks outside of radioactive decay, and its atmospheric concentration in the Southern Hemisphere is only 10–20% lower than in the North.

5.11.5 Stable Isotopes

Stable isotopes undergo no radioactive decay; D, ^{18}O and ^{16}O occur in the oceans in concentrations of about 320 mg/l HDO, 2000 mg/l $[^{18}O]H_2O$, and 997680 mg/l $[^{16}O]H_2O$. Variations in these proportions are measured by mass spectrometry and compared with the composition of standard mean ocean water (SMOW). The ratios of D/H and $^{18}O/^{16}O$ are expressed in delta units, which are parts per thousand (per mil) deviations of the isotopic ratio from these standards:

$$\delta\ 0/100 = \frac{R_{\text{sample}} - R_{\text{standard}} \times 1000}{R_{\text{standard}}} \tag{5.6}$$

where
R = isotopic ratio of interest

Differences in the isotopic composition of water samples reflect the fraction processes that occur in the hydrological cycle. The heavy isotope molecules (HDO, $[^{18}O]H_2O$) have slightly lower saturation vapor pressures than the ordinary water molecule, $[^{16}O]H_2O$; hence, when changes of state occur during evaporation and condensation, a slight fractionation takes place. For example, when evaporation occurs from an open water surface, the vapor is depleted in heavy isotopes in relation to the remaining unevaporated water, whereas when condensation occurs the initial precipitation is slightly enriched so that later precipitation becomes increasingly depleted with respect to SMOW.

Temperature is an important factor in fractionation: the lower the temperature, the greater the depletion in heavy isotopes. This influence is reflected in both latitude and altitude differences in heavy isotopic composition of precipitation (Dansgaard 1964). Back and Zötl (1975) identified four general rules for stable isotope fractionation:

1. In regions of moderate climate, precipitation is isotopically lighter in winter than in summer.
2. Precipitation becomes isotopically lighter with increasing latitude and altitude.
3. D and ^{18}O in precipitation shows a good linear correlation.
4. Enrichment of D and ^{18}O occurs in lakes because of evaporation.

The relationships that exist in the natural distribution of stable isotopes in water have been valuable in determining the recharge area and recharge seasons of groundwaters, especially where allogenic inputs are involved. Autogenic recharge is usually so well mixed in the epikarst that seasonal patterns and recharge events are difficult to identify (Yonge et al. 1985; Even et al. 1986). An application of isotope hydrology elucidates the complex relations between confined karst groundwaters, geothermal waters and groundwaters in shallow, basaltic, and alluvial aquifers (Issar et al. 1984).

5.11.6 Use of Artificial Tracers

Artificial tracers should have the following criteria to be suitable for water tracing experiments:

1. Nontoxic to the handlers, to the karstic ecosystem, and to potential consumers of the labelled water
2. Soluble in water, in the case of chemical substances, with the resulting solution having approximately the same density as water
3. Neutral in buoyancy and, in the case of particulate tracers, sufficiently fine to avoid significant losses by natural filtration
4. Unambiguously detectable in very small concentrations
5. Resistant to adsorptive loss, cation exchange, and photochemical decay; and to quenching by natural affects such as pH change in temperature variation
6. Susceptible to quantitative analysis
7. Quick to administer and technologically simple to detect
8. Inexpensive and readily obtainable

Only a few of the tracers satisfy all of the above criteria, but several do satisfy to a significant extent and are used for tracing experiments. Nontoxicity is the most important characteristic that has to be evaluated before any artificial tracer is used for tracing experiments.

5.11.7 Radiometrically Detectable Tracers

Guidelines provided by IAEA (International Atomic Energy Agency 1984) for the selection of artificial isotopes for water tracing are as follows:

1. The isotope should have a life comparable to the presumed duration of the observations. Long-lived isotopes will create pollution, and health hazards.
2. The isotopes should be resistant to adsorption by soils and rock.
3. The radioactivity emitted by isotopes should be measurable in the field. Therefore, gamma emitters are preferred in general, although β emitters are also suitable for tracing experiments.
4. The isotope must be readily available at reasonable cost.

Isotope ^3H is identified as perfect tracer because it is a part of the water molecule itself. It is also identifiable in a very low concentration, to $1:10^{15}$. However, based on the criteria 1 and 3, ^3H is eliminated. In situ detection of ^3H is not possible.

Isotopes ^3H, ^{82}Br as a bromide, and ^{35}S as a sulfate are considered to be the best isotopes for tracing experiments (Lallemand and Grison 1970).

The public health disadvantages of using radioactive isotopes may be overcome for some species by the post-sampling activation analysis technique. The method involves injection of an initially non-radioactive tracer that is later sampled and irradiated with neutrons in a reactor. If the tracer is present in the sample, it is detectable by its activity. As with the radioisotope tracers, chelation with EDTA (ethylene dinitrotetra acetic acid) forms a negatively charged complex ion that minimizes adsorption losses. Added advantages of the activation technique are that it makes possible the simultaneous use of several tracers (for example, Br, In, La) and problems of half-life can be disregarded.

Nevertheless, despite the advances of the last 25 years or so, Burdon and Papakis (1963) and Back and Zötl (1975) recommend that before turning to radioactive tracers, which are costly, hazardous, and need skilled handling by personnel supported by atomic laboratories, it is better to try to use colored or chemical tracers for tracing experiments.

5.11.8 Dyes as Water Tracers

Artificial dyes are the principal and most successful water tracers at the present time. Dyes have been used to trace undergroundwater for more than a century. The green dye fluorescein ($C_{20}H_{12}O_5$) was discovered in 1871 and first used as a karst water tracer at a stream/sink of the Upper Danube River in 1877. Tracing with fluorescein advanced considerably when it was determined that the dye is adsorbed by charcoal grains from which it may later be released by an elutant of potassium hydroxide in 5% ethanol. The field requirement is simply to suspend small mesh bags containing a few grams of granular activated carbon in a moderate current in the monitored springs. Rhodamine WT is also adsorbed and may be eluted with a warm solution of 10% ammonium hydroxide in 50% aqueous 1-propanol (Smart and Brown 1973). The great advantages of this technique are that detector bags can be changed when convenient and many sites may be easily monitored. There are no time constraints on when analysis must be undertaken, although air drying of detector bags is required if more than a few days will elapse between collection and examination. Charcoal bag detectors also work effectively in the sea should resurgences be submarine or intertidal. With careful elution and use of a fluorometer, the charcoal method may be used semi-quantitatively (Smart and Friederich 1982).

The most recent important element in dye tracing techniques has been the advent of quantitative fluorometric procedures following the work of Kass (1967) in Europe and Wilson (1968) in North America. Kass examined two dyes (uranin and eosine, yellow and red, respectively) and two orange dyes (rhodamine B and sulphorhodamine B) and showed uranin to possess the strongest fluorescence of the four and to be detectable in solutions down to 0.01 mg/m^3.

Fluorescent substances emit light immediately upon irradiation from an external source. The emitted fluoresced energy usually has longer wavelengths and lower frequencies than the adsorbed irradiation. This property of dual spectra makes fluorometry an accurate and sensitive analytical tool, because each fluorescent substance has a different combination of excitation and emission spectra (Wilson et al. 1986). Since some naturally occurring substances, such as plant leachates, phytoplankton, and some algae also fluoresce, it is important to know the background fluorescence of the water being traced before any experiments are conducted. Industrial and domestic wastes also introduce background problems. Natural substances tend to fluoresce in the green wave band; thus the use of dyes with an orange emission overcome problems of possible misidentification.

Although elutants from charcoal may be used in a fluorometer, much more significant dye concentration data are obtained from direct sampling of the spring waters. These can be taken by hand sampling or by automatic water samplers collecting at fixed intervals or taking time-integrated samples.

The most accurate quantitative water tracing is obtained by continuous fluorometry, that is, passing water from a spring or other sampling point directly and continuously through an adapted fluorometer. The shape of the dye discharge pulse is then compared with other continuously monitored variables such as discharge rate, conductivity, and temperature.

References

Atkinson TC (1985) Present and future direction in karst hydrogeology. Ann Soc Geol Belgique 108:293–296

Back W, Zötl J (1975) Application of geochemical principles, isotopic methodology, and artificial tracers to karst hydrology. In: Burger A, Dubertret L (eds) Hydrogeology of karstic terrains International Union of Geological Sciences, Series B, 3:105–121

Bakalowicz M (1977) Etude du degre d'organisation des ecoulements souterrains dans les aquiferes carbonates par une methode hydrogeochimique nouvelle. CR Acad Sc Paris, 284D, 2463–2466

Burdon DJ, Papakis N (1963) Handbook of karst hydrogeology. Institute for Geology and Subsurface Research/FAO, Athens

Dansgaard WF (1964) Stable isotopes in precipitation. Tellus 16:436–449

Deutch M, Wiesnet DR, Rango A (1981) Satellite hydrology. Fifth Annual William T. Pecora Memorial Symposium on Remote Sensing, sponsored by AWRA. Technical publication TPS81-1

DeWiest RJM (1965) Geohydrology. John Wiley, New York, 366

Even H, Carmi I, Magaritz M, Gerson R (1986) Timing the transport of water through the upper vadose zone in a karstic system above a cave in Israel. Earth Surf Proc Landf 11:181–191

Ferris JG, Knowles DB, Brown RH, Stallman RW (1962) Theory of aquifer tests. US Geol Survey Water Supply Paper 1536–E, 174 p

Fetter CW Jr (1980) Applied hydrogeology. Charles E Merrill Publishing Co, Columbus, OH, 488 p

Fontes J-Ch (1980) Environmental isotopes in groundwater hydrology. In: Fritz P, Fontes J (eds) Handbook of environmental isotope geochemistry. Vol 1:75–140. Elsevier, Amsterdam

Fontes J-Ch (1983) Dating of groundwater. In: Guidebook on nuclear techniques in hydrology. Report Series No. 91: 285–317. IAEA, Vienna

Ford D, Williams P (1992) Karst geomorphology and hydrology. New York, Chapman and Hall, p 601

Fritz P, Fontes J-Ch (eds; 1980) Handbook of environmental isotope geochemistry. Vol 1: The terrestrial environment. Elsevier, Amsterdam

Gerba CP, Wallis C, Melnick JL (1975) Fate of wastewater bacteria and viruses in soil. J Irrig and Drainage Div, Am Soc Civ Eng 101:157–174

Hemphill WR (1958) Small scale photographs in photogeologic interpretation. Photogrammetric Engineering 24(4): 562–567

Hötzl H (1996) Origin of Danube-Aach system. Environmental Geology 27(2):87–96

International Atomic Energy Agency (1981) Stable isotope hydrology. Technical Report Series No. 210. IAEA, Vienna

International Atomic Energy Agency (1983) Guidebook on nuclear techniques in hydrology. Technical Reports Series No. 91. IAEA, Vienna
International Atomic Energy Agency (1984) Isotope hydrology 1983. IAEA, Vienna
Issar A, Quijana JL, Gat JR, Castro M (1984) The isotope hydrology of the groundwaters of central Mexico. J Hydrol 71:201–224
Käss W (1967) Erfahrungen mit Uranin bei Farbversuchen. Steir Beitr z Hydrogeologie 18/19:123–340
Käss W (1969) Schriftum zur Versickerung der oberen Donau zwischen Immendingen und Fridingen (Südwestdeutschland). Steir Beitr Hydrogeol 21:215–246
Knop A (1878) Über die hydrographischen Beziehungen zwischen der Donau und der Aachquelle im badischen Oberlande. Neues Jahrb Mineral Geol Petrogr 1978:350–363
Lallemand A, Grison G (1970) Contribution à la sélection de traceurs radioactifs pour l'hydrologie. Isotopes in hydrology, Proc Symp 833–9. IAEA, Vienna
LaMoreaux PE (1991) History of karst hydrogeological studies. Proceedings of International Conference on Environmental Changes in Karst Areas. International Geological Union/Union International Sciences, Italy, pp. 215–229
LaMoreaux PE, Hughes TH, Memon BA, Lineback N (1989) Hydrogeologic assessment – Figeh Spring, Damascus, Syria. Environmental Geology and Water Sciences 13(2):77
Lloyd JW (1981) Environmental isotopes in groundwater. In: Lloyd JW (ed) Case-studies in groundwater resources evaluation. Clarendon, Oxford, 113–132
Maillet E (1905) Essais d'hydraulique souterraine et fluviale. Hermann, Paris
Meinzer OE (1923) The occurrence of groundwater in the United States, with a discussion of principles. Department of the Interior, US Geological Survey Water Supply Paper 489, Washington DC
Meinzer OE (1927) Large springs in the United States. US Geological Survey Water-Supply Paper 557, Government Printing Office, Washington 94 p
Memon BA (1995) Quantitative analysis of springs. Environmental Geology 26:111–120
Morrison A, Chown MC (1964) Photography of the Western Sahara Desert from the Mercury MA-4 Space Craft. US National Aeronautics and Space Administration Report CR-126
Newton JG (1976) Early detection and correction of sinkhole problems in Alabama, with a preliminary evaluation of remote sensing applications. Alabama Highway Department, Bureau Research and Development, Research Report No. HPR-76, 83 p
Powell WJ, Copeland CW, Drahovzal JA (1970) Delineation of linear features and application to reservoir engineering using Apollo 9 multispectral photography. Alabama Geological Survey Information Series 41, 37 p
Romero JC (1970) The movement of bacteria and virus through porous media. Groundwater 8(2):37–48
Rozanski K, Florkowski T (1979) Krypton-85 dating of groundwater. Isotope hydrology 1978. Proc Symp Neuherberg 1978, Vol. 2, p. 949. IAEA, Vienna
Salvamoser J (1984) Krypton-85 for groundwater dating. Isotope hydrology 1983:831–932. IAEA, Vienna
Shuster ET, White WB (1971) Seasonal fluctuations in the chemistry of limestone springs: A possible means of characterisizing carbonate aquifers. J Hydrol 14: 93–128
Siegenthaler U, Schotterer U, Muller I (1984) Isotopic and chemical investigations of springs from different karst zones in the Swiss Jura. Isotope hydrology 1983:153–172. IAEA, Vienna
Smart CC (1983a) Hydrology of a glacierised alpine karst. PhD Thesis, McMaster University
Smart CC (1983b) The hydrology of Castleguard Karst, Columbia Icefields, Alberta, Canada. Arctic and Alpine Res 15(4):741–786
Smart PL, Brown MC (1973) The use of activated carbon for the detection of the tracer dye Rhodamine WT. Proc 6 Internat Speleo Cong Olomouc, CSSR, Vol. 4:285–292
Smart PL, Freiderich H (1982) An assessment of the methods and results of water-tracing experiments in the Gunung Mulu National Park, Sarawak. Trans Brit Cave Res Assoc 9(2):199–212
Theis CV (1935) The relation between the lowering of piezometric surface and the duration of discharge of a well using groundwater storage. Am Geophysical Union Trans
White, WB (1988) Geomorphology and hydrology karst terrains. Oxford University Press, New York, 464
Wilson JF (1968) Fluorometric procedures for dye tracing. Techniques of water resources investigations of the United States Geological Survey TWI 03-A12, pp 31, Washington DC
Wilson JF, Cobb ED, Kilpatrick FA (1986) Fluorometric procedures for dye tracing. Techniques of water resources investigation, 03-A12, US Geol Survey, pp 34, Washington DC
Yonge CJ, Ford DC, Gray J, Schwarcz HP (1985) Stable isotope studies of cave seepage water. Chem Geol 58: 97–105
Yurtsever Y (1983) Models for tracer data analysis. In: Guidebook on nuclear techniques in hydrology. Technical Report Series No. 91:381–402. IAEA, Vienna
Zheng Y, Bai E, Libra R, Rowden R, Liu H (1996) Simulation of spring discharge from a limestone aquifer in Iowa, USA. Hydrogeology Journal 4(4):41–54

Chapter 6
Uses of Spring Water

Chapter 6
Uses of Spring Water

Lois D. George

6.1 Introduction

The many uses of spring water around the world span history. Spring water has been used for basic survival, medicinal purposes, and for man's entertainment, pleasure, and dalliance. Hippocrates and ancient Greek physicians were versed in the health benefits of mineral water therapy. In early recorded history, the Egyptians and Arabians discussed the use of mineral waters for healing the ill. Mythology and legend date the thermal springs of Bath, England, to 800 B.C. Hannibal refreshed himself with bubbling spring water at Vergeze on his way to attack Rome in 218 B.C. Therapeutic application of mineral waters was very popular in the late 1800s and early 1900s. Significant expenditures were made throughout Europe and the United States to develop lavish resorts and vacation spots at new famous spas near mineral springs.

The Romans may have initiated the uses of mineral waters, but the French are traditionally the modern developers and promoters of bottled waters. Evian was exported to the United States as early as 1905. Mountain Valley, of Hot Springs, Arkansas, has been bottled since 1871. Poland Spring water of Maine has been distributed since the mid-1800s. Spring water has become the health drink of today. Uses of spring water through time, famous springs and famous consumers of spring water, and the therapeutic attributes of spring water are summarized in this chapter. Research includes technical, nontechnical, and trade information. This chapter provides a retrospective of historical aspects of the development of spring waters, a concise summary of medicinal characteristics of spring water, and insight into the commercial enterprise of bottling water.

6.2 Bottled Water

The traditional reason for drinking bottled or mineral water was to supplement one's health. Europeans, historically the aristocratic Romans, developed somewhat of a preoccupation with consumption and immersion in mineral waters. Sipping Vichy water was associated with the wealthy and considered high fashion. The Romans may have initiated the uses of mineral waters, but the French are traditionally the modern developers and promoters of bottled waters. Bottled water has become the health drink of today. As a result of environmental and health conscientiousness, rejuvenated interest and widespread consumption of "pure waters" has led to increased "production" of bottled water.

According to Green and Green (1985), authors of *The Best Bottled Waters in the World*:

> To many Europeans, of course, this is nothing new. The French and Belgians are already drinking over 50 l (13 US gallons) a head of mineral waters annually (compared to a mere splash of 1 l in Britain). West Germans and Austrians quaff mineral water in large quantities, often mixed with white wine as a spritzer. And in Spain and Portugal it is equally customary to order a bottle of wine and a bottle of *agua minerale* to complement a good meal in a restaurant. The natural Italian reaction to any crisis, whether a strike or an earthquake, is to rush out to stock up on bread, wine, and mineral water, the basic necessities of life. An Italian going into a hospital or clinic traditionally arrives with a bottle of pure Sangemini tucked under the arm.
>
> The connoisseur will choose the right water for the right occasion, just like wine. Thus the sharp sparkle of Perrier from France or Apollinaris from West Germany makes a good aperitif, the lighter natural carbonation of Frances' Badoit, Italy's Ferrarelle and Belgium's Bru goes well with food; and the pure still waters of Evian, Vittel or Volvic from France, Panna from Italy or Font Vella from Spain are not only good table waters, but recommended for baby formulas. Chic Frenchwomen, anxious to preserve their figures, keep a bottle of the more highly mineralized Contrexeville on hand as a diuretic, while Vichy Celentins may be preferred to ease the digestion.

Currently there are hundreds of bottled water companies in the world, over 250 in the USA, and 150 in Spain, and the industry is a multi-billion dollar business. The products range from still water to sparkling water, carbonated water or spring water, all from a variety of sources around the world. The packaging and marketing of bottled water is aggressive and attractive. The text by Green and Green (1985), subtitled *The purest, most delicious, and healthful waters from Ain Sofat to Zurich*, provides historic and current scientific, medicinal, and financial information about the bottled water industry and describes mineral waters from selected sources in over 30 countries in Europe, the Middle East, southeast Asia, Russia and China, and North and South America. This text, somewhat outdated, demonstrates the longevity of consumption and the health-motivated aspects of mineral waters.

Commonly and historically in the Western Hemisphere, consumption of mineral waters has been associated with the aristocracy or the "infirm". In more recent decades, bottled water distribution has been linked to health food coops and specialty stores. Today, however, every grocery store in the United States devotes an entire section of shelves to bottled water (Fig. 6.1).

The European culture and market has a rich history of famous bottled waters. France is the leader in the bottled water industry – first in production, first in consumption, first in exports, and first in regulating the purity and identity of the mineral waters. Regulation of mineral waters began in the late 18th century when, in 1781, the first decree governing the waters was issued. In 1856 a law was passed enabling the state to declare a water source d'intérêt public, thus indicating that the waters were beneficial to health. Mineral waters were initially bottled such that those participating in treatment at the various resorts and spas in France, such as Evian-les-Bains or Contrexeville or Vittel, could continue their treatment at home or provide others with water for their ailments (Green and Green 1985).

Fig. 6.1. Bottled water of all brands and shapes are available in local grocery stores

In France, a precondition for bottling and commercial sales of mineral waters is that the waters can be consumed in unlimited quantities; the mineral content can not be of sufficient magnitude to require a daily limit as if the water were medicinal. The Greens' research indicates that approximately 50 of the 1200 known sources of mineral waters in France meet the criteria for bottling and distribution. Water from the other sources is consumed only at the spas and under medical supervision.

Spring waters, les eaux de source, are also bottled under strict regulatory requirements. However, the spring waters are not generally promoted for therapeutic properties but as refreshing waters. The spring waters are usually bottled and sold locally.

The Badoit Spring, known to the Romans, near Lyons in central France, was declared d'intérêt public in 1897, at which time the plant was producing 15 million bottles a year. In the 17th century, local doctors used and recommended consumption of the spring waters. In 1837, the water was first bottled when the spring was purchased by Auguste Saturnin Badoit. By 1859, the plant was producing over 1 million bottles a year. In 1973, Badoit was the first carbonated water to be bottled in plastic. Sales of Badoit have grown so much over the years that the company's geologists are in search of other fissures in the granite source rock to provide supply; the Ministry of Mines has established limits on the quantity of water that can be extracted annually from the presently used fissure (Green and Green 1985).

Contrexeville, one of France's top selling bottled waters, is from springs in the Vosges hills. The springs were discovered in the 18th century and in 1861 the springs were declared *d'intérêt public*. The waters which emanate from the springs are 11 °C (51.8 °F) and are high in magnesium, calcium, and sulfates. Contrexeville was purchased by the Perrier group in 1954. The demand for the product and the sophistication of the bottling facility is such that raw plastic powder is purchased for the manufacture of the bottles at the facility.

Evian is recognized as France's best selling still water. The therapeutic benefits of the spring water were first recognized in 1789 for easing the kidney stones of the Comte de Lesser, during the French Revolution. By 1815, baths and a hotel were established. With the construction of a railway to Evian in 1890, distribution of the waters began. Evian was exporting to the United States as early as 1905. The spring was declared *d'intérêt* public in 1926. The water is noted for its purity and low mineral content.

The familiar shape of the Perrier bottle was designed by one of the first owners and business developers of the source Perrier. Sir John Harmsworth, paralyzed from the waist down in an automobile accident, used Indian clubs in exercising after the accident. Harmsworth and Dr. Perrier were the "fathers" of the world-renown sparkling mineral waters. The history of the Perrier water includes the military campaigns of Hannibal and his elephants refreshing themselves in the bubbling pool at Vergeze on his way to attack Rome in 218 B.C. Archeological remains of Roman origin indicating use of the spring by the ancient Romans, and later Emperor Napoleon III granted exploitation of the spring in 1863. Napoleon designated the spring as a national treasure. Dr. Perrier and Harmsworth initiated the production and exportation of the spring waters and Perrier became famous for its sparkling character and low mineral content. The aggressive marketing of Perrier in the United States in the late 1970s resulted in great competition to the existing beverage industry, and brought about legal challenges to Perrier of false advertising, regarding its advertisement about its "naturally sparkling" character (Fig. 6.2).

The French consume 55 l (14.5 gal) of mineral water per capita annually and over 70 l (18.5 gal) including the *les eaux de source* or natural spring water. Three groups,

Fig. 6.2. **a** Source Perrier, Vergeze, France, about 15 km southwest of Nimes. One of the most popular international bottled waters. **b** Jardins De Le Fontaine at Nimes. An area of artesian springs that have been developed since Roman Times. **c** Old Roman Baths at Nimes Spring. **d** Perrier can dispenser machine for spring water

Evian, Perrier, and Vittel, control over 90% of mineral water sales. Annual incomes of $400 million, $125 million, and $110 million, respectively, are realized. Forty-five percent of the water from Source Perrier is exported to 150 countries (Green and Green 1985).

In 1853, George Kreuzberg, a farmer in the Ahr Valley, Germany, started selling spring water in earthenware bottles. The water from the Apollinaris Spring (named after St. Apollinaris by Kreuzberg) was shortly thereafter termed the "Queen of Table Waters", when a London shipowner arranged exclusive exports rights from Kreuzberg. By 1900, over 27 million bottles of Apollinaris were sold, of which over 25 million were exported. Consumption by the Germans did not exceed exports until the 1950s.

The German mineral water market is unique and in contrast to other parts of Europe for three reasons. The renown of Apollinaris with foreign consumers compared to native drinkers is in stark contrast to the history of bottled water in other European countries such as France and Italy. Sales of the German bottled waters are predominantly regional with limited export business. Furthermore, Germans' consume predominantly carbonated bottled water rather than the still waters preferred throughout the remainder of Europe. By law, German waters are divided into two categories: *Heilwasser* or health waters and *Quelle*, spring waters or table waters.

Fachingen water, licensed as a *Heilwasser*, is the favored health water by the Germans. The German poet Goethe wrote of the water in 1843 "the next four weeks are supposed to work wonders. For this purpose I hope to be favored with Fachingen water and white wine, the one to liberate the genius and the other to inspire it." In the 18th century, the water was bottled in stone jars, sealed with corks and dispatched. Today, the water is bottled in the plant adjacent to the Fachingen well. Only one percent of production is exported (Green and Green 1985).

Apollinaris, Gerolsteiner Sprudel, and Überkinger are the top three mineral waters sold. Gerolstein Spring is the source of the Gerolsteiner Sprudel water, which was first bottled in the mid-19th century. The bottles were initially stone and replaced with glass in the 1930s. Only 2.5% of production is exported. The company also bottles a *Heilwasser*, St. Gero, which competes for sales with the Fachinger.

Four springs producing four quality types of water are located at Bad Überkingen.

> The Überkinger water in the Alb Mountains of Baden Württemberg was known in the middle ages, when visitors bathed in its hot springs and drank the cooler waters. From the end of the Thirty Years War (around 1750), the drinking water was put into stone jars and sold in the locality and by 1870 a glass bottling plant and a spa with thermal treatments had developed. But it was only after 1950 that the bottling company at Bad Überkingen, armed with the communal bottle of all German mineral water producers, which can be returned to any bottler and used up to 40 times, and good modern roads, was able to take its water all over southern Germany and especially to the prosperous triangle of Stuttgart, Munich and Nuremberg (Green and Green 1985).

Italy is rich in mineral waters, with each region having at least one commercially bottled water. As a valued resource, the mineral waters belong to the Italian nation, as natural patrimony, and bottling companies are granted concessions. The ancient Romans initiated interest in and developed a way of life in "taking of water" with their preoccupation of the idea of *mens sana in corpore sano* – a healthy mind in a healthy body. Italian law covers the definition and supervision of waters for bottling and for spas. The laws required to determine water purity were enacted in 1919.

Italy's leading water groups are San Pellegrino, Sangemini, Ferrarelle, and Bognanco. Marketing and distribution of the products are hampered by the geography and topography of Italy, therefore the consumption of the three leading products and local waters is somewhat localized.

Many of the sources commercially bottled today were well known in Roman times. Ferrarelle Spring was reputed to have provided waters to Hannibal and his troops while resting before sacking a nearby Roman village. Pliny the Elder and Cicero praised the waters of Ferrarelle for its qualities – an aid to digestion, eased kidney stones, and mingles well with wine such that the Romans could drink all day and night without hangovers. Since 1925, the Farrarelle waters have been controlled by the Sangemini group. Export sales only amount to about 10% of distribution.

Acqua Panna, Italy's top bottled water, was first bottled in the 1880s. The water is piped 1128 m down the slope of Mount Gassaro in the Tuscan hills near Florence. In the early days of bottling, the water vessels were demijohns and fiaschi – straw covered glass flasks. Corks were used as stoppers until the 1930s when the present system of metal capping was initiated. Panna was the first still water in Italy to be produced in plastic bottles. Panna is one of the San Pellegrino group's premier operations.

Fig. 6.3. San Pellegrino is Italy's most renowned mineral water (label in center of picture)

Fig. 6.4.
a Mountain Valley Spring Water national headquarters visitors' center. One of the most popular waters in the United States. b Bottling works at Mountain Valley Headquarters

San Pellegrino is Italy's most renowned mineral water. The source of supply is three springs emanating from deep bedrock aquifers (396 m depth). The water is highly mineralized and emerges at the springs still warm (21 °C). Commercial bottling of the waters began in 1899. However, a reputation of "magical waters" was established in the 13th century. Later, Leonardo da Vinci was supposed to have taken the waters. The water was first analyzed for chemical content in 1782, revealing that the water is low in sodium, but with a unique combination of mineral salts. Elaborate treatment facilities were first opened in 1848, and subsequently a hotel and drinking hall and casino (Fig. 6.3). Prior to World War I, San Pellegrino was one of Italy's most fashionable spas with the list of patrons considered a social register. San Pellegrino is the only Italian water to establish substantial exports:

"Bottled water has become the fastest-growing beverage in the United States, far outstripping the soaring popularity of wine" (Green and Green 1985). Historically, a limited number of natural spring waters in the United States, such as Deer Park from Maryland and Poland Spring from Maine, were bottled at the source, and only Mountain Valley, from Arkansas, was distributed on a national scale (Fig. 6.4). Perrier's aggressive marketing of their product in the United States in the late 1970s was a catalyst to the bottled water industry and to the consumer. Sales of bottled water in the United States doubled between 1976 and 1980. Of interest:

> Those statistics, however, mask an immense difference between the American and European concept of bottled waters. The European thinks of mineral and spring waters being bottled at the source, like chateau wine, and may well drink them for their therapeutic benefit. The American is usually more concerned with the "purity" and "sterility" of the water than its actual origin (Green and Green 1985).

Processed tap water, distilled water, club soda, seltzer, or the flavored mineral waters constitute the consumption of bottled water in the United States. Green and Green (1985) characterize the American approach to bottled water as "simply to offer a pleasant, pure drinking water that is free from all traces of pollution, does not have the tang of chlorine afflicting so many municipal supplies, and is free from sodium and other minerals that worry the diet-conscious."

In some states the label "spring water" is no guarantee as to the source of the water, labeling is in the rule-making and legislative process and consumer communications are of major interest to the bottled water industry. Only half a dozen natural springs in the United States bottle and distribute water: Mountain Valley of Hot Springs, Arkansas; Poland Spring in Maine (under ownership of Great Waters of France/Perrier); Ephrata Diamond Spring and Cloister Spring in Ephrata, Pennsylvania are examples; France's Vittel has developed

Fig. 6.5. Bottled water from natural springs: Poland Spring, Maine; Mountain Valley Spring, Arkansas, and Saratoga, New York

Bartlett Mineral Springs in northern California, and Saratoga Springs, New York (Fig. 6.5).

Mountain Valley has been bottled since 1871. Sales throughout the States increased after 1908 when publisher William Randolf Hearst and gambler Richard Canfield, met by chance on a train: both gentlemen were transporting bottles of Mountain Valley for personal consumption and established a marketing joint venture. Today Mountain Valley has distributors in almost all 50 states.

Hirman Ricker of South Poland, Maine, built a granite and marble springhouse at Poland Spring in the mid-1800s. His motivation stemmed from his and other family members' improved health from drinking water from the spring. Ricker advertised that the water "Cures Dyspepsia! Cures Liver complaints of Long Standing! Cures Gravel! Drives out all humors and Purifies the blood" (Green and Green 1985). Ricker's success in sales allowed him to erect a bottling plant and subsequently build a resort hotel with golf course, which attracted droves of American and European socialites. The Ricker family bottled the spring water until 1946 when they sold the spring. Little production ensued until 1980, during the early years of the United States bottled-water boom, after Perrier's Great Waters of France purchased the spring and built a modern plant. The bottled water is sold noncarbonated, carbonated, and carbonated with natural essence flavors.

Currently the bottled-water industry in the United States is a vital, aggressive industry. In 1987, $11.5 million were spent on television advertisement (Predicasts F&S 1988). Retail bottled-water sales reached $1.8 billion in 1987, a 12% increase. The dollar value represents 4.47 billion l (Predicasts F&S 1988). In 1987, bottled water sales increased 13% while soft drink sales grew 4.6% (Predicasts F&S 1988). Records indicate that more than 8.3 billion l of bottled water were consumed in 1990 in the United States (Predicasts F&S 1991).

According to the 1992 Edition of *Bottled Water in the United States*, by Beverage Marketing Corporation (1992), the United States bottled-water market was once the fastest growing segment of the multiple beverage industry but realized only about a 0.5% increase in domestic gallonage. The maturing of the market is attributed to general national economic conditions. However, the growth in consumption in the United States, from 5.6 l annual per capita in 1976 to 33.3 l in 1991 (with incremental rise from 8.1 in 1989) are indicative of the magnitude of the market.

1988 Spas, The International Spa Guide (An International Passport to Beauty, Fitness and Health) (Bain 1988) includes some statistics and a taste rating of the most popular domestic and imported bottled waters (see Table 6.1).

6.3 Labels and Art Forms

Inherent to the production of bottled water is the packaging. Emphasis on health, purity, vitality, youth, and nature are represented on scenic labels affixed to tinted, transparent bottles of unique shapes as symbols of status and health and environmental awareness to attract young healthy consumers. Mountains and flowing waters, mythical figures and catchy logos and text allow the bottles to serve as mini billboards (Fig. 6.6).

Alone among Italian waters, a member of the Sangemini group, Ferrarelle, has capitalized on its natural light effervescence with a masterly publicity campaign, rare in the comparatively low-key world of Italian advertising. Ferrarelle was presented as the perfect compromise

Table 6.1. Taste rating and water quality data for selected bottled waters (Bain 1988)

Trade name	Country of origin	Cost per 8 oz	Total dissolved solids (ppm)	Sodium content per 8 oz	Taste rating
Bulk Still Waters					
Deep Rock Artesian	USA	$.02	600	7	5
Arrowhead Mtn. Spring	USA	$.06	99	2	4
Great Bear Natural Spring	USA	$.05	13	0	4
Mountain Spring	USA	$.04	52	1	4
Carolina Mountain	USA	$.06	55	1	3
Deer Park 100% Spring	USA	$.08	29	0	3
Natural spring waters					
Evian Natural Spring	FRANCE	$.23	400	1	5
Mountain Valley	USA	$.25	195	1	3
Poland Spring Pure Natural	USA	$.20	125	1	3
Fiuggi Natural Mineral	ITALY	$.28	120	1	3
NAYA Natural Spring Water	CANADA	$.16		0	3
Tipperary	IRELAND	$.25		0	3
Sparkling natural waters					
Ramlösa	SWEDEN	$.48	712	62	5
St. Pellegrino	ITALY	$.41	968	10	4
Peters Val	Germany	$.26	1350	39	3
Canada Dry Club S oda	USA	$.15	536	44	3
Apollinaris Natural Mineral	Germany	$.32	2250	114	3
Saratoga	USA	$.23	445	16	3
Montclair	CANADA	$.25	200	9	3
Calistoga	USA	$.23	540	29	3
Canada Dry Seltzer	USA	$.16	135	1	3
Poland Spring Sparkling	USA	$.19	99	1	3
Perrier Sparkling	FRANCE	$.26	545	4	3
Vichy Célestins	FRANCE	$.33	2987	277	3
Pedras Salgadas	PORTUGAL	$.28	1700	2	3
Vitelloise Natural Spring	FRANCE	$.36	465	13	3
Ferrarelle Natural Sparkling	ITALY	$.21	1400		2

Rating: excellent 5; very good 4; good 3; fair 2.

between still and gassy waters. The Mona Lisa and similarly well-known figures appeared in ads in three alternative versions: one with straight hair (liscia or still), one with very curly hair (gassata or fizzy), and one with hair just right, as in the original painting – the mildly effervescent Ferrarelle. Sales leapt, and Ferrarelle is now one of the few mineral waters to be asked for by name in restaurants (Green and Green 1985; Fig. 6.7).

Fig. 6.6. Catchy logos and labels used in advertising ◀

Fig. 6.7. ▼ **a** Ferrarelle Spring bottled water. Origin – pyroclastic volcanic rocks. Natural spring discharge and well development. **b** Taking the waters, Ferrarelle Spring discharge. **c** Prof. Paolo Bono shows pyroclastic core test drilling to intercept spring water. **d** Well into spring aquifer. Artesian conditions at 120 m depth, near village of Riardo, Italy; open discharge 30 m high

6.4 Water on the World Wide Web

Nearly every commodity imaginable can now be located at an Internet website, including bottled water. Most well-known springs/bottlers have constructed interesting and sophisticated homepages to describe their products, including such items as production and bottling procedures, chemical composition and source of the water, corporate information, sales, attractive graphics, games, and e-mail opportunities.

Perrier's website sports text and graphics frames in the familiar green-blue color of their famous bottle. Multiple pages exhibit colorful "Bottle Art" and limited edition posters (which can be purchased over the Internet; see Fig. 4.1). A description of Source Perrier provides information on history, the carbonation process, and the hydrogeologic setting – including a geologic cross section.

The attributes of the natural constituents lithium, potassium, calcium, magnesium, and alkalinity in Lithia Springs Mineral Water are given in Lithia Spring's website. The mineral water is marked as "natures' sports drink and stress reducer".

Discover the town secrets, European resort information, and "The Lowdown on Evian" at their website (Fig. 6.8). Crystal Springs and Labrador Spring Water have a tour guide named "Gurgle" for their website. And the Crystal/Labrador information can be accessed in either English or French. The website for Canadian Glacier is a portfolio-type arrangement with separate pages that discuss the discovery and the source of the water, protection and quality control, the natural purification process, and mineral constituents in the spring water (Fig. 6.9).

So, if you become parched while "surfing the Net", search for some refreshing spring water to quench your thirst!

Fig. 6.8.
Evian Natural Spring Water

6.5 Medicinal Values

The therapeutic value of "taking the water" was established in ancient times. The medicinal qualities result from the combination of minerals dissolved in the water. The mixture or combination is dependent on the geology of the recharge area and aquifer of the supply.

> Waters gained their reputation for specific benefits. Some highly mineralized waters, like Contrexeville in France, Fiuggi in Italy and Radenska in Yugoslavia, were famous for helping to break up kidney stones and curing urinary complaints. Others, predominantly known for their bicarbonate level, like Fachingen in Germany, Ferrarelle in Italy and Vichy Catalan in Spain, eased digestion. Still others, like Apollinaris, helped bronchial complaints or, like Evian, soothed skin diseases. The diversity of waters meant that people took the ones whose blend of minerals, or main mineral characteristic, best suited their complaint.
>
> The beneficial properties of many European waters enjoy a long history. Successive Roman emperors had a rare addiction for mineral waters both at home in Italy and on their conquests abroad. They established spas in France at Badoit and Vittel, while Julius Caesar took the warm springs at Vichy, which was known as "Vicus Cadidus" (the hot town). Pliny the Elder in his *Natural History* reported not only of the "miraculous waters" of Ferrarelle, but added that those of Spa, near Leige, Belgium, were "already famous".
>
> Testimonies to delight the modern press agent kept arriving over the centuries. Leonardo da Vinci regained his health with the water of San Pellegrino in northern Italy and Michelangelo went in to bat for Fiuggi. Tormented by kidney stones he sampled that water and soon wrote, "I am much better than I have been. Morning and evening I have been drinking the water from a spring about forty miles from Rome, which breaks up the stone … I have had to lay in a supply at home and cannot drink or cook with anything else." (Green and Green 1985)

Hippocrates and ancient Greek physicians were well versed in the health benefits of mineral water therapy. In very early recorded history, the Egyptians, Arabians, and Mohammedans discuss the use of mineral waters for healing the ill. Mythology and legend date the thermal springs of Bath, England, to 800 B.C. The springs of Ala-Shehr, the Philadelphia of the New Testament, have been used since the 3rd century B.C.

Therapeutic application of mineral waters was very popular in the late 1800s and early 1900s. Significant

Natural Mineral Balance

Canadian Glacier mountain spring water from the Canadian Rockies is the ideal water for human consumption and is a natural part of a healthy life-style. The composition of Canadian Glacier makes it the perfect replenishment and refreshment beverage for people of all ages.

Canadian Glacier has a natural balance of Calcium and Magnesium (ratio 3:1) and a neutral pH of 7.6. Canadian Glacier is also Sodium free at less than 1 part per million (PPM). This unique Balance of naturally occurring minerals is what gives Canadian Glacier water its unique, fresh, full bodied taste.

Canadian Glacier Water Composition

Cations	Anions
Calcium **50**	Bicarbonates **150**
Magnesium **17**	Sulfate **37**
Potassium **ND**	Chloride **ND**
Sodium **ND**	Nitrates **ND**

ND = Not Detected

Above measurements are in MG/L or PPM

Fig. 6.9. Canadian Glacier advertises on the World Wide Web

expenditures were made throughout Europe and the United States to develop lavish resorts and vacation spots at the famous mineral springs. The activities at such establishments were focused on various types of therapy, both internal and external, and were distributed through exotic hotels and lodges, drinking halls, and baths and pools. Additional attractions included casinos and outdoor activities such as golf and tennis. The image of high society and romance is easy to conjure in association with these resort establishments.

Technical publications contemporary to the hay day of the spas detail the scientific and medical theory of the therapy of mineral waters. Several works, such as Fitch (1927) and Moorman (1873) were written by physicians with direct experience in the administration of mineral waters for medical purposes. The technical basics, i.e., the attributes of the waters in aiding various ailments, are still applicable.

Fitch (1927) writes of medical hydrology as "the comprehensive designation of the whole science which treats of the several factors comprising the medicinal use of water in all its solutions." Hydrotherapy is the scientific use of water as a therapeutic agent and treatment of disease in various ways and in various gradations of temperature from ice to steam. Balneotherapy is the science of baths and bathing, including their effects in the treatment of diseases. Loutrotherapy is the knowledge and application of carbonated baths and crounotherapy is the internal administration (oral consumption) of mineral waters in the treatment of disease.

What is the magic of the minerals? Some mineral waters result in cathartic action (purgative), others in diuretic action (increase in urination), and some a diaphoretic action (increase in perspiration). The nature and extent of the reactions are dependent on the predominant chemical constituents and the magnitude of concentration.

Various diseases have been treated by the administration of mineral waters. A listing of generic diseases includes (see also Table 6.2):

- Diseases of metabolism
  - Obesity
  - Diabetes
  - Gout
  - Rheumatism
- Diseases of the alimentary tract
  - Chronic constipation
  - Chronic diarrhea
- Diseases of the liver
  - Hyperemia
  - Jaundice
- Diseases of the urinary tract
  - Kidney stones
  - Dropsy
- Diseases of the circulatory system
  - High blood-pressure
  - Cardiac palpitation
  - Cardiac disease
  - Anemia
- Diseases of the respiratory system
  - Rhinopharyngitis
  - Pharyngitis
  - Laryngitis
  - Emphysema
- Fever and infection

For the French *la médecine douce* (alternative medicine) is widely accepted and is often paid for by social security. Doctors can prescribe "taking the waters" to aid kidney and digestive ailments, rheumatism and arthritis; physiotherapy and massage with waters is sometimes recommended after accidents.

... people drink from a selection of different waters, more varied in mineral content and temperature than the water selected for bottling. At Vichy, apart from the cool water of Source de Celestins, which is bottled, there are the hot waters of Source Hospital, Source Chomel, and Source Grande Grill, and the cooler waters of Source du Parc and Source Lucas; the precise quantities of each, to be drunk at carefully charted intervals during the day, are prescribed individually. The busiest time in the drinking halls of Contrexeville, Evian, Vittel or Vichy is between 11:30 a.m. and noon, when a ceaseless *passaggiata* of people, equipped with glass drinking mugs marked in centiliters and housed in little wicker baskets, line up at the *gryphons* through which the water flows ceaselessly. Everyone carefully fills up to the recommended level, and then goes to sit in the sun, sipping judiciously and chatting with friends. (Green and Green 1985)

The *Heilwasser* Fachingen is well documented in German medical literature. Technical papers have been written describing the effects in curing infantile vomiting, eliminating kidney stones and preventing recurrence, its use in treating ailments of the gastro-intestinal tract, liver and gall bladder. The water is used regularly in German hospitals as the high levels of calcium and magnesium make it a useful diuretic.

The skin of the average male is about 1.4 m^2 in area. Bathing in hot or cold waters has many benefits to the condition of the epidermis and the circulation of blood through the multilayered system of the human skin. The primary benefits of application of heat include: dilation of blood vessels, excitation of nerves, quickening and facilitation of respiration, decrease in acid production in the stomach, and increased production of urea. The application of cold slows deepens respiration, increases red blood cells, excites kidney action, increases production of stomach acids and activity, and diminishes activity in skin tissue.

6.6 Mineral Exploration Using Artesian Well Spring Water

Hot springs or gas exhalations from geyser areas often indicate either mineral deposits in the formative stage or evidence of existing minerals. Some solids resulting from the discharge and evaporation of hot springs are the source of some economic minerals. In some cases, springs occur in the lodes themselves, e.g. Comstock Lode. Spring waters, seeps or effluence from mining operations can also present a very serious environmental problem due to the emission of high concentrations of iron, sulphur, arsenic and other sulfates.

It is not uncommon in mines at depths of 1 000 ft or more to observe substantial water penetration with temperatures of 110 °F or more. Examples of this are Telluride, Colorado, Cripple Creek, Colorado and others.

Springs and wells often discharge water in mineralized zones that contain a wide variety of chemical constituents, including metals such as lead, zinc, copper, gold, and silver. In areas of mineralized water, the local vegetation commonly picks up traces of these organics and inorganics. A rather unique example is from Cyprus where tree roots extend into a mineralized copper zone and flecks of native copper occur in sections of the tree trunks (Beyschlag et al. 1916).

A newsletter of the Virginia Geological Survey contains the article, *Ancient Warm Springs Deposits in Bath*

Table 6.2.
Medical attributes of various mineral waters

Chemical character	Medicinal characteristics	Comments
Carbonated or bicarbonated	Stimulates digestive secretions, neutralizes hyperacidity, antacid and diuretic effects	
Sodium carbonate of bicarbonate	Neutralizes acidity, increases metabolism, elimination of uric acid and uric acid deposits	
Potassium carbonate and bicarbonate	Diuretic, antacid, and antilithic effects, elimination of uric acid and uric acid deposits	Similar to above but more effective
Magnesium carbonate and bicarbonate	Mild laxative effect, antacid, elimination of uric acid and aric acid deposits	
Calcium carbonate and bicarbonate	Constipates and decreases secretions of respiratory, digestive and urinary tracts	
Iron bicarbonate	Increases hemoglobin content in blood; increases temperature, pulse and body weight; stimulates appetite	
Hydrogen chloride/saline	Increases flow of urine and excretion of uric acid	
Hydrogen sulfide/saline	Diuretic and slight purgative, stimulate excretion of bile,	
Sodium/hydrogen chloride	Brine baths to increase action of the skin, increase production of gastric juices, improve appetite	Usually diluted when taken internally
Calcium chloride	Diuretic effect and increases perspiration and excretion of bile	
Sodium, magnesium, chloride	Laxative to purgative depending on dosage	
Calcium sulfate	Relieves kidney irritation, diuretic effect, increases kidney function	
Sodium bromide and sodium iodide	Stimulate activity of lymphatic system, elimination of metallic poisons in system	
Arsenic	Increases appetite and digestion, stimulate secretion of gastro-intestinal mucous membrane	
Free carbon dioxide	Increases flow of saliva and intestinal fluids, increases digestive function, diuretic and antacid effect	
Radioactive water	Activates enzymes; increases autolytic, glycolytic, diastaltic, urolytic, pancreatic, peptic and lactic acid fermentation	Inhalation and oral administration

and Rockingham Counties, Virginia (Nolde and Giannini 1997) and provides an example of the occurrence of metals in spring water in the White Springs area of Virginia. The article is summarized as follows:

Natural springs are sites at which groundwater flows to the surface and issues freely from the ground. Groundwater is insulated from fluctuations in air temperature and therefore approaches the mean annual temperature for the area. A spring whose flow is at least 9 °F (5 °C) higher than this mean temperature and lower than 100 °F (38 °C) is classified as a warm spring. A spring whose flow is warmer than 100°F is classified as a hot spring. The mean annual air temperature in northwestern Virginia ranges between 48 ° and 54 °F. Reeves (1932) has shown that, in this area, water temperatures in springs range from 55 ° to 106 °F. From a sample of 85 springs, 72 temperatures fall between 55 ° and 75 °F.

Two models have been proposed for the temperature of warm springs. These are: shallow circulation above a still-cooling pluton and deep circulation of groundwater which is structurally controlled. Heat flow studies (Costain and others, 1976) have shown no evidence for the existence of a shallow pluton. Costain and others (1976) have shown no evidence for the existence of a shallow pluton. Costain and others (1976) and Perry and others (1979) reported the geothermal gradient in the area to be about 1.5 °F/100 ft (10 °C/km). They concluded that the warm springs in Virginia are not associated with a cooling pluton. Instead they attribute the warm springs to deep circulating groundwater. Assuming a constant gradient remains at depth, water would have to circulate to a depth of about 26 000 feet to be heated to the temperature observed at the surface. Since basement lies about 32 500 feet below the surface (Harris, 1975), deep circulation in the sedimentary section is possible.

In this report, the term "sinter" was restricted to deposits consisting dominantly of silica minerals formed at the surface by deposition from thermal waters. Spring deposits consisting dominantly of carbonate minerals are called travertine.

6.7 Spas of Renown

As discussed previously, springs in the Eastern Hemisphere have been formally developed as spas since the early ancient history. Historical writings place Socrates, Hannibal, Roman aristocracy, Michelangelo, and royalty at the famous springs and spas of their time including the hot springs of Bath, England, Vichy in France, and Fiuggi in Italy.

The Karlovy Vary (Carlsbad), the largest and currently the best known Czechoslovak spa, has been fashionable as a thermal spring since the 14th century. During the 19th century the area was an architectural and cultural center. Twelve major springs are located at Karlovy Vary. The spring water is of hydrocarbonate-sulfate-sodium-chloride composition with a temperature of 71 °C. The Thermal Spa Hotel at Karlovy Vary offers spas facilities, spa programs and treatment and the latest balneotherapeutic equipment, medical supervision and laboratory tests.

The "cure" at Vichy dates back to Roman times. The thermal waters from 12 springs are often the sodium bicarbonate type and are comprised of a combination of therapeutic metals such as iron, lithium, fluorine, iodine, and radon gases. The *Grand Etablissement Thermal* include the Institute De Vichy. The famous spa facilities include hydrotherapy and individualized thermally based medical programs approved by the institute and recreational spa facilities.

Fiuggi, near Rome, provided restorative powers from spring water to Michelangelo. Current facilities at the spa provide thermal showers, an array of spa treatments and a specialty of application of hot phytoplankton mud. Montecatini, near Florence, is acclaimed for numerous thermal springs and nine spas. The spring water is radioactive, of chloride-sulfate and sodium bicarbonate chemical character and typically is 24–34 °C. Carbonic baths, sulphur baths, oral balneotherapy, various irrigations, skin treatments, and radioactive treatments are offered at the spas. Many celebrities, including Sophia Loren and Princess Grace, have come to Montecatini for spa treatments.

Fitch (1927) wrote "we have in our own country every variety of mineral water found in other parts of the world, and without fear of contradiction we can safely assert that our American waters are just as good, just as potent, just as efficacious, and not only equal to any in the world, but in some instances, superior." Fitch described many of the greater American spas as directed by eminent physicians and equipped with the proper equipment of treatment and bathing establishments which make it "no longer necessary for physicians to subject their patients to long, tedious, expensive and, in many instances, hazardous sea voyages to secure treatment at a foreign mineral spring or spa. At the turn of the century many grand hotels and establishments for health and recreation were constructed at natural springs throughout the United States.

Legend has it that Ponce de Leon sought the Fountain of Youth in the Ozark region of the country and that De Soto and his explorers visited the thermal springs of Hot Springs National Park in the 16th century. Today, Hot Springs, Arkansas, remains as one of the renown area of spas in the Western Hemisphere. The bathing establishments were initially operated under the rules and regulations of the Secretary of the Interior.

Radioactive spring waters from Manitou Springs, Colorado, and French Lick Springs in Indiana provided "striking improvement" of musculoskeletal conditions when first established as spas.

Other vintage springs, such as White Sulphur Springs at Sharon Springs, New York; Saratoga Springs, New York; Hot Springs, South Dakota; and White Sulphur Springs, West Virginia still provide hydrotherapy in conjunction with other health and beauty activities.

References

Bain JH (1988) 1988 Spas. The international spa guide, 1st edn. An international passport to beauty, fitness and health. B.D.I.T., Inc., Flushing, New York

Beverage Marketing Corporation (1992) Bottled water in the United States, 1992 Edition. New York, pp. 1–53

Beyschlag F, Vogt JHL, Krusch P (1916) The deposits of the useful minerals and rocks – their origin, form, and content. Macmillan and Co., Ltd.; Vol. II, London, England, pp 515–1262

Costain JK, Keller GV, Crewdson RA (1976) Geological and geophysical study of the origin of the warm springs in Bath County, Virginia. U.S. Department of Energy, TID-28271, 184 p

Fitch WE (1927) Mineral waters of the United States and American spas. Lea & Febiger, Philadelphia and New York

Green M, Green T (1985) The best bottled waters in the world. Simon & Schuster, Inc., New York, 172 pp

Harris LD (1975) Oil and gas data from the Lower Ordovician and Cambrian rocks of the Appalachian basin. U.S. Geological Survey Miscellaneous investigations Series 917-D

Moorman JJ (1873) Mineral springs of North America, how to reach, and how to use them. J. B. Lippincott & Co., Philadelphia, 294 pp

Nolde JE, Giannini WF (1997) Ancient warm springs deposits in Bath and Rockingham Counties, Virginia. In: Virginia Minerals, Vol. 43, No. 2, May 1997, Publisher quarterly by the Division of Mineral Resources, 900 Natural Resources Drive, Charlottesville, VA 22903, p. 1

Perry LD, Costain JK, Geiser PA (1979) Heat flow in western Virginia and a model for the origin of thermal springs in the folded Appalachians. Journal of Geophysical Research 84/B12: 6875–6883

Predicasts F&S (1988) Industries and products: pollution control 4950. Predicasts, Cleveland, Ohio, 1:917

Predicasts F&S (1991) Industries and products: water supply and use 4940. Predicasts, Cleveland, Ohio, 1:881

Reeves F (1932) Thermal springs of Virginia. Virginia Geological Survey Bulletin 36:56

Chapter 7

In Search of a Uniform Standard For Bottled Water

CHAPTER 7
In Search of a Uniform Standard For Bottled Water

Karen L. Bryan

7.1 Introduction

Bottled water, as distinguished from drinking water derived through a piped water distribution system ("tap water"), has unique characteristics that justify its regulation apart from tap water. First, the bulk of bottled water is used specifically for human consumption through drinking, cooking, and food preparation. Second, the perception that bottled water meets or exceeds the standards of tap water is widespread and is supported by the fact that people commonly pay a higher unit cost for bottled water than for tap water, from 500 to 1 000 times greater, depending on the brand of the bottled water and local market conditions. Third, the time and vessels in which bottled water and tap water are carried to the consumer are different. Unlike tap water, which may be exposed to a variety of conduit materials during its fairly constant time of travel to the consumer, the residence time of bottled water in a single container made of a single material can be days, weeks, months, or years. Fourth, bottled water has the potential for widespread distribution far beyond its point of origin, including different cities, regions, and countries. Lastly, consumers of bottled water are subject to a myriad of individual labeling possibilities, whereas there are no such labels for tap water.

In response to the fact that drinkable water is essential to humans, and the fact that bottled water is ubiquitous, would it not be useful to have a uniform standard for a quality of bottled water that can be consumed safely by everyone throughout their lifetime, rather than the diverse and inconsistent patchwork of regulations and directives that currently prevail worldwide?

7.2 Social, Economic, Cultural, and Environmental Impacts on Bottled Water Use

To understand why there are so many systems of regulations and directives, or absence of regulations altogether, it is essential to consider bottled water use in the context of local or national environmental, social, economic, and cultural conditions. Drinking bottled water, including natural mineral water, is an important part of many cultures of the world. "Taking the waters", not only to get a refreshing, better tasting water than the local water supply, but for curative reasons as well, is a longstanding tradition in many places. In Europe, for example, for hundreds of years, some bottlers of mineral water shipped sealed jugs made of clay all over the continent. The products were in high demand and were known in European courts of royalty.

In contemporary Europe, different types of water are marketed, each strictly defined by regulations. Mineral water is treated separately from any other type of water, and it is not judged by standards for tap water or for other forms of bottled water. Mineral water is not drinking or processed water, it is not seltzer or club soda, and it is not distilled or demineralized water, rather mineral water is water from a spring or a well that contains minerals naturally present in the water as it flows from the source. European mineral water has a high, upscale image. It is frequently taken with a meal. It is considered by many to be an integral part of the daily European lifestyle.

In Ireland, bottled water consumption in 1995 was 36.7 million l; and in the United States the consumption has increased more than fourfold, from 487.7 million gal in 1979 to 2 014.2 million gal in 1991. These enormous sales of bottled water around the world have been attributed by consumers to a variety of factors, including better taste and real or perceived health benefits, including an absence of contaminants. In a 1990 survey, *Californians' Views On Water* (The Field Institute 1990), about half of the consumers surveyed said that they drank bottled water because it tasted better than water from the tap, about one-fourth gave safety and health reasons, and one-fourth believed that bottled water was free of contaminants. Similar results were reported by a nationwide survey, *Public Attitudes Towards Drinking Water Issues* (Audit and Surveys, Inc. 1985).

Although bottled drinking water is an important segment of the bottled water industry, it is not the only part. The bottled water industry includes specialty waters too, such as distilled waters used for pharmaceutical purposes, mineral waters and carbonated spring waters used for beverage purposes, and other bottled waters that are not consumed as a primary source of drinking water.

7.3 Regulatory Environment

7.3.1 United States of America

In the United States, two primary federal laws protect the public from contaminants in drinking water: the Safe Drinking Water Act (SDWA) and the Federal Food, Drug and Cosmetic Act (FFDCA). The Food and Drug Administration (FDA), an agency of the US Department, Health and Human Services, is responsible for protecting the public from unsafe food or drink. The role of FDA in the regulation of bottled water stems from the classification of bottled water as "food" under the FFDCA. Under Section 201(f) of the FFDCA, food means "articles used for food or drink by man or other animals and components of such articles." By this definition, water is considered a food.

The SDWA of 1974 gave the Environmental Protection Agency (EPA) federal jurisdiction over drinking water. EPA is responsible for public water systems and sets primary and secondary water quality standards for them. Primary water quality standards establish legal maximum levels for certain contaminants in drinking waters in an effort to protect human health. Secondary water quality standards set recommended maximum levels for contaminants related to taste, odor, and other aesthetic concerns (see Table 7.1, 7.2).

A June 1979 Memorandum of Understanding (1979 agreement) between the FDA and EPA outlines the areas of authority for both agencies concerning water. Based on the express provisions of Section 410 of the FFDCA, FDA retains authority over bottled drinking water. The jurisdiction of FDA starts where water enters a food manufacturing or processing establishment. The 1979 Agreement states that all substances in water used in the preparation or processing of food are "added substances" and are subject to the provisions of the FFDCA. However, no substances added to the public drinking water system before the water enters the food processing establishment are to be considered a food additive.

7.3.2 Federal Food, Drug and Cosmetic Act: Regulatory Scheme of FDA over Foods Generally

The FDA, is responsible for protecting the public from unsafe food or drink. Bottled water is one of many foods that the FDA monitors to insure compliance with the FFDCA and other statutes. Specifically, Sections 402 and 403 of the FFDCA serve as the pillars for enforcement.

Section 402 addresses adulterated food. Food, including bottled water, is forbidden under Section 402(a)(1) from containing any substance in amounts that may be injurious to health. Water quality standards are not necessarily enforced by FDA. Under Section 402(a)(1), food is adulterated if it contains a poisonous substance which may render it injurious to health. Under Section 402(a)(4), food is adulterated if it has been prepared, packed, or held under insanitary conditions whereby it may become contaminated with filth or rendered injurious to health. By comparison, SDWA which controls tap water, contains no such prohibition and no such enforcement provisions with respect to public water supplies. The EPA can protect tap water consumers through enforcement of primary drinking water standards (maximum contaminant levels, MCLs, or required treatment techniques) which are set through rule-making by the EPA but only for specific contaminants. Unlike bottled water quality standards set by the FDA, tap water MCLs must be set by the EPA in light of a balancing of consumer safety against the cost and the feasibility for the tap water industry of meeting the maximum contaminant levels. The FDA is under no such restrictions in providing quality assurance for bottled water.

Section 403, the second pillar of enforcement of the FFDCA, addresses misbranding of foods. Food, including bottled water, is misbranded if its labeling is false or misleading in any way, or if the label does not bear the common or usual name of the food. Also, the label must provide a list of ingredients if the food is fabricated from two or more ingredients. The term "labeling" includes labeling on the product in any written, printed or graphic material accompanying the product. FDA regulations detail the food labeling requirements, such as prominence of statements in nutrition labeling requirements.

Table 7.1. FDA's Bottled Water Standards and EPA's Pubic Drinking Water Standards

Substance or property	FDA quality standards	EPA primary standards	EPA secondary standards
Inorganic chemicals (mg/l)			
Arsenic	0.05	0.05	
Barium	1.0	1	
Cadmium	0.01	0.01	
Chloride	250.0		250
Chromium	0.05	0.05	
Copper	1.0		1
Fluoride	1.4 – 2.4[a]	4.0	2.0
Iron	0.3		0.3
Lead	0.05	0.05	
Manganese	0.05		0.05
Mercury	0.002	0.002	
Nitrates	10.0	10	
Phenols	0.001		
Selenium	0.01	0.01	
Silver	0.05	0.05	
Sulfate	250.0		250
Total dissolved solids	500.0		500
Zinc	5.0		5
Organic chemicals (mg/l)			
Total trihalomethanes	0.10	0.10	
Vinyl chloride	b	0.002	
1,1-Dichloroethylene	b	0.007	
1,2-Dichloroethane	b	0.005	
1,1,1-Trichloroethane	b	0.20	
Carbon tetrachloride	b	0.005	
Trichloroethylene	b	0.005	
Para-dichlorobenzene	c	0.075	
Benzene	b	0.005	
Endrin	0.0002	0.0002	
Lindane	0.004	0.004	
Methoxychlor	0.1	0.1	
Toxaphene	0.005	0.005	
2,4-D	0.1	0.1	
2,4,5-TP silvex	0.01	0.01	

[a] FDA has two fluoride quality standards 1.4–2.4 when fluoride has not been added to the water, and 0.8–1.7 when fluoride has been added to the water.
[b] FDA proposed these standards on 6 July 1990, but as of 31 December 1990, had not adopted them.
[c] FDA deferred adopting this primary standard because EPA was considering developing a stricter secondary standard. FDA wants bottled water standards to address aesthetic concerns such as unpleasant odors, and thus believes bottled water quality standards should encompass EPA's stricter secondary standards.
[d] In mg/l.
[e] As measured on the pH scale, whose values run from 0 to 14, with 7 representing neutrality.
[f] Turbidity units.
[g] Color units.
[h] Odor threshold number.
[i] Most probable number of coliform organisms per 100 ml when using the multiple-tube fermentation test method.
[j] Total coliform-positive results per monthly samples for systems that test fewer than 40 samples per month.
[k] Picocuries per liter.

Table 7.1. *Continued*

Substance or property	FDA quality standards	EPA primary standards	EPA secondary standards
Physical characteristics			
Corrosivity			Noncorrosive
Foaming agents			0.5[d]
pH			6.5 – 8.5[e]
Turbidity	5[f]	5[f]	
Color	15[g]		15[g]
Odor	3[h]		3[h]
Microbiological standards			
Coliforms	2.2/100[i]	1[j]	
Radiological standards			
Gross alpha	15[k]	15[k]	
Combined radium 226 and 228	5[k]	5[k]	

[a] FDA has two fluoride quality standards 1.4–2.4 when fluoride has not been added to the water, and 0.8–1.7 when fluoride has been added to the water.
[b] FDA proposed these standards on 6 July 1990, but as of 31 December 1990, had not adopted them.
[c] FDA deferred adopting this primary standard because EPA was considering developing a stricter secondary standard. FDA wants bottled water standards to address aesthetic concerns such as unpleasant odors, and thus believes bottled water quality standards should encompass EPA's stricter secondary standards.
[d] In mg/l.
[e] As measured on the pH scale, whose values run from 0 to 14, with 7 representing neutrality.
[f] Turbidity units.
[g] Color units.
[h] Odor threshold number.
[i] Most probable number of coliform organisms per 100 ml when using the multiple-tube fermentation test method.
[j] Total coliform-positive results per monthly samples for systems that test fewer than 40 samples per month.
[k] Picocuries per liter.

Adulteration or misbranding of food in interstate commerce, including bottled water, is prohibited under Section 301. Sections 302, 303 and 304 provide for injunctions, fines and/or imprisonment and seizure, respectively, for violation of Section 301. Imported bottled water may be detained if it appears that it is adulterated or misbranded. Voluntary recalls, regulatory letters, and notices of adverse findings are other approaches used by FDA to obtain compliance.

The use of food and color additives in foods requires pre-market authorization through a petition and a regulation process to establish safe conditions of use before an additive can be used. Food additives include food-contact materials that have a reasonable expectation of becoming a component of food, so-called "indirect additives". FFDCA covers thousands of compounds and uses for food-contact materials, from polycarbonate and polyethylene resins used to fabricate bottles, to materials for compounding rubber for gaskets in bottle processing plants. It is the responsibility of the manufacturer to insure that all additives to food and food-contact materials (processing plant food-contact equipment) are authorized by food additive regulations, or are otherwise legal and generally recognized as safe.

The Fair Packaging and Labeling Act (FPLA) provides an additional legal framework for the labeling requirements for food products, including bottled water, packaged for sale to consumers.

7.3.3 Federal Food, Drug and Cosmetic Act: Regulatory Scheme of FDA over Bottled Water Specifically

Section 401 of the FFDCA provides for issuance of regulations establishing standards of identity, standards of quality and standards of fill when such standards would promote honesty and fair dealing in the interest of consumers. Standards for bottled water were first set by FDA in November 1973. Although the FDA quality standards were essentially the same as the 1962 Public Health Service (PHS) Drinking Water Standards, FDA pointed out

Table 7.2. Specific dates of actions of EPA and FDA

Contaminant category	Date EPA finalized	Date FDA proposed	Days elapsed
Initial standards for 6 organic chemicals, 10 inorganic chemicals, turbidity, and microbiological contaminants	12/24/75	06/21/76[a]	180
Initial standard for radioactivity	07/09/76	01/04/77	179
Initial standard for total trihalomethanes	11/29/79	06/13/80	197
Revised standard for fluoride[b]	02/27/87	09/16/88	567
Initial standards for 8 volatile organic chemicals	07/08/87	07/06/90[c]	1094
Revised standard for total coliforms	06/29/89	[d]	550[d]

[a] FDA adopted EPA's organic and inorganic chemical standards and already had turbidity and microbiological contaminant standards.
[b] On Sept. 18, 1988, FDA proposed adopting EPA's revised fluoride standard. FDA's proposed fluoride standard was not final as of Dec. 31, 1990.
[c] FDA proposed adopting all 8 volatile organic chemical standards except the para-dichlorobenzene standard for which EPA is considering further action. FDA's proposed standards were not final as of Dec. 31, 1990.
[d] FDA has not announced any action on this standard as of Dec. 31, 1990.

that EPA would be revising drinking water regulations periodically and that it was the intention of FDA to keep the FDA quality standard for bottled water compatible with EPA quality standards. Although the quality standards for bottled water have been amended many times since November 1973, the promulgation of new, up-to-date quality standards for bottled water generally has trailed behind EPA's actions and the needs of the public.

Regulations of FDA, pertaining to bottled water, appear in Parts 103 and 129 (21CFR Parts 103 and 129). They provide for microbiological, physical, chemical, and radiological quality standards of bottled water. Bottled water that does not meet the numerical quality standards must be labeled as substandard, even if the standard is not health-based or if consumption of the product is safe. This prohibition applies regardless of the cost to industry. Further, it is illegal to sell a bottled water product if it contains unsafe levels of any compound (even compounds not specifically regulated with numerical quality standards).

When the SDWA of 1974 was enacted, the SDWA legislation added Section 410 to the FFDCA. Section 410 of FFDCA obligates FDA, whenever EPA prescribes interim or revised national primary drinking water regulations under Section 1412 of the Public Health Service Act (PHSA) (the SDWA [42 U.S.C. 300f through 300j-9]), to consult with EPA concerning those regulations. In 21 U.S.C. 349, FFDCA, further requires that within 180 days FDA must either promulgate amendments to its regulation applicable to bottled water or publish in The Federal Register its reasons for not amending the regulations. To comply with 21 USC 349, FDA reviews new EPA regulations and determines whether the new or revised primary drinking water regulations or treatment technology regulations can be incorporated into the quality standards for FDA without further modification. The last revisions were effective 13 May 1996.

Historically, FDA considered mineral water and soft drinks to be inherently different from other types of bottled water which were promoted as substitutes for tap water. Mineral water was therefore exempt from quality standards. FDA believed that the mineral content of water would exceed certain existing secondary water quality standards, such as that for total dissolved solids. This is so even though Section 103.35 (a)(1) of the FFDCA defined "bottled water" as "water that is sealed in bottles or other containers and intended for human consumption."

FDA has now developed a standard of identity for bottled water and has recodified the standard of quality for bottled water to include mineral water and ingredient uses of the project. FDA exempted mineral water from certain physical and chemical allowable levels.

FDA regulations require that all bottled water be produced in food plants that comply with special Good Manufacturing Practices (GMP) for handling, processing and distributing food products. GMPs were originally tailored to address the following:

1. Bottling under sanitary conditions
2. Collection of samples by bottlers for bacterial analysis
3. Collection of samples by bottlers for chemical analysis
4. Quality control to assure bacteriological and chemical safety of water
5. Protection of water source from contamination
6. Frequent checks on processing equipment
7. Sanitation and protection of containers prior to filling and capping
8. Coding on labels of bottled water

GMPs were designed to establish a comprehensive set of good manufacturing practices for bottled water products, including bottled mineral water, and to serve as a corollary to the standard of quality. In 21 CFR Part 129.3(b), FDA defines bottled drinking water to include "all water which is sealed in bottles, packages, or other containers and offered for sale for human consumption, including bottled mineral water". Thus, the GMPs of FDA found in 21 CFR Part 129, as well as the strong protection of the FFDCA itself, have always applied to mineral water.

As in the case of quality standards, GMPs have been amended several times. GMP regulations do not apply to tap water.

To insure that bottlers meet Federal regulations, FDA uses a multi-pronged approach. First, FDA requires bottlers to use water sources (wells, springs, public drinking water systems, etc.) that have been inspected, tested and approved by appropriate regulatory organizations, such as EPA or state agencies. Second, FDA inspects domestic bottling plants for proper operating practices in cleanliness. Third, FDA requires bottlers to test their source water and bottled water periodically to insure compliance with bottled water quality standards. Finally, FDA tests selected samples of domestic source waters and domestic and imported bottled waters for contaminants.

FDA does not have sufficient resources to check every bottler and inspect and sample all food products every quarter, or even every year. The regulatory context under which the FDA operates places the responsibility on bottlers to insure that their products are safe, wholesome and truthfully labeled in full compliance with the FFDCA. It is the responsibility of the bottled water plant operator to run sufficient quality control checks on the ingredients, during the processing and on the finished product, to insure that every lot complies with statutes and regulations. For bottled water, this means that manufacturers need to analyze their water as often as necessary for all potential contaminants that may render the water product injurious to health, whether or not FDA has set a specific maximum level for the contaminants. Self-regulatory efforts by bottled water manufacturers are therefore an important adjunct to the quality control efforts by the federal government.

7.3.4 Current USA Food and Drug Administration Regulations

The current status of regulations in the United States is taken from the *Federal Register* dated Monday, 13 November 1995. In its application, appropriate agencies in each state will incorporate the Federal regulations as a minimum requirement for bottled waters. In summary, these regulations are as follows:

The Food and Drug Administration (FDA) is establishing a standard of identity for bottled water. At the same time, the agency is recodifying the standard of quality for bottled water. FDA is revising the definition for bottled water in the quality standard to include mineral water and ingredient uses of this product. In addition, FDA is defining "artesian water", "groundwater", "mineral water", "purified water", "sparkling bottled water", "spring water", "sterile water", and "well water". FDA is exempting mineral water from certain physical and chemical allowable levels. FDA is taking these actions, in part, in response to a petition submitted by the International Bottled Water Association (IBWA). FDA finds that the regulations will promote honesty and fair dealing in the interest of consumers as well as the interests of the regulated industry.

Effective Date: 13 May 1996. The Director of the Office of the Federal Register approves the incorporations by reference in accordance with 5 U.S.C. 552(a) and 1 CFR part 51 of certain publications at 21 CFR 129.35(a)(3)(ii), 129.80(g), and 184.1563(c), effective 13 May 1996.

7.3.5 Regulatory Environment at the State Level and Self-Regulation in the United States of America

The jurisdiction of FDA pertains to interstate commerce only, leaving the regulation of intrastate products to individual states. The International Bottled Water Association (IBWA), a private association representing all segments of the bottled water industry in the United States of America, including many foreign bottlers who export to the USA, wanted uniform bottled water regulations to be promulgated to govern intrastate commerce. IBWA took their case to the states (Table 7.3).

IBWA estimates that 60 to 70% of bottled water is shipped intrastate. Although individual states within the United States of America are responsible for regulating

Table 7.3. States with Bottled Water Regulations

Alabama	Alaska	Arizona
Arkansas	California	Colorado
Connecticut	Delaware	District of Columbia
Florida	Georgia	Hawaii
Idaho	Illinois	Indiana
Iowa	Kansas	Kentucky
Louisiana	Maine	Maryland
Massachusetts	Michigan	Minnesota
Mississippi	Missouri	Montana
Nebraska	Nevada	New Hampshire
New Jersey	New Mexico	New York
North Carolina	North Dakota	Ohio
Oklahoma	Oregon	Pennsylvania
Rhode Island	South Dakota	Tennessee
Texas	Vermont	West Virginia
Wisconsin	Wyoming	

intrastate state bottlers, they are not required to mirror FDA regulations. Just as topographic, hydrologic, and geological conditions vary from state to state, the members of IBWA found regulations of individual states differed greatly from one another and from the FFDCA. The IBWA has therefore advocated a unified bottled water standard for the public through its support of the Model Code for Bottled Water.

A program of self-regulation instituted by the IBWA has steadily grown over the past 20 years. In the USA, as a requirement of membership in the IBWA, each bottler member, domestic and now foreign as well, that produces bottled water for sale in the USA, must undergo an annual plant inspection by an outside, non-profit, independent entity, the National Sanitation Foundation (the Foundation). The Foundation is nationally recognized in the field of plant audits, consumer product certification, food sanitation and inspection. Foreign bottle members are not obliged to utilize the Foundation, but may instead utilize an independent inspection entity that is recognized in the country of export.

FDA bottled water quality standards are used as a yardstick for compliance in these annual, independent inspections. Inspection also relies on a comprehensive plant technical manual prepared by the IBWA that provides a comprehensive, uniform, guideline on health and sanitation procedures for bottled water plants. EPA sampling and analytical criteria are used for source water and product water samples taken during the inspection.

Each plant of the member company and each type of bottled water sold is covered in the annual inspection requirement. The IBWA pays for the inspection out of its annual budget. The inspection and a passing score are mandatory. If a plant of a member company fails any of the numerous critical test items, or if it receives an overall failing score, the entire member company fails the test. The plant must correct all deficiencies. The plant must be reinspected and pass a second inspection within 90 days if the company is to remain a member company of the IBWA. Member companies participating in this self regulatory program account for over 85% of the volume of water sold in the United States.

As part of a pro-regulation strategy, the IBWA petitioned FDA to revise the standards of quality for bottled water in Section 103.35 and the bottled water GMP regulations to expand the scope of these regulations to include mineral water in the standards of quality and to regulate more closely the labeling, production, and distribution of bottled water in order to promote honesty and fair dealing in the interest of consumers, as well as the interest of regulated industry. This was accomplished in 1996. Still, the lack of uniformity as to bottled water requirements in the USA has been fostered in the past, in part, by the absence of clear, up-to-date, comprehensive FDA regulations. As a result, there is still some confusion and inconsistency which is carried over to the regulatory frameworks of the states and the efforts of IBWA to self-regulate member companies.

7.4 European Economic Community

In keeping with the 1964 impetus to create a united Europe, the treaty of Rome was written to create a "united economic home market". The objective was to set compatible standards for food and other commodities. The differences in regulations governing the bottled water market were hindering the free movement of natural mineral water and creating disparate competitive situations. Of course, this directly affected the establishment and functioning of the common market, at least as it related to natural mineral waters. It took 16 years to set a European standard for the natural mineral water "bottled water market". In 1980, a Directive was prepared by the Council of the European Communities, EC-Directive 80/777, to address the different regulations in the member countries. The Directive, effective in 1984, established a unified quality standard in labeling requirements for natural mineral water throughout the member states. In contrast to the United States, where there must be compliance with both federal and individual state regulations for bottled water, the Directive set a single regulation for the European community. If a mineral water is recognized as complying with the Directive in one member state, then it can be sold in any member state under this seal of quality.

Natural mineral water, as defined by the Directive, means "micro-biologically wholesome water originating in an undergroundwater table or deposit and emerging from a spring tapped at one or more natural or bore exits – the underground origin of such water is protected from all risk of pollution."

Natural mineral water can be clearly distinguished from ordinary drinking water by its nature. The former is characterized by its specific mineral content and trace elements, which in the original state, at the point of emergence, remain constant.

Article 4 of the Directive states:

1. Natural mineral water, in its state at source, may not be subject to any treatment or addition other than: (a) the separation of its unstable elements, such as iron and sulphur compounds, by filtration or decanting, possibly preceded by oxygenation, in so far as this treatment does not alter the composition of the water as regards the essential constituents which give it its properties; (b) the total or partial elimination of free carbon dioxide by exclusively physical methods; (c) the introduction or reintroduction of carbon dioxide under the conditions laid down in Annex 1, section III. In particular, any disinfection treatment by whatever means and, subject to paragraph 1(c), the addition of bacteriostatic elements or any other treatment likely to change the viable colony count of the natural mineral water shall be prohibited.

The bottom line is that if a source begins to give off polluted water, it is not permissible for a manufacturer to treat the water to make it potable. Rather, the manufacturer is required to shut down the source until the cause of the pollution can be identified and corrected, or cease doing business.

The Directive states that natural mineral water must be assessed from properties favorable to health, including some of the following points of view:

1. Geological and hydrological
2. Physical and chemical
3. Microbiological, and
4. If necessary, pharmaceutical, physiological and clinical

The Directive further lists the methods and requirements for testing of mineral waters and the parameters to which mineral waters must adhere.

Since the adoption of the Directive, the European Council has further endorsed the principal of developing Europe-wide voluntary standards that establish state-of-the-art specifications and codes. The aim is to remove differences of a technical nature that could cause technical barriers to trade, either among the national standards of the European community or among measures applied at a national level to certify national conformity. It is expected that the EEC will contract with European regional standard-setting bodies, i.e., the European Committee for Standardization (CEN), to develop voluntary standards in the areas of water supply and water analysis. These multi-country efforts towards international standardization are becoming increasingly important in view of an ever-changing global economy which is challenging outdated trade barriers. Competitiveness is the watchword.

7.5 The World Health Organization

The World Health Organization (WHO) published recommended *Guidelines For Drinking Water Quality* (Guidelines) in 1984, which were revised in 1993 (Table 7.4, 7.5). These Guidelines supersede both the European (1970) and the International (1971) standards for drinking water which heretofore were used worldwide in the establishment of national drinking water quality standards

Table 7.4. Guideline values for health-related inorganic constituents

Constituent	Guideline value (mg/l)
Arsenic	0.05
Cadmium	0.005
Chromium	0.05
Cyanide	0.1
Fluoride	1.5
Lead	0.05
Mercury	0.001
Nitrate (as N)	10.00
Selenium	0.01

Table 7.5. Guideline values for health-related organic contaminants

Contaminant	Guideline value (μg/l)
Aldrin and dieldrin	0.03
Benzene	10.0
Benzo[a]pyrene[a]	0.01
Chlordane (total isomers)	0.3
Chloroform[c,d]	30
2,4-D	100
DDT (total isomers)	1
1,2-dichloroethene[a]	10
1,1-dichloroethene[a,c]	0.3
Heptachlor and heptachlor epoxide	0.1
Hexachlorobenezene[e]	0.01
Gamma-HCL (lindane)	3
Methoxychlor	30
Pentachlorophenol	10
2,4,6-trichlorophenol[a,b]	10

[a] The guideline values for these substances were computed from a conservative, hypothetical, mathematical model that cannot be experimentally verified and therefore should be interpreted differently. Uncertainties involved are considerable and a variation of about two orders of magnitude could exist.
[b] The threshold taste and odor value for this compound is 0.1 g/l.
[c] Since the FAO/WHO conditional ADI of 0.0006 mg/kg body weight has been withdrawn, this value was derived from the linear multi-state extrapolation model for a cancer risk of less than 1 in 100 000 for a lifetime of exposure.
[d] The microbiological quality of drinking water should not be compromised by efforts to control the concentration of chloroform.
[e] Previously known as 1,1-dichloroethylene.

and drinking water quality control programs. The purpose of the Guidelines is to provide a sound base for the development of national and local drinking water quality standards. The Guidelines specifically note that "bottled water must be at least as good in bacterial quality as unbottled potable water" and that it "should contain an acceptable level of chemical contaminant(s)." The obvious possibilities for wide-spread transmission of disease through the use of bottled water indicate a special need to ensure that bottled water use meets the same standards as for drinking water. In effect, WHO has recommended that bottled water should meet or exceed drinking water quality standards of a country in all respects.

Guidelines apply to bottled water but do not apply to natural mineral waters, which are regarded by WHO as beverages rather than drinking water. Standards for mineral waters are separately governed by The Codex Alimentarius Commission, which is an international body established in 1962, and is charged with implementing Joint FAO/WHO Food Standards. The standards and codes of practice are developed according to a procedure that is designed to respond to and unite the views of government, consumer, and the food industry. Approximately 150 countries are members of the Commission. Codex Alimentarius Standards exist for collecting, processing, and marketing natural mineral waters.

WHO regards aesthetic and organoleptic characteristics of bottled water subject to a wide variety of individual preference, as well as social, economic, and cultural considerations. Although guidance is given by WHO on the levels of substances that may be aesthetically unacceptable, no guideline values have been set for substances where they do not represent a potential hazard to health.

7.6 WHO Microbiological and Biological Aspects of Bottled Water Quality

WHO Guidelines for bacteriological quality of bottled water are the same as tap water supplies including both treated and untreated water entering the distribution system, as well as water within the distribution system. WHO deems it extremely important that the source and the processing of water be adequately protected from any possibility of contamination by the coliform group and pathogenic organisms.

WHO Guidelines mandate that bottled water must be at least as good in bacterial quality as unbottled potable water and must be free of coliform organisms.

7.6.1 WHO Chemical Quality

The concern for chemical contaminants, whether organic or inorganic, is based on the potential for adverse health effects after prolonged exposures. The objective of the WHO Guidelines values is "to define a quality of water that can be consumed safely by everyone throughout their lifetime".

According to WHO, the Guideline values represent the judgment of committees of the leading experts around the world. They are based on scientific criteria defined by dose-response relationships, analytical data on the frequency of occurrence and concentrations commonly occurring in drinking water, and toxicological evidence.

7.6.2 WHO Radioactive Materials

WHO Guidelines values are based on an assumed per capita daily intake of 2 l of water for 1 year and are calculated on the basis of the metabolism of an adult. Guidelines are based on the conservative assumption that only the most toxic radionuclides are the cause of gross radioactivity and take into account radiation due to both man made as well as natural radioactivity. The Guidelines represent a value below which water can be considered potable. Because groundwater is both the most frequent source of bottled water and groundwater contains significant radioactivity much more frequently than surface water, WHO Guidelines address the radioactivity of water in water quality standards for bottled water.

7.6.3 Labeling

Clear labeling with sufficient information for the consumer to make an informed choice is deemed good practice to protect the consumer, particularly in view of indefinite storage times at temperatures that are usually warmer than tap water. WHO Guidelines advocate that the date of bottling, lot numbers and the concentrations of chemical constituents are all items that should appear on labels.

7.7 Conclusions

Clearly bottled water, including mineral water, has unique characteristics that justify its regulation separate and apart from tap water. Its very presence in the market place as a commodity to be purchased at a premium rather than a service to be provided to the general public places a special burden on bottlers to deliver a product better than tap water and a burden on the consumer to be informed as to the many types of and differences among bottled waters.

In maintaining a balance between normal market forces and regulations, it is critical to be aware of the position of the consumer, who may become confused or disenfranchised by inconsistent and unfamiliar regulations. Consumers want and need clear, useful information. Only then can they make informed decisions to purchase the type of bottled water that fits their need. Thus, in an economy where bottled water may be produced on one side of the world and sold on the other, it is imperative for the consumer and the bottler alike that there be a uniform, globally accepted standard.

Governments, private industry, and interested third parties alike must respond by promoting a truly international and collaborative effort to develop and implement comprehensive and consistent global regulations that address all types of bottled water. Technically anchored rules and regulations for the many phases of the bottled water process, with clear definitions for all bottled waters of the world that allow for cultural diversity, would help to clarify a sea of murky national, state, and self-imposed regulations. Uniform, worldwide regulations would usher in a new era of better understanding and fair trade for all.

References

Allen HE (1990) Chemical quality of bottled water. Proceedings of the Bottled Water Workshop. A Report Prepared for the Use of the Subcommittee on Oversight and Investigations of the Committee on Energy and Commerce

Audit and Surveys, Inc. (1985) Public attitudes toward drinking water issues, Princeton, New Jersey

Cech I (1990) US House of Representatives, bottled water: truth in advertising and selected water quality issues. Proceedings of the bottled water workshop. A report prepared for the use of the subcommittee on oversight and investigations of the committee on energy and commerce

Codex Alimentarius (1982) Codex standards for natural mineral waters and edible ices and ice mixes. Food and Agriculture Organization of the United Nations, Volume XII. World Health Organization, Rome

Codex Alimentarius (1985) Recommended international code of hygienic practice for the collecting, processing, and marketing of natural mineral waters. CAC/RCP 33–1985, 14 p

Codex Alimentarius (1991) Codex standard for natural mineral waters (European Regional Standard), Codex Stan 108-1981 (amended 1985, 1991). Codex Alimentarius Vol. 11-1994

Deal WF (1991) Statement of international bottled water association. A report prepared for the use of the subcommittee on oversight and investigations of the committee on energy and commerce

Department of Health and Human Services, Food and Drug Administration (1993) Federal Register, Part II, , 21 CFR Part 103

Department of Health and Human Services, Food and Drug Administration (1995) Federal Register, Vol. 60, No. 218, Rules and Regulations, 21 CFR Parts 103, 129, 165, and 184; Beverages: Bottled Water

Ferstandig R (1990) Consumers and bottled water. Proceedings of the Bottled Water Workshop. A report prepared for the use of the subcommittee on oversight and investigations of the committee on energy and commerce

Fricke M (1992) Bottled water of the USA – definitions and market. Verlag W. Sachon, D 8948 Schloss Mindelburg, pp. 11–16

Fricke M (n.d.) Natural mineral waters, curative-medical water and their protection. Lang Str. 118, D-3490 Bad Driburg

Fricke M, Griesing K-H (1994) Natural mineral water – drinking water, where's the difference? Der Mineralbrunnen 5(94): 6 p

General Accounting Office (1991) Food safety and quality, stronger FDA standards and oversight needed for bottled water. Report to the Chairman, Subcommittee on Oversight and Investigation Committee on Energy and Commerce, House of Representative

Guidelines for Canadian Drinking Water Quality, 1987, Federal-Provincial Subcommittee on Drinking Water of the Federal-Provincial Advisory Committee on Environmental and Occupational Health, Canadian Minister of National Health and Welfare

Harker TL (1988) Letters, 20 January. Center for Food Safety and Applied Nutrition; Food & Drug Administration

Harker TL (1988) Letters, 25 March. Center for Food Safety and Applied Nutrition; Food & Drug Administration

Harker TL (1990) Letters, 12 September. Center for Food Safety and Applied Nutrition; Food & Drug Administration

Harker TL (1991) Letters, 28 February. Center for Food Safety and Applied Nutrition; Food & Drug Administration

Harker TL (1991) Letters, 17 May. Center for Food Safety and Applied Nutrition; Food & Drug Administration

Harris TL (1986) Groundwater resources, control and management. Vintage Press, Inc., New York, 173 p

IBWA (1992) Model code for bottled water. Alexandria, Virginia

IBWA (1993) Model code for bottled water – regulation questions and answers. Alexandria, Virginia

O'Donnell WJ (1990) Sanpellegrino. Proceedings of the bottled water workshop. A report prepared for the use of the subcommittee on oversight and investigations of the committee on energy and commerce

Shank FR (1991) Statement before the subcommittee on oversight and investigations committee on energy and commerce, house of representatives

The Council of the European Communities (1980) Council Directive of 15 July 1980 on the approximation of the laws of the member states relating to the exploitation and marketing of natural mineral waters. Official Journal of the European Communities No. L:229/1

The Field Institute (1990) Californians' views on water

Troxell T (1990) Role of FDA in regulatory bottled water. A report prepared for the use of the subcommittee on oversight and investigations of the committee on energy and commerce

US Bottled water market gallonage by segment 1976–1991. International Bottled Water Association, Alexandria, Virginia

World Health Organization (1993) Guidelines for drinking-water quality. vol. 1: Recommendations, 2nd edn. The World Health Organization, Geneva, Switzerland, 188 p

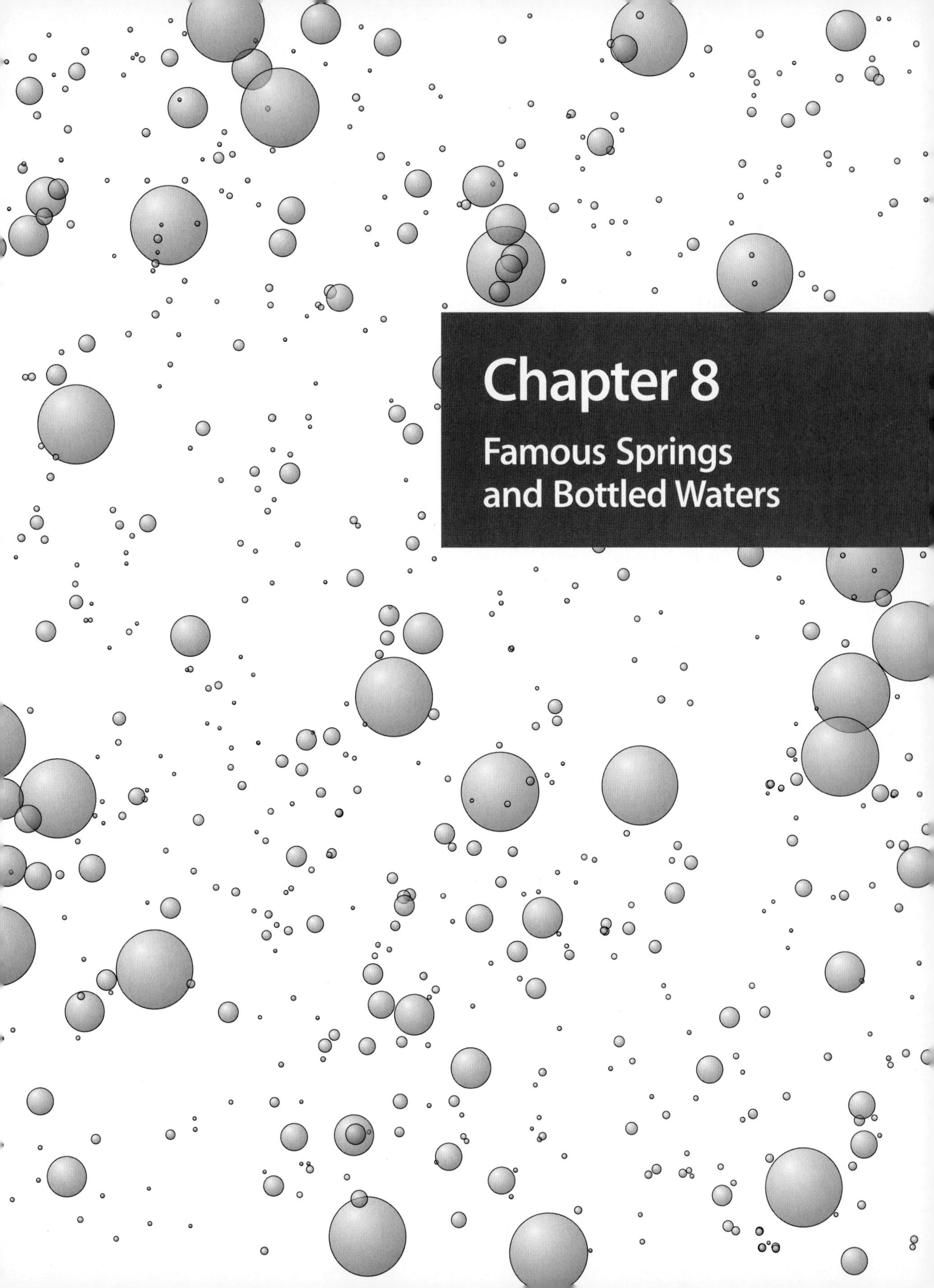

Chapter 8

Famous Springs and Bottled Waters

CHAPTER 8
Famous Springs and Bottled Waters

Philip E. LaMoreaux
(Individual authors are given for each spring subsection)

8.1 Introduction

Grandmother used to think of bottled water as a heating pad to keep her feet warm on a cold night. Today, few Americans would have trouble relating to the terms bottled water, spring water, sparkling water, or mineral water, all of which now come in vessels of different shapes, sizes, colors, and labels that appear on supermarket shelves. They are premium products in today's marketplace. Many of the worlds famous springs supply water for bottling and for sale.

Historically, the use of spring waters dates back to the earliest civilizations. Legend reports that Hannibal, after crossing the Alps, rested his troops and elephants at a spring near Nimes, France, a spring now known as Les Bouillens, the source of the original *Perrier* water. The principal source of water at the ancient Biblical city of Palmyra in Syria is the spring called Efca. It is on the east flank of Gebel Muntar, south of the gate to Damascus. The spring provides 5 000 m^3 of water a day. The water is warm, 33 °C. It is also sulphurous and radioactive and it is believed by some to have curative powers. The spring bubbles up in a grotto that is the termination of a lengthy underground canal, reported to be at least 600 m long. The last 50 m are straight, the more interior part of the channel curving, with two or three branches, possibly a form altered by man. Similar tunnels which tap underground water occur at Shobek, Kirhareshet in Moab, Lachish, Jerusalem, Gibeon, and Megiddo, as well as at Hazor and Gezer.

Dr. Dora P. Crouch, in her book, *Water Management and Ancient Greek Cities* (Crouch 1993), describes a unique and very interesting relationship between karst, water supplies (springs), and the location and development of ancient Greek cities. She relates that in ancient Greece, cities were located based on the perennial availability of water in areas suitable for agriculture, proxinity to trade routes, aesthetic landscapes, and their defensibility. Her chapter, Karst: The hydrogeological basis of civilization, describes the early historical interrelationship between springs, water supplies and development. It is very interesting reading.

In Europe, the famous springs at Bath, England, are among the earliest to have been developed and have an international reputation based on a long history of use of the warm water. Legend implies the water was used for curing the disease of leprosy suffered by Prince Bladud (who later became the mythologic god-king and father of King Lear) and his pigs about 863 B.C. When the Romans came with their enthusiasm for hot springs, the small town of Bath became a fashionable resort for rest and recuperation. For a period of 4 centuries, the hot springs were famous throughout the Roman Empire. The Royal Mineral Water Hospital was established at Bath by the British Parliament in 1739 and it evolved into a modern treatment facility using the thermal water for physical therapy in the treatment of rheumatic diseases.

More than 25 centuries after Prince Bladud discovered the curative power of the springs at Bath, the popularity of springs and baths generated the desire for duplication in the United States, and one of the early popular spas was named Bath (now Berkeley Spring, West Virginia). Native Americans used those springs for hundreds of years for medicinal purposes, especially for treating rheumatism. American Indian tribes from the Great Lakes to the Carolinas, including the Six Nations, Delawares, Tuscaroras, and Catawbas came to the springs. In 1748, George Washington was at Bath, West Virginia, as part of a surveying party to establish the western limit of Lord Fairfax's land grant. Thereafter, Washington visited these springs often and in 1761 wrote, "I think myself benefitted by the water and I am not without hope of their making a cure for me." He had a cottage built at the springs which can be considered the first summer White House. Similarly, Franklin Delano Roosevelt had a cottage at Lithia Springs, Georgia, USA, and even ordered a special study made of the spring by

O. E. Meinzer, Chief of the Groundwater Branch of the US Geological Survey.

In this early period of hot or mineral spring use in the United States, it was generally accepted that medical value could be obtained from drinking the water from mineral springs, or more from the mineral bath. As a result, spas and mineral springs were developed at many places in the USA. This was also a time when the study of mineral water for bottling was a topic of active research. Some great scientists, including English chemists Joseph Priestley (1733–1804) and William Henry (1774–1836) were involved in mineral water development and production. The Schweppes Company, the earliest bottler of carbonated water, regards Joseph Priestley as "the father of our industry". Jacob Schweppe's association with Priestley in developing "aerated" waters for commercial sale began in Geneva in 1783. Carbonated waters were developed to imitate the popular and naturally discharged waters from other famous springs. Priestley suggested that carbonated water might be useful in curing or preventing sea scurvy. In 1781, Thomas Henry, father of William Henry (who developed Henry's Gas Law in 1802), described a prescription of making carbonated waters for use at sea. His recipe, slightly modified, is as follows: to every gallon spring water add 1 scruple of magnesia, 30 grains of Epsom salt, 10 grains of common salt, and a few pieces of iron wire or filings. The operation was then to proceed impregnating the water with fixed air. If intended for keeping, the water must be put into bottles closely corked and sealed.

In America, the renowned geologist, Benjamin Silliman, Professor of Chemistry and Pharmacy at Yale University, also known for his research on oil from the first producing well in the USA (the Drake Well), began to produce bottled soda water on a commercial scale in 1807. He most likely developed his interest in carbonated water during his trip to England, Scotland, and Holland in 1805. In Philadelphia, Joseph Hawkins was issued the first US patent for the preparation of imitation mineral water in 1809. However, as would be expected, making mineral water was not accepted by the producers of natural mineral water from springs. In 1806, S. R. Stoddard published his book, *Saratoga Springs, Its Mineral Waters*, and wrote "Artificial mineral waters are, if possible, avoided. Nature can only be imitated, never equalled." The 122 springs in Saratoga Springs are heavily charged with carbon dioxide gas and contain various proportions of bicarbonate, chloride, calcium, sodium, and magnesium. The first iodine discovered in the United States was in Saratoga water in 1828. These springs made Saratoga one of the most fashionable watering places in North America. It was widely recognized in the United States as "the Queen of Spas". However, it lost much of its aura of respectability with the introduction of horse racing in 1863 and the opening of the casino in 1867.

The discovery of many springs towards the end of the 18th century led to the development of spas in the eastern states and many spas opened along the Appalachians. Westbound settlers found mineral springs in every state except North Dakota. The Rocky Mountains were particularly rich in mineral waters. Wyoming, richest of all with 2 244 springs, also has the largest hot spring in the world, the Big Horn Spring, with a flow of 60 million l a day. In 1886, the US Geological Survey (USGS) listed 2 822 American spring localities with a total of 8 843 springs, of which 223 were sources of commercial mineral waters.

By the middle of the 19th century, bottling of mineral waters was a well-established American industry that provided water for table and medical use, some of which were exported to European cities. Saratoga Springs was a major source of bottled water during the 19th and early 20th centuries. By 1900, many large wells were drilled for the production of carbon dioxide to be used in the manufacture of soda pop and other drinks. The artificial well pumping severely decreased the spring flow. To preserve the natural resources, the Saratoga Park-Reservation was established. Through the authority of State ownership, it was possible to preserve this mineral resource by closing many wells and carefully regulating the flow of water and gas from the producing wells. Hydrogeologists played a key role in the legal and legislative battle that led to this early conservation measure.

By recognizing the value of mineral waters, the United States government built its Army and Navy General Hospital in Hot Springs, Arkansas in 1882. The Hot Springs area was the first land to be preserved as a federal reservation. Through the height of its popularity, Hot Springs consisted of 17 concessionaire bathhouses and was host to many famous visitors.

Thermal springs have played a major role in developing the National Park System. Of the current 51 parks, 13 contain hot springs. Yellowstone National Park is among the largest and its establishment in 1872 made it the first land to be set aside specifically for a park. Hydrogeologists made a significant contribution to understanding the hydrology and geochemistry of the thermal system. Thermal springs were also instrumental in establishing the small national parks. The Platt National Park consisted of 1.42 acres in southern Oklahoma and was established as a reservation in 1906. The recreation area contains 32 springs, each having slightly different chemistry, of which 18 have high levels of sulphur, four have high levels of iron, three have bromide, and seven have no significant mineralization. The two largest springs discharge about 5 million gal/d.

In recent years, springs have taken on even greater importance for humans with the increasing popularity of commercially bottled water as a result of public awareness regarding the importance of the quality and reliability of drinking water. Bottled spring water produced by hundreds of different bottled water companies around the world is marketed to and consumed by people in virtually every nation of the world. The companies involved are making an increased number of requests for hydrogeological and geochemical information and advice. Although in modern times in the Unites States this industry has not received much professional consideration from hydrogeologists and medical professionals, people have given the terms "spring", "spring water", and "mineral water" special meanings. Many bottled water products are being labeled with terms such as "pure spring water", "mountain spring water", and "natural spring water", implying that "spring water" carries a special connotation of high quality, good taste and healthiness.

Spring and bottled water development spans history over the world. An international standard definition of spring water is badly needed. Water appropriately termed as "spring water" must be clearly associated with a groundwater source feeding a spring or group of springs. In a news article in US Water News (December 1994), the National Spring Water Association released a position on bottled water, "what does not flow naturally is not a spring". They correctly state that many bottled waters labeled as "spring water" come from wells. This issue is discussed more in Chap. 7.

One of the most active groups involved with present day research and case histories of springs and thermal and mineral waters of the world is the Commission on Mineral and Thermal Waters of the International Association of Hydrogeologists (IAH). The Commission meets once a year in a well organized meeting. Technical, as well as administrative discussions are held and each year the meeting is described in detail by the Executive Secretary. The minutes of the Commission are included in an annual book titled *Internal Communications* and represent research work by the members of the Commission on Mineral and Thermal Waters. These papers are not just case histories but contain some very detailed research results. Unfortunately, the annual meeting books that contain detailed descriptive material about the geology, hydrology, and geochemistry of springs, mineral and thermal waters, geothermal energy, and the environment are not widely distributed except to the membership. This material comprises a very excellent source of reference material for those doing research on mineral, thermal and bottled waters of the world.

During the last half of the 20th century, the earliest quantitative studies on spring discharge and quality were carried out. Some of the most detailed studies were undertaken at Huntsville Spring in Madison County, Alabama, USA. This work included detailed studies of satellite imagery, sequential air photography, geological mapping, well and spring inventories, test drilling and establishment of monitor wells and springs, water quality studies and detailed pumping tests at springs and wells. A number of reports were published on these studies. One of the most detailed, *Geology and Groundwater of Madison County* is by Mamberg. A final report, *An Environmental Atlas* integrated all of this information in the first atlas of its type (LaMoreaux 1975).

Extensive studies have also been made of a number of springs in the United States: Silver Springs, Florida, discharging from the Ocala; and Hot Springs in Hot Springs, Arkansas. Probably the most detailed examination of geologic and hydrologic controls for springs are the studies of the Edwards Aquifer in Texas as described in an unpublished report of the US Geological Survey (1994). Comal Spring, one of the largest springs in Texas issues from the Edwards Formation. It is hydrogeologically connected in a system that provides water to municipalities including the city of San Antonio, Texas, the military, and extensive agricultural and industrial use. Minimum requirements for spring flow have been established to protect the environment.

A rather detailed study emphasizing the importance of geologic structural and stratigraphy with detailed water quality information, pumping test and recession curve analyses has been completed for Figeh Spring near Damascus, Syria (LaMoreaux et al. 1989).

While preparing material for this book, the Senior Editor requested well-known scientists/hydrogeologists around the world to provide information about the most interesting springs in their countries. These scientists, all members of the International Association of Hydrogeologists (IAH), are a part of a membership of about 2 000, representing over 69 countries. IAH was organized in 1956 with several Commissions. Those that are directly involved with research about springs are: the Commission on Mineral and Thermal Waters, the Karst Commission, and the Commission on Groundwater Protection. The following series of selected reports about springs will provide a background on different kinds of springs and how these springs have been developed, how the water is used, historical background, and some legal aspects of spring development and protection.

References

Crouch DP (1993) Water Management and ancient greek cities. Oxford University Press, Inc., New York, NY, 380 p

Green M, Green T (1985) The best bottled waters in the world. Simon & Schuster, Inc., New York, 172 pp

LaMoreaux PE (1975) Environmental geology and hydrology, Huntsville and Madison County, Alabama. Geological Survey of Alabama, Atlas Series 8, Geological Survey of Alabama, University, AL 35 486, 118 p

LaMoreaux PE (1996) Special issue, springs and bottled waters. Environmental Geology 27(2), 142 p, Springer-Verlag Heidelberg

LaMoreaux PE, Hughes TH, Memon BA, Lineback N (1989) Hydrogeologic assessment – Figeh Spring, Damascus, Syria. Environmental Geology and Water Sciences 13(2): p. 77

US Geological Survey (1994) Edwards aquifer bibliography, draft copy. US Geological Survey, Water Resources Division, Texas District, 8011 Cameron Road, Austin, TX 78 753

US Water News (1994) What doesn't flow naturally isn't a spring

W. Schmidt

8.2 Silver Springs, Florida, USA

8.2.1 Introduction

Silver Springs has long been considered the largest freshwater spring in Florida when considering the long-term average discharge. The spring is a classic example of discharge from the extensive Floridan aquifer system which extends from southern Alabama through Georgia into South Carolina, and underlies all of Florida. The springs have been commercially developed since the late 1800s including the world famous glass bottom boats.

8.2.2 Location

Silver Springs is just northeast of the city of Ocala, Florida at lat. 29°12'57" N, and long. 82°03'11" W. From downtown Ocala on US Highways 301 and 441, the springs can be found by driving about 6 miles east on State Route 40. At the large sign for the entrance to the nature theme park, turn right into the parking area. The address is, 5656 E. Silver Springs Blvd., Silver Springs, Florida 34 488.

8.2.3 Background and Description

Silver Springs, which is the headwater of the Silver River, is one of Florida's original tourist destinations. As early as 1878 the now world famous glass bottom boats were first in operation. The springs currently are a commercially developed tourist attraction. Attractions include: a jungle cruise, a jeep safari, a lost river voyage, animal shows, a petting zoo, and the famous glass bottom boats, along with numerous gift shops and restaurants. Several movies have been filmed in the spring pools over the years. These include six original *Tarzan* movies and the television series *Sea Hunt*.

The springs are located along the western edge of the Oklawaha River Valley. The spring run (the Silver River) flows into the Oklawaha River in east central Marion County after winding through dense cypress swamps for about 8 km. The Oklawaha in turn flows north and then east until it empties into the St. Johns River in Putnam County. The St. Johns flows into the Atlantic Ocean at Jacksonville.

There are numerous spring vents from which the water flows in the vicinity of the head pool. The main pool area is about 250 ft in diameter (Fig. 8.1), and it is deepest in the eastern portion. Here, springflow emerges from a cavern located below a limestone ledge at a depth of 25 to 30 ft. The cavern opens towards the west and is 5 ft high and 135 ft wide. Cave divers report that the cavern extends east and northeast to a large room that extends about 65 ft from the opening where it narrows downward to small orifices in the cavern walls and floor. Maximum water depth in the open pool area is about 30 ft, inside the cavern maximum depth measured is 55 ft. From the cavern opening across the open pool area to the southwest the bottom gradually shallows to a depth of about 5 ft. In this area the water flows out over the lip of the limestone ledge, then flows east down the Silver River (Fig. 8.2 and 8.3).

Along the spring run there are numerous smaller springs in the bed and along the edges of the river. These are all within 3 500 ft of the main spring vent. Most open-

Fig. 8.1. Aerial view of Silver Springs from the southeast (Rosenau et al. 1977)

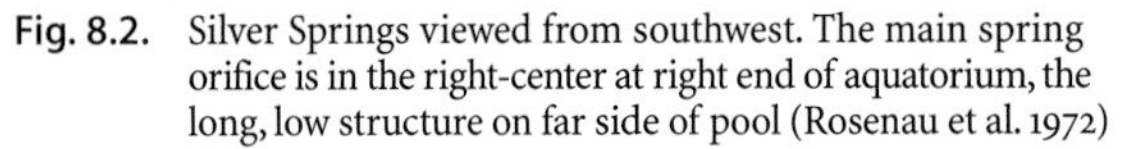

Fig. 8.2. Silver Springs viewed from southwest. The main spring orifice is in the right-center at right end of aquatorium, the long, low structure on far side of pool (Rosenau et al. 1972)

Fig. 8.3. Underwater view west from inside cavern at main orifice of Silver Springs. Photograph taken in January 1970 by L. I. Briel (Rosenau et al. 1977)

ings are about 30 ft deep, and allow the limestone in the bottom of the opening to be clearly seen. Often, the spring water discharging along the various fractures in the limestone forms sand boils above the vent. The majority of the flow originates within the main pool. Along the run, water depths vary from 6 to 30 ft, with the deeper locations resulting from the scour of the bottom by the turbulent water flow. The bottom of the Silver River is often covered with aquatic plants and the banks are lined with subtropical vegetation. The bottom is typically easily visible through the clear spring water. The temperature of the water measured a few feet below the surface in the main pool consistently ranges between 23 and 24 °C (73 and 74 °F). Divers have reported differences in temperature within the large cavern from two different inflow points, one from the floor and one from the wall. One measured 22.5 and the other 24.5 °C (72 and 75 °F). It is postulated that the warmer water originates from a deeper source.

A vast number of fractures and solution channels in the limestone of the Floridan aquifer system supply the water to Silver Springs. It has been estimated that the flow to the springs comes from a drainage area of about 730 miles^2 (Fig. 8.4) and also results from local rainfall.

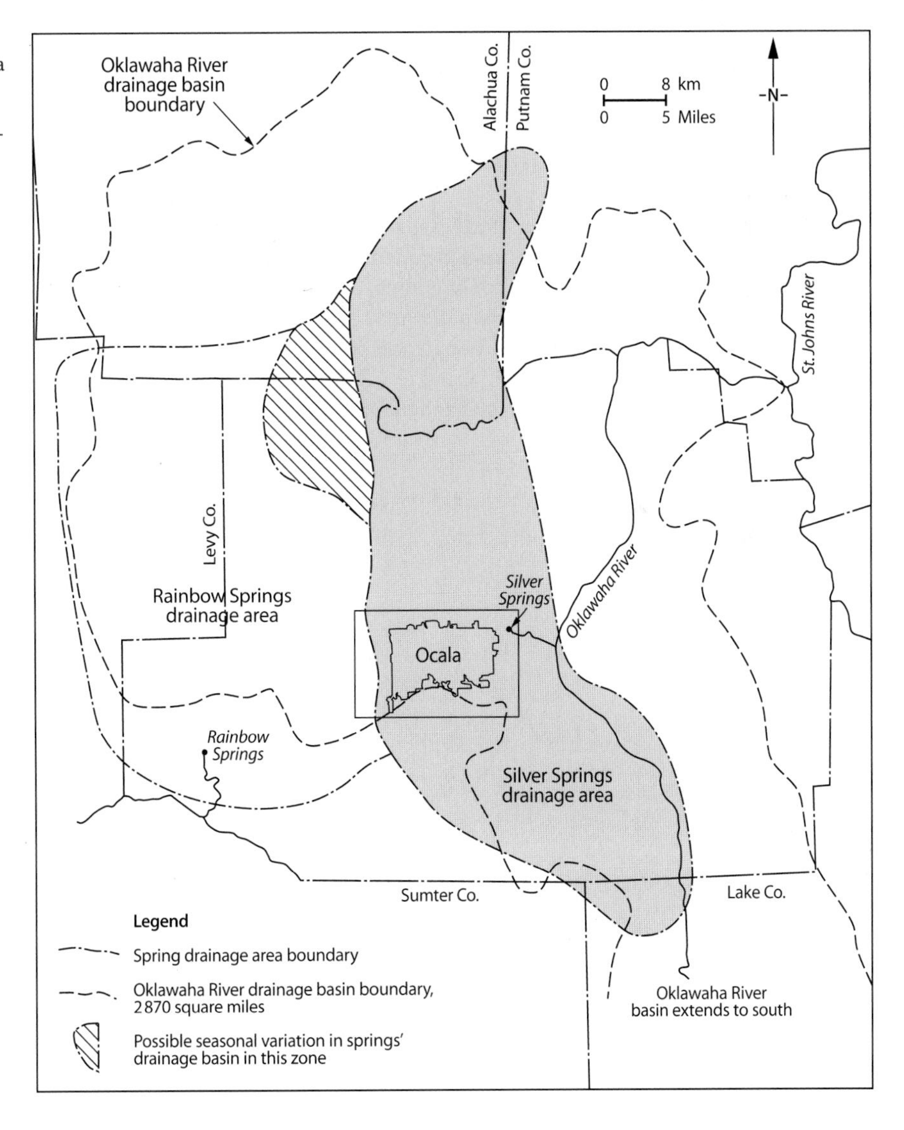

Fig. 8.4. Drainage basin of the Oklawaha River with drainage basins of Silver and Rainbow Springs superimposed (Lane and Hoenstine 1991)

8.2.4 Discharge and chemistry

Some early discharge measurements were recorded by the US Geological Survey as early as 1906. Measurements were sporadic for the next 26 years.

From October 1932 through current times, the US Geological Survey has computed daily discharge rates for the spring. The maximum flow was recorded on 7 separate days in October of 1960, at 1 290 ft^3/s. The highest annual mean also occurred in 1960 at 1 058 ft^3/s. The minimum recorded flow occurred on 12 February 1991, at 517 ft^3/s, and the lowest annual mean occurred in 1956 when 583 ft^3/s was calculated. Figure 8.5 compares rainfall and groundwater levels with the spring discharge from 1981 through 1986. The commercial attraction claims the springs discharge 550 million gal of water each day.

Water quality analyses have also been carried by the USGS. Earliest published data was collected in 1907. Routine analysis has been carried out at least through May 1991 (USGS 1996). Measurements have included: turbidity, transparency, color, specific conductance, dissolved oxygen, biochemical oxygen demand, pH, dissolved carbon dioxide, alkalinity, bicarbonate, carbonate, nitrogen (general, organic, ammonia, and as nitrite and nitrate), phosphate, phosphorus (total and organic), carbon (total and inorganic), hardness, calcium, magnesium, sodium (dissolved and adsorbed), potassium, chloride, sulfate, fluoride, silica, arsenic, cadmium, copper, iron, lead, manganese, nickel, strontium, zinc, aluminum, coliform (total and fecal), streptococci, solids (suspended, residue, dissolved), mercury, and sediment (suspended and discharging). Table 8.1 compares some water quality measurements between the years 1946 and 1972.

In general, the water discharging from the springs is hard, reflecting the dissolved carbonate rocks such as limestone and dolomite comprising the Floridan aquifer system.

8.2.5 Hydrogeology

The Floridan aquifer system is the principal source of water for this area, and indeed the source of water for the springs. In this vicinity, the nearsurface limestone which comprises the Floridan aquifer system is composed of the Ocala Limestone Formation. The Ocala Limestone is an upper Eocene granular limestone (packstone to wackestone with limited occurrence of grainstone). The unit is often soft and friable with abundant large foraminifera, mollusks, echinoids, and coral fossils.

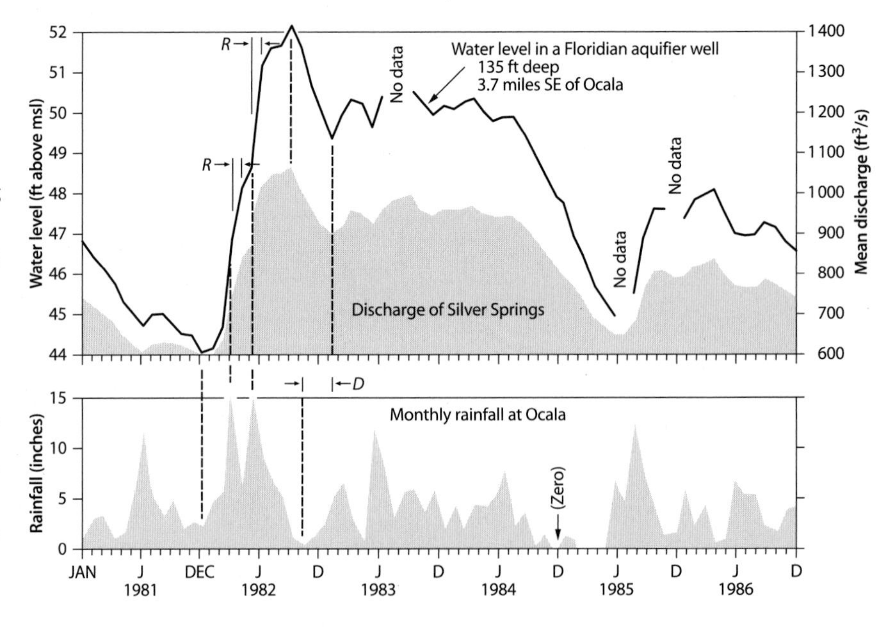

Fig. 8.5. Graphs showing relationships between rainfall, groundwater levels in the Floridan aquifer system and discharge of Silver Springs. Note that all these graphs display very similar curves, which indicates a strong cause-effect relationship. Note the very short recharge lag-time (*R*) between major rainfall events and a rise in the water level in the Floridan aquifer system well, usually a few days. The discharge lag-time (*D*) is longer and represents the delays due to the water's travel through recharge, storage within the aquifer, and eventual discharge at Silver Springs; this can vary from a few days to a few weeks. Data from US Geological Survey, Water Resources Data, 1981–1986 (Lane and Hoenstine 1991)

Table 8.1. Water quality for Silver Springs for 1946 and 1972. Units are in mg/l unless otherwise noted (Rosenau et al. 1977)

Date of collection	21 Oct. 1946	16 Sep. 1972
Nitrite (NO_2 as N)	–	0.00
Nitrate (NO_3 as N	0.29	2.6
Calcium (Ca)	68	68
Magnesium (Mg)	9.6	9.3
Sodium (Na)	4.0	4.3
Potassium (K)	1.1	0.2
Silica (SiO_2)	9.2	9.8
Bicarbonate (HCO_3)	200	200
Carbonate (CO_3)	0	0
Sulfate (SO_4)	34	39
Chloride (Cl)	7.8	8.0
Fluoride (F)	0.1	0.2
Nitrate (CO_3)	1.3	–
Dissolved solids		
calculated	–	242
residue on evaporation at 180 °C	237	246
Hardness as $CaCO_3$	210	210
Noncarbonate hardness as $CaCO_3$	–	42
Alkalinity as $CaCO_3$	–	170
Specific conductance (S/cm at 25 °C)	401	420
pH (units)	7.8	8.1
Color (platinum cobalt units)	4	0
Temperature (°C)	–	23.5
Turbidity (JTU)	–	0
Biochemical oxygen demand (BOD, 5-day)	–	0.1
Total organic carbon (TOC)	–	8.0
Organic nitrogen (N)	–	0.37
Ammonium (NH_4 as N)	–	0.03
Orthophosphate (PO_4 as P)	–	0.14
Total phosphorus (P)	–	0.14
Strontium (Sr)	–	500 µg/l
Arsenic (As)	–	0 µg/l
Cadmium (Cd)	–	0 µg/l
Chromium (Cr^6)	–	0 µg/l
Cobalt (CO)	–	0 µg/l
Copper (Cu)	–	0 µg/l
Lead (Pb)	–	2 µg/l

The Floridan aquifer system is extremely variable lithologically. These changes are a composite result of a number of factors including: original deposition, subsequent diagenesis, structural features and preferential dissolution of limestones, dolostones, and evaporites. The near surface terrain of north-central Florida is extensively altered by karst processes. In the area are numerous springs, caves, sinkholes, and other karst features. The post-depositional processes acting on the limestone are the primary reason the Floridan aquifer system is as prolific as it is. The increased porosity and permeability from these dissolutional and diagenetic processes have resulted in one of the most prolific and extensive aquifer systems in the world. The Floridan aquifer system functions as the principal artesian aquifer from the coastal plain of Alabama, throughout the Florida Platform, beneath the coastal plain of Georgia, and into South Carolina. There are off-shore spring discharges from this aquifer beneath both the Atlantic Ocean and the Gulf of Mexico as well as hundreds of springs dotting the Florida uplands. Florida is home to at least 27 first magnitude springs that discharge water from this aquifer (Rosenau et al. 1977).

References

Lane E, Hoenstine RH (1991) Environmental geology and hydrogeology of the Ocala area, Florida. Florida Geological Survey, Special Publication No. 31, 71 pp

Rosenau JC, Faulkner GL, Hendry CW Jr, Hull RW (1977) Springs of Florida. Florida Bureau of Geology, Bulletin No. 31, 461 pp

US Geological Survey, Water Resources Division (1996) Water quality data printout for Silver Springs, FL

J. B. W. Day

8.3 Springs of Great Britain

Predictably, in a country such as Britain, with its preponderance of consolidated, sedimentary, mainly fissure-flow aquifers, there is a very large number of springs, many of which are, or have been, used for public supply. Migratory springs are a feature of the British (Ur. Cretaceous) chalk, the most important British aquifer. The chalk's low specific yield and high capillary moisture retention together give rise to very considerable water level fluctuations in wells (more than 33 m in some areas) of the unconfined water table. Along the gentle dip slopes of the chalk (North and South Downs of southern and southeastern England) springs may migrate laterally for several miles, giving rise to seasonal streams locally known as "bournes" or "lavants". However, springs such as at Duncton, West Sussex, at the base of the much steeper scarp slopes of the chalk, form point sources, the flows from which tend to be relatively steady; such springs commonly supply and are the original reason for the existence of many of the small towns and villages that nestle along the bases of the chalk scarps of Sussex and Kent.

Where the chalk forms coastal cliffs, a number of springs break out at the base of the cliff between high and low tide levels; there are major chalk coastal springs, for instance, at St. Margaret's Bay (Kent) and at Arish Mells, east of Lulworth Cove, Dorset. Such springs are not used for direct supply (their salinity is usually too high) but are indicators of the presence of local reserves of groundwater for possible future development.

Perhaps one of the most (historically) famous of chalk springs is the Chad Well, to the north of London in Hertfordshire, and the original source of the capital's piped water supply. In the days before the chalk of the London Basin became heavily overdeveloped, with accompanying decline in water levels, the Chadwell Spring is reported to have had an average flow of 9–13 million l per day (million/d).

By far the largest chalk spring (or group of about 20 springs) occurs at Bedhampton, Hampshire, and has an average flow of about 90 million/d, with minimum and maximum flows of 64 and 164 million/d, respectively. The springs drain a catchment of about 64 km^2 and supply, or partly supply, the city and surrounding conurbation of Portsmouth. This group of springs discharges along an axis at the head of a tidal inlet, at a height of a few meters above mean sea level, by an anticline, with accompanying syncline. A few kilometers to the east, similar hydrogeological conditions result in another smaller group of springs at Fishbourne, used to supply the city of Chichester.

Farther to the north, in Lincolnshire, the Lincolnshire Limestone of Jurassic Age forms a locally important aquifer which dips gently eastward beneath overlying clays. In south Lincolnshire, and within the artesian overflow area that results, individual spring flows of more than 20 million/d have been recorded.

During the latter part of the last century, it became fashionable to "take the waters" at a number of spa towns throughout England, Scotland and Wales, where mineral or thermal waters occurred. Perhaps the most famous of these is the English town of Bath where thermal waters have been utilized for bathing since Roman times; the waters are said to have health-giving (or restorative) properties. However, not all the clients of such spas had their health restored. In 1766 the 31-year-old 17th Earl of Sutherland visited the Roman baths at Bath, contracted typhoid, and died while being nursed by his countess, who herself contracted the disease and died shortly afterwards. The warm and highly mineralized waters at Bath occur in beds of Triassic Age, but are believed to have their origin in the underlying Carboniferous Limestone, which crops out in the Mendip Hills some 19 km distant. The nearby spa town of Cheltenham is also well known for its therapeutic waters which issue from Jurassic limestones.

In recent years, there has been a dramatic growth in the bottled water industry in the United Kingdom, all the more marked, perhaps, because usage of bottled water seems to have lagged behind that of continental Europe. Perhaps the British, in their dogged insularity, put a greater faith in the purity and quality of their public water supplies than their continental counterparts, but whatever the reason, nowadays the supermarket shelves all carry a good selection of "still" and "carbonated" "natural mineral water". Most is genuine spring water, but in at least one case the water is derived from a well, although, of course, it is none the worse for that. In England, perhaps one of the best known sources occurs at Malvern, where a number of springs issue from rocks within a Pre-Cambrian inlier. Individual springs here have low but relatively constant flows and are suitable for bottling. An analysis (1985) of water from the Grimeswell Spring indicate the constituents shown in Table 8.2.

Table 8.2. Analysis of the water from the Grimeswell Spring (mg/l)

Compound or parameter	Concentration (mg/l) or value
Calcium carbonate	83
Magnesium carbonate	15
Magnesium sulphate	41
Magnesium chloride	37
Sodium chloride	34
Sodium nitrate	6
Potassium nitrate	4
Silica	10
pH	6.0

Malvern water is said to have been bottled since 1851 and is probably the first natural mineral water to have been so treated.

Another well-known English source of bottled water (from the north of England, Derbyshire Dome region) comprises a group of thermo-mineral springs at Buxton emerging from the Carboniferous Limestone. These springs were known to the Romans, who named the town "Aquae Arnemetiae" (or "the waters of the goddess of the grove"); a recent analysis of water from the St. Ann's Spring shows Table 8.3.

In Scotland there are many springs which issue from rocks of widely differing ages. Some are used for bottling and may be advertised as particularly well suited for mixing with that other well-known and widely drunk Scottish liquid – whiskey, the Gaelic "usquebaugh", the water of life. One leading supermarket chain has its own "Scottish Spring" in the Ochil Hills above the Perthshire Village of Blackford, and another, the "Caledonian Spring" in the Campsie Fells above Lennoxtown Village in Stirlingshire. Results of analyses from these springs are shown in Table 8.4.

These waters may be sold in the natural (still) form or may be artificially carbonated.

Table 8.3. Analysis of water from the St. Ann's Spring (in mg/l)

Compound or parameter	Concentration (mg/l) or value
Calcium	55
Magnesium	19
Sodium	24
Potassium	1
Bicarbonate	248
TDS (total dissolved solids) at 180 °C	280
Chloride	42
Sulphates	23
Nitrates	<0.1
Iron	0
Aluminum	0
pH	7.4

Table 8.4. Analyses of the "Scottish Spring" and the "Caledonian Spring" gave the following results (in mg/l)

	Scottish Spring	Caledonian Spring
Calcium	48	22
Magnesium	13	5
Potassium	1	0.3
Sodium	65	6.8
Bicarbonate	180	77
Sulphate	12	12
Nitrate	3.5	2
Fluoride	0.03	0.07
Chloride	60	10
Silicate	8	5.7
TDS at 180 °C	264	95
pH	7.6	4.6

R. T. Sniegocki †

8.4 The Waters of Hot Springs National Park, Arkansas, USA – Their Nature and Origin

8.4.1 Introduction

The springs of Hot Springs National Park, Arkansas, USA, issue from the plunging crestline of a large overturned anticline, along the southern margin of the Ouachita anticlinorium, in the Zigzag Mountains. The flow of the hot springs is highest in the winter and spring. The dissolved solids of the waters range from 175 to 200 mg/l. Based on carbon-14 analyses, the major portion of the water is 4 400 years old.

8.4.2 Hot Springs – Their Flow, Chemical Characteristics and Age

Rocks in the vicinity of the hot springs range in age from Ordovician to Mississippian. The rocks – cherts, novaculites, sandstones, and shales – are well indurated, folded, faulted, and jointed. The springs emerge from the Hot Springs Sandstone Member of the Stanley Shale near the anticlinal axis, between the traces of two thrust faults that are parallel to the axis of the anticline.

The combined flow of the hot springs ranges from 2 838.750 to 3 595.750 m^3/d (32.9 to 41.6 l/s). The flow of the springs is highest in the winter and spring and is lowest in the summer and fall. The temperature of the combined hot-springs waters is about 62 °C.

The radioactivity and chemical composition of the hot-water springs are similar to that of the cold-water springs and wells in the area. The dissolved-solids concentrations of the waters in the area generally range from 175 to 200 mg/l. The main differences in the quality of hot water, compared with nearby cold groundwaters, are the higher temperatures and the higher silica concentrations of the hot springs. Cold waters in the area generally range from 15.0 to 26.8 °C. The silica concentrations of cold waters range from 2.6 to 13.0 mg/l, whereas the silica concentration of the hot springs is about 42 mg/l. The high silica concentration of the hot springs is due to the increased solubility of silica in hot water. The silica concentration of the hot springs indicates that the maximum temperature of the hotspring water is no more than a few degrees higher than the temperature at which the springs emerge.

Tritium and carbon-14 analyses of the water indicate that the water is a mixture of a very small amount of water less than 20 years old with a preponderance of water being about 4 400 years old. The deuterium and oxygen-18 concentrations of the hotspring waters are not significantly different from those of the cold groundwaters.

The presence of radium and radon in the hotspring waters has been established by analysis. Analyses done in 1973 showed the radium concentration to be 2.1 pCi/l. Analyses made in 1953 of the radon gas, a radioactive decay product of radium, ranged from 0.14 to 30.5 nCi/l.

Mathematical models have been employed to test various conceptual models of the hot-springs flow system (Bedinger et al. 1979). The geochemical data, flow measurements, and geological structure of the region support the concept that virtually all the hotspring water is of local, meteoric origin. Recharge to the hotspring artesian-flow system is by infiltration of rainfall in the outcrop areas of the Bigfork Chert and the Arkansas Novaculite. The water moves slowly to depth, where it is heated by contact with rocks of high temperature. Highly permeable zones, related to jointing or faulting, collect the heated water in the aquifer and provide avenues for the water to move rapidly to the surface.

References

Bedinger MS, Pearson FJ Jr, Reed JE, Sniegocki RT, Stone CG (1979) The waters of Hot Springs National Park, Arkansas – their nature and origin. Geological Survey Professional Paper 1044-C: C1

Uri Kafri

8.5 Main Karstic Springs of Israel

8.5.1 Introduction

A series of karstic springs in Israel belongs either to the western (Mediterranean) or eastern (Rift Valley) watersheds. Most of them are presently managed or diverted. Salinities range from very fresh through brackish to very saline waters.

The main karstic springs of Israel drain karstic, hard carbonate aquifers of Jurassic to Eocene age. Among these, the largest regional aquifer is that of the Cenomanian-Turonian aquifer which builds the mountain crest of Israel.

The different springs belong either to the western watershed, issuing along the foothills, or to the eastern watershed, in connection with the Rift Valley base level. The discharge of most of the big springs has been diminished and is presently managed by nearby exploitation wells or by diversion canals. The data given here are therefore historical, prior to the management of the springs.

8.5.2 Western Watershed

8.5.2.1 *Rosh Ha'ayin Springs*

These springs used to drain a large watershed of the Cenomanian-Turonian aquifer of central Israel, the average discharge being over 200×10^6 m^3/yr. Average chlorinity of the waters is about 200 mg/l.

8.5.2.2 *Taninim Springs*

Draining the Cenomanian-Turonian aquifer of the Samaria and Carmel Mountains, the average discharge is approximately 25×10^6 m^3/yr. The waters are brackish and the chlorinity is up to 1 000 mg/l, due to the mixture of freshwaters with seawater and/or deep-seated brines.

8.5.2.3 *Afeq Springs*

They drain the Cenomanian-Turonian aquifer of western Galilee Mountains with average historical discharges of approximately 50×10^6 m^3/yr. The average chlorinity, several hundred milligrams per liter, is a result of mixtures between freshwater and intruding seawater.

8.5.2.4 *Kabri Springs*

They drain the Cenomanian-Turonian aquifer of the western Galilee Mountains with average historical discharges of approximately 10×10^6 m^3/yr. Due to the low chlorinity, below 20 mg/l, they were used in the past to manufacture bottled mineral waters.

8.5.3 Eastern Watershed

8.5.3.1 *Jordan River Sources*

These include the three big springs, namely the Banias, Dan, and Hazbani Springs, with a cumulative average discharge of approximately 500×10^6 m^3/yr. These springs drain the large Jurassic karstic aquifer of Mt. Hermon to the Jordan River which flows southward through Lake Tiberias to the Dead Sea. The chlorinity of waters is very low, about 20 mg/l.

8.5.3.2 *'Einan Springs*

They drain the western Galilee Cenomanian-Turonian aquifer at a historical average discharge of approximately 20×10^6 m^3/yr, with a chlorinity of a few tens of milligrams per liter. At present, the waters are managed and pushed upward to the mountain area.

8.5.3.3 *Lake Tiberias Saline Springs*

They drain the eastern Galilee Cenomanian-Turonian and Eocene aquifers, along the western shores of Lake Tiberias and at the bottom of the lake, some of which are thermal. The onshore known cumulative average discharges are approximately 30×10^6 m^3/yr, with chlorinities ranging between those of freshwaters to those of Mediterranean waters. Salinity is attributed to the mixture of freshwater with either deep-seated brines, and/or seawater that emerges along the Rift Valley border faults. The brackish waters are presently diverted to prevent salination of the lake waters.

8.5.3.4 *Bet Shean Valley Spring*

This drains the Cenomanian-Turonian and Eocene aquifers to the Rift base level. The cumulative average discharge is approximately 85×10^6 m^3/yr, and the chlorinities range from a few hundred to 1 000 mg/l due to a mixture with saline end-members.

8.5.3.5 *Dead Sea Saline Springs*

A few springs along the western fault escarpment of the Dead Sea drain the Cenomanian-Turonian aquifer, with average discharge of a few million cubic meters per year. These waters are partly mineral and thermal and are used as spas. The chlorinity, up to a few thousand milligrams per liter, is a result of mixture of freshwater with calcium chloride concentrated brines.

A. S. Issar · Philip E. LaMoreaux

8.6 Kadesh Barnea

The spring mentioned frequently in the Bible as a camp of the tribes of Israel was Kadesh Barnea. From here the mission headed by Joshua and Caleb went to find the promised land and from here the first attempt to enter the promised land was made. Kadesh Barnea is an ideal place for the concentration of people in the middle of the desert.

A large spring issuing from the limestones of Eocene rocks emerges at the lower slope of an anticline of the Ramon, which is the highest mountain of the Negev (at an altitude of 1 000 m above m.s.l) and receives more precipitation. The water infiltrates into solution channels in the limestone rocks of Middle Eocene age until it reaches impermeable chalk layers of Lower Eocene age and marls of Paleocene age, on which a regional perched water table is formed (Fig. 8.6). In addition to the main spring of Kadesh Barnea, special geological conditions create many small springs and seeps from the rocks in this area. One spring still has the name Ein Qadis. The outlet of the narrow valley of Kadesh Barnea is into a broad valley at the small town of Al-Quseime. Here, many small brackish springs seep from the ground, supporting shrubs, palms, and grasses. Today the spring of Ein Qudeirat flows at about 40 m^3/h. This quantity may be sufficient for supplying drinking water for a few tens of thousands of people provided they water their stock from the brackish springs in the lower valley. Realizing that a more humid period would have caused an even larger flow, one can understand the rationale in making Kadesh Barnea the pivot point for the wandering tribes.

It is more difficult to understand the story about the need to strike the rock to get water at Kadesh. Was the story based on another place in order for the tribes to move to a more permanent location or did the spring of Kadesh also fail, thus necessitating excavation into the rock? According to records known to date, this spring did not dry up even after a series of a few dry years, although its flow did diminish. One possible explanation is that the last stages of the establishment at Kadesh Barnea were the start of a dry period, which caused the springs to dry up, requiring the digging of wells. The same dry period forced the Israelis to try and break out of the desert into the more humid lands. The less conservative approach suggests that this story was attrib-

Fig. 8.6.
Aerial photograph of the oasis of Ein Qudeirat (Kadesh Barnea). The spring outflows (at the *upper left* of the picture) from the limestones of Eocene age which builds the surrounding hills. The ancient 'tel' can be seen in the *upper part* (photograph by Scientific Survey, Sinai)

uted to this place by the writers and editors of the Bible during some later period. (Editor's note: as previously mentioned, the purpose of the study by Issar was not to explain biblical and mythological phenomena. Thus, it can be stated that with a better knowledge of the Bedouins' methods of excavating for water in the crystalline rocks, or the story of Moses' striking of the rock, the spring or well study can be used to describe a hydrogeological situation at Kadesh Barnea, Ayun Musa or in Wadi Fieran.)

The first attempt of the Israelis to enter Canaan by force and their first defeat was from their base at the foot of the mountains near the springs at Kadesh Barnea. The Israelis were beaten back by the "King of Arad". The interpretation of archeological excavations at the site definitely identified the ancient town of Arad indicating that this town was flourishing as a strong fortified city during the lower part of the Early Bronze Age. It was abandoned at about 4600 B.P. (2600 B.C.), and was not rebuilt until the Iron Age. Thus, during the Upper Bronze Age, the assumed period when the Israelis arrived, the town was abandoned (Aharoni and Amiran 1964). In Issar's opinion, there is no controversy between the story of the Bible and the archeological findings. In general, the history of this site as revealed by the excavations agrees with the paleoclimatic information for the area (Issar 1990). The city flourished during the Early Bronze Age which coincides with the wet climate period which continued from the Chalcolitic period until its abandonment at about 4600 B.P. This is in agreement with the desiccation of the climate, which can be seen on the ^{18}O curve of the core samples taken from the bottom of the Sea of Galilee.

The significance of the knowledge regarding well or spring locations becomes an important archaeological tool. For example, while the city of Arad was still occupied, a deep shaft was excavated into the shallow groundwater table. Although the archeologist Ruth Amiran (pers. comm. with Issar), who excavated the site, maintains that the well was developed in the Iron Age. Issar disagreed with this conclusion because the archeological determinations did not take into account that a humid period could have caused the water table to rise or springs to flow and thus a shallow well could have provided sufficient water for the town's water supply. During the Iron Age, when methods of digging wells improved, the well could have been enlarged in diameter and deepened.

At Arad, this well provided a supply of water in times of siege. As the area of recharge for this water table is rather limited, it seems probable that a series of dry years and the concentrated building in the zone of recharge caused the depletion of the source of water supply and the desertion of the city by most of its people. Spells of wet years during the Upper Bronze period caused the water table to rise and, as the sites of wells or springs were well known to the semi-nomads who dwelt in the surrounding area due to tradition passed from father to son, and they would again gather around it. Thus, it can be understood that a site such as Arad with a well or springs, like other sites in the vicinity, were settled.

At some stage, the Israelis decided to avoid a direct assault across the desert border, where too strong an opposition was encountered, and to circumnavigate it. They tried to march through the Land of Edom, most probably the present Central Negev but were denied free passage. They then pressed southward toward the Gulf of Elat, crossing somewhere north of the gulf, to reach the more humid heights of the mountains of Moab and Ammon where they avoided war by abstaining from harassing the local inhabitants whom they considered kin. They moved northward, defeated the Amorites inhabiting the Gilead and Bashan heights and were ready to cross the Jordan. After they had crossed the Rift Valley and proceeded up the mountains of Trans-Jordan, *no more complaints of thirst were reported.* During this part of their journey the tribes experienced an event connected with the excavation of a water well, an occasion that prompted the composing of a special hymn:

> Spring O Well, sing ye unto it. The princes digged the well, the nobles of the people digged it, by the direction of the lawgiver, with their stave. (Numbers 21:17,18)

In conclusion and significant about these stories of Biblical wells and springs is that the hydrogeologist using the tools of his trade is able to contribute scientifically to the archaeological history of an area.

References

Aharoni Y, Amiran RA (1964) Archeology 17:43–53
Issar A (1990) Water shall flow from the rock. Hydrogeology and climate in the lands of the Bible. Springer-Verlag, New York

C. R. Aldwell

8.7 Mallow Springs, County Cork, Ireland

8.7.1 Introduction

Because of its copious and reliable rainfall Ireland has an abundance of springs. Many of the larger ones issue from the Carboniferous limestone that occur in over 40% of the country. The spring water is mainly a calcium bicarbonate type with a temperature of about 10 °C. In the 18th century, warm and cold springs were developed as spas in various parts of Ireland. The popularity of these springs was short and most were in major decline by 1850. Today, only one cold water spa at Lisdoonvarna, Co. Clare is still operating. Springs in Ireland were places of religious significance for the pre-Christian Druidic religion. In the Christian period they became holy wells, under the patronage of various saints, and cures for many different ailments were attributed to water from these wells or springs.

Ireland is located in a seismically stable region with no extensive deep aquifers or hot springs (Fig. 8.7). The average thermal gradient in Ireland is 20 °C/km and lukewarm water springs of 20–23 °C occur only in two areas. One such area is at Mallow in north County Cork.

8.7.2 History of Mallow Springs

A number of springs rise in an area known locally as Spa Glen. Since warm springs are rare in Ireland it is to be expected that they would have been a source of awe and reference in historic times. With the advent of Christianity one of the Mallow Springs was dedicated to the national apostle of Ireland – St. Patrick. The largest of the group is known as Lady's Well.

The belief that the spring had medicinal properties stems from the work of Dr. Rogers of Cork. Called to treat a patient in Mallow in 1727, he learned that the water from the spring was the only liquid she was able to retain. Her subsequent recovery was attributed to the spring, and Dr. Rogers invited J. Rutty from the Bristol Spa to visit Mallow. Rutty was very excited by what he observed and in his book *Mineral Waters of Ireland*, published in 1757, he wrote with obvious enthusiasm of the valuable medical properties of Mallow water. He compared it favorably to the English springs at

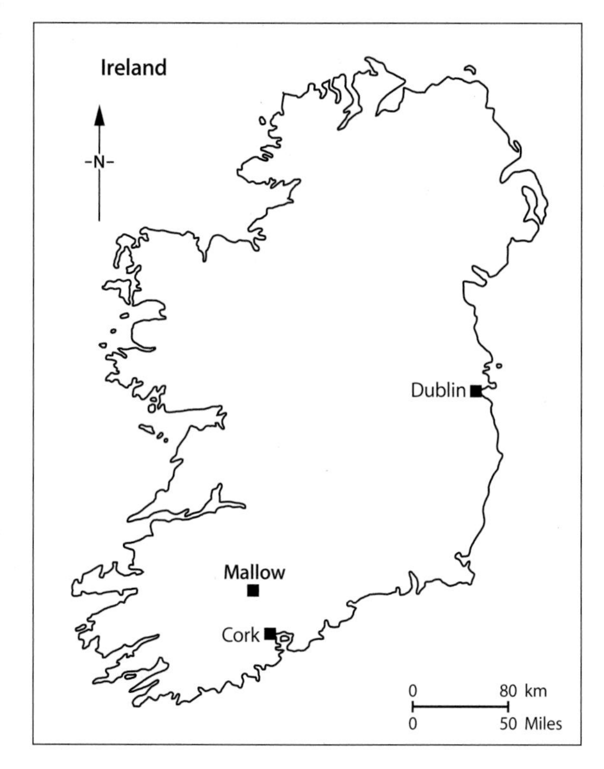

Fig. 8.7. Location Map

Bristol and Buxton, noting "there is one peculiar advantage in the use of this and other waters endued with the like small degree of heat, that it is attended with less danger and requires less caution than the hotter baths such as Aix-la-Chapelle, Bath, etc." Rutty (1757) quotes Rogers as reporting a wide range of cures including: disorders of the stomach and skin; respiratory problems such as catarrh and coughs and asthma; urinary disorders; and diabetes. Thanks largely to Rogers, by 1735 Mallow was famous throughout Ireland as a spa. Moreover, because of its mild climate, pleasant scenery, and social life, Mallow rapidly developed to become a fashionable vacation resort. Spring and Autumn were the most important times for visitors. From about 1775, a decline in popularity appears to have begun. This was in part due to changes in fashion, but the succeeding 19th century was one of the most turbulent and traumatic in Irish history. Social upheaval reached its peak in the Great Irish Famine (1845–1848). Death and disease (cholera, typhus, etc.) stalked the land and the confidence and morale of Irish society were severely shaken.

In any event, by the 1850s Mallow Spa had ceased functioning as a resort. For more than 100 years it remained unnoticed except by local people and an occasional scientist. One such visit was made in 1876, and a paper read to the Royal Irish Academy the following year on the gaseous contents of the Mallow Spa (Plunkett and Studdert 1883). In the 1960s one of the springs was made into a children's bathing pool.

In 1973 came the oil crisis, and a renewed interest in alternative sources of energy began. In 1975 the EC commenced a series of 4-year geothermal R&D programs. Mallow was again examined as a part of the 1979–1983 Irish contribution to the EC program. This provided for a combined study involving University College Cork, the Geological Survey of Ireland, and the Cork County Council. Various test boreholes provided additional information. Above all, Cork County Council, prompted by the determined enthusiasm of Deputy County Engineer Michael O'Brien, set out to try and provide a modern use for this historic spring.

8.7.3 Geological setting of Mallow Springs

The rocks of the Mallow area comprise Devonian sandstones of terrestrial origin and shales with Carboniferous limestones, sandstones and mudstones. The geological structure of Mallow is a complex series of major thrusts associated with the Hercynian orogenic belt of southern Ireland. The source of the warm water is considered to be the Dinantian Reef limestone (Bruck et al. 1986). This rock is a pale, locally highly silicified micrite and at Mallow is estimated to be 1200 m thick (Bruck et al. 1986). The fracturing of the limestone resulting from the tectonic activity provides ample opportunity for the upward movement of warm water to the springs.

8.7.4 Flow and Temperature of the Springs

The springs were monitored regularly by University College Cork for three years (1980–1982). Lady's Well, the largest spring, had a mean flow of 617 l/min. The discharge of the complex was thought to be about double that of Lady's Well (Bruck et al. 1986). The mean temperature over the three-year period was 19.5 °C, with a range of 17–22 °C. The warmest temperatures were recorded in summer and lowest in winter, roughly co-

Table 8.5. Chemical constituents (ppm) of water from Lady's Well, Mallow, compared to cold control spring (Bruck et al. 1986)

	Lady's Well	Cold control spring
Conductivity	466	534
Ca	73	114
Mg	11	5
Na	15	15
K	2	2
Si	4	3
HCO_3	234	289
SO_4	17	16
Cl	27	34
NO_3	3	9
PO_3	0.02	0.02
F	0.06	0.1

inciding with seasonal variations in air temperature (Shannon Airport 30-year mean monthly average air temperature, coldest month, January, 5 °C; hottest month, July, 15 °C).

8.7.5 Chemistry of Mallow Springs

The chemistry of the springs was investigated by University College Cork over the period September 1981 – January 1983. The results of the analyses for the principal parameters of water from Lady's Well and a cold control well are given in Table 8.5.

The results of the analyses indicate that the water is a calcium bicarbonate type and similar to the local groundwater in the limestone aquifers. The main differences are the lower calcium, bicarbonate, and nitrate concentrations in water from Lady's Well.

Isotopic and gas analyses from Lady's Well showed $^4He \times 10^7$ at about 170 and tritium (TU) of 11. These results are interpreted as showing deeper circulation and longer residence time than usual for Irish groundwater.

Results of silicon geothermometry tests suggest an aquifer temperature at Mallow of 30–40 °C at a depth of about 1100 m.

8.7.6 Use of Mallow Springs – Present and Future

In the 1980s, a renewed interest awakened in the use of Mallow's warm water, prompted by the EC Geothermal Program (Aldwell et al. 1985). Shallow exploration boreholes penetrated large amounts of 20 °C water. With heat pumps, this water is heating the local municipal swimming pool. The Spa House, in which the original spa well is located, has been completely renovated and there are plans to open it to the public (Fig. 8.8).

At the same time, the Cork County Council has ambitious plans for a major development of Mallow based on the warm water resource and the town's position as the crossroads of the province of Munster. The proposal is for a natural energy park. Included in the plans are: (1) space heating using heat pumps, (2) aquaculture and fish hatcheries for native and exotic species, (3) greenhouses and hydroponics, (4) tourism – gracious living holidays and an exhibition center with associated educational courses, and (5) a school of natural energy. Mallow Springs once again is set to assume importance after a gap of nearly 150 years.

Acknowledgements

References for this paper came from work of the late David Burdon, hydrogeologist, who lived some 10 km from Mallow Spring, and Peter Bruck, Professor of Geology in University College Cork, in interpreting the hydrogeology of Mallow Spring. I also acknowledge the vision and persistent efforts of Deputy County Engineer Michael O'Brien in gaining recognition for the resource potential of the warm waters of Mallow and who assisted in the preparation of this paper

References

Aldwell CR, Burdon DJ, Peel S (1985) Heat extraction from Irish groundwater. Memoirs of 18th Congress IAH, pp. 79–94
Bruck PM, Cooper CE, Cooper MA, Duggan K, Wright DJ (1986) The geology and geochemistry of the warm springs of Munster. Ir J Earth Sci 7:169–194
Plunkett W, Studdert L (1883) Report on the solid and gaseous constituents of the Mallow Spa in the county of Cork. Proceedings of Royal Irish Academy Vol. 3, Series II, Science Pt. 1, pp 75–78
Rutty J (1757) Mineral water of Ireland – book V, of the warm waters of Ireland and particularly those of Mallow. National Library of Ireland, Ref. J. 61338, pp. 286–304

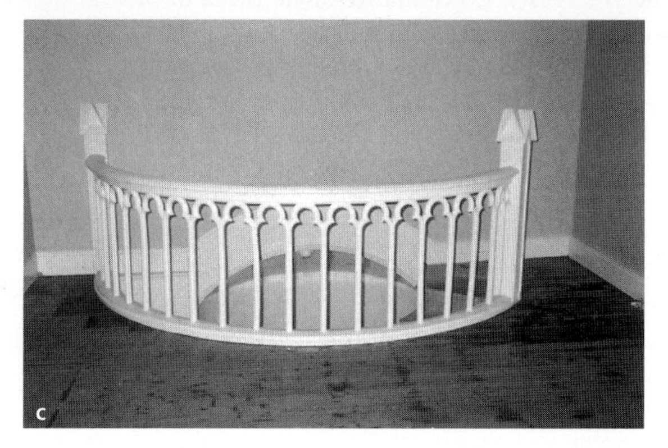

Fig. 8.8.
a Spa House. **b** Mallow and Lady's Well.
c Entrance down to Spa House

D. Drew

8.8 Lisdoonvarna Springs and Spa Wells, County Clare, Ireland

8.8.1 Introduction

Lisdoonvarna is a small town with a resident population of ca. 700, which is seasonally increased severalfold by tourists. It is located in the northwest of County Clare, western Ireland about 10 km from the Atlantic seaboard and adjacent to the karstic Burren plateau. Lisdoonvarna has two claims to fame: it is the only currently operative Spa in Ireland, and it is the centre for a long-established match-making festival held each September (postharvest) at which bachelor farmers are found partners by skilled match-makers. These two features of Lisdoonvarna are not wholly unconnected even though the September festival now attracts many young people from outside of the area who may not patronize the spa waters.

8.8.2 Geological Setting

In the area of the Lisdoonvarna Springs, the Visean limestones that comprise the Burren plateau dip beneath the younger Upper Carboniferous Namurian strata in the vicinity of Lisdoonvarna. It is on these latter strata (predominantly shales, siltstones and sandstones) that the flow of mineral springs are located (Dwyer 1876). The Namurian strata are less than 20 m thick in the area, and many of the rivers have been cut down to the level of the underlying limestone. The geologic column in the Lisdoonvarna area includes: (1) ribbed beds (sandy shales, the youngest), (2) Goniatite Shale (8–14 m of black shales), (3) Phosphate Shale (0–3 m of dark grey-black shales with calcite and dolomite, very hard and compact), (4) Cahermacon Shale (0–3.5 m of black shale with some phosphate and pyrites), (5) base of Namurian, (6) top of Visean; and (7) Burren Limestones (the oldest).

In some outcrops in the area the phosphatic shale rests directly upon the limestone, the Cahermacon Shale being absent. The shales are rich in organic matter as well as iron sulphide and concretions of calcium and magnesium carbonate, while in addition to phosphate the Phosphatic Shale also contains silica, sodium, manganese, uranium, and aluminum. The strata are almost horizontal with a 1–2° dip to the southwest in places. The soils of the area are acidic and heavily leached with extensive peat deposits.

8.8.3 Hydrology and Chemistry of the Springs

The Namurian rocks form a broadly dissected plateau in west County Clare. In the Lisdoonvarna area, a series of streams have eroded deeply into the physically weak strata – probably in postglacial times – to create a series of deep gorges. It is the existence of these gorges that have allowed the mineral springs to develop, groundwater in the shales emerging at or close to the floor of the gorges at numerous points but in particular at the level of the Phosphate Shale. It appears that the Phosphate Shale represents an effective barrier to the vertical movement of groundwater and that lateral flow predominates above the stratum (Fig. 8.9). Figure 8.10 illustrates the angled joints through the Clare Shales that serve as the main mode of transmission of groundwater.

The Namurian rocks are conventionally regarded as being unproductive in groundwater, with effective porosities of 1–2% at most. However, it is apparent that a limited circulation of groundwater is possible, almost certainly in joints and secondarily between the laminae of the shale bands and it is this limited recharge (5% of rainfall) that feeds the springs and wells.

The spa wells that are recognized are simply a small proportion of the total number of groundwater outflows in the area and have presumably been developed because of their chemical and discharge characteristics (Mapother 1871). During the 19th century, six separate

Fig. 8.9. Incised gorge showing groundwater flow directions

Fig. 8.10. A 3-m section through the Clare Shales showing angled joints, the main mode of transmission of groundwater

Fig. 8.11. The Lisdoonvarna Pump Room and Sulphur Baths on the banks of the Gowlaun River. The Sulphur Well waters overflow into the river at this site

springs, each with a distinct chemical characteristic, were exploited, but at present only three sites are operational, the Magnesia and Iron wells located only a few metres apart and the Sulphur Well at the further end of the town (Flinn 1888). Only limited chemical data are available, and they suggest that none of the springs have very high mineral concentrations except for iron (ca. 4.5 mg/l in the Iron Well and 3 mg/l in the Magnesia Well), sulphate (150 mg/l in the Iron Well) and total hardness (344 and 300, respectively, in the Iron and Magnesia wells).

The Sulphur Well has a low level of total dissolved solids (TDS) but contains an average 4.4 cc/l of H_2S gas with a range of 2.5–7.5. None of the springs are thermal; the temperature of the waters is approximately the mean annual air temperature (10–11 °C) for the locality.

As would be expected from the very limited permeability of the rock and from the restricted catchment areas for each spring, outflows are very small indeed: ca. 400 l/h for the Sulphur Well, 80 l/h for the Iron Well, and 40 l/h for the Magnesia Well. It is remarkable, given these small discharges, that the town was able to develop spas.

8.8.4 History of the Spa Waters

Although Lisdoonvarna did not officially become a spa until 1845 and its heyday was during the later years of the 19th century, its mineral waters were known and were used at least a century earlier. For example, Dutton (1808) observed: "Lisdoonvarna has been long celebrated for its virtues, particularly in obstructions and some find it beneficial after winter's drinking of bad whiskey from private stills; strongly ferruginous, and of an astringent taste and strong smell, but not fetid."

During the late 19th century and into the early years of this century, individual spring waters were used for particular purposes. For example, the Copperas Well was used (externally) for skin diseases, the waters from the Magnesia Well were used as table waters in the hotels, while the sulphur water was used for bathing and for drinking. For many years in the late 19th century a female named Biddy the Sulphur dispensed free glasses of sulphur water to visitors.

The Victorian pump room, baths, and pumping equipment still function at the Sulphur Well, but the facilities have been expanded in recent years to include saunas, a solarium, etc. (Fig. 8.11). Although sulphur water is sold both by the glass and in bottles, the mineral springs are still relatively undeveloped in comparison with spas elsewhere in Europe (Luke 1919).

References

Dutton H (1808) Statistical survey of Clare. Graisberry and Campbell, Dublin, 369 p

Dwyer P (1876) Handbook to Lisdoonvarna and its vicinity. Hodges, Foster and Co., Dublin, 86 p

Flinn E (1888) Irish health resorts and watering places. Kegan Paul and Co., London, 175 p

Luke TD (1919) Spas and health resorts of the British Isles. A. & C. Black, London, 318 p

Mapother ED (1871) Lisdoonvarna Spa and some other Irish watering places. Fannin and Co., Dublin, 117 p

H. Idris

8.9 Springs in Egypt

8.9.1 Introduction

Descriptions of springs in Egypt deal only with natural springs producing potable water. No water from natural springs in Egypt is bottled, and thermal springs are not identified. Egyptian standards state that the total dissolved solids in potable water should not exceed 1000 ppm, except in Siwa, where the only available source for water for human consumption is from springs that have water containing more than 2000 ppm total dissolved solids (TDS). Six natural springs in Egypt are typical examples for Sinai and the Western Desert: Ain Furtaga in the southern pre-Cambrian province of Sinai Peninsula; Ain El Gudeirat in the sedimentary plateau of north Sinai; and Ain El Bishmo, Ain El Bousa, and Ain El Gabal in the Western Desert Oases of Bahariya, Kharga, and Dakhla. The springs in the Western Desert discharge from the Nubian Sandstone aquifer system. The sixth spring, Ain El Arayes, is in Siwa Oasis.

8.9.2 Springs of Sinai

8.9.2.1 *Ain Furtaga*

Ain Furtaga is the largest spring in the southern province of Sinai. It is in the downstream part of Wadi Watir 15 km from the Gulf of Aqaba. Wadi Watir incises the Pre-Cambrian igneous mass in an area of perpendicular parallel dikes. The dikes, acting as barriers to inundation of water from floods, and water percolates into the sorted aggregates sedimentation in the Wai flood plain during floods of short duration. In Wadi Watir, the alluvial fill is characterized by the presence of considerable quantities of silt and clay material filling between coarser aggregates. Wadi Watir and Wadi Ghazal merge and continue their course to the Gulf of Aqaba. At the merging point, Wadi Watir sedimentary fill is eroded and refilled with ag-

Table 8.6. Chemical analyses for water samples from some springs in Egypt

Spring	Electr. conduct.	TDS	pH	Temp.	Cations concentration				Anions concentration		
	(S/cm)	(ppm)		(°C)	Units	Mg	Ca	Na + K	HCO_3	SO_4	Cl
Ain Furtaga	2010	1450	7.9	24	ppm	18	135	173	166	415	185
					EPM[a]	1.48	6.74	7.47	2.72	8.64	5.22
					Z	9.43	42.96	47.10	16.41	52.11	31.48
Ain Gudeirat	2000	1440		23	ppm	53.38	85.77	312.28	222.7	221.9	568.09
					EPM[a]	4.39	4.28	13.50	3.65	4.62	16.02
					Z	19.8	19.3	60.9	15.0	19.0	66.0
Ain El Bishmo, western outlet	260	186		31.6	ppm	8	0.60	0.61	45	3	37.0
					EPM[a]	0.66	4.28	13.50	0.74	0.06	1.04
					Z	35.3	19.3	32.6	40.2	3.3	56.5
Ain El Bishmo, eastern outlet	300	206		28.9	ppm	13	10	16	54	10	42.0
					EPM[a]	1.07	0.5	0.7	0.88	0.21	1.18
					Z	47.2	22.0	30.8	38.8	9.2	52.0
Ain El Gabal	240	168		40	ppm	5.4	16	33	37	48	22.0
					EPM[a]	0.44	0.80	1.44	0.61	1.00	0.62
					Z	16.9	29.2	53.9	27.0	45.1	27.9
Ain El Bousa	417	292		31	ppm	17.5	22.4	54	95	10	96
					EPM[a]	1.44	1.12	1.74	1.55	0.2	2.7
					Z	33.5	26	40.5	34.8	4.5	60.7
Ain El Arayes	3300	2112		27	ppm	108	150.87	487.5	182.88	485	773.24
					EPM[a]	8.24	7.53	20.62	3.12	10.12	21.80
					Z	22.64	20.69	56.66	8.90	28.88	62.22

[a] EPM = equivalent per million.

gregates containing less silty material, a more favorable groundwater aquifer system.

In Ain Furtaga in Wadi Watir, about 7 km upstream of the merging site of Wadi Ghazal, Wadi Watir issues in an area where groundwater in highly fractured basement zone feeds the alluvial wadi fill and discharges to the surface where the fill is intercepted by a dike. Due to the steep slope in this reach of the wadi floor and the presence of low permeability silty material, groundwater in the valley fill moves upwards and discharges to the surface of the wadi. The spring water flows on the surface for 7 km and then disappears as it percolates back into the fractured basement. The spring discharge is an estimated 2 500 m^3/d, with a total dissolved solids content of 1 450 ppm. The chemical analysis of Ain Furtaga water shown in Table 8.6 indicates that this spring water is of the sodium sulphate type. Water from Ain Furtaga is being considered for development as a source of potable water for future tourism.

8.9.2.2 *Ain El Gudeirat*

Ain El Gudeirat, in Wadi El Gudeirat near El Qosaima Village in the northeastern part of Sinai is near Egypt's eastern international borders (Fig. 8.12). The spring is on the axis of a small syncline gently plunging west to outcrops of Eocene Limestone which overlie Paleocene shales. Wadi El Gudeirat incises the limestone formation to the contact between the two formations.

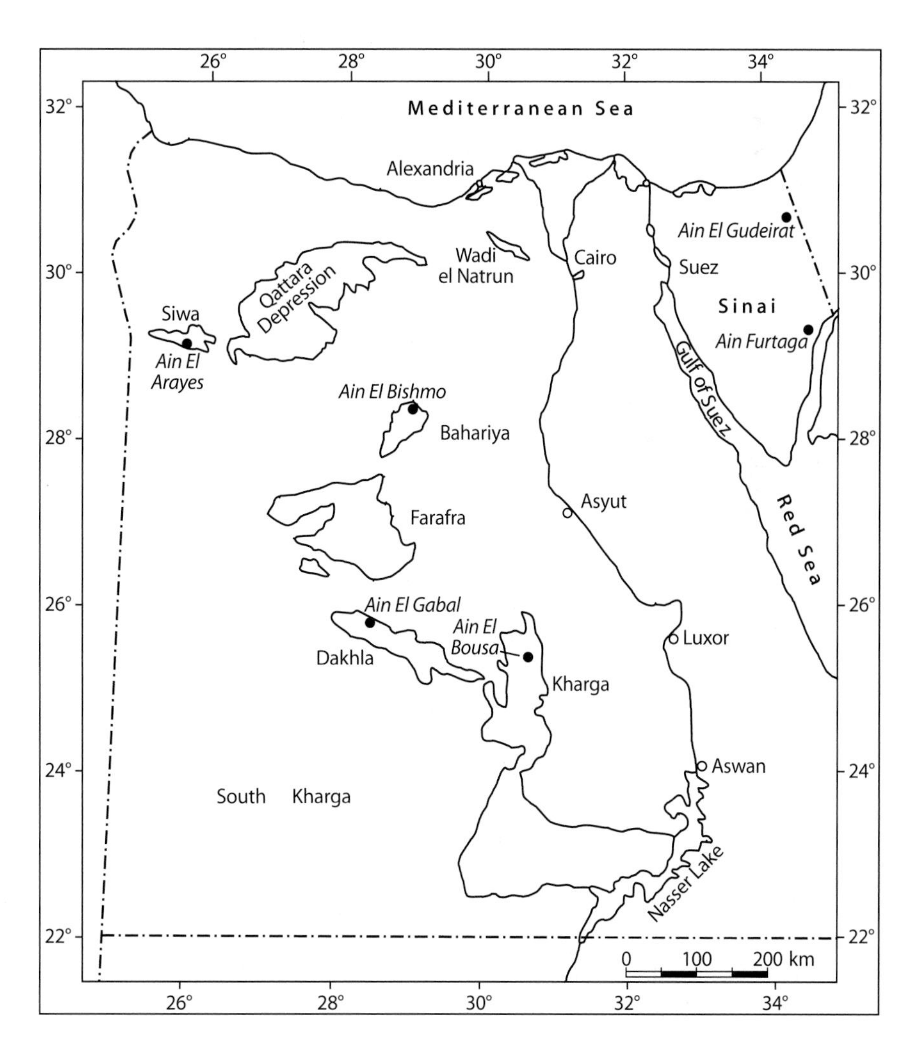

Fig. 8.12. Location map: *1* Ain Furtaga; *2* Ain El Gudeirat; *3* Ain El Bishmo; *4* Ain El Gabal; *5* Ain El Bousa; *6* Ain El Arayes

Ain El Gudeirat issues from the lowermost part of the highly fractured limestones at a daily rate of 1 500 m^3. The spring water flows in a small channel and is used to irrigate several hundred feddans (1 feddan = 1.04 acre) of olive trees and is a source of water supply for the local villagers.

El Gudeirat water has a total dissolved solids content of 1 440 ppm. Chemical analyses shown in Table 8.6, indicates that the water from this spring is of sodium chloride type. Groundwater age dating studies indicate that the age of water from Ain El Gudeirat is 14 000 years B.P., indicating that the recharge to this spring is late Pleistocene.

8.9.3 Springs in the Western Desert

In the Western Desert, a number of springs emerge from the Nubian Sandstone aquifer along through faults or joint zones. These structures provide the avenue for the upward movement of groundwater from deep sandstone beds under artesian pressure. These springs issue in desert depressions where the ground surface elevation is low and the piezometric head of the aquifer is high enough for groundwater to discharge at the land surface. This spring water has been utilized for centuries by inhabitants of the oases for agriculture. They were mentioned by Herodotus over 2 500 B.P.

8.9.4 Nubian Sandstone Aquifer System in the Western Desert, Egypt

Al Sahara Al Gharbiya, the Western Desert of Egypt, is a typical desert region, one of the very driest and hottest climates in the world. Precipitation ranges from zero to about 15 mm/yr. The relative humidity has a mean value of 4% in winter and 27% in summer. The absolute maximum temperature is 50 °C while the minimum is zero.

Sediments of Paleozoic, Mesozoic, and Tertiary ages dip gently to the north and overlie the Precambrian Basement Complex. Sediments become progressively younger in age from the southern boundary to the shores of the Mediterranean Sea (Fig. 8.13). Superimposed on this great northward dipping monocline are various structural trends, including anticlines, synclines, and folds and faults.

The Western Desert Nubian aquifer system underlies the Western Desert. It extends to the high massif of Ennedi, Eridi, and Tibesti in the southwest; to the northern confines of Darfour and Kordofan of Sudan in the south; to the Tibesti-Sirte structural high in the west; and to the Red Sea pre-Cambrian Mountain range in the east.

This aquifer consists of a thick sequence of coarse, clastic sediments of sandstone with intercalations of sandy clay, shale, and clay beds. The more clayey impermeable beds restrict the vertical movement; thus aquiferous zones are formed within the Nubian Sandstone that constitutes a regional single aquifer complex. It increases in thickness from south to north. It has a thickness of 800 m in Kharga Oases, 1 500 m in Dakhla Oases, and about 1 800 m in Bahariya Oases. At Farafra Oasis, the total thickness of this complex is estimated to be 2 800 m, and is 3 500 m thick at Desouky, as determined from an exploratory oil well south of Siwa.

Based on the piezometric map in Fig. 8.14 (modified after Ezzat and Abu El Atta 1974), groundwater moves in general from the Eridi and Ennedi region on the borders of Chad Basin, in a southwest-northeast direction, discharging mainly from springs, wells, and diffuse seepage flow into the depression areas of Kharga, Dakhla, Farafra, Siwa, and Qattara.

The evaluation of groundwater resources of the Nubian aquifer system indicated that 1 020 million m^3/yr can be exploited in the Western Desert, New Valley Oases in Egypt. At present, about 600 m^3 are being used annually for irrigation. Centuries ago, this Nubian Sandstone aquifer was the source of many springs. In recent generations, hundreds of wells tapped this very large aquifer system, lowering the pressure surface. Today most water in the area is from wells.

The results of test drilling in the Western Desert Farafra Oasis (Fig. 8.15) to 1 200 m depth, indicate that three Nubian aquifer zones have been penetrated with different piezometric.

To illustrate the natural springs in this hydrogeological setting of the Western Desert, a number of springs at different Oases are described.

8.9.4.1 *Ain El Bishmo*

Ain El Bishmo is in Bawiti, the capital of Bahariya Oases (Fig. 8.12). Folding and faulting controls the topography of Bahariya Oases Depression, which is surrounded by steep escarpments. Bahariya Oases originally depended on natural springs for water supply. Springs that owe their existence to these structures. A pattern of cross faulting creates conditions favorable for the upward movement of groundwater under artesian pressure to the land surface. Ain El Bishmo, an example of this type of spring, is in the northeastern part of the Oases near the El-Bawiti fault. Flowing water moves upwards along the fault zone. After hundreds of years of well development, a general lowering of the piezometric pressure has occurred in the oases and resulted in a decline in flow. In an attempt to rehabilitate the spring, the local inhabitants have sank casings at two locations along the fault. Both attempts were successful, and groundwater now flows at two of these outlets.

The western of the two spring discharges water from a deeper zone and is locally called El Sokhna "the hot". The eastern outlet produces water from a shallow zone, has a lower temperature, and is called El Barda "the cold". The chemical analyses of water discharging from the two outlets at Ain El Bishmo are shown in Table 8.6. The water has total dissolved solids ranging between 186 and 206 ppm.

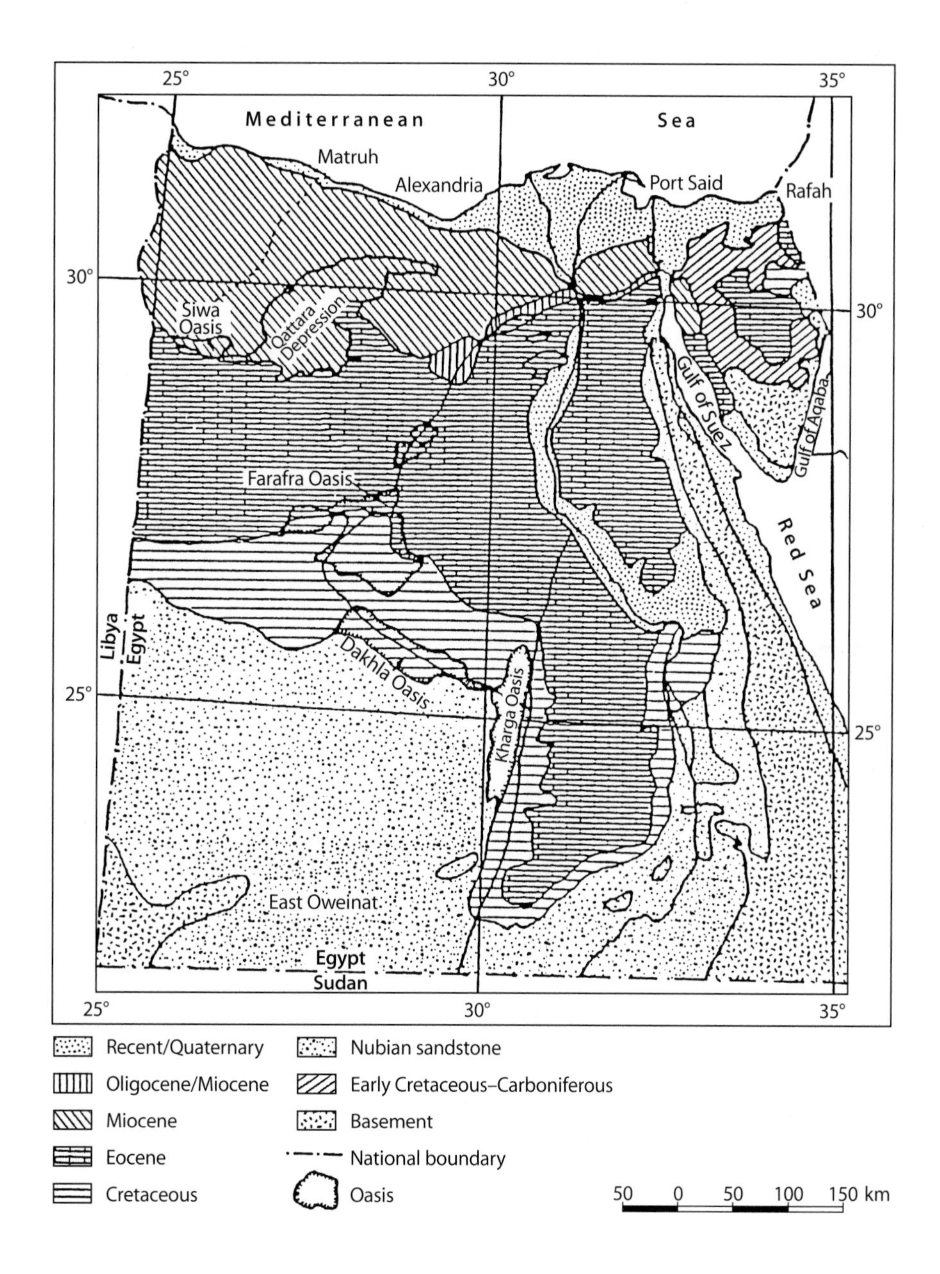

Fig. 8.13.
A geological map of Egypt

8.9.4.2 *Ain El Gabal*

Ain El Gabal is in the northwestern part of Dakhla Oases (Fig. 8.12). The Dakhla Oases Depression is a major syncline dominated by a series of minor anticlines and synclines. The Dakhla Oases is characterized by the occurrence of numerous large springs. The spring Ain El Gabal is one of the most famous and is still flowing, while many of the other springs have ceased due to a general lowering of piezometric pressures caused by the drilling and production from wells. The spring issues from a faulted zone with a water temperature of 40 °C.

Table 8.6 shows the chemical analyses of Ain El Gabal water, which has a low total dissolved solids (168 ppm) and is of sodium sulphate type.

Ain El Gabal is well known for its therapeutic uses. People in Dakhla Oases suffering from rheumatic pains bathe in this water for relief. The water has an obvious sulphur odor. Ain El Gabal is also famous as the traditional bath for new brides on their wedding day. Brides go to the spring accompanied by their female friends and bridesmaids. Elder women follow to perform hair dressing and make-up. It is believed that the bath ceremony gives the bride a happy, successful, and long married life. It is also believed that the ceremony is a good omen for the other women for a speedy wedding and a family of their own.

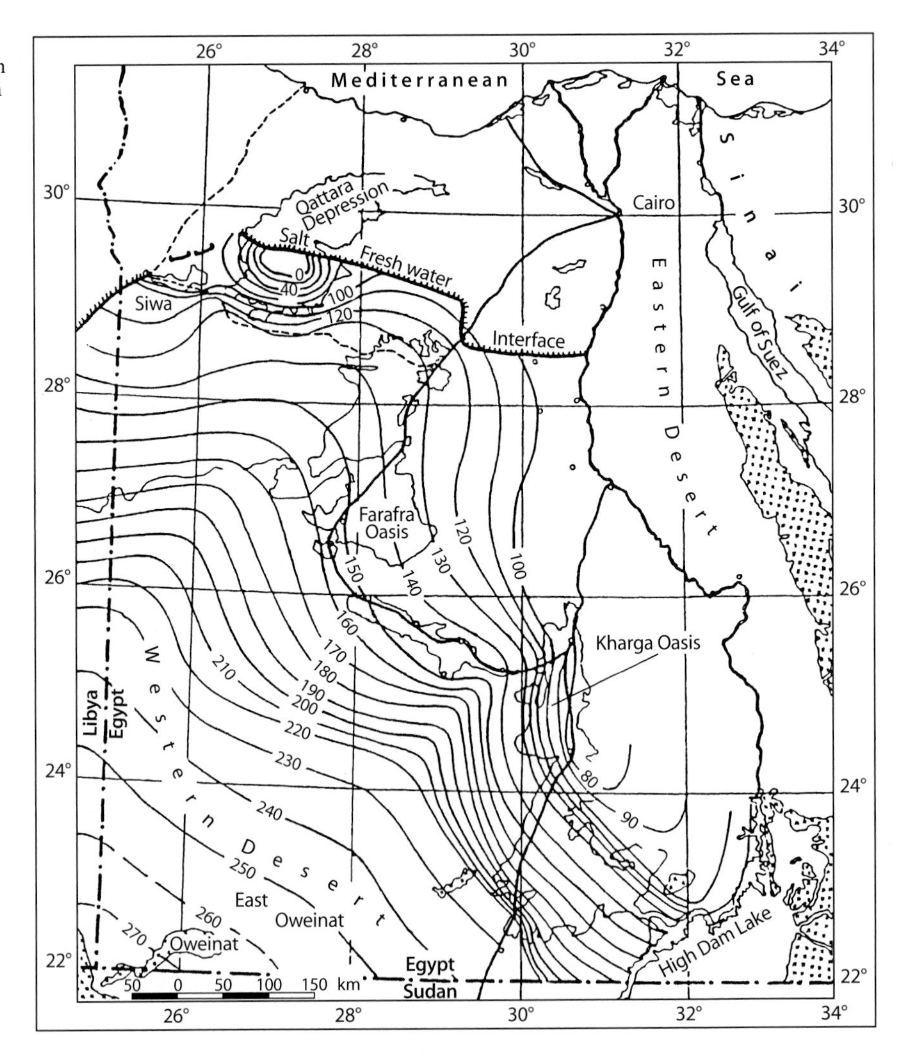

Fig. 8.14.
Piezometric map of the Nubian aquifer system (modified from Ezzat and Abu El Atta 1974)

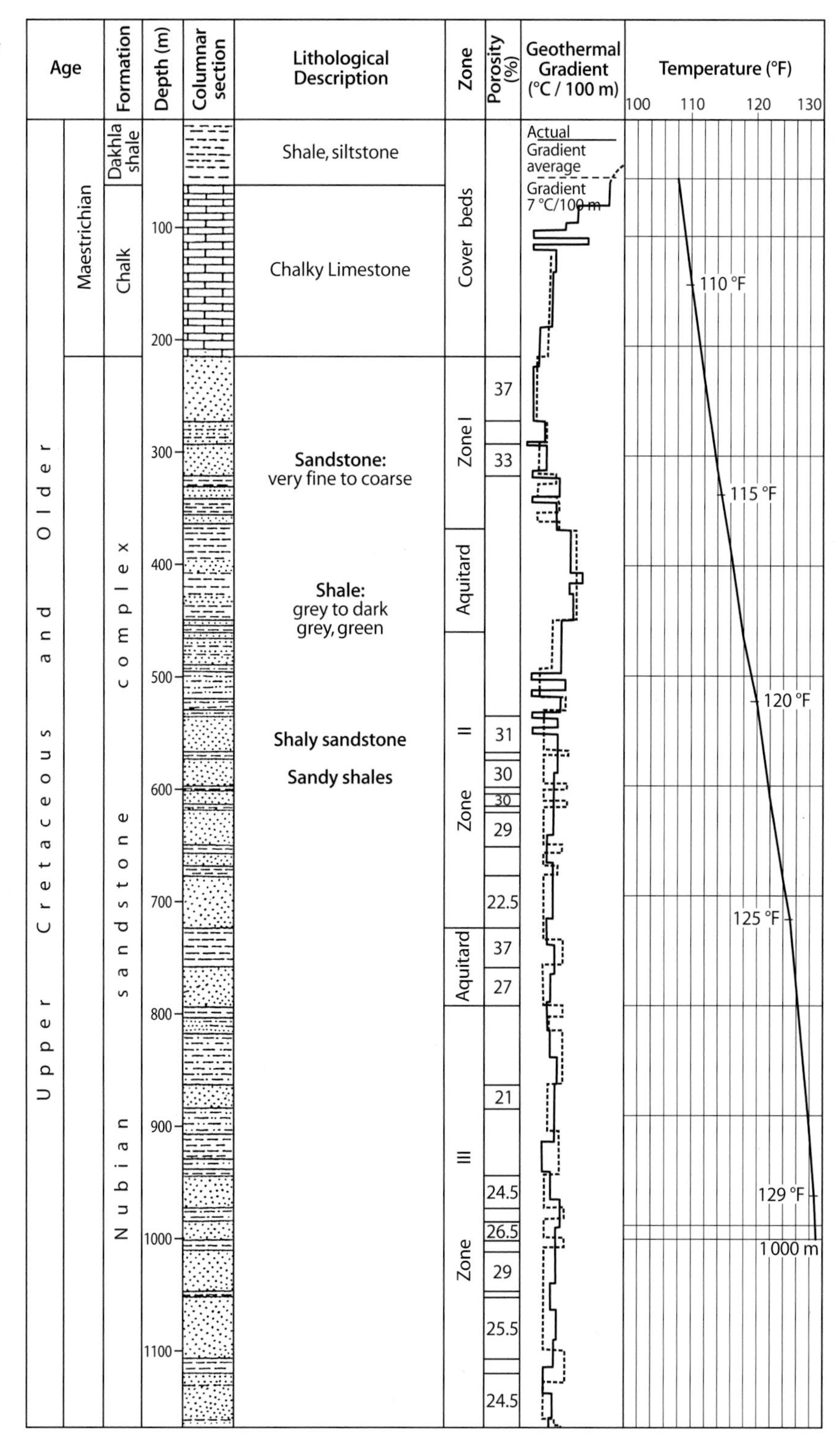

Fig. 8.15.
El Sheikh Marzouk exploratory well at Farafra Oasis

8.9.4.3 *Ain El Bousa*

Ain El Bousa is in Kharga Town, Kharga Oases (Fig. 8.12). The Kharga depression is oriented with its long axis north and south and is surrounded by a steep, step-like escarpment in which is exposed Upper Cretaceous and Eocene limestone, chalk, and shale. The floor of the depression is covered by thick variegated shales that confine the underlying Nubian Sandstone aquifer.

Kharga Oases is an eroded anticline intersected by a N–S system of faults at the contact with the eastern plateau. The fault forms an impermeable barrier to lateral flow of groundwater. A north-south fault cuts across the northern end of the oases where most springs are located.

Ain El Bousa is one of the rare springs still flowing in Kharga Oases and discharges water low in TDS. The results of the chemical analyses of water from Ain El Bousa are given in Table 8.6. The water is of chloride sodium type with a total dissolved solids content of 290 ppm.

8.9.5 Springs of Siwa Oasis

Siwa Oasis occupies the westernmost depression in the Western Desert. It is perhaps the most famous and remote oasis in Egypt and springs provided the water supply from ancient times. It is here that Alexander the Great went to meet the "Great Oracle". Siwa is near the Libyan-Egyptian border, about 250 km south of the Mediterranean Sea. It is the furthest Egyptian depression from the Nile Valley. It is about 80 km long and ranges in width from 9 to 28 km covering an area of about 1000 km^2. Four salt lakes occur in Siwa. The Marmarika Limestone plateau forms the northern boundary of this depression, while on the south it is bounded by an elevated plain. The elevation of the depression floor ranges from –10 to –18 m.

Springs in western Siwa issue from the Miocene limestone, and in the eastern part from the Eocene limestones. Water discharges upward from the underlying Cretaceous Nubian Sandstone aquifer. The occurrence and magnitude of the fractures in the limestone control the distribution of the springs, their flow rates and mineral content.

In recent years, drilling of exploratory oil wells has revealed that the upper zone of the Nubian Sandstone contains freshwater. Water wells drilled to the upper Nubian sandstone during the 1990s provide freshwater that was recharged to the aquifer 30 000 years B.P., based on radioactive dating.

8.9.5.1 *Ain El Arayes*

Among the 200 existing springs in Siwa Oasis, Ain El Arayes, "The Brides Spring", is most famous. In Arabic, Arayes is the plural of Arousa. Before the Islamic era, the spring was called in the Siwan unwritten language "Tamous".

The spring is in the eastern vicinity of the town of Siwa. Its water emerges from fractures and solution springs from limestone at a daily rate of 485 m^3 with a temperature of 27 °C. The chemical analysis of the spring water, shown in Table 8.6, indicates the water from the spring has a total dissolved solids content of 2 112 ppm.

Ain El Arayes owes its fame to being the spring that brides use on their wedding day. Also, women who are widowed go to the spring after 40 days of mourning following their husbands' death. During the 40 days, the widow is detained in a house. She is not to see or be seen by anyone. Her food is dropped from a hole in the wall. Siwan myths tell us that the widow is transformed to a "ghoula," which in Arabic is the feminine of ghoul. Ghoula is a harpy similar to the harpies of Greek and Roman myth. Although the Egyptian harpy is a rapacious monster that has a woman's face with untidy hair and claws, it does not have wings. The Siwan myths assure that anyone who sees or is seen by El Ghoula (The Harpy) will become "bewitched" and consequently will die or be transformed into a mad person. Such myth is to ensure no contact with the widow. At the end of the 40-day mourning period, the village announcer declares that El Ghoula may leave her home and go to Ain El Arayes for bathing and be transformed into a normal woman. Everyone must clear the road, houses and shops along the route to the spring are closed. On her way back home, the widow is no longer a ghoula.

In Upper Egypt in remote areas, a widow stays barefooted at home for forty days after the death of her husband. In Sudan, widows stay at home for the same period of time with the female relatives of the deceased husband. They do not use beds or chairs during this mourning period. The treatment of Siwan women is mild and the author has witnessed a lady Sudanese Cabinet Minister refraining from going to her office for the mourning period in honor of that tradition.

References

Ezzat MA, Abu El Atta A (1974) Regional hydrogeologic conditions, El Wadi and El Gedid Area (New Valley). Part I of groundwater series in the Republic of Egypt. Ministry of Agriculture and Land Reclamation, Cairo

P. Milanović

8.10 Ombla Spring, Croatia

8.10.1 Introduction

Ombla Spring is on the Adriatic coast near the town of Dubrovnik. The spring discharges at sea level. To eliminate the influence of the tide, a small dam was constructed 50 m downstream of the spring outlet. The spring water overflows the dam crest at an elevation of 2.40 m. Since 1897 the spring water has been used for the water supply for Dubrovnik.

8.10.2 Outlines of Geology

The spring outlet originates on a reverse fault, along which the Mesozoic carbonate complex (dolomite and limestone) overthrusted the autochthonous Eocene flysh sediments (Fig. 8.16 and 17). The carbonate complex is strongly fractured. A number of faults cut the limestone and dolomite rock mass to a depth of a few hundred meters. The surface is extremely karstified. More than 100 swallow holes, caves, and shafts were investigated in the catchment area. The flysh zone represents a hydrogeological barrier, which at this location is eroded to the sea level.

Fig. 8.16. Ombla Spring: position of underground dam axis

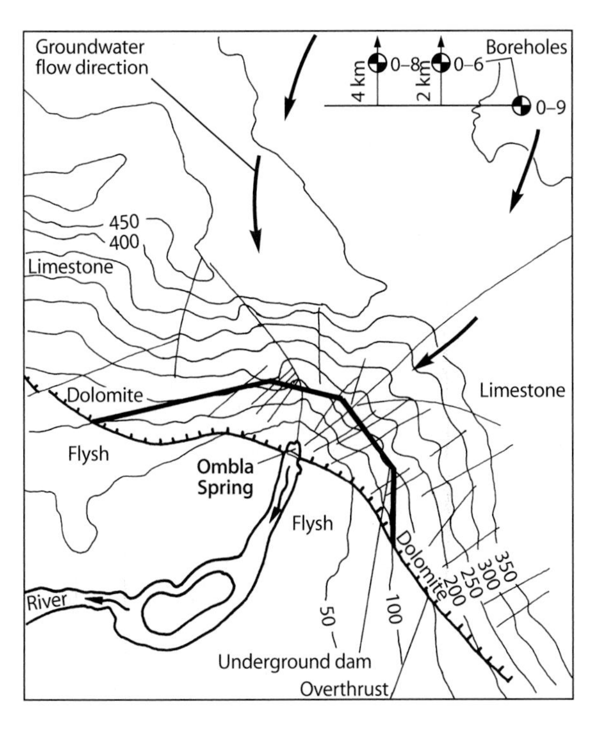

8.10.3 Hydrological Setting

The average yearly discharge under natural conditions is $Q_{av} = 33.8\ m^3/s$. After construction of Hydrosystem Trebišnjica, which influences a part of Ombla catchment, the discharge was reduced to $Q_{av} = 24.4\ m^3/s$. The recorded minimum discharge is 2.3 m^3/s and maximum is 112.5 m^3/s. The temperature of the spring water varies over the year from 12.0 to 14.8 °C. The catchment area is about 600 km^2. The aquifer recharge is partially influenced by the sub-catchment in the hinterland with an area of 1 630 km^2.

8.10.4 Main Features of Ombla Karstic Aquifer

The investigation of the karstic aquifer included drilling of 15 deep boreholes, many dye tests, plus speleological and complex geophysical investigations. A relatively low total porosity and extraordinary transmissivity (large karst conduits) result in very rapid fluctuations of the water table in the Ombla aquifer.

The amplitudes of water level fluctuation can be as much as 200 m. After heavy precipitation, the water table and discharge of Ombla Spring respond in less than 4 h (Fig. 8.18). In some precipitation events, the water table rises more than 90 m within 10 h (piezometer 0–8, Fig. 8.18), and at times the discharge increases from 20 to 88 m^3/s. The deep Ombla karstic aquifer functions as a hydraulic system under pressure. The average groundwater flow velocities during the period when the aquifer is not completely saturated (dry period of year) are about 3 cm/s, but in the period when it is fully saturated, the velocity is about 7 cm/s.

8.10.5 Characteristics of the Spring Zone

The investigation of the spring was aided by the construction of 19 piezometric boreholes, the study of 600 m of galleries, the construction of 28 boreholes drilled from galleries (Fig. 8.19), and numerous geophysical methods.

Fig. 8.17.
Ombla Spring: cross section

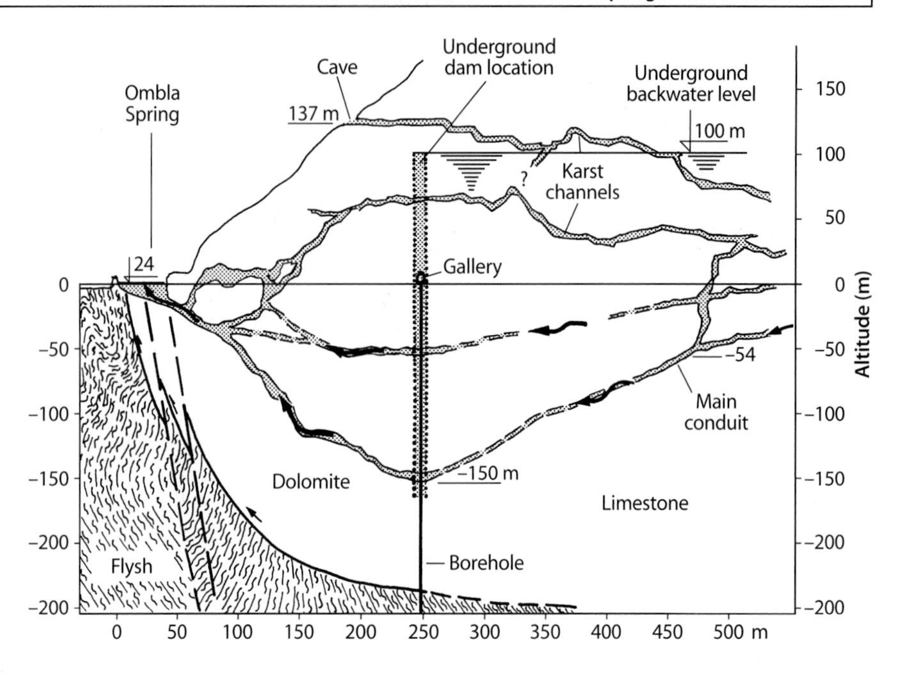

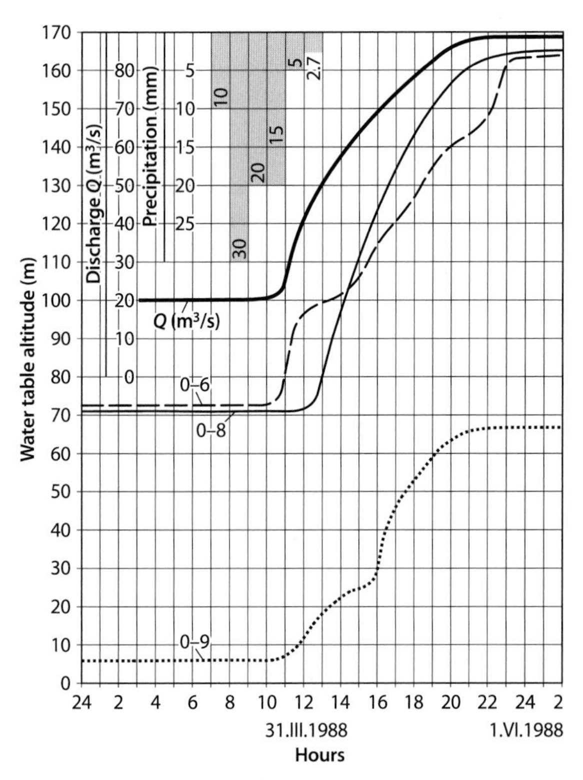

Fig. 8.18. Water table fluctuation graphs, spring discharge graphs, and precipitation data

Geophysical methods included: gravimetry, thermometry, different electrical methods, seismicity (cross-hole, reflection, refraction), and borehole radar. About 3 km of karst channels were investigated by speleologists and divers.

The flysh hydrogeological barrier is between the karstic aquifer and erosion base level (sea coast). The thickness of the barrier is around 800 m, and the vertical depth is more than 300 m. There is no possibility for deep filtration below the flysh barrier toward the bottom of the Adriatic Sea. Four different levels of karstic channels indicate the stages of karst aquifer evolutionary process. Many stalactites and stalagmites are present in the two upper levels and indicate that some of the karst channels have been inactive for a long time.

Presently, the main water circulation occurs through the deep conduits. The deepest part of the siphon in the system was discovered by drilling, thermometry, and borehole radar at a depth of 150 m below sea level and about 200 m upgradient from the main spring outlet. The siphon has been explored by speleologist divers to a depth of 54 m below sea level. That part of the siphonal channel is about 50 m upgradient from the spring outlet. A large cave was discovered by the divers (24 m below sea level and 10 m behind the outlet). The length of the cave is 80 m with a width of 10–20 m. The upper part is above sea level.

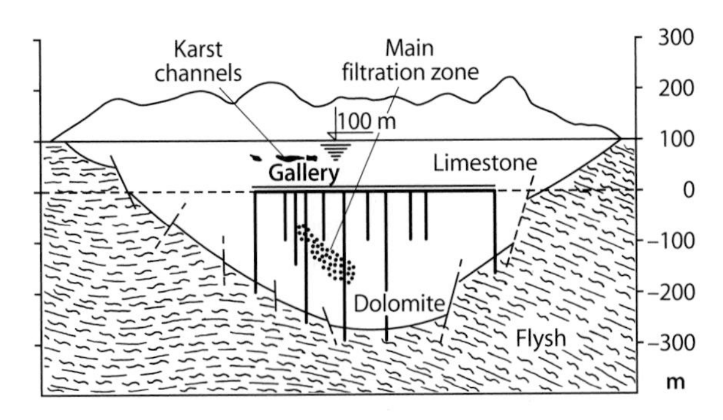

Fig. 8.19. Cross section along the underground dam axis

The limestone and dolomite rock mass in the spring zone is very solid. The cracks are mostly compressional, indicating that filtration occurs only through the large diameter conduits. The surrounding rock mass is mostly impervious.

8.10.6 Possibilities for Underground Dam Construction

Ombla Spring discharges a large quantity of water to the sea. There is, however, the potential for utilizing the water for energy and additional water supplies by constructing an underground dam for groundwater storage. Geostructural and hydrogeological conditions indicate the idea is feasible as the preferable position of impervious flysh makes it possible. Laterally, from the spring along both sides, the barrier to the flysh is formed in such a way that the open profile has the form of a "V", with the spring at the lowest point.

The position of the flysh and the elevation of lateral overlying springs limit the underground dam at about 100 m above mean sea level (maximum 130 m). Due to this height and the required overburden of the alignment, the archlike water-permeable structure must be located at least 200 m behind the spring outlet.

The underground power plant could be downstream of the water-permeable structure and upstream of the discovered cavern. The water from the plant would be conveyed to the cavern, then through the natural siphon, and finally discharged into surface flow.

The structure (underground dam) could be a combination of grout curtain and plugged karst channels. Upstream from the plugging, all karst channels should be connected by shaft collectors and then connected further up-gradient to the underground plant.

Compared with conventional structures (surface dams and reservoirs) underground dam, power plant, and storage are, from all environmental aspects, preferable. Destruction of the landscape is minimal, as are river valley deterioration and microclimate alterations, the quality of water is better, and a break in the dam would not be as catastrophic as a similar event for a conventional surface dam.

References

Milanović PT (1979) Hidrogeologija Karsta, I Metode Istrazivanja, Godina izdanja, 302 p

S. Wohnlich

8.11 The Spa of Baden-Baden, Germany

8.11.1 Introduction

The famous spa of Baden-Baden is in the northern Black Forest area not far from the upper Rhine Valley in southwestern Germany. The spring discharges a natural sodium chloride thermal water (68 °C), occurs in an extremely mild climate and pleasant landscape, which made Baden-Baden one of the most favored spas in Germany. Baden-Baden was originally used by the Romans, is a former capital of the German state of Baden (which means bath or spa), and its name now is included in the name of the state of Baden-Württemberg. Baden-Württemberg is the only area in Germany where a geologic phenomena is part of the name of the state.

8.11.2 History

In the 1st century A.D., the Roman settlement of Aquae can be traced next to the thermal district within the old town of Baden-Baden. In the 2nd and 3rd century A.D., the town became the center of a larger administration unit at the rim of the Black Forest. Below the market place of Baden-Baden, remnants of the Roman baths have been excavated. These so-called "Kaiserthermen" (Emperors thermal bath) were used for curing Roman soldiers. After the destruction at about A.D. 260, the ruins were covered by 6-m-high sinter terraces related to the thermal springs.

The medieval town of Baden-Baden was constructed on top of the Roman ruins. The thermal springs were used for physical treatments of ill people from the Middle Age to recent times. The baths were frequently visited by European royalty and nobles. From 1845 to 1970 Baden-Baden was the most important spa in Europe. In 1863 the three emperors of Austria, Russia, and France stayed at Baden-Baden. Most of the hotels and villas in Baden-Baden are remnants from this period when Baden-Baden was the luxury spa of Europe.

8.11.3 Geological Setting

The Baden-Baden depression is in the northwestern part of the Black Forest. It consists of sediments and volcanics of Carboniferous and Permian age. It is underlain by Devonian greywacke and a metamorphic series of the Black Forest. The Baden-Baden depression is surrounded by granites to the south and east. In the north, sandstones of Triassic age occur, and to the west the tectonic structure of the Rhine Graben. Tectonically, the depression is subdivided into two parallel synclines with a northeast to southwest trend. The synclines are separated by the Battert anticline, which crosses beneath the center of town.

In the center of the anticline, greywacke and metamorphic rock of Devonian age outcrop west of the town center. The Carboniferous rock consists of granite (Friesenberg-Granit) and terrestrial sediments. The latter consist of sandstones, conglomerates, and schists, which include coal seams. In the lower Permian, volcanic rocks occur not far south of Baden-Baden (porphyry and por-

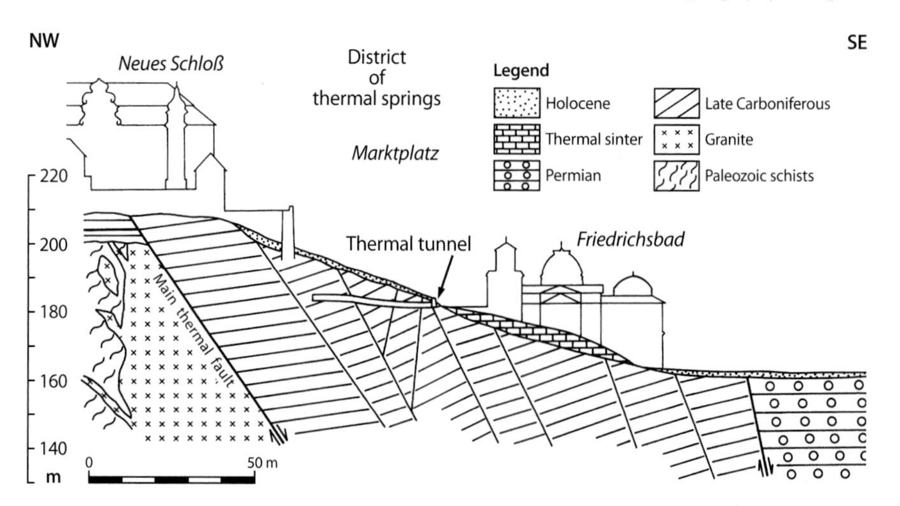

Fig. 8.20. Geological NW-SE section through the main discharge areas of the thermal springs of Baden-Baden (according to Metz 1975)

phyry tuff with sandy schists and arkoses). The upper Permian is formed by a series of conglomerates and claystone, grading into the later sandstone environment of the lower Triassic (Buntsandstein).

8.11.4 Hydrogeology

In the city area of Baden-Baden, the Carboniferous series are predominant; in particular, the New Castle and the Old Town are situated on these rocks. These rocks are intensely fractured by southwest to northeast trending faults. Along these faults, mineral thermal water circulates and rises to the surface at 20 historic springs located downhill of the New Castle, but approximately 30 m above the valley floor. It is assumed that the faults in the main valley must be clogged. The hottest spring (68.9 °C) is at the highest elevation. Until the middle of the last century, most of the springs discharged freely. Prior to the construction of the old steam bath, discharge from three of the springs at the market place were impounded. In the period 1868–1871, the major thermal springs at Florentiner Berg were diverted into two tunnel systems (Fig. 8.20). In 1965, additional thermal water was located by drilling two wells north of the spring district. These wells, 305 m and 511 m deep, penetrated artesian thermal water along a fault between Paleozoic schists and the granite.

The total discharge from all Baden-Baden springs is about 8.9 l/s; the wells produce as much as 3.4 l/s. The temperature of the natural springs varies between 54.0–66.1 °C, and the temperature and mineralization have reportedly been constant since the start of measurements in 1894.

The water from Baden-Baden Springs is a sodium chloride type (Table 8.7). This mineralization has its origin in the evaporative sediments in the Rhine Graben, where the sediments occur within 1 200 m of the Tertiary deposits just west of Baden-Baden. These salty waters from the Rhine Graben are mixed with groundwater from the granite areas during migration to Baden-Baden. Owing to a high geothermal gradient in this area (19–20 m per 1 °C), a circulation depth of approximately 500 to 600 m is adequate for the temperatures observed.

The water has a radioactive content of up to 22.1 nCi/l. The latter is referred to sinters containing magnesium and uranium (similar to the mineral "Reissacherit"), which occurs along the fault zones. The sources of these radioactive minerals may be from sedimentary deposits within the Permian sandstones.

References

Carlé W (1975) Die Mineral- und Thermalwässer von Mitteleuropa. Stuttgart, 643 pp
Metz R (1975) Mineralogisch-landeskundliche Wanderungen im Nordschwarzwald, 2nd edn. Moritz Schauenburg Verlag, Laar

Table 8.7. Mineralization of the Friedrich-Quelle in 1954. At 68.8 °C it is the hottest spring in the Baden-Baden area (analysis from Carlé 1975)

Constituent	Concentration (ppm)	Constituent	Concentration (ppm)
Li	8.71	F	1.37
NH_4	0.72	Cl	1 438.0
Na	823.3	Br	2.80
K	82.1	J	0.011
Cs	1.18	SO_4	162.4
Mg	5.02	HPO_4	0.16
Ca	137.49	$HAsO_4$	0.565
Sr	3.98	NO_3	1.24
Fe	0.89	HCO_3	154.19
Mn	0.02	H_2SiO_3	156.4
Zn	0.003	HBO_2	6.57
Co	0.006	H_2TiO_3	0.04
Al	0.02	TDS	2 987.18
		CO_2	13.30

J. Gunn

8.12 Source of the Shannon

The Shannon is the longest river in the British Isles (280 km) and drains an area of about 15 530 km^2. The source of the river is Shannon Pot, a karst rising in County Cavan, which is one of the most famous springs in Ireland. Water tracer experiments have shown that the rising spring an immediate area of about 12.8 km^2 on the slopes of Cuilcagh Mountain, of which about 60% is underlain by limestone. However, two sinks 10–11 km east of the rising and ca. 200 m higher have also been shown to be hydrologically connected during high flow conditions. This suggests that Shannon Pot may once have had a substantially larger catchment area.

The Shannon's traditional source is Shannon Pot (Log na Sionnainne or Lag na Sionna in Gaelic), a karst rising in County Cavan in the Irish Republic (Fig. 8.21). The rising pool is about 16 m in diameter and has been explored by divers to a depth of 9.5 m, where water emerges from a 2-m-wide, but impenetrably narrow, fissure (Elliot and Solari 1972; Fig. 8.22). The spring is one of the most famous in Ireland and there is a local tradition in the area that many hundreds of years ago, a terrible plague ravished the countryside claiming many lives. One day a saint, wandering in the vicinity in search of drinking water, fainted. As he regained his senses, he could hear the trickle of water running from the Pot. He drank his fill and since then the spring has never gone dry. Local folklore also has it that Poteen (illicit pot-still whiskey) was once brewed by the side of the stream that drains the Garvagh Lough about 2.2 km to the northeast (Fig. 8.21). A sudden warning of the approach of Crown Revenue Officers led to the still being hastily dumped into the stream, which washed it into the sink at Pollnaowen (Fig. 8.21). A few days later, so the tale goes, the still was recovered intact from Shannon Pot! Garvagh Lough is also the site of the first recorded water tracer experiment in Ireland (Hull 1878, p. 172). Hull notes that "Mr. S. B. Wilkinson proved by experimenting the truth of this statement, having thrown hay or straw into the little lough which, on disappearing, has come up in the waters of the Shannon Pot." However, there is

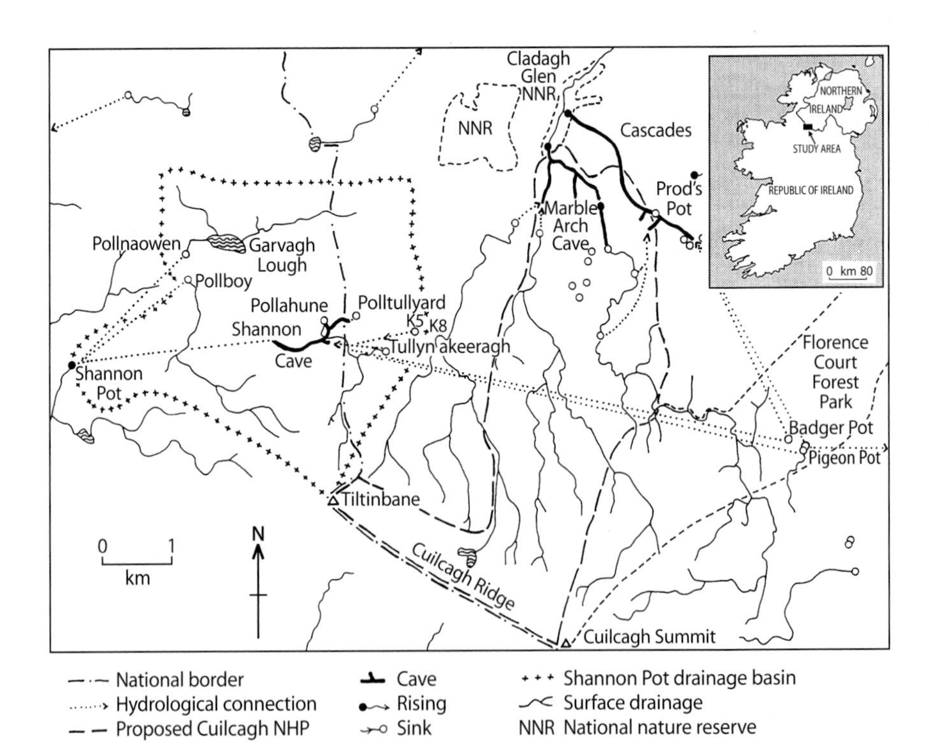

Fig. 8.21. Cuilcagh Mountain and the Shannon Pot drainage basin

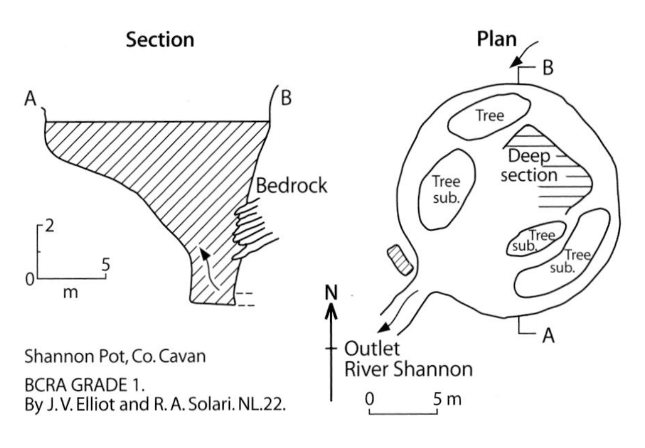

Fig. 8.22. Shannon Pot, County Cavan (after Elliot and Solari 1972)

some doubt as to whom this experiment should be attributed since Wilkinson and Cruise (1886) simply note that "people living in the district state that, on one occasion when sacks of chaff were emptied where the small river which drains the Lough disappears, after a short interval the chaff reappeared in the Shannon Pot." Further confusion is added by the fact that Wilkinson and Cruise cite Hull as the source of their information! However, Gunn (1982) demonstrated that the water from Garvagh Lough does indeed flow to Shannon Pot, a travel time of 240 m/h being recorded during a period of high flow. The sink-rising gradient is 22 m/km which is typical of the mid-reaches of the catchment (Table 8.8). The catchment boundary was constructed on the basis of these tracer experiments together with geological and topographical information, and the catchment area is estimated as 12.8 km^2, of which about 60% is underlain by limestone.

During the period 31 March 1979–26 March 1980, 50 water samples were collected from the Shannon at the rising pool under a variety of flow conditions and analyzed for calcium and magnesium. The long-term annual precipitation and evapotranspiration were estimated from data supplied by the Northern Ireland Meteorological Service and a preliminary limestone erosion rate of 74 t/(km^2/yr) for the whole basin was determined (Table 8.9).

Further water tracer experiments were performed on the eastern flank of Cuilcagh Mountain in 1994–1995, and it was determined that two stream sinks at Pigeon Pot and Badger Pot (Fig. 8.21) also provide drainage to Shannon Pot, but only under high flow conditions.

Table 8.8. Summary of connections established by water tracing in Shannon Pot catchment

	Average linear velocity (m/h)	Sink-rising gradient (m/km)
Leegeelan Lower	120	24
Gowlan 1	130	25
Pollnaowen[a]	240	22
Pollboy	(80–250)	23
Garvagh 1	60	29
Pollahune	<90	36
Tullynakeeragh	105	40
Derrylahan 1	50	35
Derrylahan 5	140	70
Polltullyard		39
Killykeeghan 5		33
Killykeeghan 8		32
Pigeon Pot[a]	100	20
Badger Pot[a]	85	20

[a] Traced during a period of high flow.

Under normal to low flow conditions, the two sinks drain to the Cascades Cave System to the north (Fig. 8.21) and also to Gortalaughany Rising to the east. The connection during high flow is probably via Shannon Cave (Fig. 8.21), which contains a misfit stream, substantially smaller than the large and impressive passage it occupies. The upgradient passages in Shannon Cave terminate in boulder chokes and it is likely that a further, large cave passage is present beneath Cuilcagh Mountain along the 8 km that separates the sinks from the known cave. The objectives of the ongoing research are to locate this passage and to test the hypothesis that Shannon Pot is the outlet for a large, ancient drainage system that has been partially cut off.

The reason why Shannon Pot should be regarded as the source of the Shannon, as opposed to any of the river's other headwater tributaries, has been lost in time, although it probably relates to the mystical nature of the rising pool. However, tracer experiments have shown that at least sinking streams drain to Shannon Pot and two of these have some claim to being the ultimate

Table 8.9. Preliminary solutional erosion rate estimate – Shannon Pot catchment

Data			
	Basin Area (km^2, 60% underlain by limestone)		12.8
	Limestone density (g/cm^3)		2.7
	Mean ionic concentrations in spring water (n = 50):	Ca Mg	25.5 1.5
	Mean ionic concentrations in rain water (n = 5):	Ca Mg	1.0 0.3
	Mean annual precipitation (mm)		1 500
	Mean annual evapo transpiration (mm)		350
	Mean annual runoff (mm)		1 150
Computations			
i	Solute load evacuated Ca 25.5 × 1 150 × 12.8 = 376 360 kg Mg 1.4 × 1 150 × 12.8 = 20 608 kg		
ii	Solute accessions in rainfall Ca 1.0 × 1 500 × 12.8 = 19 200 kg Mg 0.3 × 1 500 × 12.8 = 5 760 kg		
iii	Total solutional erosion (375 360 – 19 200) × 2.4973 + (20 608 – 5 760) × 3.4676 = 941 t		
iV	Erosion rates a) Whole basin 941/12.8 = 74 t/(km^2 yr) b) Area underlain by limestone 941/7.7 = 123 t/(km^2 yr) = 43 m^3/(km^2 yr)		

source of the River Shannon. If distance from the rising pool is considered to be the primary criterion, then the stream which sinks at the Pigeon Pots across the international border in County Fermanagh is the ultimate source, as it is the farthest known sink which is hydraulically connected to Shannon Pot (10.6 km in a straight line). Alternatively, if height above sea level is considered to be more important, then the highest flow in the catchment is an unnamed stream whose source is a minor spring at about 400 m a.s.l. on Tiltinbane, the western end of Cuilcagh Mountain.

The stream follows a steep course down the mountain but bifurcates before sinking at Tullynakeeragh (in County Fermanagh, Northern Ireland) and at Pollahune (in County Cavan, Republic of Ireland). Although small, the stream may be unique in having a sink in two different countries. In fact, the two sinks at Tullynakeeragh and Pollahune supply different branches of Shannon Cave, although only the Pollahune sink can be entered at present.

Acknowledgements

Thanks are due to Fermanagh District Council, which funded the research; to all those who helped with the field work, particularly Paul Hardwick, Robbie Pattison, Clare Walker and Richard Watson; to Judith Ayre for typing the paper; and to Cathy Gunn who drew the figures. This paper is a contribution to the work of IGU Commission 92.CO5 Environmental Changes and Conservation in Karst Areas.

References

Elliot JV, Solari RA (1972) Shannon Pot. Cave Diving Group Newsl 22:10

Gunn J (1982) Water tracing in Ireland: a review with special reference to the Cuilcagh karst. Ir Geogr 15:94–106

Hull E (1878) Physical geology and geography of Ireland. Dublin, Ireland

Wilkinson SB, Cruise RJ (1886) Explanatory memoir to accompany sheet 56 of the maps of the Geological Survey of Ireland. Geological Survey of Ireland, Dublin

J. R. Vegter

8.13 The Fountains of Pretoria

8.13.1 Introduction

The occurrence of groundwater in South Africa is to a very large degree limited to a surficial zone of weathered and fractured hard rock formations. Primary aquifers include narrow strips of alluvium along certain river stretches and to Cenozoic coastal deposits. Coupled with a rainfall that is well below the world average, South Africa is therefore poorly endowed with large springs which are almost totally confined to karst areas.

8.13.2 Background

The principal occurrences of carbonate rocks are the chronostratigraphically equivalent Proterozoic Chuniespoort and Ghaap Groups in the Transvaal and Northern Cape Province (Fig. 8.23). The Chuniespoort dolomitic strata are overlain by a thick succession of clastics of the Pretoria Group and are underlain by a thin band of the Black Reef Quartzite Formation overlying Arhcean granite gneisses south of Pretoria, and other formations elsewhere.

The Chuniespoort Group, which dips northwards in the Pretoria area at about 20°, is over 1 000 m thick (Fig. 8.24). It consists of four formations: chert-free micritic or recrystallized dolomite (bottom and third unit) alternating with chert-rich dolomite composed of alternating beds, bands, and laminae of chert and dolomite (second and top unit).

The chert-free and chert-rich dolomite units weather differently. Whereas deeply penetrating dikes in the former occur only on well spaced discontinuities, dissolution occurs on many more joints and bedding planes in the alternating chert and dolomite sequences. There is a considerable widening of passages below chert ceilings.

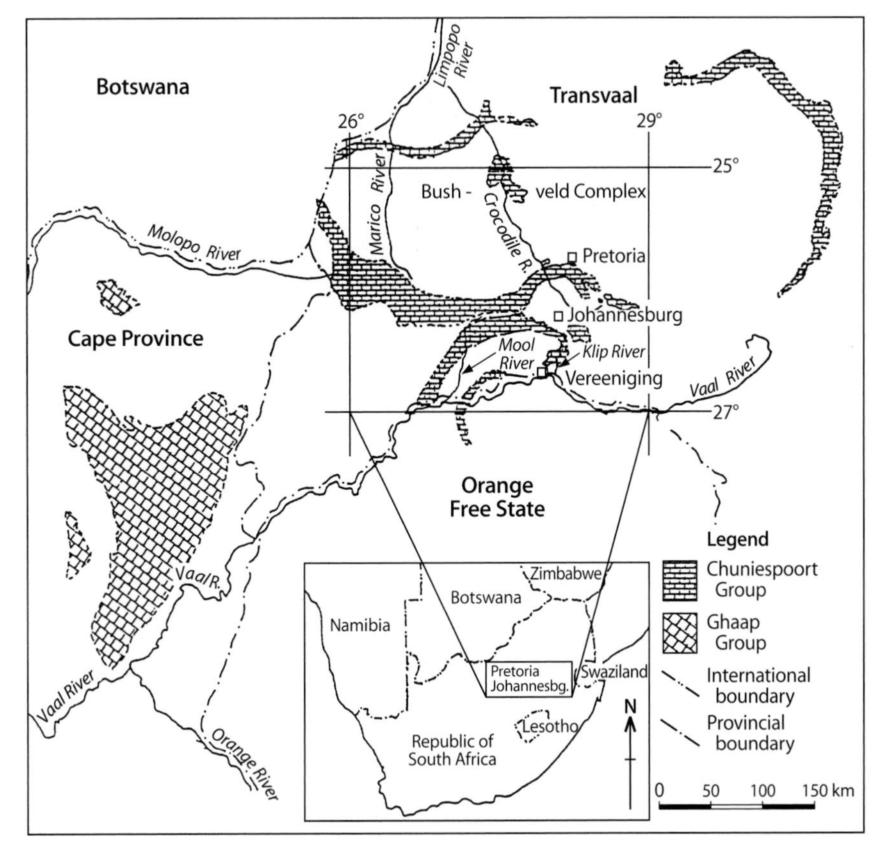

Fig. 8.23. Location map showing the distribution of carbonate rocks

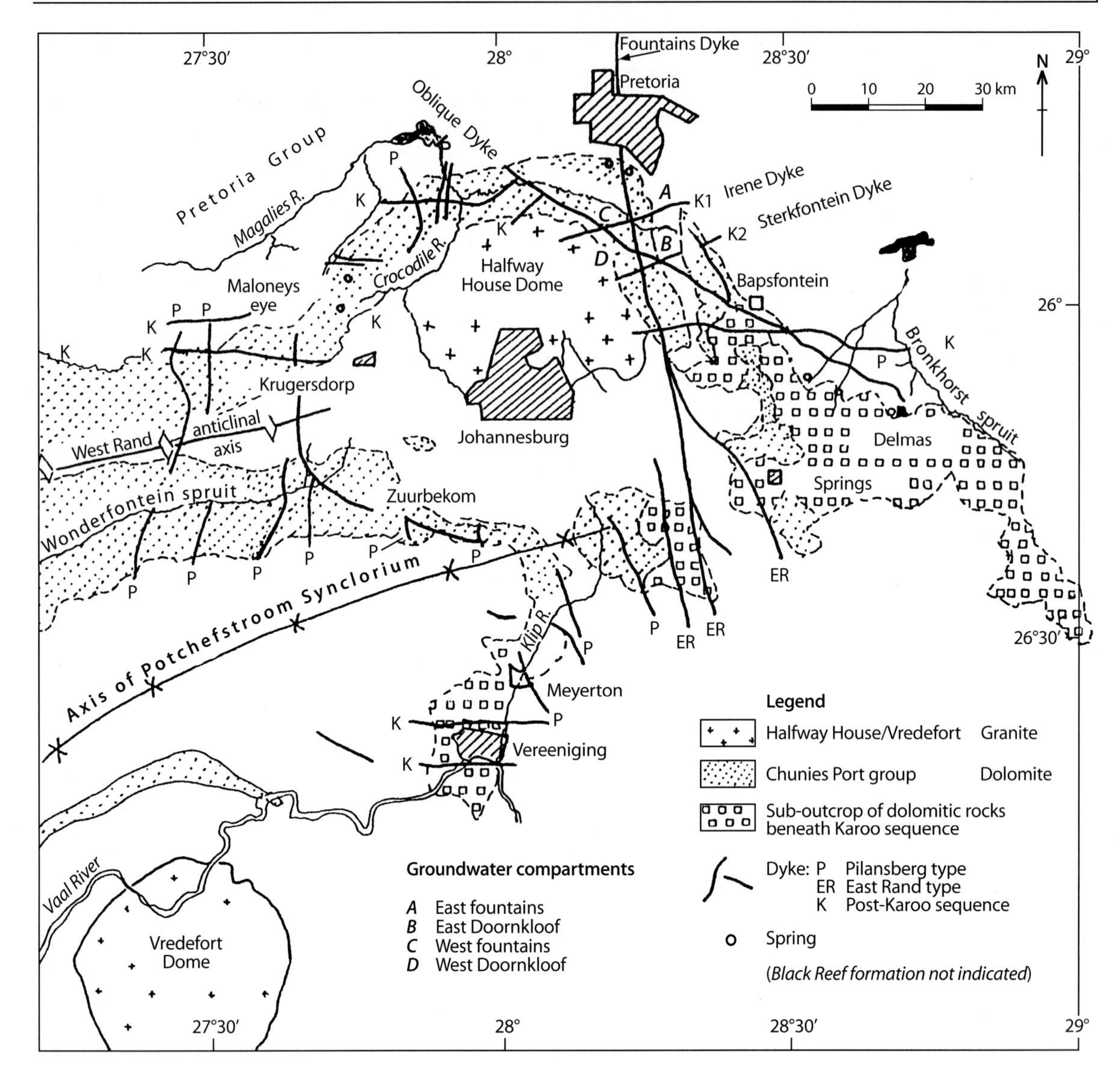

Fig. 8.24. Partitioning of Chuniespoort dolomitic strata by dikes of different ages

The dolomitic aquifers consist of residual dissolution products, chert fragments and wad (a cellular fabric of iron and manganese oxides) and hydroxides, silica and an underlying zone of cavernous to fractured dolomitic rock. The residual products have thicknesses of up to 100 m.

A characteristic feature is the network of dikes of different ages which have intruded the dolomitic strata. The dikes and sills (where present; not shown in figures) have a profound influence on the hydrologic regime by acting as barriers to groundwater movement; thus dividing the strata into separate hydraulic units or compartments. Subsurface flow may, however, occur between compartments at gaps in dikes, where faults displace them and where weathered and fractured dike rock extends to below the groundwater level. Most springs issue on or near contacts within dikes in the underlying Black Reef quartzite or the overlying Pretoria Group clastics. Flows range from less than 0.001 to about 3 m^3/s.

8.13.3 The Pretoria Fountains

Although by no means the largest, the two springs known as "The Fountains" have been selected as most noteworthy for South Africa. The city of Pretoria, the administrative capital of the Republic of South Africa, owes its birth in 1855 to these springs, which form the headwaters of the Apies River.

Initially, the flow of the springs was diverted by means of a weir into furrows running down the sidewalks. This crude distribution system was superseded by pipelines in 1890. By 1928 the discharge from The Fountains could no longer meet the city's requirements, and spring and surface water had to be developed further away from the city, first from Rietvlei Dam on the Hennops River and, subsequently, after the Second World War, from the Vaal River. In 1992, The Fountains supplied 7.3% of the city's average requirements.

8.13.4 Hydrogeological Setting

The hydrogeologic setting is illustrated by a simplified map (Fig. 8.25) and a north-south section of the piezometric surface (Fig. 8.26) east of the Fountains dike. A section west of the dike is very similar. The compartmenting role of the Irene, Sterkfontein, Oblique, and

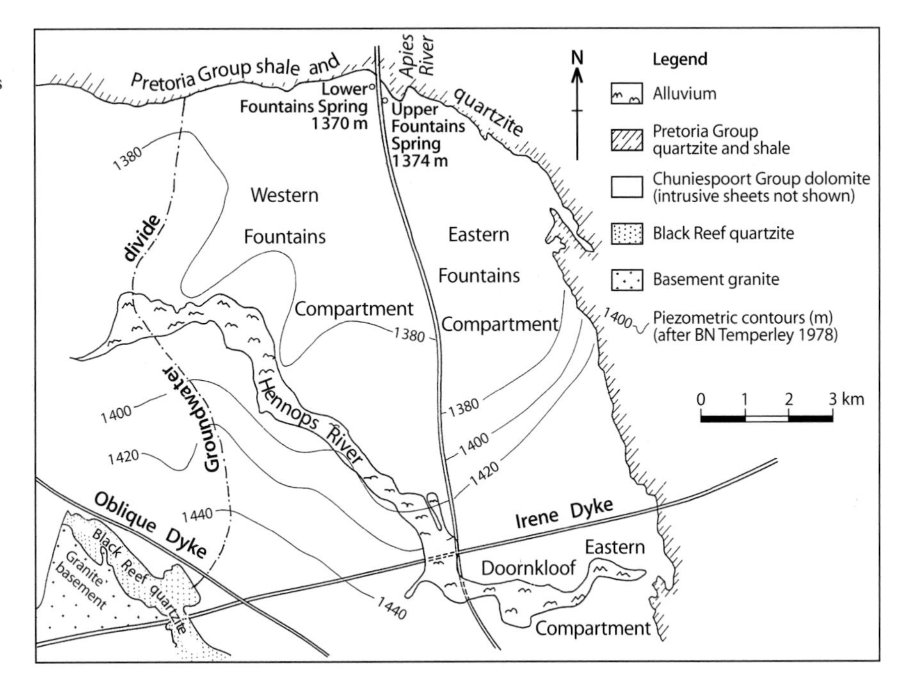

Fig. 8.25. Simplified hydrogeological map of the Pretoria Fountains

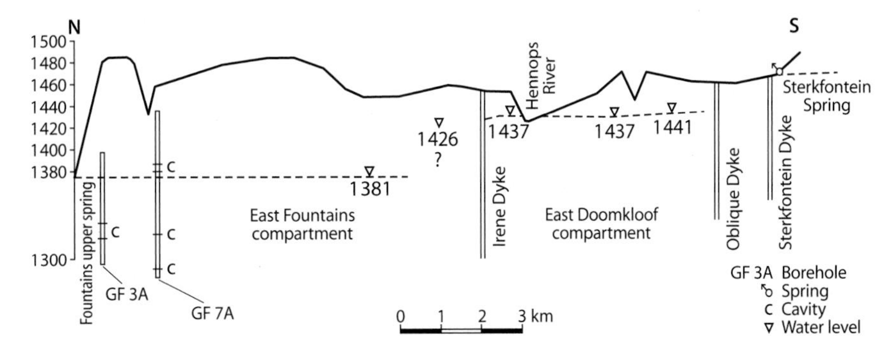

Fig. 8.26. Schematic piezometric section *A* east of The Fountains dike (adapted from Robertson and Kirsten 1985)

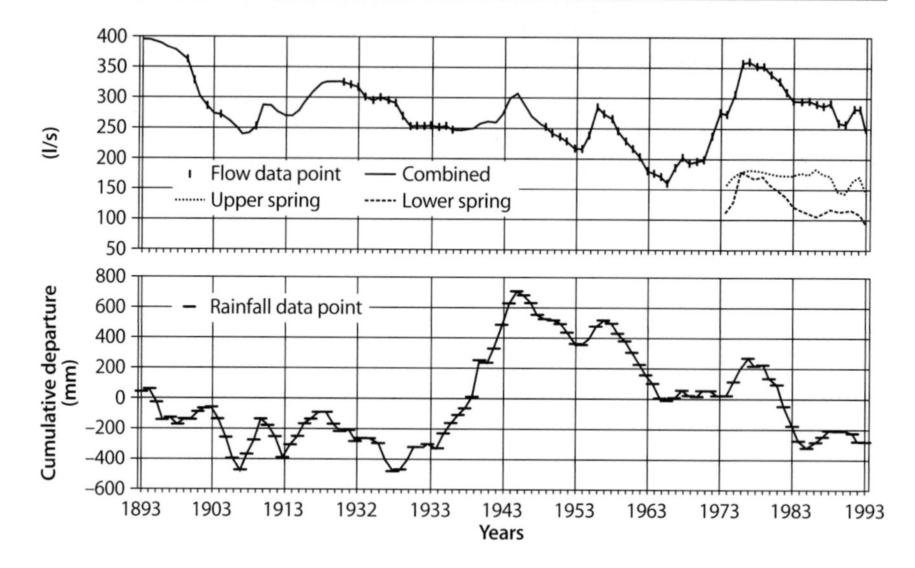

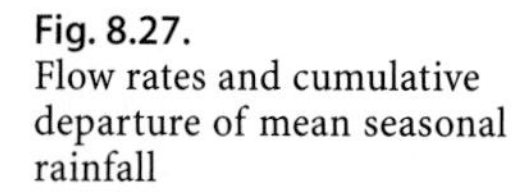
Fig. 8.27.
Flow rates and cumulative departure of mean seasonal rainfall

Fountains dikes should be noted as well as the stepped and low hydraulic gradients that are typical of the Transvaal karst. Note also that the east and west Fountains compartments straddle catchments of the Apies and Hennops Rivers.

The upper (eastern) and lower (western) springs (piezometric difference about 4 m) are approximately 500 m apart on either side of the impervious Fountains dike. They are about 200 m south of the line of junction between the overlying Pretoria Group shale and quartzite and the dolomitic strata. In this area the strata are cut by many dip and oblique faults and The Fountains dike follows one of them.

8.13.5 Flow Characteristics

Although the flows from the springs have been gaged separately over the past 24 years, data available for past years are for combined flows only. In Fig. 8.27 graphs of the yearly mean combined flow rate in l/s and the smoothed (moving 3-year average) cumulative departure from the long-term mean rainfall (of the hydrologic year October to September) are shown. The mean summer rainfall is about 675 mm. In general, both graphs have the same pattern. The fact that recharge and spring flow depend on other factors besides total seasonal rainfall is clearly demonstrated by differing spring responses to cumulative departures from mean rainfall. Note that the flow has been interpolated for the period 1892 to 1908 from some widely separated discharge figures, the accuracy of which may be questioned. The flow has been extrapolated for the periods 1908 to 1919 and 1935 to 1948 for which no data are available. Over the past century the combined flow has fluctuated between 160 and 197 l/s. The dissimilar behavior of discharge between 1969 and 1992 of the upper and lower springs is also noteworthy.

8.13.6 Hydrochemistry

Analysis of cuttings from drill holes penetrating the different stratigraphic horizons has shown that in more than 80% of the samples, the CaO : MgO ratio exceeds the theoretical value for pure dolomite, i.e., 1 : 0.72 by weight. The dolomitic strata are calcium rich (Vegter and Foster 1992).

It has been recognized by Bond (1946) that the CaO : MgO ratio in water from dolomitic springs in the Transvaal is lower than in the dolomite, in spite of being calcium rich. This means that between the dissolution process by rain percolating through the dolomite and its reappearance at the springs, part of the calcium is deposited as calcite and aragonite in caves above the water level (Martini 1973). The variation in chemical composition with time is illustrated in Table 8.10. The spring water is essentially a calcium magnesium bicarbonate water.

Table 8.10. Variation in chemical composition of dolomitic springs in the Transvaal

	Dates of samples				
	Feb–April 1937	Jan 1940	Aug 1980	July–Dec 1985	July–Oct 1990
Total dissolved solids (mg/l)	209	210	272	312	260
Total alkalinity as $CaCO_3$ (mg/l)	198	192		199	200
Total hardness as $CaCO_3$ (mg/l)	200	204		218	222
Ca	40.3	36.5	34.5	41.4	43.2
Mg	23.7	27.3	26.1	32.7	31.0
Na	5.9		7.9		
Cl	1.7	1.7	13.9	25.1	19.5
SO_4	(3.0)?	0.05	4.5	6.9	7.8
NO_3		0.04	1.9	10.6	8.4
PO_4			0.03	0.0	0.40
SiO_2	12.8	70.2	31.7		
CaO:MgO ratio	1:0.70	1:0.88	1:0.78	1:0.90	1:0.90
Reference	Cilliers (1953)	Bond (1946)	Water Affairs Laboratory	Municipal laboratory	Municipal laboratory

Composition of combined flow except for the Jan 1940 analysis which is of one of the springs.

8.13.7 Remarks

Over the past 100 years the land use of the groundwater catchment areas of the upper and lower springs has changed from mixed farming to largely urban. This change has occurred at an accelerating pace from about 1950 to present. Because of the complicating effect of long-term fluctuations, it is uncertain whether spring flow is gradually declining. As far as is known, little, if any, groundwater is being abstracted from boreholes. The effect of human activities is best illustrated by the increase in chloride, sulphate, nitrate and phosphate. Judging by the ratios 1:82 and 1:100 between mean volume groundwater recharged annually and volume of water in storage for two compartments on the Far West Rand (Vegter and Foster 1992), the full effect of ongoing groundwater pollution will probably be realized only after a lengthy period of time.

Acknowledgements

Thanks are due to the Director, Geohydrology, Department of Water Affairs, the Chief Engineer, and the Head of the Municipal Laboratories of Pretoria for making reports and data available.

References

Bond GW (1946) A geochemical survey of underground water supplies of the Union of South Africa. Geol Surv SAfr Mem 41

Martini J (1973) Chemical composition of spring water and the evolution of Transvaal karst. S Afr pelaeological Assoc

Robertson S, Kirsten (1985) Investigation into the groundwater supply potential of the Pretoria dolomites. Unpublished phase 2 technical report for the Department of Water Affairs

Vegter JR, Foster MBJ (1992) The hydrogeology of dolomitic formations in the Southern and Western Transvaal. In: Hydrogeology of selected karst regions, IAH Vol. 13. Verlag Heinz Heise, Hannover, Germany, pp. 355–376

S. Yamamoto

8.14 Springs of Japan

8.14.1 Introduction

There are numerous springs in Japan. The largest and most famous ones are volcanic and karstic springs. Springs in the areas of Mount Fuji, Mount Aso, and Akiyoshidai are discussed here as sources of water used for drinking, irrigation, fish cultivation, industrial water, and sightseeing.

Recently, the Japanese government determined by vote "hundreds of selected significant valuable water" sources. Figure 8.28 shows the distribution of these waters. The source of all water at these locations is from springs, except for three.

8.14.2 Mount Fuji Area

There are about 180 springs around Mt. Fuji. Oshino (northern), Inogashira (western), Fujimiya and Yoshi-

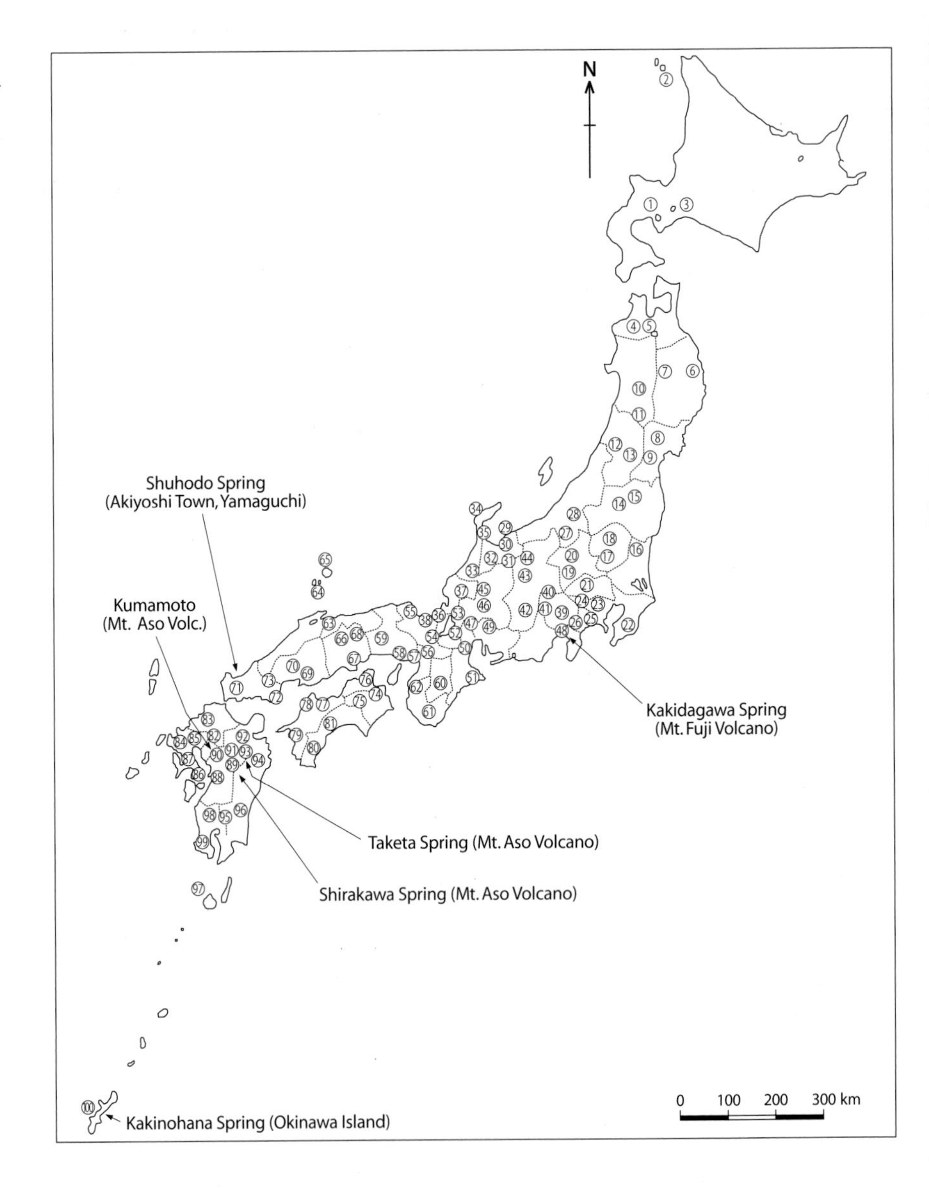

Fig. 8.28. Location of sites of 100 valuable waters in Japan

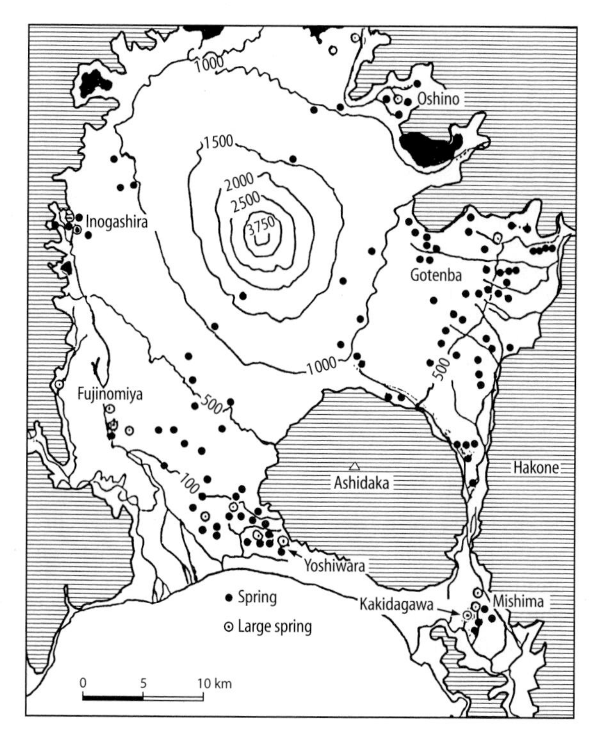

Fig. 8.29. Distribution of springs in the Mount Fuji area

wara (southern), and Mishima Springs are very famous (Fig. 8.29). Spring water is used for drinking, irrigation, fish cultivation, and industrial water and as a sightseeing attraction.

Mt. Fuji, with a peak of 3776 m, consists of a triple-layered structure of superimposed cones. It is on Misaka Miocene groups of green tuff and diorite and is underlain by volcanoes of Mt. Ashidaka and Mt. Komitake. It is thinly covered by New Fuji deposits of pyroclastic ash and small-scale lava sheets. The older deposits are called Kofuji, which means older Fuji and the youngest are called Shin Fuji (New Fuji). Large springs gush out from the lower unconformity portions of New Fuji lava flow.

Mishima Spring is very famous historically. A folk song notes, "White snow on the top of Fuji melts away and flows down to Mishim. It supplies girl's cosmetic water. She stops passengers at Mishima." Large springs are located at Mishima shrine and Kakidagawa with a discharge of 800×10^3 m^3/d, of which 312×10^3 m^3/d are used for drinking, irrigation, and industrial water.

The temperature of this water is 15 °C; pH is 6.8–7.0; and *EC* is 100–170 μS/cm. Based on tritium measurements, the age of this water is about 60 years.

8.14.3 Mount Aso Area

There are more than 1500 springs on and around the Aso Volcano. Springs near Kumamoto City along Kiyama River, on the central cone of the caldera and around Takeda City are the most famous. These are used for drinking, irrigation, and fish cultivation. The total amount of discharge is estimated at 1300×10^3 m^3/d. A spring near Kumamoto City discharges at 800×10^3 m^3/d.

The Aso deposits consist of four tephras: Aso 1, Aso 2, Aso 3, and Aso 4. Aso 1 correlates to the Togawa lava and supplies most of the water to this spring. The temperature of this spring ranges from 14 to 16 °C; pH ranges from 6.6 to 7.0; and *EC* is 90–210 μS/cm.

8.14.4 Akiyoshidai Area (Karstic Region)

Akiyoshidai is a famous karstic region in Japan. There are many doline, uvale, polije, and related springs. Among these Shuhodo is the most famous spring in Japan which issues from a cave. This cave is inhabited by many rare animals. The temperature of the water from this spring is 14.3 °C, with a pH of 7.2 and an *EC* of 160 μS/cm. The water is used for irrigation and fish cultivation.

References

Meinzer OE (1923) The occurrence of groundwater in the United States. US Geological Survey Water-Supply Paper 489, 321 pp

J. Vrba

8.15 Thermal Mineral Water Springs in Karlovy Vary

8.15.1 Introduction

In the western part of the Czech Republic about 130–180 km west of the capital of Prague, in an area of about 300 km^2, several dozen mineral springs occur from various origins, with water of different chemical characteristics, temperatures, and levels of carbonation and radioactive intensity. Mineral waters are widely utilized, in particular for spa treatment of a broad range of ailments as well as for bottling (curative and table waters), industrial uses of carbon dioxide, evaporation for the salts dissolved in them and the thermal waters are used for local heating.

8.15.2 Mineral Waters in Western Bohemia

The best-known spa resorts in western Bohemia include Marianske Lazne, Frantiskovy Lazne, Jachymov, and Karlovy Vary. Thermal (40–73 °C), carbonated (75 l/s of gaseous CO_2 and 400–1 000 mg/l of free CO_2 dissolved in water) mineral waters rich in dissolved solids (TDS 6 400 mg/l) with a total discharge of 30 l/s help cure diseases of the stomach and the digestive tract. About 85 000 patients are treated every year in Karlovy Vary during 3- to 4-week stays. Thousands of visitors use the spas' sport and recreational facilities for short relaxation stays.

8.15.3 Historical Background

Discharging from the Tepla riverbed (tepla means warm in Czech), the Karlovy Vary thermal springs were known by the Slavonic tribes that settled in that region in the 6th century A.D. Artifacts of the Stone Age settlements have also been found near Karlovy Vary. The never-freezing mineral springs issuing into the Tepla River were known as "vary" (boiling, in Czech). The town of Karlovy Vary was founded by Czech King and Roman Emperor Charles IV in 1348, who conferred on it special privileges to be used for its thermal waters. The following beautiful romantic legend associated with the foundation of the town:

King Charles set out from the Loket Castle (about 12 km from Karlovy Vary) to hunt in the surrounding deep forests. However, the hunt was not successful. At dusk, upon returning back to the castle, the royal entourage beheld a stately deer. They started to chase it. The deer tried to flee from the hunters and their dogs; it was pursued to the top of a steep rocky cliff and jumped into an abyss below. When the hunters climbed down into the deep valley they saw geysers of hot springs. "These are boiling waters" said the Royal Burgrave, "I wanted to show them to your Royal Highness for a long time but the deer came first." The king, enchanted by the beauty of the natural sight, said: "I will found a city on this place, and a castle to bear my name. They will enjoy all the royal privileges and will return health to those who ail. Let the rock from which jumped the deer that has led us to the hot springs be called the Deer's Jump for eternal memory." The statue of a deer on the rocky outcrop over the town, and the name of Karlovy Vary (Charles's Hot Springs) shall always remind us of their history.

The first written reports on physicians' recommendations to use the Karlovy Vary mineral waters for curative purposes date back to the early 16th century, and sound amazingly modern. Doctor Vaclav Payer recommends alternating drinking of thermal water and bathing in it, plus a diet. Fifty years later, and throughout the Baroque period, the medical community's views of mineral water cures were quite controversial. Physicians prescribed drastic dosages: drink 60 cups of water a day and spend 12 h/d in the thermal water baths until "the skin cracks and the disease is allowed to leave the body." Treatment of patients as we know it today was introduced by Dr. Becher as late as 1750.

The spa's importance began to grow from the mid-17th century. In the early 18th century, Karlovy Vary was already a social center popular with European rulers, nobility, and aristocracy (Fig. 8.30). New bath houses, hotels, a theater, and a colonnade were built there. Historical documents list 247 patients in 1756, 5 000 patients in 1848, and 70 000 patients in 1912. Books kept by spa houses register repeated calls and stays of geniuses of the world's music, literature, and arts: Bach, Beethoven, Goethe, Schiller, Chopin, Wagner, Dvorak, Brahms, Paganini, Tchaikovski, etc. After 1945, a tradition was started in Karlovy Vary. International music and film festivals and scientific congresses were held there.

The first reports on the chemical composition of the Karlovy Vary Springs date back to 1522. The first analysis, on a par with current methods, was made by Dr. Becher in 1749. The first credible measurements of the mineral waters' discharge (yield >2 200 l/min) were from the same period.

Fig. 8.30. Unknown artist of the Dresden School: view of Karlovy Vary from the east (colored engraving, 1815)

The capture of mineral springs at Karlovy Vary has undergone considerable development. It was complicated by the carbonated mineral water frequently bursting through both inside and outside the Tepla riverbed. Initially (1570), thermal springs were accumulated in shallow collection holes in the bed of the river and water passed through open troughs to the bath houses. To achieve the required head, the water was allowed to expand and was collected in shallow boreholes, 3–5 m deep, on the Tepla banks. The boreholes quickly filled with incrustations from the mineral water and had to be redrilled. Also, the thermal water kept breaking out around the boreholes as well as in the Tepla riverbed. This happened frequently during the 18th and 19th centuries and caused damage to buildings and roads and caused a temporary decline in the yields from boreholes and springs. A burst in 1809 was described by laymen as an earthquake.

A new technique of capturing carbonated thermal water was introduced in 1912 when R. Kampe installed flow-over towers for the water. At the head of a borehole, free carbon dioxide is separated from the water and fanned away. The thermal water is then piped to spa houses, containing only the dissolved CO_2. The high level of the Karlovy Vary thermal water's vulnerability, prompted extensive hydrogeological investigations, which resulted in a novel concept of collection. In 1981 and 1982, three inclined collection wells were drilled, thereby intercepting thermal water at a depth of more than 8 m for the first time in history. Thermal water accumulates at a depth of 44–88 m, which is the level of CO_2 evasion, thereby bringing to an end – after 500 years – the era of shallow development of the Karlovy Vary thermal springs. In 1984–1986, deep boreholes also intercepted the source of water to the small springs.

8.15.4 Geologic and Hydrologic Setting

The Karlovy Vary region lies in the western part of the Bohemian Massif, which forms the easternmost part of the European Hercynian folding (Meso-Europe) and covers Bohemia and the western part of Moravia. The development of the Bohemian Massif has been affected by major geotectonic cycles – Cadomian orogeny (which consolidated the base of the Bohemian Massif), Caledonian, Hercynian (characterized by disintegrating the Bohemian Massif into blocks of different development), and Alpine (neoid tectogenesis manifested in rejuvenation of the Epihercynian platform and vulcanism associated with Saxonian tectonics).

Of decisive importance for the genesis of mineral waters (cold, hot, carbonated) was the Ohersky Rift, which came into being by the resumption of neoid tectogenesis in an ancient mobile zone. The rift separates two important geological units of the Bohemian Massif: the Ore Mountains block (an area of intensive Hercynian tectogenesis) and the Tepla-Barrandien block (a stable Varisan intermontane block). The Ohersky Rift is delineated by two rift faults of a northeast-southwest direction and a distinct central fault along the longitudinal axis of the rift (Kopecky 1971).

The Ohersky Rift is part of the European-African system and has all the characteristics typical of an intercontinental rift (Kopecky 1971). In terms of hydrogeology and genesis of mineral water, the important factors are an asymmetric terraced trough with a markedly alkaline magmatism, the existence of a gravimetric minimum, an intensive thermal flow, migration of neoid tectonic and volcanic activity in the direction of the structural axis (from northeast to southwest), and allied occurrence of CO_2 in the aftermath of active volcanism. Since the early 20th century, various authors have been pointing out the relation between the occurrence of CO_2 and neoid volcanism. The genesis of regional distribution of mineral waters in the Bohemian Massif in relation to old neoid-rejuvenated structural-tectonic lines was first noted by Vrba (1964; Fig. 8.31).

The Karlovy Vary thermal mineral waters are genetically linked to the Karlovy Vary Pluton as part of a complex of Hercynian intrusive rocks (Carbon-Perm) of granitoid composition. The Karlovy Vary Pluton is oriented crosswise to the Ohersky Rift and is composed of porphyric, biotitic, two-mica, and muscovitic granite to granodiorite, in some places heavily weathered to kaolin (to a depth of 50 m) and hydrothermally transformed (along the fracture systems). Pluton emerged in two intrusions separated in time – the older "montane granite" some 300 million years ago, and the younger "Ore Mountains granite" 250 million years ago. Thermal waters are associated with the latter. The Karlovy Vary Pluton surfaces include a large outcrop covering about 1 000 km^2.

The occurrence of Karlovy Vary Springs, concentrated in a short stretch of the Tepla Valley thermal zone, is predisposed tectonically and associated with the intersections of faults of two directions. The west-east to WSW-ENE direction, which corresponds to the Ohersky Rift, is of transregional importance. Of local importance is a system of faults in the NNW-SSE direction (azimuth 330°, inclined 70–80° towards southwest), which is referred to as the Karlovy Vary Thermal Spring Line. Geophysical investigations identified the depth of the Mohorovičić discontinuity as 30–32 km, the thickness of the granite pluton at about 10 km, a high geothermal activity of 80–90 mW/m^2 (Cermak 1979), and the highest negative values of Bouguer anomalies in the Bohemian Massif.

Fossil and recent carbonated mineral waters have been identified in the thermal zone. The tectonically predisposed shatter belts are as much as 6 m thick, the rocks in them are disturbed by strong cataclastic to mylonitic deformations. They were subsequently silicified and filled with chert veins containing small crystals of barite, which is evidence of their hydrothermal origin. On the surface of the thermal zone there are layers of different types of sinters, as much as 8 m thick, including aragonite, a chemically almost pure calcium bicarbonate that crystallizes rhombically over 50 °C and, characteristic of Karlovy Vary, pisolite (colloquially called pea-stone in Czech).

The fractures with active inflows of thermal water are open; as determined in the instance of an inclined investigation borehole, and terminated at a depth of 133 m due to a massive influx of thermal water. A sound log (length 1 050 mm, width 36 mm), sunk in the borehole passed through the footwall of the borehole and into an open fracture to a depth of 370 m. The direction of the open fracture corresponded with the Karlovy Vary Thermal Spring Line.

The discharge of the Large Springs is controlled, according to Vylita and Pecek (1981), by a fault system of the Karlovy Vary Thermal Springs Line (NNW-SSE) at the points where it crosses deep faults genetically asso-

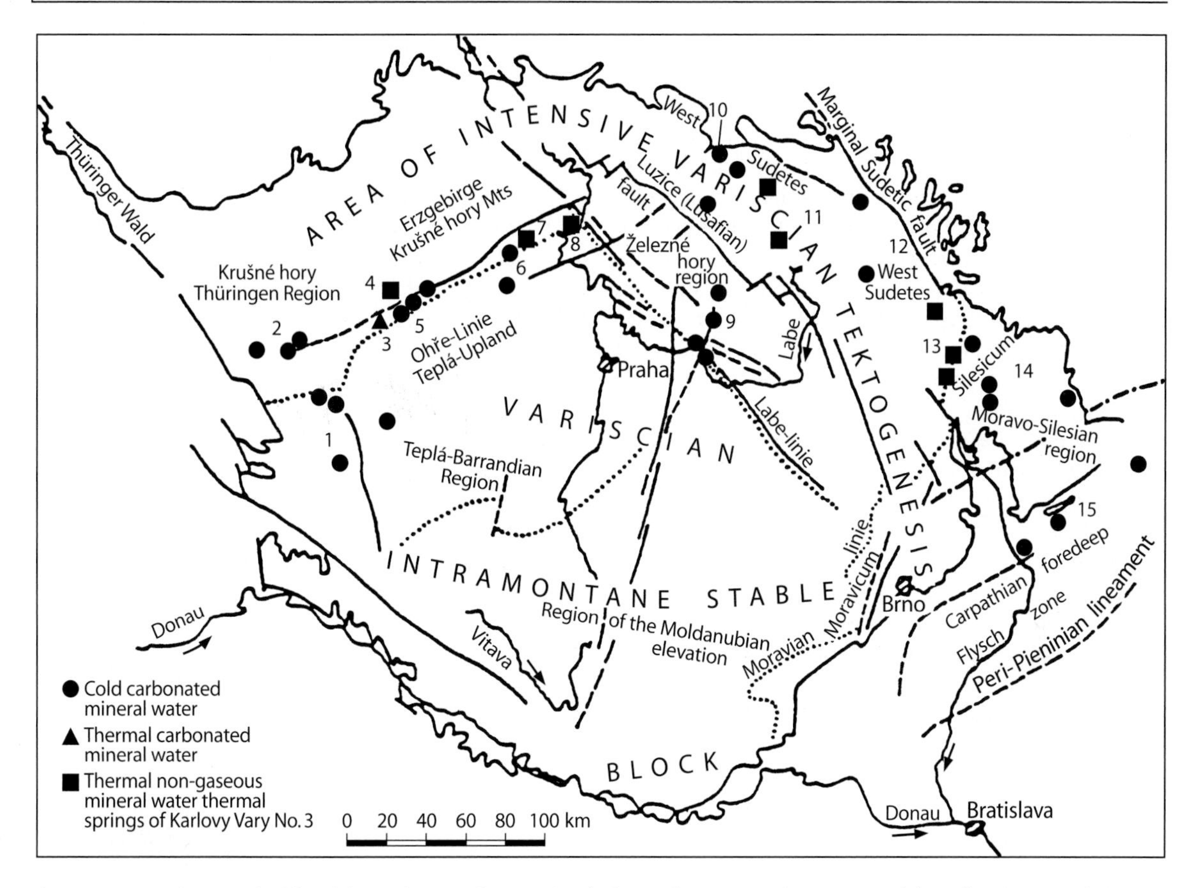

Fig. 8.31. Distribution of cold and thermal mineral waters in relation to the geotectonic structure of the Bohemian Massif

ciated with the Ohersky Rift (NE-SW). Investigations have identified outflow of the springs over an open, steeply inclined fault having a NNW-SSE direction. The surfacing of the thermal water is mobilized by carbon dioxide (the gas-lift effect). Due to high pressures, gaseous CO_2, which rises from the depths, completely dissolves upon contact with water. It escapes from water only in the last stage of the thermal water rise when a change in pressure and temperature triggers its escape at a depth of 60–80 m. The process of evasion accelerates as it nears the surface, whereby the mass of the water-gas mixture (1:3) decreases, which positively affects the discharge level and yield from interceptor wells. The springs' high overpressure had ruled out the possibility of thermal water mixing with a cold water aquifer. Nevertheless, its effects were eliminated with development of a collection system for the thermal water installed in 1981–1982.

Isotope analysis of sinter deposits and studies of the morphological development and terraces of the Tepla River show that the emergence of the Karlovy Vary Springs can be estimated at between 40 000 and 250 000 years ago.

Since the first half of the last century, the group of so-called small thermal springs attracted geologists' attention by its linear arrangement. These springs form a peripheral area to the large spring field. Discharges of thermal water are concentrated in a tectonically disturbed zone up to 150 m broad, which runs through the Karlovy Vary granite in the Tepla Valley along an azimuth of 326°. According to the latest investigations, the central part of the disturbed zone constitutes a tectonic trough (Vylita and Pecek 1981). In the tectonic zone, granite is extremely disturbed tectonically, is silicified, and contains numerous chert veins. The differences in the temperature and yield of the small springs is attributed to the different outflow paths of mineral water in the tectonically disturbed subsurface zone and the extent to which thermal water mixes with the shallow groundwater system drained by the Tepla River. In tectonically fractured granite, a two-layer aquifer with fracture permeability has formed. The upper aquifer is composed of cold water with shallow circulation, that overlays the bottom aquifer with deep-circulating thermal waters. There is a close relationship between the two aquifers. In the course of increased recharge of the upper aquifer during periods of precipitation, the thermal aquifer is contaminated and subsequently contaminates the shallow collecting holes that collect water from the small springs utilized for drinking cures. For this reason, in 1984–1986, the small springs were intercepted in boreholes at different depths to eliminate the possibility of their contamination from water in the upper shallow aquifer.

8.15.5 Methods of Development and Use

Development of the large springs, known as the Karlovy Vary Thermal Spring Field, was started nearly 500 years ago (about A.D. 1500). At that time the springs discharged to two shallow holes on the right bank of Tepla. Gradually, the number of holes increased to ten but their depth had never exceeded 8 m under the riverbed. Thermal water was captured near the surface, in the caverns of an aragonite layer. The elevation of discharges of these springs gradually rose by 4 m. To prevent uncontrolled discharges in the Tepla riverbed, the flow of thermal water was sealed using various methods. However, with the growth of the spa resort and, in particular, erection of a new colonnade over the springs in 1975, shallow collection of springs was fraught with many technical problems. It was only the deep capture of the large springs in 1981–1982 that enabled the sealing off (by means of injecting, under pressure, a mixture of clay and cement) of uncontrolled discharges of thermal water into the riverbed. These injections could not be used for the shallow capture because the mixture penetrated into the shallow wells through caverns. Into the layer of aragonite were drilled so-called regulating boreholes to relieve the pressure and flow of thermal water and gas upon the aragonite layer and the colonnade's foundations sunk under the level of the river's bottom. Relieving the aragonite plate was essential because accurate measurements had demonstrated its rise by 25 mm over 3 years.

The deep captation of the thermal springs had been preceded by several years of comprehensive geotectonic, petrological, hydrogeological, atmochemical, hydrochemical and geophysical investigations, and the drilling of several investigation boreholes.

On the basis of this investigation, four inclined (14.5–20.5°) capture wells were constructed which intersected at different depths (48–88 m) the tectonic fault zone through which carbonated thermal water rises to the surface (Fig. 8.32). The spring, with an overflowing fountain of thermal water and gas, was intercepted in an inclined well at 14.5°, azimuth 314°, depth 71 m, discharge level 12 m over the terrain. The discharge from the four wells of the large springs is 1700 l/min and 4795 l/min of carbon dioxide. The wells were designed with anticorrosion screens and cement on their outer perimeter to a depth of about 16 m. When drilling penetrated a fracture with thermal water, a strong jet of water always spurted out, carrying clastic rocks up to

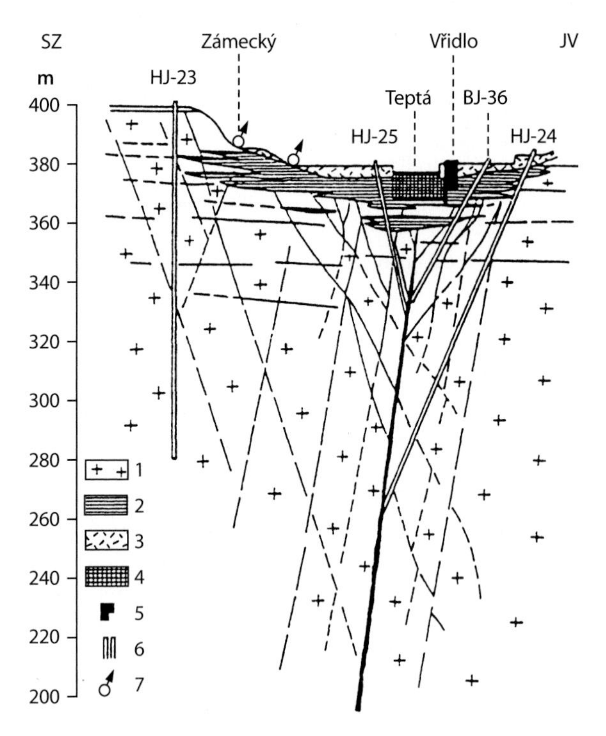

Fig. 8.32. Scheme of the discharge path of Karlovy Vary thermal mineral water (after Vylita and Pecek 1981): *1* granite; *2* aragonite; *3* Quaternary deposits and stockpiles; *4* old sealing of the riverbed; *5* old shallow captation boreholes; *6* new inclined investigation and captation wells; *7* thermal springs

150 mm in diameter (in the 1–2 h prior to closing of the borehole, several cubic meters of this material were ejected). The drilling required special equipment to keep discharges of thermal water and gas fully under control.

Capture of the large springs in deep wells and injections into the Tepla riverbed have resulted in an increase in and stabilization of the yield of thermal water and gas and a profound reduction in their vulnerability. The ratio of water and gas improved from 1:1.24 to 1:1.33. Formerly, it had been 1:3 because thermal water was captured at approximately the depth of CO_2 evasion (Vylita and Pecek 1981).

Since the 15th century, the large springs have traditionally been used for baths and drinking for various cures. Later, evaporation for salts was introduced and carbon dioxide started to be used in industry. Souvenirs are made of polished aragonite, and the famous "petrified" Karlovy Vary roses are created by immersing them in the wells.

Capture and therapeutic application of the group of small springs began as late as the 1770s when local physician Dr. Becher introduced them in drinking cures. The small springs' yield is substantially lower. Depending on the complexity of their rise to the surface, they have varying temperatures, TDS, and CO_2 levels. The discharges from the springs were taken from shallow wells. The latest modifications of the wells were made before 1912. Shallow captation resulted in an increase in contamination of mineral waters, and closing down of the springs for several days was normal. For this reason, in the case of the small springs, comprehensive investigations resulted in a new concept of captation in wells of different depths (the original shallow wells have been preserved). The new wells have helped stabilize mineral water yields and temperatures, as well as volumes of gas. Only natural overflow from the boreholes is utilized, and the effects of the shallow cold aquifer have been completely eliminated. Small springs, and recently also mineral water from the wells that have replaced them, are used only for drinking cures. Some of the small springs have traditional Czech names.

Table 8.11 indicates that the yield and, particularly, the temperature of the mineral water from boreholes have increased in comparison with the earlier shallow captation.

8.15.6 Chemical Character and Temperature

The Karlovy Vary siliceous carbonated thermal waters are the hottest mineral waters of the Bohemian Massif. Their chemical type is uniformly Na-SO_4-C_1-HCO_3, their TDS more than 6 g/l, temperatures range from 40 to 73 °C, free carbon dioxide dissolved in water ranges from 400 to 1 000 mg/l. Major cations and anions are listed in Table 8.12.

Carbon dioxide dominates in gas content (95–99% vol). Among the remaining gases, nitrogen occurs at a level of tenths to units of percentage volume, and oxygen at thousandths to units of percentage volume. Argon and helium are at very low levels.

Table 8.11. Comparison of average values of yield, temperature and CO_2 level of some small springs from shallow captation holes and wells

Name	Temperature (°C)		Yield (l/min)		Free dissolved CO_2 (mg/l)		Depth (m)
	Captation holes	Wells	Captation holes	Wells	Captation holes	Wells	Wells
Karel	42.6	61.4	2	20.0	692	660	70.0
Vaclav	59.4	64.5	13.4	2.5	539	579	9.6
Libuse	41.2	62.9	5.1	3.8	610	814	17.6
Skalni	52.9	58.4	3.6	4.2	687	821	20.0
Rusalka	49.1	61.0	2.4	5.1	624	618	7.8
Mylnsky	52.9	54.8	2.6	3.4	704	682	24.0
Trzni	49.6	64.0	7.2	9.6	763	836	38.0
Sadovy	47.4	45.0	9.0	10.0	1 020	1 400	37.0
Svobody	61.4	64.6	8.5	9.5	566	755	38.6

Table 8.12. Principal ions in Karlovy Vary thermal water[a]

Cations	mg/l	m/val	mval%	Anions	mg/l	m/val	mval%
Na	1701.00	73.985	85.52	Cl	612.34	17.270	19.56
K	78.90	2.018	2.33	NO_2	0.00	0.000	0.00
NH_4	0.11	0.006	0.00	NO_3	0.50	0.008	0.00
Mg	47.42	3.900	4.51	HCO_3	2127.13	34.860	39.49
Ca	124.25	6.200	7.17	CO_4	0.00	0.000	0.00
Mn	0.09	0.003	0.00	SO_4	1722.46	17.932	40.62
Fe	1.34	0.048	0.06	PO_4	0.32	0.003	0.00
Li	2.47	0.356	0.41	F	5.20	0.274	0.31
H	0.00	0.000	0.00	OH	0.00	0.000	0.00

[a] SiO_2: 76.20 mg/l; TDS: 6499.73 mg/l; pH: 7.45; BOD: 1.54 mg/l; Conductivity: 8100; ^{222}Rn: 5–50 Bq/l; mval: total amount of substance in milligrams per liter which corresponds to 1 gramatom of hydrogen; mval%: percentage proportion of total cation/anion content in 1-l volume.

8.15.7 Thermal Water Protection

Thermal water protection should be understood in terms of internal and external protection. Internal protection of thermal water against construction activities (laying of foundations) and local pollution sources was addressed successfully by introducing a deeper captation of the water in boreholes. External protection is of principal importance, as kaolin and lignite are mined from open-pit mines near the spa resort. Protective measures are defined in terms of area and implemented in three degrees of protection zones, as well as in terms of depth (extraction of kaolin and lignite is restricted to a certain depth derived from the discharge level of the thermal water). This restriction on mining is important, as formerly hot and carbonated groundwater broke out several times in kaolin and lignite mines.

The regulation that defines the extent of protective zones, and the Mining Act specifically list the activities that may be undertaken in protection zones, including the recharge areas.

References

Cermak V (1979) Thermal flow in Czechoslovakia. Geological Survey Institute, Prague
Kopecky L (1971) The Ore Mountains – Charecka Zone and Rynsky Fault and their importance for deep structures. In: Investigation of deep geological structures of Czechoslovakia. Brno, pp 157–185 (in Czech)
Vrba J (1964) Origin and occurrence of carbon dioxide and carbonated mineral waters, Varisan Platform, Central Europe. Econ Geol 59:874–882
Vylita B, Pecek J (1981) Investigation of the Karlovy Vary Springs structure. Geol Surv J 161–614 (in Czech)

P. Bono · C. Boni

8.16 Mineral Waters in Italy

8.16.1 Introduction

The use of spring water as a drinking, therapeutic, and ornamental resource has historical origins with the Romans. The most ancient regulations on "mineral waters" was enacted in Italy long before the union (1870). In Italy, consumption of "mineral waters" as a beverage is increasing. In 1992 the home market was about 6200 million liters, with a per capital consumption of 108 l. There are 233 national mineral water brands and there are at present 181 spa units (Fig. 8.33–8.35, Table 8.13).

8.16.2 Bottled Waters in Italy

There is a large variety of physical, chemical, and medicinal characteristics of mineral waters, many of which were recognized by the Romans. Two mineral waters that have a long historical tradition are described. The selected springs originate from aquifers with different hydrogeological conditions.

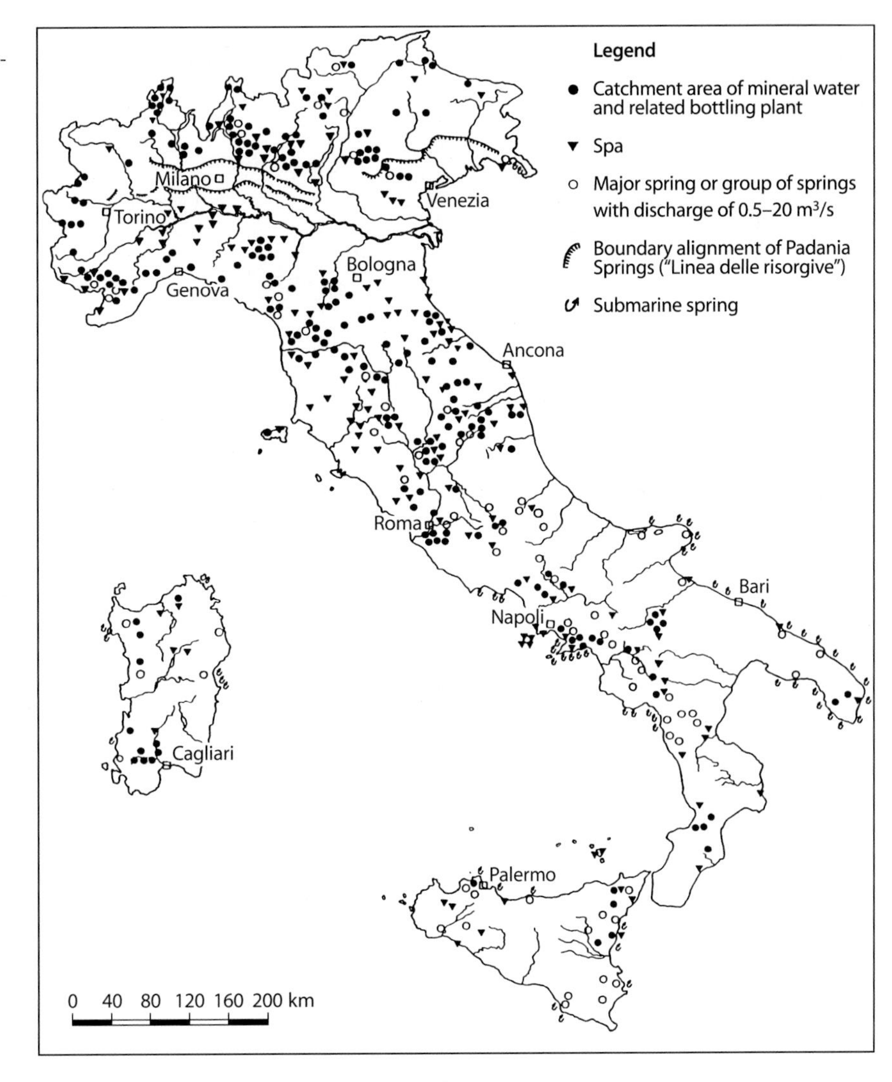

Fig. 8.33. Location of mineral water catchment areas and spa plants in Italy

8.16.2.1 *Ferrarelle Mineral Water*

These springs are along the southeastern edge of Roccamonfina stratocone volcano (Campania region) close to the northern limits of the Monte Maggiore Mesozoic karst unit. The springs are near Ricardo (Caserta Province), the nearest municipality to the Ferrarelle catchment and bottling plants.

The Savone River crosses the Quaternary volcanic deposits bordering the Monte Maggiore carbonate unit, a morphological step to the gentle slopes of the volcano. The river and the aquifer in volcanic terrains form a system on which linear springs occur along the main drainage network. The springs were known by the Oscian and Etruscan people, and certainly quenched the thirsts of the invading Greeks, Carthaginians, Sannites, Romans, Goths, and Longobards.

Historical and archaeological finds show unmistakably that the Romans improved this water resource and constructed several catchments and thermal plants in the area. Vitruvio (1st century B.C.) mentions different acidulous waters in the area which were thought to have special therapeutic power. Later the famous naturalist Plinio the Old (A.D. 27–79), Cicero, and Horace refer to the same spring water, keeping alive the tradition

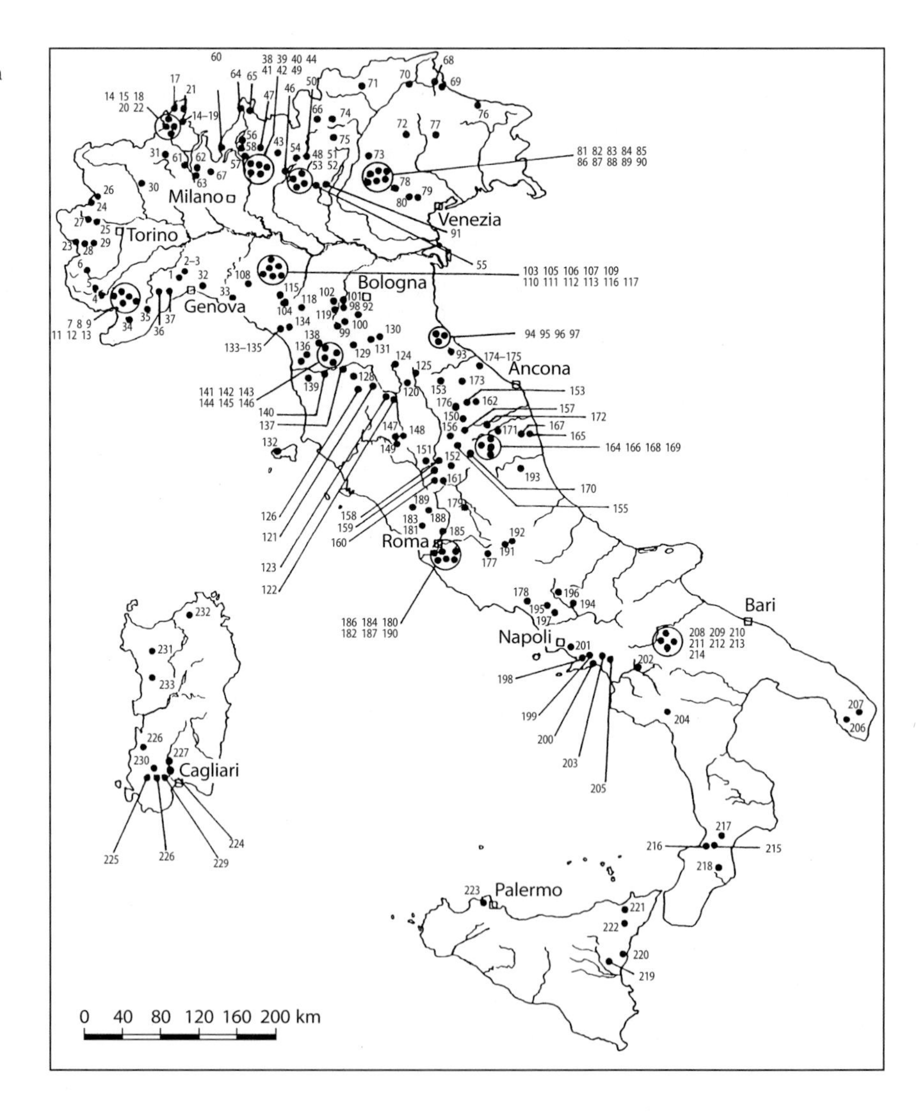

Fig. 8.34. Mineral water spring location and reference number (see Table 8.13)

but not the splendor that existed after the end of the Roman Empire.

The stratigraphy of the catchment area is represented by Quaternary volcanoclastic sequences with heterotic continental deposits which overlie a dissected karst bedrock structurally referred to as the Monte Maggiore unit. The spring area is along a fault zone of two main tectonic systems with NW and SE trends. The Roccamonfina Quaternary graben and the related volcanic activity provided such a schematic structural frame. The carbonate bedrock is intensely fractured and karstified, while the volcanic cover has a remarkable anisotrophy due to heterotic facies or phreatomagmatic pyroclastic flow and ashfall sequences. The recharge area is on the southern slope of the stratocone volcano, while groundwater flow occurs in karst and volcanic reservoirs are hydraulically connected. A remarkable CO_2 content and calcium and silica concentrations characterize the chemical facies of Ferrarelle groundwater (Table 8.14).

8.16.2.2 *Fiuggi Mineral Water*

The town and springs of Fiuggi are in the middle of the Italian peninsula, 100 km south of Rome, in a wooded depression on the southern slope of the Simbruini

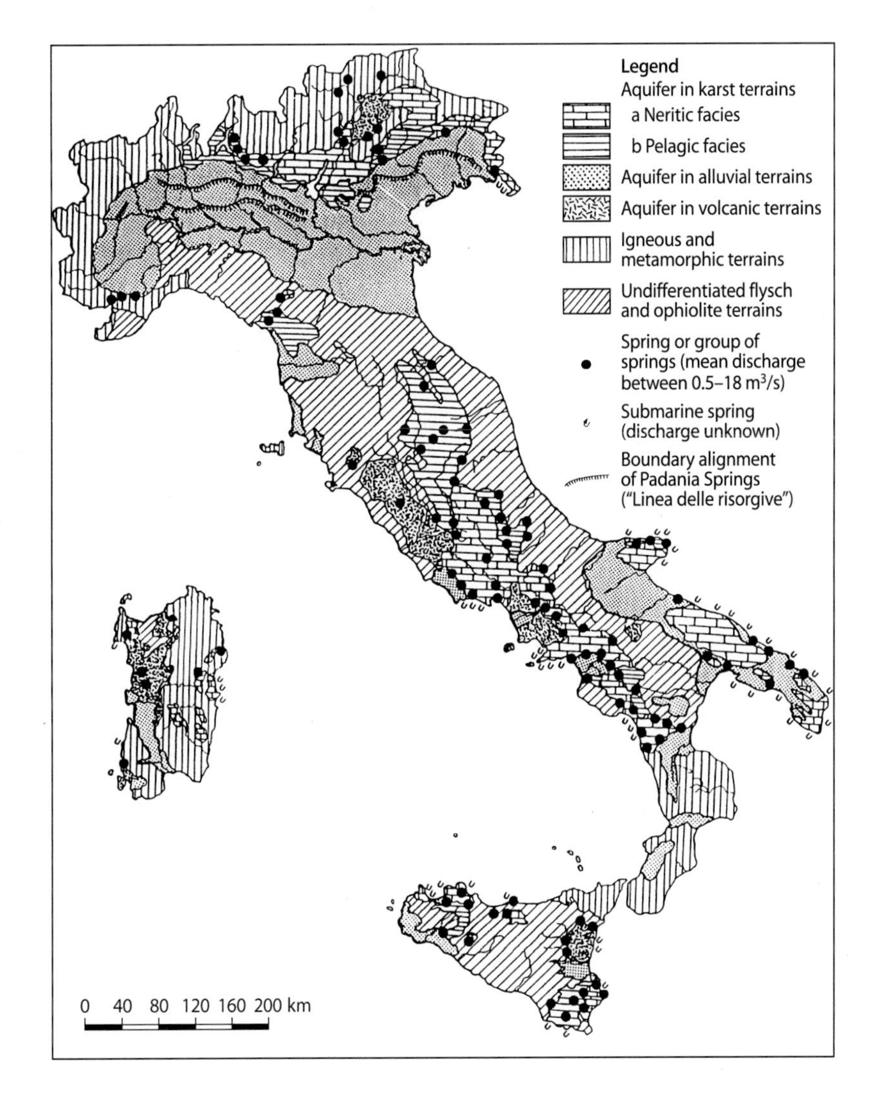

Fig. 8.35. Regional aquifer scheme of Italy

Table 8.13. Mineral water in Italy and main water characteristics (data from Annuario Acque *Minerali Italiane* 1991–1992, ed. Laus, Milano)

Brand name	Location refer. no.	Spring altitude (m)	TDS 180 °C (mg/l)	pH	Production (mio. l/yr)
Piemonte region					
Augusta	1	240	496		20/30
San Rocco	2	456	392	7.25	10 20
Sovrana	3	240	366	7.32	20/30
Abrau	4	570	132	7.05	<10
Camorei	5	585	276	7.60	10 20
Coralba	6	743	202	7.50	20/30
Garbarino	7	720	111	7.01	70/100
Lurisia	8	1 400	34	6.65	70/100
Nuova Gareisa	9	595	126		10 20
Roccolo	10	600	84		<10
Sorg. della Rocca	11	1 500	126	7.36	
San Bernardo	12	1 000	54	7.70	150/200
San Bernardo (Rocciaviva)	13	1 150	42		150/200
Alpia	14	760	53		10 20
Ausonia	15	680	780	5.76	<10
Buvera	16	238	54	7.70	10 20
Crodo Lisiel	17	510	226	7.70	70/100
Gaudenziana	18	900	96	7.85	<10
Gioiosa	19	293	182	7.55	20/30
San Lorenzo	20	700	2 282	6.14	<10
Valle d'Oro	21	500	2 160		70/100
Vigezzo	22	875	30		10 20
Alpi Cozie	23	900	33	6.50	10 20
Plan della Mussa	24	1 432	39	6.75	10 20
Pic	25	400	186	7.68	10 20
San Grato	26	800	1 139	6.80	<10
San Michele	27	1 400	35	6.80	10 20
Sparea	28	600	24	6.38	10 20
Valmora	29		23	6.70	10 20
Lauretana	30	1 050	14	6.10	20/30
Valverde	31	780	37		10 20
Liguria region					
Madonna della Guardia	32	680	101	7.20	10 20
Santa Rita	33	750	591	8.10	30/50
S. Vittoria	34	1 380	183	7.45	20/30
Bauda Calizzano	35	980	35	6.30	20/30
Fonte del Lupo	36	398	31	5.90	20/30
Vallechiara	37	398	34		20/30
Lombardia region					
Bracca	38	400	725	7.20	50/70
Flavia	39	400	259	7.34	50/70
Gaverina	40	500	575	7.44	50/70
Limpia	41				200/300
Lombardia region (cont.)					
Orobica	42	400	344	7.24	20/30
Pineta	43	700	189	7.60	10 20
Pracastello	44	358	827	7.22	200/300
Prealpi	45	480	330	7.23	20/30
Primula	46	400	368	7.28	50/70
Stella Alpina	47	900	50	7.13	20/30
San Carlo Spinone	48	400	422	7.15	50/70
San Pellegrino	49	358	197	7.71	200/300
Boario	50	217	535	7.35	200/300
Castello di Vallio	51	296	265	7.58	10/20
Linda	52	65	343	7.10	50/70
Sole	53	150	371	7.10	<10
San Silvestro	54	500	1 355	7.05	20/30
Tavina	55	70	350	7.10	50/70
Chiarella	56	760	166	7.70	30/50
Daggio	57	1 935	48	7.25	70/100
Gajum	58	481	200	7.70	30/50
Leonardo	59	900	60	7.60	70/100
Paraviso	60	907	229	7.45	<10
Sant'Antonio	61	313	162	7.90	150/200
San Francesco	62		134		150/200
San Luigi	63	627	219	7.50	10/20
Bernina	64	625	37	7.35	30/50
Frisia	65	435	99	7.05	30/50
Levissima	66	1 848	68	7.75	>300
Frida	67	271	222	7.75	10/20
Trentino region					
Acqua Imperatore	68	1 175	755		10/20
Lavaredo	69	1 340	1 381	7.60	10/20
Plose	70	1 830	26	6.50	10/20
San Vigilio	71	1 540	21	6.20	<10
Idrea	72	1 250	2 422	7.40	<10
Levico-Casara	73	1 640	37	6.80	20/30
Pejo	74	1 394	85	6.60	100/150
Surgiva	75	1 100	38	6.00	20/30
Friuli region					
Goccia di Garnia	76	1 370	71	8.08	30/50
Veneto region					
Vena d'Oro	77	452	194	7.70	10/20
Vera	78	60	157	7.80	>300
Guizza	79	20	285	7.40	>300
San Benedetto	80	20	271	7.50	>300
Acquachiara	81	805	125	7.95	70/100

Table 8.13. *Continued*

Brand name	Location refer. no.	Spring altitude (m)	TDS 180 °C (mg/l)	pH	Production (mio. l/yr)
Veneto region (cont.)					
Alba	82	530	36	6.00	10/20
Azzurra	83		377	7.00	<10
Beber	84	725	136	7.70	10/20
Dolomiti	85	640	274	7.65	70/100
Lissa	86	630	180	7.70	10/20
Lonera	87	450	100	8.00	<10
Lora	88	800	154	8.10	150/200
Margherita	89	260	2358	6.30	<10
Regina	90	450	1634	6.40	<10
Balda	91	65	298	7.50	10/20
Emilia-Romagna region					
Cerelia	92	680	382	7.33	20/30
Bonora	93	190	433	7.40	<10
Fontesana	94	150	558	7.15	10/20
Galvanina	95	150	570	7.05	10/20
Sacramora	96	2	563	7.23	30/50
San Giuliano	97	2	571	7.25	30/50
Fonte del Parco	98	900	265	7.01	10/20
Monte Cimone	99	936	109	7.88	20/30
Monteforte	100	716	366	7.25	
San Daniele	101	800	223	7.53	10/20
Tre Fontane	102	800	123	7.38	10/20
Aemilia	103	165	620	7.11	30/50
Ducale	104	950	54	7.80	20/30
Fontechiara	105	164	461	7.65	100/150
Fontenova	106	447	181	7.20	10/20
Lidia	107	200	540	7.00	100/150
Lynx	108	1015	163	7.45	30/50
Madonna della Mercede	109	159	609	7.18	30/50
Montinverno	110	600	615	7.33	<10
Pergoli	111	165	2889	6.45	<10
Riviana	112	300	819	7.81	<10
Rocca Galgana	113	258	251	8.04	10/20
Sant'Andrea	114	200	609	7.65	100/150
S. Moderanno	115	865	364	7.03	10/20
Varanina	116	225	577	7.10	<10
Verdiana	117		350		30/50
Ventasso	118	1006	159	7.77	20/30
Vis	119	590	676	7.65	<10
Toscana region					
Fontemura	120		318	7.40	10/20
Leona	121	144	217	7.00	<10
Perla	122	330	633	7.80	10/20

Brand name	Location refer. no.	Spring altitude (m)	TDS 180 °C (mg/l)	pH	Production (mio. l/yr)
Toscana region (cont.)					
Santa Flora	123	330	625	7.80	10/20
Sapore di Toscana	124	1200	76		<10
Verna	125	960	144	7.50	20/30
Cintoia	126	400	306	7.15	20/30
Fontepatri	127	200	563	7.25	<10
Ilaria	128	30	1988	7.70	
Lentula	129	600	18	7.60	10/20
Palina	130	580	167	7.75	10/20
Panna	131	292	126	7.65	>300
Napoleone	132	375	64	5.75	<10
Amorosa	133	500	22	5.70	20/30
Fonteviva	134	500	49	5.70	20/30
San Carlo	135	300	559	5.76	<10
Corona	136	6	176	6.60	20/30
Generosa	137	140	921	7.11	<10
Pieve	138	42	247	6.70	<10
San Leopoldo	139	125	506	6.54	<10
Tesorino	140	98	741	7.05	20/30
Uliveto	141	12	1000	6.41	200/300
Vallicelle	142	35	488	7.05	10/20
Regina	143	19	16224		<10
Silva	144	850	91	7.80	20/30
San Felice	145	205	199	7.53	<10
Tettuccio	146	29	7171	6.80	<10
Acqua Santa	147	500	3280		<10
Fucoli	148	500	2600	6.80	<10
Sant'Elena	149	550	447	7.40	<10
Umbria region					
Flaminia	150	500	205	7.35	30/50
Fonte Tullia	151	550	227		20/30
Misia	152		249	7.55	<10
Motette	153	700	147	7.40	20/30
Rocchetta	154	536	174	7.78	20/30
Sanfaustino	155	350	1076	6.05	20/30
Santo Raggio	156	424	529	7.29	<10
Sassovivo	157	600	200	7.59	<10
Vasciano	158	400	1440	6.15	<10
Amerino	159	390	407	7.48	<10
Fabia	160	330	402	7.00	200/300
Sangemini	161	380	1059	6.10	200/300
Marche region					
Frasassi	162	322	319	7.40	30/50
San Cassiano	163	325	212	7.70	30/50

Table 8.13. *Continued*

Brand name	Location refer. no.	Spring altitude (m)	TDS 180 °C (mg/l)	pH	Production (mio. l/yr)
Marche region (cont.)					
Fonte del Gallo	164	638	463	7.80	<10
Fonte di Palme	165	160	460	7.21	<10
Madonna dell' Ambro	166	638	372	7.30	<10
Palmense del Piceno	167	120	578	6.73	<10
Preistorica	168	700	385	8.00	<10
Tinnea	169	700	216	7.65	70/100
Roana	170	1492	103		10/20
San Giacomo	171	539	489	7.22	<10
Santa Lucia	172	228	549	7.19	<10
Cinzia	173	801	289	7.40	<10
Orianna	174	12	465	7.40	<10
Petra Pertusa	175	118	261	7.31	10/20
Val di Meti	176	600	318	7.35	20/30
Lazio region					
Fiuggi	177	590	108	6.50	70/100
San Marco	178	141	1848	6.20	<10
Cottorella	179	400	297	7.30	<10
Appia	180	110	673		30/50
Claudia	181	143	788	5.80	50/70
Egeria	182	20	621		30/50
Giulia	183	143			50/70
Meo	184	200	216	6.67	<10
Regilla	185	768	113	7.60	<10
S.M. Capannelle	186	20	689	5.90	10/20
San Pietro	187	146	673	5.70	10/20
Acqua di Nepi	188	227	508	5.60	30/50
Mineral Neri	189	350	248	6.40	<10
Acetosa San Paolo	190	20	2270		10/20
Abruzzo region					
Santa Croce	191	800	182		20/30
Sponga	192	830	182	7.40	20/30
Santa Reparata	193	589	608	7.15	<10
Campania region					
Telese	194	55	1675	6.38	<10
Ferrarelle	195	111	1597	5.80	>300
Lete	196	300	976	5.94	30/50
Santagata	197	98	1035	6.00	30/50
Acetosella	198	0	820	6.20	<10
Acqua della Madonna	199	6	790	6.80	20/30

Brand name	Location refer. no.	Spring altitude (m)	TDS 180 °C (mg/l)	pH	Production (mio. l/yr)
Campania region (cont.)					
Faito	200		319	7.48	50/70
Vesuvio	201	139	1730	6.30	10/20
Don Carlo	202	99	602		70/100
Irno	203	50	893	6.60	
Santo Stefano	204	650	212	7.50	10/20
Vitologatti	205	50	1796	6.37	10/20
Puglia region					
Eureka	206	97	355	7.45	20/30
Para vita	207		510	7.10	<io
Basilicata region					
Cutolo-Rionero	208	656	637	5.80	50/70
Gaudianello	209	640	1189	5.90	100/150
Itala	210	450	515	5.70	<10
La Francesca	211	656	680	5.96	50/70
Toka	212	450	1970	6.10	10/20
Traficante	213	491	1067	6.50	20/30
Visciolo	214	656	560	6.10	50/70
Calabria region					
Certosa (F. Camarda)	215	400	77	6.68	<10
Certosa (F. Perna)	216	400	59		<10
Madonnina della Calabria	217	750	79	6.46	<10
Mangiatorella	218	1200	581	6.39	50/70
Sicilia region					
Acquarossa	219	551	1330		<10
Pozzillo	220	12	1136	6.70	30/50
Ciappazzi	221	12	1530	7.27	50/70
Fontalba	222	920	119	6.80	30/50
Acquabaida	223	550	342	7.00	20/30
Sardegna region					
Giara	224	71	230	6.65	20/30
Levia	225	400	276		150/200
Pura	226	400	245		150/200
Sandalia	227	71	1479	6.87	50/70
Sattai	228	160	249	6.40	10/20
Sant'Angelo	229	400	241		150/200
San Giorgio	230	400	263	6.20	150/200
San Martino	231	270	2967	6.62	10/20
Smeraldina	232	900	171	6.82	10/20
Santa Lucia	233	350	1227		20/30

Mountains. Archaeological remains demonstrate that Fiuggi Springs were known and exploited during the Roman period (4th century B.C.). With the fall of the Roman Empire, the region became a part of the Roman dukedom, and later a feudal holding of the Pope; in the 1500s the area became a feudal holding of the Colonna family.

Fiuggi Springs have been mentioned several times in Vatican documents. Pope Bonifacio VIII and the great Michelangelo Buonarroti drank at Fiuggi Springs to treat calculuses. More recently, a number of popes, statesmen, and artists spent time at Fiuggi Springs to be treated for calculuses and gout.

Although Fiuggi Springs are in a depression area within a karst ridge, they are not fed by a karst aquifer, but by an outcrop of volcanic tufa, erupted by the Albano Volcano, that lies on lacustrine sediments that fill the karst depression. The tufa deposits outcrop over an area of a few square kilometers and supply some small springs; the main spring being named after Pope Bonifacio VIII. The entire outcrop of tufa is covered by a flourishing forest of chestnut.

The springs are surrounded by a huge park and it is possible to drink Fiuggi waters and enjoy the cultural and recreational activities. Spring discharge, which amounts to only 203 l/s, is inadequate to meet the increased demand for mineral waters in Italy. During the 1970s, to meet the consumer's requirements, a well field was built near the springs which doubled the production of mineral water. The physical and chemical characteristics of Fiuggi water are its acidic pH and very low mineral content (Table 8.15).

Table 8.14. Chemical analysis of "Ferrarelle" water (ions, TDS and dissolved gas in mg/l)

Parameter	Value	Parameter	Value
T (°C)	16	SO_4^-	4.1
pH	6.0	PO_4^-	0.4
EC (μS/cm)	1 800	NO_3^-	4.1
Na^+	51	HCO_3^-	1 604
K^+	49	SiO_2	78.5
Mn^{2+}	0.2	NH_3^-	ass.
Sr^{2+}	0.3	Oxidability (O_2 cons.)	0.7
Li^+	0.1	Alkalinity (ml HCl)	263
Ca^{2+}	441.8	TDS (180 °C)	1 462.9
Mg^{2+}	19.5	CO_2	2 060
Cl^-	20.6	O_2	0.5
F^-	0.3		

Table 8.15. Chemical analysis of "Fiuggi" water (ions and TDS in mg/l)

Parameter	Value	Parameter	Value
T (°C)	13	Mg^{2+}	4.9
pH	6.50	Cl^-	12.6
EC (μS/cm)	159	F^-	0.1
Na^+	6.5	SO_4^-	5.1
K^+	4.6	PO_4^-	0.14
Sr^{2+}	0.2	HCO_3^-	61
Ba^{2+}	0.06	SiO_2	19.0
Ca^{2+}	16.0	TDS (180 °C)	107.7

H. Hötzl

8.17 Origin of the Danube-Aach System

8.17.1 Introduction

The Swabian Alb formed by an Upper Jurassic carbonate sequence is the most extensive karst area of Germany. The western part is crossed by the upper Danube River. It represents an old, mainly Pliocene, drainage system which is now restricted by the young Rhine system. The low base level of the upper Rhine Graben causes a strong headward erosion. Since the Upper Pliocene, the Danube has lost more than 90% of its headwaters. The underground Danube-Aach karst system of the Western Alb represents the last capture of the Rhine, leading periodically to a complete loss of water in the upper Danube. The seepage of this water, together with the huge karst catchment area, supplies the strong discharge of the Aach Spring, forming the largest spring of Germany with an average discharge of 8.5 m^3/s.

Germany has several karst areas in the outcrops of Paleozoic, Mesozoic, and Tertiary limestone and gypsum that vary in size and importance. The largest closed karst area is the Swabian Alb between the Neckar and the upper Danube Rivers (Fig. 8.36). It is a part of the South Germany cuesta landscape and is formed by the outcrop of the Upper Jurassic carbonate sequence. This flat mountain ridge east of the Black Forest forms the northern margin of the Molasse Basin, and to the southwest it passes into the Swiss Jura Mountains. The Alb extends from SW to NE more than 200 km, and the width is about 40 km. The cuesta faces northwest, and its steep escarpment, the so-called Albtrauf, rises from 600 m in the foreland to 1 000 m a.s.l., from where the Alb plain dips gently to the southeast.

This carbonate, flat-topped ridge exhibits moderate to intensive karstification. The most pronounced morphological feature is the widely spaced network of dry valleys, which are in sharp contrast to the dense drainage network in the clayey foreland of the Alb. Numerous karst depressions, dolines, cave systems, and sinter formations underline the karst. Active sinkholes and karst springs demonstrate the hydrogeologic efficiency of the underground drainage system discharging towards the Danube River.

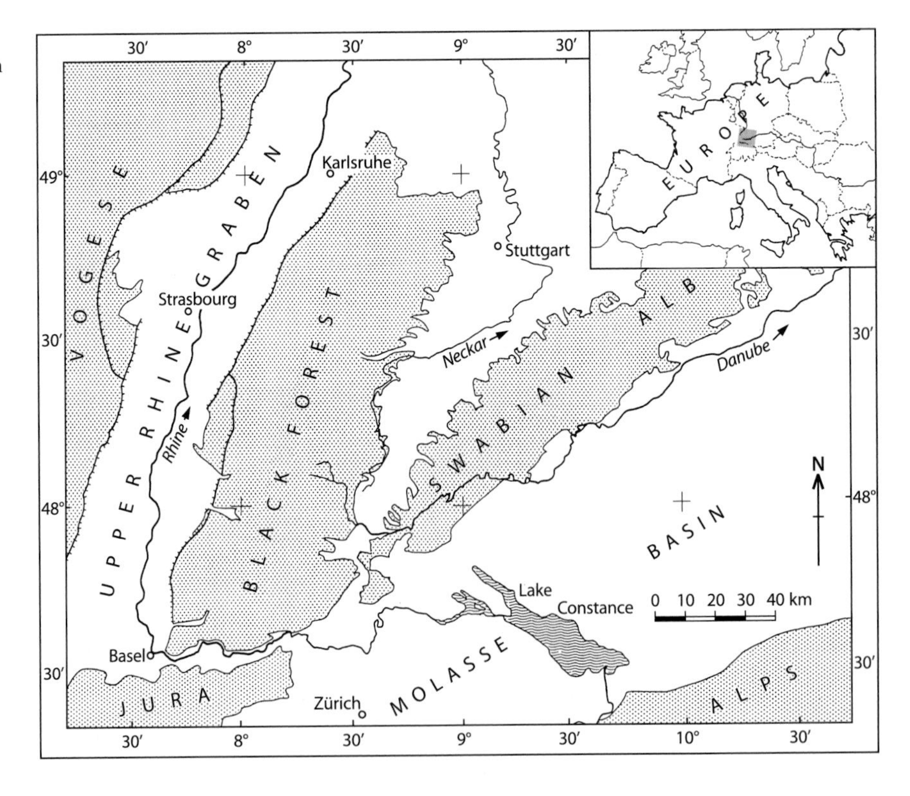

Fig. 8.36. Location of the Swabian Alb in southwest Germany

The young, dynamic morphology is determined by the relative position of the Danube and Rhine River drainage systems. Favored by the deep base level in the upper Rhine Graben (Fig. 8.36), the Rhine River gradually captures more of the old Danube drainage area. This affects not only the surface drainage but also the groundwater and has led to one of the most spectacular karst phenomena of Germany, the "Donauversickerung", the loss of the Danube water in the western part of the Swabian Alb to the Aach Spring, a tributary of the Rhine system. The hydrogeologic details of this phenomena and its development are described herein.

8.17.2 Geological Setting

The Swabian Alb is part of the main tectonic block of southern Germany (Geyer and Gwinner 1984, 1986). This area is associated with one of the Variscian orogenic zones, which was folded, overthrusted, metamorphosed, and intruded by granitic plutons. During the Upper Carboniferous and Permian, the area was uplifted and eroded forming the "basement" which now crops out with granite and gneiss in the Black Forest.

The new epicontinental sedimentary cycle started during the Lower Triassic with a sandy continental sequence. Marine carbonates with an evaporitic intercalation succeeded during the Middle Triassic. In the Upper Triassic, continental influence prevailed. During the Jurassic, clayey, marly, and carbonaceous sediment sequences were deposited. The most conspicuous sequence is the more than 300-m-thick Upper Jurassic carbonates. These consist of light colored, regularly bedded limestone and marl which pass partly into massive algal-sponge limestone with tower-like structures.

At the end of the Jurassic, the area of the Swabian Alb became subaerial and vast parts have remained so. Strong weathering and karstification occurred on the flat carbonate landscape especially during the Cretaceous and Early Tertiary with the warm and humid climate. The topography of the recent landscape originated from the tectonic events of the Middle and Late Tertiary. In the south, the overthrusting of the Alps caused the foredeep basin of the molasse. With the table-like uplifting of the Black Forest in connection with development of the upper Rhine Graben, the Swabian Alb tilted southeastward, toward the Molasse Basin. The Molasse Basin then controlled further orientation of the drainage pattern.

8.17.3 Development of the Danubian Drainage Pattern

With the tilting of the Swabian Alb during the mid-Tertiary, a system of consequent streams was formed. These pre-Danube rivers followed the slightly dipping surface toward the Molasse Basin (Schreiner 1974). After the filling of the basin, mainly due to the sediments derived from the Alps in the south, the sea retreated eastward in the Late Miocene. The drainage channels followed the retreating sea, and the early Danube was formed during the Late Miocene and Early Pliocene (Fig. 8.37). It comprised a huge catchment area, including that of the Aare, along with vast parts of the western Swiss Alps (Wagner 1961; Villinger 1986). The early Danube River drained more than 20 000 km^2 in one section of the Swabian Alb, which now has a drainage area of 900 km^2 (Hötzl 1973).

During the Pliocene, tectonic displacement and additional uplifting of the northwestern part of the Alb caused a gradual movement of the Danube toward the southeast. Later, it started to dissect the Upper Jurassic limestone, and by the Middle Pliocene it had reached a base level of 200 m below the old surface. The remnants of this old meandering valley are represented as meander-scar terraces about 50 m above the recent valley floor (Wagner 1961). Only the main tributaries within the Alb followed this deep dissection. Because of the increasing karstification, larger areas of the new Alb high plains began draining into the subterranean discharge systems. Vertical tectonic displacements during the Pliocene (Illies 1965) caused additional critical changes in the drainage pattern outside the Alb. In the west, the Saonne River, a tributary of the Rhône River, captured – by eastward erosion – one of the main headwaters of the early Danube, the Aare River. Therefore, the main part of the former upper Danube catchment along with vast parts of the Swiss Alps was draining westward through the Burgundian gate directly to the Mediterranean Sea. In the Late Pliocene, the Danube catchment area was reduced gradually but still included the Alpine Rhine and the early Eschach River with its catchment in the central Black Forest, now part of the Neckar system (Fig. 8.37).

The final phase of development of the drainage area was caused by the activation of the upper Rhine Graben in the Late Pliocene (Illies 1965). The downwarping of the graben basin coupled with additional uplifting of the graben edges, Black Forest and Vogese, produced a new, deeper base level with a short flow path to the North Sea. This was the birth of the new Rhine discharge sys-

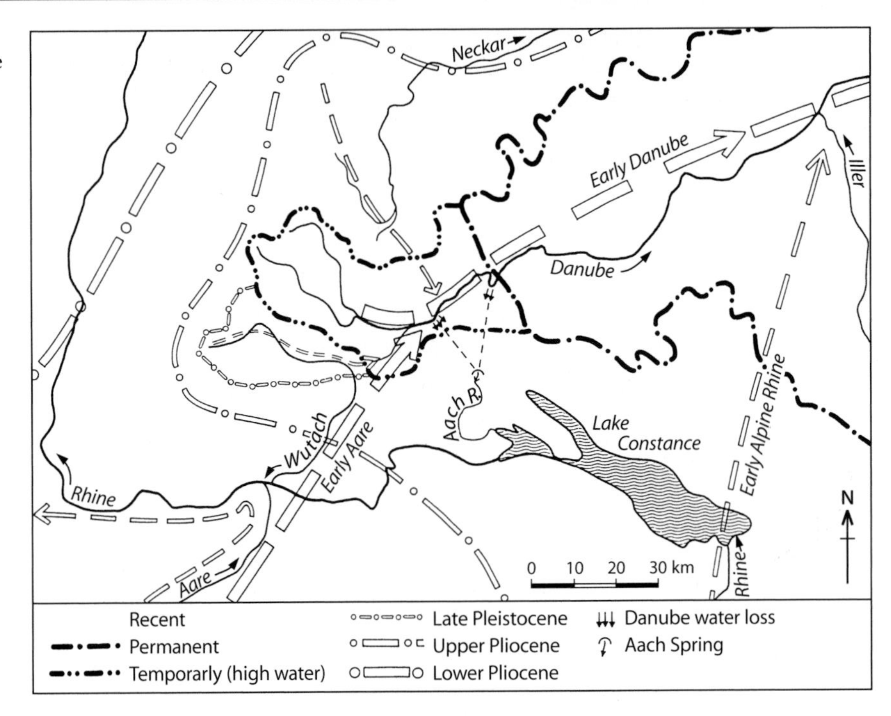

Fig. 8.37. The constriction of the Danube catchment area by the Rhine River system

tem. By the Late Pliocene, the Aare River was diverted to the north into the upper Rhine Graben. Due to the increased headward erosion of the Aare and the Neckar Rivers, the Rhine system was prograding to the east and capturing more of the Danube headwaters.

Three decisive steps may be mentioned. In connection with the first big glaciation, the Danube glacial, about 1.5 million years B.P., the diversion of the Alpine Rhine toward the west to the upper Rhine took place (Villinger 1986). The next important step was the deep scouring of the Hegau Basin during the Mindel-Riss interglacial, approximately 450 000 to 400 000 years B.P. The Mindel gravels are above 600 m a.s.l., and the Riss moraine is at 450 m a.s.l. (Schreiner 1974). This deep erosion also removed the impermeable Molasse cover on the Jurassic limestone in the Hegau Basin. Therefore, a new base level was developed for the limestone of the Western Swabian Alb, leading finally to the subterranean Danube-Aach system (Hötzl 1973). The third step began only 20 000 years ago. At the maximum stage of the Würm glaciation, melting waters of the Feldberg glacier caused the diversion of the Feldberg Danube, now called Wutach, southward to the Rhine (Wagner 1961). After the loss of these headwaters, the Danube retained only the relatively small headwaters of the Breg and Brigach Rivers encompassing 482 km^2 in the Black Forest. Their junction now forms the "young Danube".

8.17.4 The Underground Danube-Aach System

8.17.4.1 *Karst Hydrogeologic Overview*

The Western Swabian Alb (Fig. 8.38) is obliquely crossed by the deep Danube Valley cutting the Jurassic limestone into a northern and a southern Alb high plain. For both karstified regions, the Danube originally functioned as collecting stream. From this, the Hegau Alb, the high plain south of the Danube between Immendingen and Fridingen, is drained by the Rhenish Aach River, which is supplied mainly by the Aach Spring, the largest spring in Germany. The underground drainage area of the Aach Spring extends headward up the Danube Valley. Due to the deeper karst water level, sinking of river water occurs in several places, which periodically causes complete loss of the Danube water. The underground connection to the Aach Spring defines the Danube-Aach system.

The lost Danube is one of the most impressive karst features of this region. The water loss first occurs at the Immendingen weir (Fig. 8.38 and 8.39), where the strati-

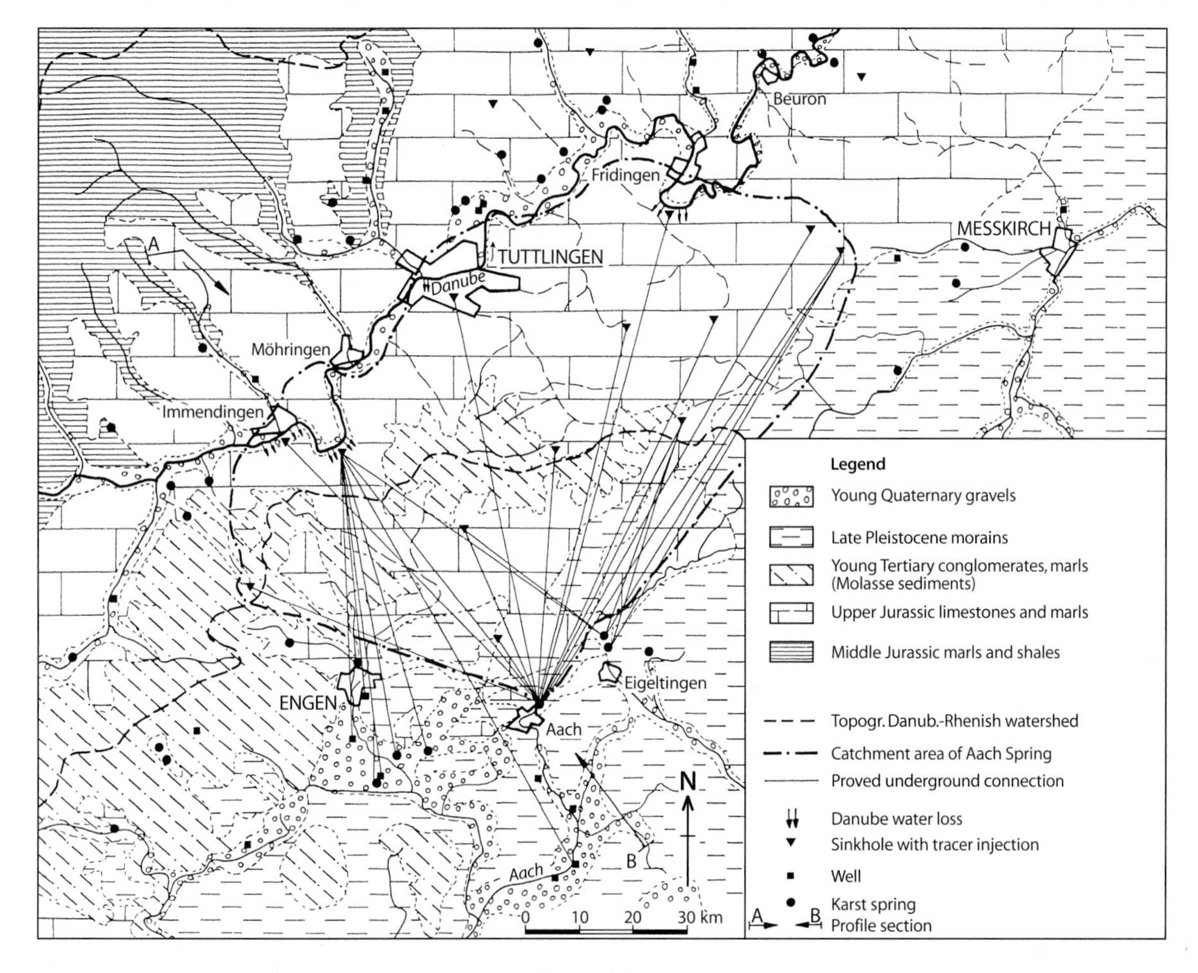

Fig. 8.38. Hydrogeological map of the western Swabian Alb and of the Hegau Basin in the south

fied Oxfordian limestone submerges below the valley floor along a flexure zone and the river comes in direct contact with permeable rocks. Several sinkholes have been developed along the river bank. The main water loss occurs 1.5 km downstream along the undercut slope of the Brühl curve between Immendingen and Möhringen, where the water is sinking into the open joints of the Oxfordian limestone or disappearing through the gravel of the river bottom. During dry periods, the water level was measured at 30 m below the river bed. The loss amounts to the entire river discharge under low water conditions so that the river bed downstream from the Brühl curve is dry (Fig. 8.40). On average, this occurs about 150 days per year, mainly during the summer and autumn. Downstream from the Brühl curve, the Danube is renewed during the dry period by small tribu-

Fig. 8.39. The Danube at the Immendingen weir with sinkholes on the right bank and the well-stratified Oxfordian limestone behind

Fig. 8.40. The Brühl curve of the Danube between Immendingen and Möhringen; on the right bank the undercut slope with the main section showing the downstream Danube completely dry

Fig. 8.41. The Danube at the Fridingen loop with sinkholes on the left bank within the Kimmeridgean limestone

taries and spring discharges coming from the northern part of the Alb plain. Further water losses of the Danube occur in the valley loop of Fridingen 40 km downstream where the water submerges into the joints and bedding planes of the Kimmeridgean limestone (Fig. 8.41). The water lost from the Danube emerges in different springs of the Hegau Basin (Fig. 8.38). The distances between the river sinkholes and the springs range from 8 to 19 km. The decisive point is the deep topographic level of the Hegau Basin. The Aach Spring, the dominating karst spring with an outflow at 475 m a.s.l., lies 175 m deeper than the Danube at the Brühl curve (distance: 11.7 km).

The hydrogeology is illustrated on the geologic profile (Fig. 8.42). The base of the karstification is formed by the Oxfordian marl and the underlying marl and slate of the Middle Jurassic. The overlying carbonate sequence acts as one unique, hydraulically connected karst aquifer. Due to disruption by faults and the existence of massive limestone, the intercalated Kimmeridgean and Tithonian marl have minor influence as aquitards. The loss of Danube water occurs for the upstream reach into the Oxfordian limestone and for the downstream segment into Kimmeridgean limestone. Independently from this, the groundwater emerges in the Hegau Basin from the Tithonian limestone, thus confirming the hydraulic connection of the whole karst body.

Outside the catchment area of the Danube-Aach system, the Danube is still the collecting river of the discharges from the remaining parts of the Western Alb (Batsche et al. 1970). Based on the position of the base level below or above the karst base, two types of the classical German karst are illustrated there, flat and deep karst. For example, there are important karst springs east of Tuttlingen just on the northern slope of the Danube Valley, where the karst base at the Oxfordian marl crops out a few meters above the valley floor. On the other side east of Fridingen, where the karst base is below the valley floor, karst springs occur on both margins of the valley.

8.17.4.2 *Hydrographic Conditions*

Due to the karstification, the river pattern of the Western Alb is poorly developed and restricts rivers coming from outside of the karst area. This is valid for the main river, the Danube, as well as for tributaries coming from the northern region, while hardly any significant tributaries exist from the orographic right side. Therefore, the discharge of the upper Danube River is mainly influenced from outside. Its water regime follows the hydrologic events of the Black Forest with its prevailing winter precipitation creating high discharge during the snowmelt in April. In the Alb proper, the summer rains are heavier, but because of high evaporation, most of the small rivers have their maximum discharge in spring time as a result of the snowmelt. The precipitation rate varies from 1 000 mm/yr at the Albtrauf in the northwest to approximately 750 mm/yr in the southeast.

The Danube at the contact with the karst area drains a catchment area of 766 km^2 resulting in a mean yearly discharge of 12.5 m^3/s, with a maximum discharge of nearly 200 m^3/s. Downstream to Beuron, where the Dan-

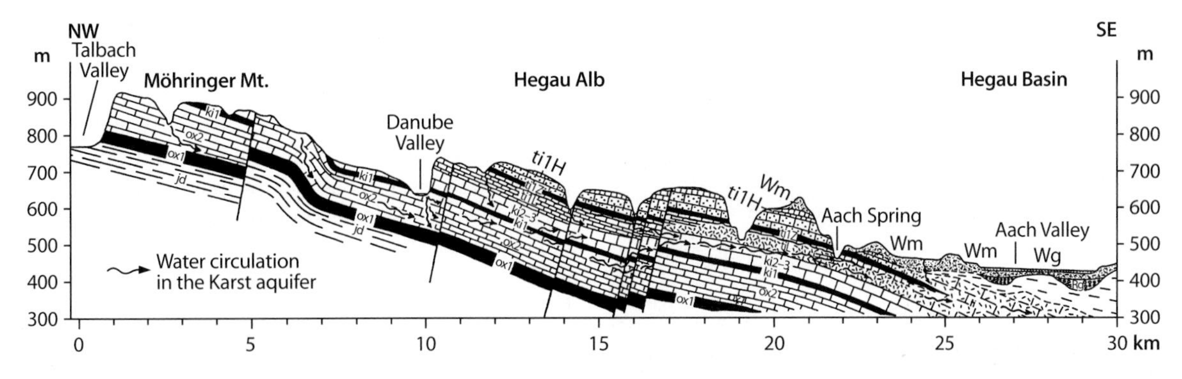

Fig. 8.42. Geological section through the western Swabian Alb with the Hegau Basin in the southeast, according to Batsche et al. (1970) with supplements north of the Danube by Hötzl (1973). *Wg* Würm gravels, *Wm* Würm moraine, *ti 1H* Lower Tithonian limestones, *ti 1Z* Lower Tithonian marl, *ox 2* Oxfordian limestone, *ox 1* Oxfordian marl, *jd* Middle Jurassic marl and shale

ube is again the main receiving stream, the catchment area increases to 1320 km^2 with an average discharge of 10.7 m^3/s. The loss in discharge of the Danube to the Rhenish Aach system averages annually approximately 7 m^3/s. Discharge measurements compared the hydrographs of water level measurements (Türk 1932; LfU 1980) have shown that the relationship between the river and the underground system is complicated. It depends on the karst water level and the recharge rate in the karstic catchment area. The loss of discharge of the river is limited by different inflow capacities along the river banks, the discharge capacity of the underground system, and the restricted outflow capacity of the Aach Spring (Hötzl and Huber 1972).

At low water levels, a flood in the Danube may cause an increase in the amount of surface water loss to the groundwater system. The filling of underground channels with water from the karst plain sinking through fine fissures toward the main channels creates a damming effect. In one area, a delay of ten days was determined with the aid of tracers for the movement of water compared with the normal time of two days from a sinkhole on the Alb plain to the Aach Spring. The rising of the karst water table aids the runoff from the fine fissures into the preferential channels, so that the seepage loss from the Danube is reduced, but even if there is no local recharge, the throughflow capacity is largely limited by the outflow system. Comparison of hydrographs has shown that at a discharge of the Aach Spring of more than 20 m^3/s, the outflow becomes independent and approximately stagnant in spite of further increases in Danube flood stage (LfU 1980). In the case of a large amount of local recharge, this can completely stop the infiltration of the river water and can even lead to karst discharge along the same river section.

The average discharge of the Aach Spring is about 8.5 m^3/s. The highest rate observed was 24.1 m^3/s, the lowest rate was 1.3 m^3/s. The water emerges from two parallel adjacent joints from a depth of 17 m (Fig. 8.43). There the water comes through a narrow passage behind which a wide, slightly ascending cave occurs, which has been explored by divers more than 600 m to the north (Hasenmayer 1972; Schettler 1991). Beside the main spring, there are several smaller springs nearby with discharges as much as 1 m^3/s, which have been included in the discharge rates above (Hötzl and Huber 1972). One of these smaller springs, located only 5 m from the main joints, is of special interest. Water outflow occurs only during total discharge of more than 5 m^3/s, and when the total discharge is less than 3 m^3/s it becomes a swallow pool (Käss and Hötzl 1973).

The water balance for the Danube-Aach system was calculated from available hydrologic data and runoff measurements (Hötzl and Huber 1972). For the Aach Spring, about one third of the discharge, 2.7 m^3/s, was estimated as coming from its own catchment area. The other two thirds, about 5.8 m^3/s, are from the karst area north of the Danube, and from the more upstream part of the Danube. The remaining 1 m^3/s, which is unaccounted for at the Beuron discharge station can be assumed to be a part of the discharge from the small karst springs near Engen and Eigeltingen and as recharge to the gravel aquifers of the Hegau Basin. Similar results were obtained with new data by Villinger (1977) and by Ganz and Villinger (LfU 1980), showing more clearly the variation over succeeding years due to the changing hydrologic conditions.

Fig. 8.43. The Aach Spring discharging from Tithonian limestones in the Hegau during low water in the autumn. The main spring is situated in the central notch, while on the orographical right side a smaller outlet along a flat horizontal crack can be seen

8.17.5 Tracing Experiments

In the first note in the literature on the sinking of the Danube water, the opinion was expressed that there could be a connection to the Aach Spring in the Hegau (Bräuninger 1719). The first documented tracer experiment was performed with aniline red by Ten Brink, a spinning industrialist from the Aach River in 1869 (Käss 1969). The experiment failed, probably due to the high adsorptivity and low concentration of the color. The experiment was repeated by Ten Brink together with Knop from Karlsruhe in 1877. Shale oil, fluorescein (this was the first hydrological use of this now very well-known groundwater tracer), and sodium salt were injected successively within 14 days. All three tracers reappeared within 2.5 days in the Aach Spring. The use of sodium salt as a tracer was the first quantitative experiment. By gravimetric analysis, Knop could prove that 18 530 kg of the original 20 000 kg of injected tacer reappeared in the Aach Spring (Knop 1878).

Several later tracer experiments were performed mainly with salt tracers in sinkholes along the Danube as well as from the high plain. The results of these experiments provided the first information on the underground flow velocity and the catchment area of the Aach Spring. A combined tracer experiment with simultaneous tracer injections (fluorescein dyes, salts, colored spores, and nonpathogenic germs) in eight different places was performed by an international group of scientists (Batsche et al. 1970). For the first time, the reappearance of water from the Danube was identified at several springs and wells in the Hegau Basin. The results with the overlapping fan-like dispersion of the tracers are shown in Fig. 8.38. It could be confirmed that not only karst springs but also the gravel aquifers in the Hegau are in part recharged (approximately 1 m^3/s) by water from the Danube. Based on these results, it is now possible to outline the underground catchment area of the Aach Spring. It comprises an area of about 250 km^2 in comparison with a topographic drainage basin area of 10 km^2 (Hötzl 1973). The flow velocity from the sinkholes at the Danube to the Aach Spring ranges from 160 to 435 m/h for the first reappearance of the tracer and from 120 to 300 m/h for the maximum concentration of the tracer. The flow rates from sinkholes on the Alb plain are slower and vary between 100 to 300 m/h for the first detection and between 50 to 200 m/h for the detection of the maximum concentration.

8.17.6 Hydrochemistry of Karst Water

The chemical characteristics of the karst water is shown by the analyses of water from springs in the Danube Valley, which have their catchment area exclusively within the karstified limestone. The relative distribution of the ions is shown in the last line of Fig. 8.44. As expected, there is a predominant concentration of calcium and bicarbonate that constitute 90% of the cations and anions. The Ca^{2+}/Mg^{2+} ratio of 18 is extremely high. The alkali ions and sulfate constitute less than 5% of the dissolved solids. The average content of the ions given as the sum of cations (= the sum of anions) is about 5.48 meq/l.

The Danube water from upstream of the Swabian Alb has a different composition due to different geology and environmental factors (Hötzl 1973). Alhough the dominant ions are calcium and bicarbonate, there are elevated concentrations of magnesium, sodium, chloride, and sulfate. Yet the total mineralization is less than that of the groundwater (Fig. 8.44). At the second control station of the Danube downstream from the sinking section of the river near Fridingen, the composition is typical for karst water. This pattern can be recognized initially from the average concentration values (Fig. 8.44) and becomes more clear when considering the yearly variation where, for instance, the bicarbonate values rise during the periods when the flow in the upper part of

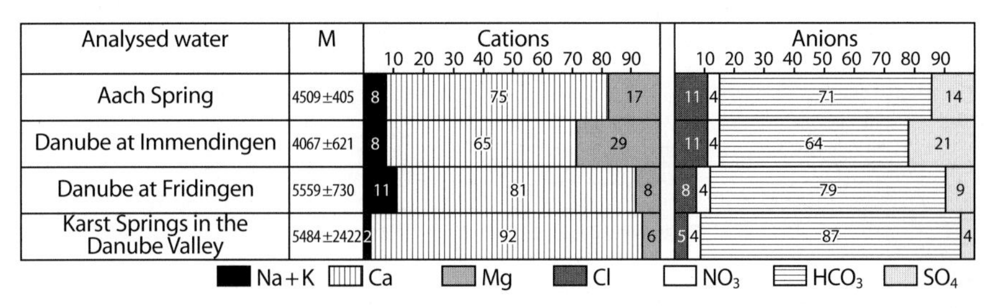

Fig. 8.44. Chemical composition of water from the Danube-Aach area showing the relative ionic distribution. In *column M* the average values of the mineralization as the sum of cations are given in μeq/l

the Danube ceases (Fig. 8.45). During these dry periods, the Danube water at Fridingen is supplied only from groundwater discharge from the karst area of the Swabian Alb.

The water discharged at Aach Spring represents a mixture of water from the upper Danube and groundwater. The mixing ratio determined from the chemical components is approximately 2:1, with the major component being the Danube water, which coincides with the value calculated from the water balance. The influence of the Danube becomes especially evident when studying the seasonal variations of chemical components (Fig. 8.45). While the noncarbonate components of the Aach Spring (number 8 in Fig. 8.45) follow the upper Danube water (number 2 in Fig. 8.45), the bicarbonate confirms the changing ratios due to the recharge conditions (Fig. 8.45) in the Aach catchment area. The bicarbonate curves illustrates the effect that the large amount of melt water coming from the Black Forest has on this component. From March through May the bicarbonate content decreases in the lower part of the Danube too (number 4 in Fig. 8.45). In spite of high water in the Danube, the proportion of karst water in the Aach Spring is increased due to the high recharge in the karst catchment area, which contributed to a rise in the bicarbonate content. Similar results were also obtained from tritium measurements (Batsche et al. 1970). They confirmed that the greatest similarity between components in waters in the Danube and Aach occurs during August to November, when the karst water level is low; and greatest difference occurs during the spring, when the karst level is high. This is in contrast to the downstream part of the Danube, where the greatest difference occurs in autumn, due to the complete loss of the river in the upstream section.

Based on the increased concentration of the dissolved components during the underground flow of the Aach Spring, the amount of limestone dissolved was calculated at 11 350 t/yr or 6.370 m^3/yr (Schreiner 1978). Some authors tried to interpret this in terms of the expansion of the storage capacity or the widening of the preferential cave system. In reality, most of the dissolved carbonate comes from the soil and subsurface zone, while solutional processes along the preferential paths and previously existing cave systems are rather minor due to the high velocity of flow.

8.17.7 Evolution of the Danube-Aach System

Because of the corrosional processes that form the karstification of limestone, time markers to date the time of origin are difficult to determine. Therefore, several indirect indications and relationships with other geological and climatological processes are required to date the gradual development of a karst system. In the upper Danube, several cycles of alternating phases of sedimentation and erosion have occurred that are well dated. They enable us not only to reconstruct the development of the surface drainage pattern but also to compare it with the development of the underground discharge systems. Based on this, we know that the Danube-Aach system is relatively young, its initial stages dating back to the mid-Pleistocene. The whole drainage pattern of the Pliocene Danube from the Alps was oriented on the northward-dipping alluvial plain to the baseline of the Early Danube. With the downward cutting of the Danube into the Jurassic limestone plate during the mid-Pliocene, a new karstification cycle developed, but the accompanying processes were only oriented toward the Danube. The decisive change of the drainage pattern for the area south of the Danube occurred with the deep scouring of the Hegau Basin during the mid-Pleistocene. The new southward-oriented drainage channels from the Hegau Alb cut through the impermeable sheet of the Molasses sediments down to the Jurassic limestone. A new base level was formed nearly 200 m deeper than that of the Danube Valley (Schreiner 1974). Thus, the

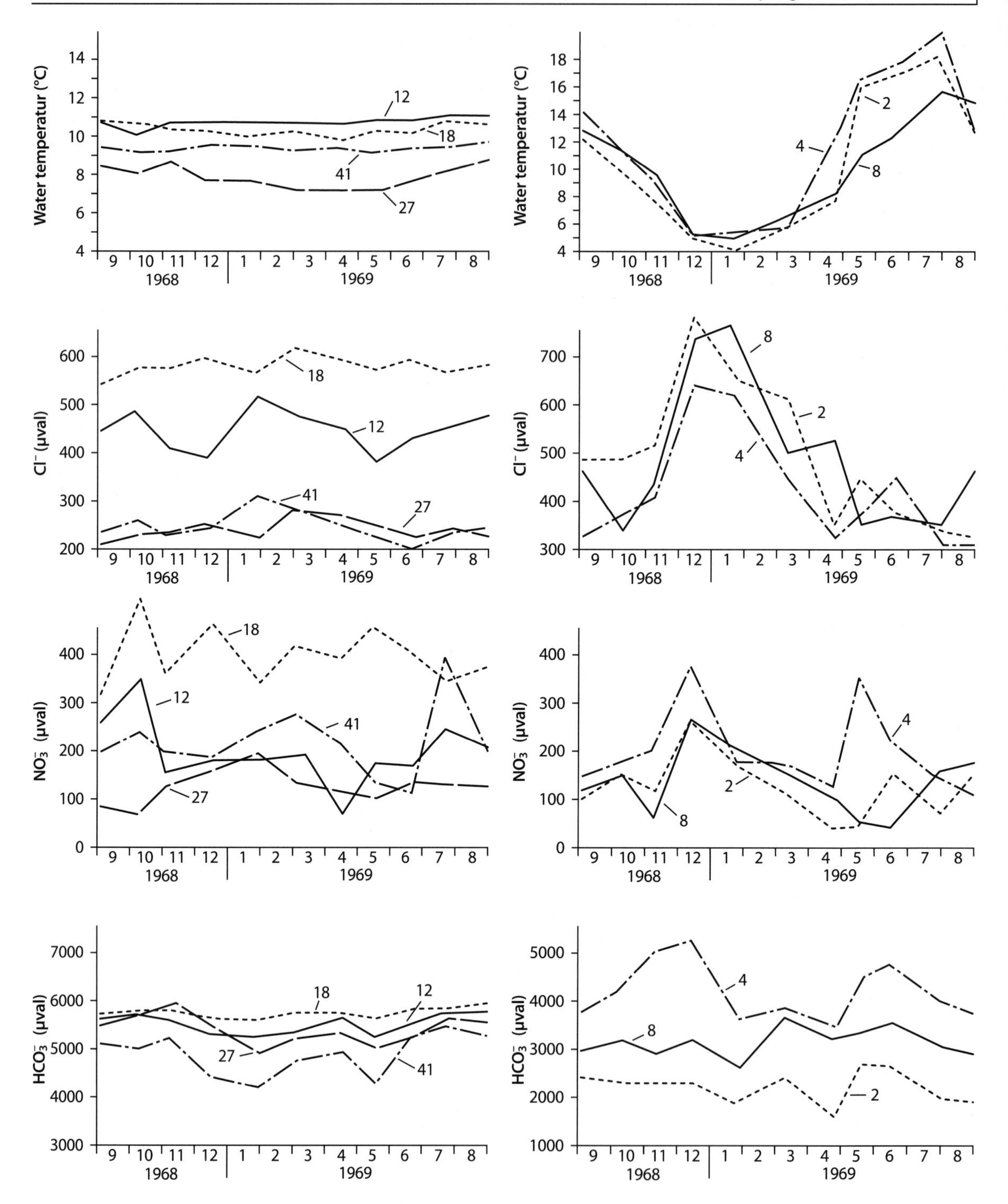

Fig. 8.45. Seasonal variations of some chemical components of the water from the Danube-Aach area. *Left column* Spring water from karst (*41* Wulfbachquelle, *12* Kressenlochquelle), Tertiary conglomerate (*27* Dorsteltal) and Quaternary gravel (*18*). *Right column* Danube at Immendingen (*2*), Danube at Fridingen (*4*), and the Aach Spring (*8*)

limestones of the Hegau Alb could drain southward and the newly beginning karstification of them followed that direction. This was the initial stage of the Danube-Aach system, which, at that time, was only an early Aach system and can be dated 400 000–450 000 years B.P.

The thick ice sheet of the Riss glacial, reaching up the Hegau Alb to an elevation of more than 650 m a.s.l. temporarily stopped the further development of the karst, but with the melting of this ice sheet even deeper erosion channels were formed. Their bottoms are about 50 m below the recent valley floor. One of the main interglacial valleys could be traced by geophysical surveying up to about 5 km south of the recent Aach Spring (Schreiner 1978). It can be assumed that the dewatering of the karst during the last interglacial period with preferential water paths was oriented toward this deep exposure of the limestone, where the Aach Spring was originally situated, which was being covered by gravels and moraines of the last glacial event. An indication that this former karst outlet is still partially functioning can be determined from the results of the tracing experiments of 1969. An artesian well tracer from the Danube introduced 18 km away was detected 8 days after injection. The short travel time can only be explained by the well-defined karst conduits (Batsche et al. 1970). This underground drainage system was blocked once again by the huge ice sheets of the Würm glaciation, which covered the whole Hegau Basin up to an elevation of nearly 600 m a.s.l. The recession of the ice cover started soon after its maximum extension at 20 000 years B.P. The recession stages in the Hegau are very well marked by end moraines as well as by partly refilled meltwater channels in front of them. One of these channels, which is dated from 18 000 to 16 000 years B.P., was cut into the limestone so deeply that one of the former preferential flow paths leading to the pre-Würm main groundwater outlet 5 km south was opened. Thus, a new main groundwater outlet, the Aach Spring, was formed for the drainage system which had been developed since the mid-Pleistocene.

The time when the headward progression of underground erosion reached the Danube Valley located 11 km away could not have been dated until now. The first mention of the complete loss of Danube water in historic documents was made only 300 years ago and the first observation of the complete water loss of the Danube occurred in the middle of the last century, indicating a very recent inclusion of the Danube Valley into the underground southward-directed drainage system. This verifies the interpretation that the underground connection of the Danube with the Rhenish Aach system is of Holocene age (Hötzl 1973). Based on the well-developed underground karst system or from the large relief difference, other authors consider that the inclusion of the Danube is older than the Würm glaciation (Schreiner 1974; Villinger 1977).

References

Batsche H, Bauer F, Behrens H, Buchtela K, Dombrowski HJ, Geisler R, Geyh MA, Hötzl H, Hribar F, Käss W, Mairhofer J, Maurin V, Moser H, Neumaier F, Schmitz J, Schnitzer WA, Schreiner A, Vogg H, Zötl J (1970) Kombinierte Karstwasseruntersuchungen im Gebiet der Donauversickerung (Baden-Württemberg) in den Jahren 1967–1969. Steir Beitr Hydrogeol 22:5–165

Bräuninger FW (1719) Fons Danubii primus et naturalis oder Die Urquelle des weltberühmten Donaustroms. Tübingen, 388 pp

Geyer OF, Gwinner MP (1984) Die Schwäbische Alb und ihr Vorland. Samml Geol Führer 67

Geyer OF, Gwinner MP (1986) Geologie von Baden-Württemberg. 3. Aufl. Schweizerbart, Stuttgart, 472 pp

Hasenmayer J (1972) Tauchabstiege in die Quellhöhle der Aach. Geol Jahrb C2: 351–357

Hötzl H (1973) Die Hydrogeologie und Hydrochemie des Einzugsgebietes der obersten Donau. Steir Beitr Hydrogeol 25:5–102

Hötzl H, Huber W (1972) Über die Hydrogeologie und wasserwirtschaftliche Nutzung der Aachquelle (Baden-Württemberg, BRD). Geol Jahrb C2:359–382

Illies JH (1965) Bauplan und Baugeschichte des Oberrheingrabens. Ein Beitrag zum "Upper Mantle Project." Oberrhein Geol Abh 14:1–54

Käss W (1969) Schriftum zur Versickerung der oberen Donau zwischen Immendingen und Fridingen (Südwestdeutschland). Steir Beitr Hydrogeol 21:215–246

Käss W, Hötzl H (1973) Weitere Untersuchungen im Raum Donauversickerung-Aachquelle (Baden-Württemberg). Steir Beitr Hydrogeol 25:103–116

Knop A (1878) Über die hydrographischen Beziehungen zwischen der Donau und der Aachquelle im badischen Oberlande. Neues Jahrb Mineral Geol Petrogr 1978:350–363

LfU-Landesanstalt für Umweltschutz, Baden-Württemberg (1980) Donau und Aach, Bericht 1980. Karlsruhe, 436 pp

Schettler H (1991) Die Aachhöhle. Südkurier, Konstanz, 163 pp

Schreiner A (1974) Erläuterungen zur geologischen Karte des Landkreises Konstanz mit Umgebung 1:50 000. 2. ber. Aufl. Stuttgart, 286 pp

Schreiner A (1978) Erläuterungen zu Blatt 8119 Eigeltingen. Geol Karte 1:25 000 Baden-Württemberg, Stuttgart, 82 pp

Türk W (1932) Wesen und Wirken der Donauversickerung (als Manuskript gedruckt). Karlsruhe, 64 pp

Villinger E (1977) Über Potentialverteilung und Strömungsversuche im Karstwasser der Schwäbischen Alb (Oberer Jura, SW-Deutschland). Geol Jahrb C18:3–93

Villinger E (1986) Untersuchungen zur Flußgeschichte von Aare-Donau/Alpenrhein und zur Entwicklung des Malm-Karsts in Südwestdeutschland. Jahrb Geol Landesamt Baden-Württemberg 28:295–362

Wagner G (1961) Zur Flußgeschichte von oberer Donau und oberem Neckar. Jahresber Mitt Oberrheinischer Geol Ver 43:93–98

P. Bono · C. Boni

8.18 Water Supply of Rome in Antiquity and Today

8.18.1 Introduction

In ancient Rome, water was considered a deity to be worshipped and most of all utilized in health and art. The availability of a huge water supply was considered a symbol of opulence and therefore an expression of power. Springs were the principal sources of water for ancient Rome.

The countryside around Rome offered a spectacular view: it was adorned with an incalculable number of monuments, temples, and villas and it was crossed by sturdy aqueducts with magnificent arcades. The aqueduct as a superelevated monumental work is a typical concept of Roman engineering, although it is possible to recognize that the inspiration and the basic ideas came from the Etruscan technology. The Etruscans did not construct real aqueducts, even though they built hydraulic works like irrigation channels, drainage systems, dams, etc. The Greeks had also constructed similar hydraulic structures, before the Roman influence. Interesting aqueduct remains are found in Rome, Segovia (Spain), Nimes (France), and Cologne (Germany).

8.18.2 Ancient Water Supply of Rome (700 B.C.–A.D. 500)

Rome initially used the water of the Tiber River and water from wells and many small springs that existed inside its town area, such as Acque Lautole, Acque Tulliane, Fonte Giuturna, Fonte Lupercale. Since the 4th century B.C., Rome gradually built aqueducts. The aqueducts conveyed water from springs many kilometers away from Rome. The water passed through underground tunnels and over huge arched bridges that maintained the slope of the flow until they reached the outskirts of Rome where "water castles" distributed the water for public (baths and fountains) and for private uses (Fig. 8.46). Most aqueducts were in the area east of Rome, except one located in the north. Water from all eastern aqueducts was collected in the Porta Maggiore area, called by Romans "ad Spem Veterem" (Fig. 8.47 and 8.48).

The first aqueduct was built in 312 B.C. During the subsequent 600 years, ten more aqueducts were built. The last one was completed in the 3rd century A.D. With completion of construction, there were Aqua Applia, Anio Vetus, Aqua Marcia, Aqua Tepula, Aqua Julia, Aqua Virgo, Aqua Alsietina, Aqua Claudia, Anio Novus, Aqua Traiana, Aqua Alexandriana.

8.18.3 Aqua Applia

No remains are left of the first great Roman aqueduct constructed in 323 B.C. It was entirely underground because of the war against Sannites. Therefore its route is almost unknown. Appius Claudius Crassus (later called Caecus) and Caius Plautius (called Venox) identified the spring sources of water. The aqueduct and the coeval consular road were named after Appius and called Applia. The aqueduct was 16.5 km long, and three main restoration works were carried out by: Quinto Marcio in 144 B.C. to eliminate unauthorized connections by citizens, Agrippa in 33 B.C., and Augusto in 11–4 B.C. to collect more springs and to build a new aqueduct 9.4 km long called Applia Augusta. The original catchment area is not exactly located, however, it is the area east of Rome on the northern slope of Albano Volcano near Pantano Borghese, the ancient Lake Regillo. Total discharge was recorded by Frontino (Sextus Jiulius

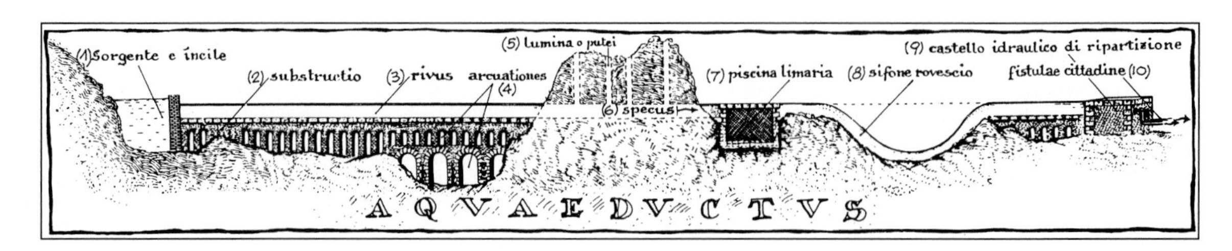

Fig. 8.46. Section of Roman aqueduct: *1* Spring and inlet; *2* Bearing wall; *3* Open channel; *4* Arcades; *5* Shafts; *6* Underground channel; *7* Settling basin; *8* Reverse siphon; *9* Main reservoir; *10* Water pipes

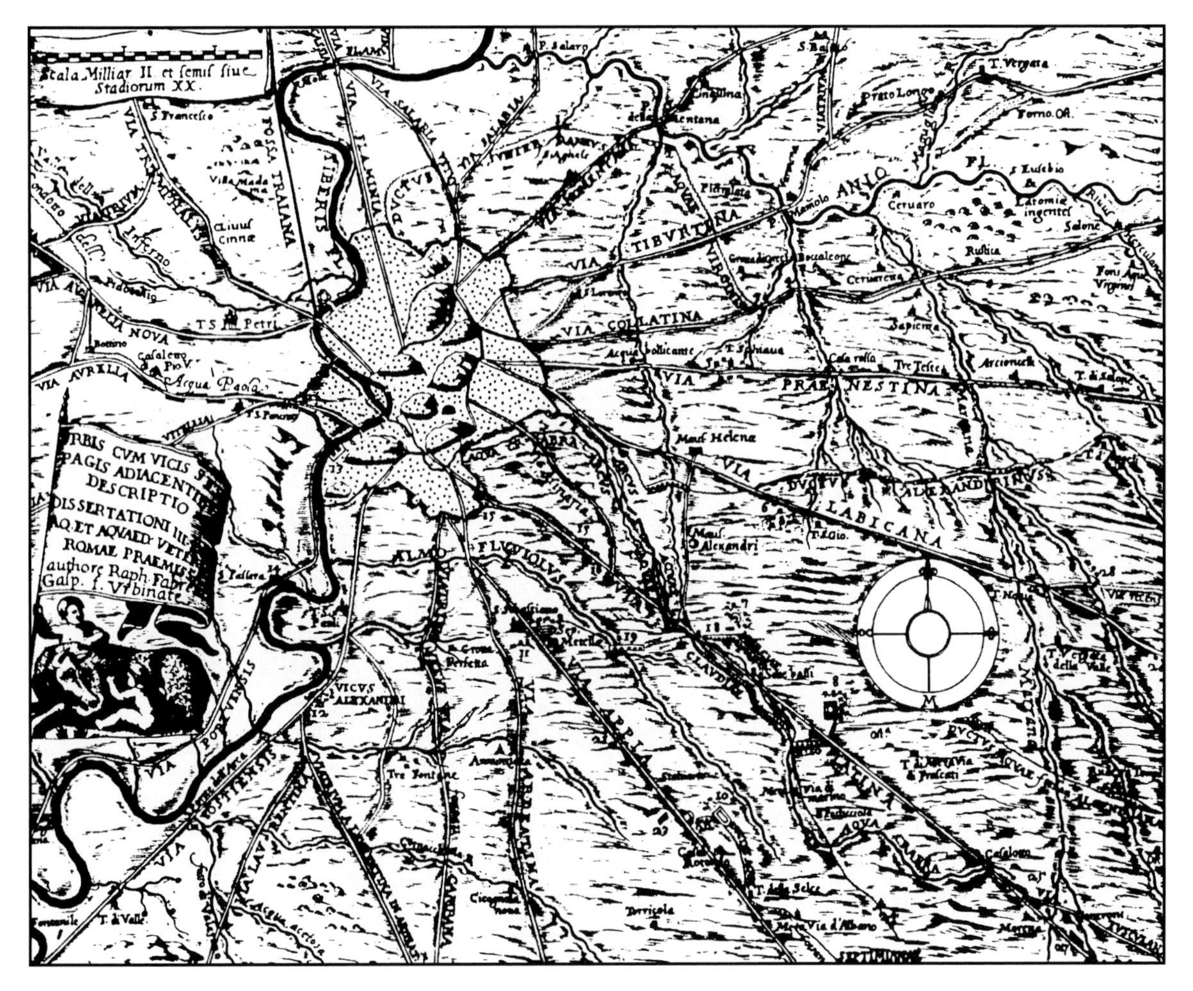

Fig. 8.47. Outline of Roman aqueducts near the city (*dotted area*). Only the 14th District of Rome (Regio XIV) is located along the right bank of Tiber River

Frontinus "curator aquarum", i.e., head of Roman aqueducts or water magistrate, lived at the time of Emperors Domitiano, Nerva, and Traiano). He wrote a fundamental treaty on Roman aqueducts in imperial times to which we often refer, *De Aquaeductu Urbis Romae* A.D. 97–103 Quinaria: a Roman discharge unit equal to 41 472 m^3/d (480 l/s) at the main reservoir near Rome (ad Spem Veterem) was 75 686 m^3/d (876 l/s).

8.18.4 Anio Vetus

Frontino dates the beginning of works for Anio Vetus aqueduct to 272 B.C. It was built by M. Curio Dentato and L. Papirio Cursore with the plunder obtained from the victory over Pirro (Punic Wars). The springs have not been located precisely but are certainly karst springs that were East of Rome, along the Aniene (Anio) River, and not far from the Agosta springs collected later by the Anio Novus aqueduct.

Anio Vetus aqueduct is 64 km long, mainly underground. It has many lumina (shafts), and it follows along the left bank of the Aniene River to Rome. According to Frontino, the aqueduct was restored by Quinto Marcio Re 127 years after its construction, by Menenio Agrippa in 33 B.C., and by Augusto, who provided it with mileage stones. The spring discharge was 182 394 m^3/d (2.11 m^3/s) according to Frontino.

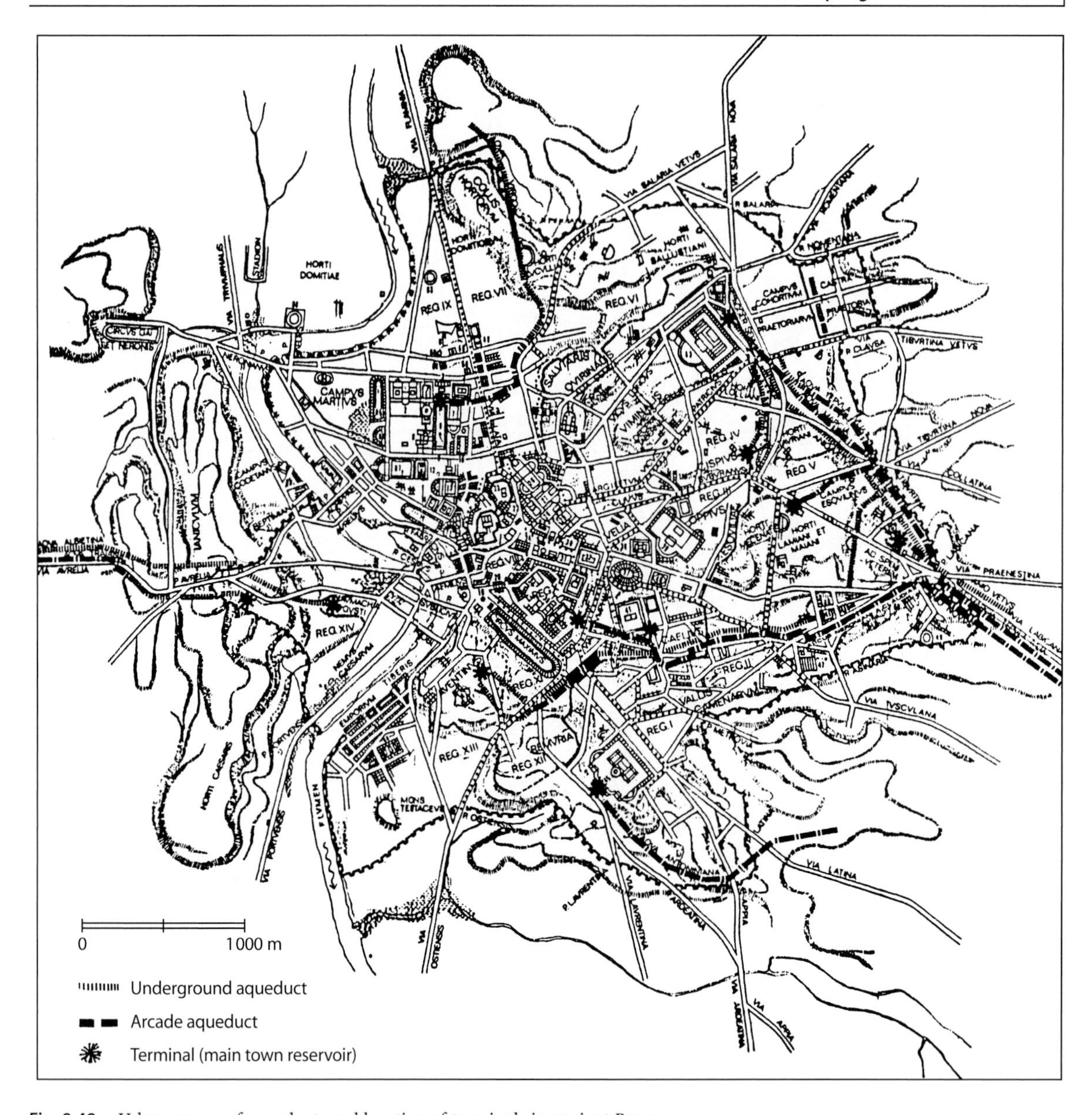

Fig. 8.48. Urban course of aqueducts and location of terminals in ancient Rome

8.18.5 Aqua Marcia

In 144 B.C., the Roman Senate charged Praetor Quinto Marcio Re to restore Anio Vetus and Aqua Appia to prevent undue connections by unauthorized citizens. Meanwhile, the population of Rome had grown and water requirements were constantly increasing. Q. Marcio Re was also charged therefore with building a new aqueduct to ensure more water of good quality. The new aqueduct was supplied by karst springs in the valley of Aniene River. The quality of this water was greatly praised by ancient authors such as Plinio the Old for being fresh and healthy.

The aqueduct worked perfectly for over one century. In 33 B.C. Agrippa carried out the first of many repair works. In 11–4 B.C., Augusto modified the aqueduct structures by collecting more springs, thus doubling the water discharge. In A.D. 79 Tito and later Adriano and Severi restored and kept the aqueduct working. In A.D. 212–213, Caracalla collected new springs (Aqua Antoniniana) to increase the discharge as needed by the huge thermal baths. Diocleziano was ordered to collect new karst springs, promoting the renewal of the entire structure of the aqueduct and its terminal. More restoration work occurred during the time of Arcadio, Onorio, and Popes Adriano I, Sergio II, and Nicola I.

Frontino states that the aqueduct was 91.33 km long. The water ran 80.28 km through underground tunnels and arched bridges (aquae pensiles) (Fig. 8.49). The springs discharged 194 504 m^3/d (2.25 m^3/s), although 30% of it was lost before reaching Rome due to unauthorized private connections, one of the earliest documentations of water theft.

8.18.6 Aqua Tepula and Aqua Julia

Consuls C. Servilio Cepione and L. Cassio Longino in 125 B.C. promoted the catchment of Aqua Tepula. These springs were in the Albano volcanic area near Marino-Castel Savelli. The quality of the water, according to Frontino, was rather poor due to its temperature of 16–17 °C. The name *tepula* means lukewarm. At this source the water from many small springs was also collected by aqueduct.

In 35 B.C., during Agrippa's rule, many additional springs were identified near Grottaferrata and aqueduct restoration was carried out to mix with water from Aqua Tepula to improve its taste and physical characteristics. There was a piscina limaria (settling basin) downstream for the mixing of the water from the two groups of springs. The water was then channeled in two pipes, underground and overland on the Marcia arched bridges about 10 km to Rome. The distance to Tepula Springs to Porta Maggiore (ad Spem Veterem) was about 17.8 km.

Frontino states that the Tepula Springs discharge was 7 550 m^3/d (87 l/s) supplemented with 10 108 m^3/d (117 l/s) drawn from Aqua Marcia and Anio Novus aqueducts. The total discharge of Julia-Tepula Springs at the settling pool was 47 952 m^3/d (555 l/s). Before reaching Rome, Julio aqueduct received about 75 l/s from Aqua Claudia.

8.18.7 Aqua Virgo

Frontino and Plinio the Old wrote a story about a young girl (virgo) who showed the location of some springs to Roman soldiers. Therefore the aqueduct was named after her, however, more probably the name of virgin is due to the purity of the water, which has been praised by the poet Marziale. The Virgo aqueduct is the only one that operated from the time of Augusto up to the present. The aqueduct is underground in volcanic rocks. It reached Agrippa's thermal baths, near Trevi and Navona square fountains that are fed by the Aqua Virgo.

The construction of the aqueduct was ordered by Agrippa and its inauguration took place on 9 June 19 B.C. It was mainly supplied by Salone Springs and its discharge was 99 519 m^3/d (1 150 l/s) according to Frontino. Along its route lateral drainage tunnels branch off 210 l/s of the total discharge. The springs were at the northern border of Albano Volcano, east of Rome, in a marshy area near Aniene River. Restoration works were carried out during the time of Emperor Tiberio, A.D. 36–37; Claudio, A.D. 46–47; and Constantino at the beginning of the 4th century. The slope of the aqueduct tunnel is 4.2 m over a distance of 19 km (0.22%).

8.18.8 Aqua Alsietina

The quality of this water was poor. At the time, however, there was no option to supply the 14th district of the city since that district (Trastevere) is on the right bank of Tiber River, opposite the terminal of the major aqueducts of the town. The water was conveyed to Rome in 2 B.C. mainly to supply the monuments built by Augusto near the Gianicolo. The surplus of water was used to supply the imperial and private gardens and Trastevere fountains (Fontino).

Naval battles or shows (naumachia) were performed in a large pool supplied by Alsietino aqueduct and located in a huge park area where a monumental complex was also built (Fig. 8.50). The elliptical basin (whose axis were 533 and 355 m long) was 1.5 m deep with a storage capacity of 200 000 m^3 of water.

The spring catchment was in the volcanic area of the Sabatini Mountains at the border of Martignano Lake (Lacus Alsietinus), north of Rome. The water of a lake was diverted at an altitude of 207 m by a tunnel. Augusto

Fig. 8.49. **a–c** Roman aqueduct arcades (arcuationes) were first built in volcanic tuff (tophus) or in travertine squared blocks (opus quadratum). **d, e** Later on during Emporer Silla, the technology was improved by means of brick tiles, which were used to curtain the wall (opus latericium) and to strengthen the arches

Fig. 8.50. Emperor Augusto's naumachia (naval battle basin)

Fig. 8.51. **a** Anio Novus aqueduct near Rome. **b** Emperor Claudio's aqueduct near Rome

probably ordered the catchment to ensure the regular level of the lake. To keep the discharge of the aqueduct constant (16 257 m^3/d), water from Bracciano Lake was collected as well (Frontino). The aqueduct was 32.77 km long, built mainly as a tunnel through volcanic rocks. The water was conveyed 529 m on arcades which underwent great restoration work and structural modification at the time of Traiano (A.D. 109) and in the 18th century during the time of Pope Benedetto XIV.

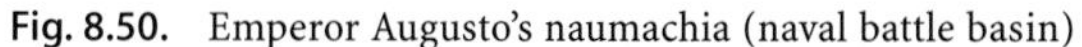

8.18.9 Aqua Claudia

Caligola started the work of two new aqueducts in A.D. 38: Aqua Claudia and Anio Novus. The works were completed in A.D. 52 by Claudio (Fig. 8.51). Historical writings (Tacito) report, however, that in A.D. 47 water from Aqua Claudia was already distributing water in Rome. Anio Novus is supposed to have been finished 5 years later.

Structural modifications and maintenance works were completed by Vespasiano (A.D. 71), Tito (A.D. 81), and Domiziano. Also, several works of consolidation are due to Adriano, Settimio Severo and Diocleziano. During the Gothic War (A.D. 537), the aqueduct was seriously damaged and restored during Belisario's time. Pope Adriano I in A.D. 776 repaired the aqueduct but its previous discharge had been considerably reduced.

The springs of Aqua Claudia were karst springs (fons Caeruleus, fons Curtius, fons Albudinus, Aqua Augusta) that were located along the right bank of Aniene River, east of Rome and not far from the Marcio aqueduct inlet. Frontino records the spring discharge at the catchment area of 191 190 m^3/d (2 213 l/s). At Spem Veterem, in Rome, this discharge was reduced to 1 341 l/s due to unauthorized connections existing along the aqueduct. Both Aqua Claudia and Anio Novus reached Porta Maggiore (Rome).

From the main reservoir (castellum aquae), the water from the two aqueducts was distributed to all 14 districts of Rome (called Augusto regions) through 92 subsidiary reservoirs (Frontino). The Aqua Claudia aqueduct was 69.75 km long, of which 54.5 km of tunnels and 15.2 km trough arcades were originally built in tuff (tophus).

8.18.10 Anio Novus

The construction of the Anio Novus aqueduct was begun by Caligola in A.D. 38 and completed by Claudio about A.D. 50. Historically, the Anio Novus and Aqua Claudia aqueducts are closely related. Important modifications to both aqueducts were made by Traiano (A.D. 109), and several maintenance and restoration projects are recorded up to the 4th century.

According to Frontino, the aqueduct Anio Novus was built in opus reticolatum (tuff wall) and opus latericium (brick wall). The catchment area was near the Aqua Claudia Springs along Aniene River, from which the aqueduct collected part of its natural discharge, mainly represented by karst groundwater. A large settling basin (piscina limaria) was built near the banks of the river as surface water was muddy during major floods.

To improve the resource quality, some springs of Rivus Hercolanus were collected and drawn to the same aqueduct, and Frontino states that the water quality had the same standard as the celebrated water from Aqua Marcia. Total discharge of Anio Novus is 196 490 m^3/d (2 274 l/s); the aqueduct was 86.876 km long, 73 km of which was in tunnels.

8.18.11 Aqua Traiana

The aqueduct was built by Traiano in A.D. 109–110 to supply the 14th district of Rome (Trastevere) with good, drinkable water. The district was previously supplied by the poor-quality water of Aqua Alsietina. Few data are available about Aqua Traiana as Frontino died and the following civil servants in charge (curatores aquarum) did not register data about new aqueducts.

Many springs were collected to supply the aqueduct. They were scattered in the volcanic area north of Bracciano Lake, northeast of Rome. Although many of the springs have not been located, it is believed that they are those that appear in an 18th century map showing the aqueduct of Pope Paolo V. The Aqua Traiana aqueduct was 32.5 km long. Most of it was underground and partially on arcades. It supplied mainly the Trastevere area and Traiano baths on Colle Oppio. The aqueduct was partially utilized by Pope Paolo V in 1608 and its name at that time changed to Aqua Paola.

8.18.12 Aqua Alexandriana

Emperor Alessandro Severo (A.D. 222–235) decided to build the aqueduct given his name during the last years of his reign. It was the last great aqueduct built in Rome in ancient times. The Emperor intended to supply water to the Campo Marzio baths built by Nerone in A.D. 6 and restored in A.D. 227. The catchment area of the springs is in the volcanic area east of Rome near ancient Gabi, not far from Aqua Appia Springs. During the Gothic War (A.D. 537), the aqueduct was destroyed to cut the water supply of besieged Romans. In 1585 the springs were collected anew to supply Felice aqueduct built by Pope Sisto V. This aqueduct was 22 km long, 8 km in tunnels and 14 km on arcades. Repair works are recorded up to A.D. 500. The Alexandriana and Vergine aqueducts were kept in use during the Middle Ages.

8.18.13 Water Potential and Use in Antiquity

Aqua Virgo and Aqua Traiana-Paola are the aqueducts built during Roman times which are still in use. Total discharge of the ancient aqueducts (excluding Aqua Traiana and Aqua Alexandriana, whose data are missing) was 24 360 quinariae (1 010 258 m^3/d [11.69 m^3/s]). The population of Rome at the end of the 1st century A.D. was about 500 000; consequently, a mean of 1 550 l/d per capita (Fig. 8.52).

At the beginning of the 4th century A.D., the large monuments of Rome supplied by aqueducts are: 11 large baths, 856 public baths, 15 monumental fountains, 1 352 fountains (nimpheos) and basins, and 2 naumachiae (naval battle basins). According to Frontino, water consumption included: 17.2% by the emperor; 38.6% by citizens; and 44.2% by public services. Today, Rome is supplied with 1 987 200 m^3/d (23 m^3/s) of groundwater, mainly from karst aquifers. Its population is 3.5 million, with a per-capita water availability of 500 l/d, including industrial uses.

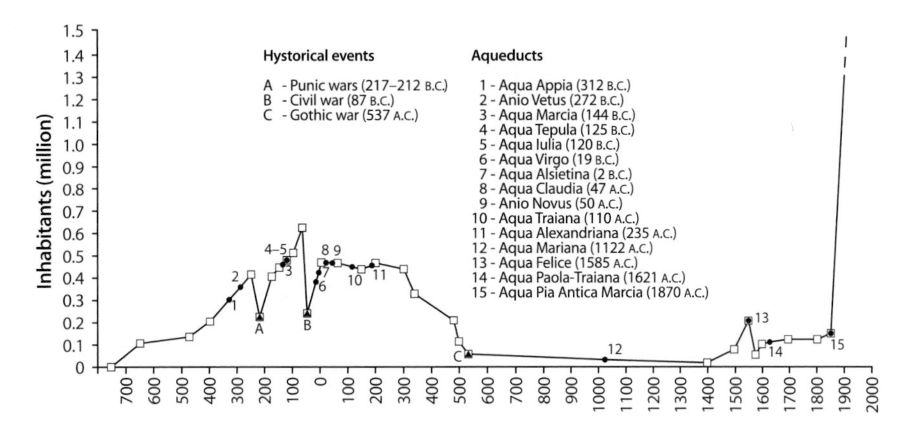

Fig. 8.52. Demographic trend of Rome and chronology of aqueducts

8.18.14 Decline and Revival of Rome Aqueducts (A.D. 500–1800)

The siege of Rome by Vitige, the king of Goths in A.D. 537, marks the decline of the water supply systems built by Romans. The Goths destroyed the aqueducts to cut water supplies to the besieged city. After a short restoration period by the Byzantine governments, Rome underwent political and economic decay. The aqueducts were neither reconstructed nor maintained and most of the arcades collapsed.

By the end of the Middle Ages, only Aqua Virgo was still in operation. Although the population had decreased considerably, the available water supply was insufficient. Therefore, people used water from the Tiber River and from wells drilled in the urban area. During this time, a great number of Roman buildings, villas, and monuments were torn down to recover bricks and materials for new construction. This practice went on until modern times.

In 1122 Pope Callisto II planned to build the aqueduct Mariano, recovering ancient Roman catchment areas (Aqua Julia and Aqua Tepula). The new water line cannot be compared to the magnificence of the Roman aqueducts. However, many windmills and workshops were located along its country and urban routes; therefore it had economic importance from the Middle Ages to modern times.

In 1570 Pope Pio V restored Aqua Virgo. Pope Sisto V in 1585–1590 collected Aqua Alexandriana's Springs to build the Felice aqueduct, while in 1607 Pope Paolo V Borghese restored Aqua Traiana (later called the Paolo aqueduct). In 1870, Pio IX completed the construction of the Pio aqueduct collecting some of the springs of the ancient Aqua Marcia.

8.18.15 Modern Water Supply of Rome

From the end of the papal rule (1870) to 1938, the water supply of the city was administered by the Municipality of Rome. By then, ACEA, which was established during the Fascist period as an electricity production company, took over the management of the water supply of Rome up until the present (Fig. 8.53). In 1964 the ACEA was given responsibility for the water supply of all communities belonging to Rome Municipality. The water supply requirement for Rome at present is about 23 m^3/s, 86% of which is from karstic springs. It is interesting that the water quality standard does not require chemical treatment for human consumption except for chlorination as a preventive measure of organic pollution (Table 8.16).

8.18.16 Marcio Aqueduct

This aqueduct conveys spring water of the ancient aqueducts: Anio Vetus, Aqua Marcia, Aqua Claudia, and Anio Novus, all destroyed by the Goths in A.D. 537. At the time of Pio IX in 1870, a new aqueduct was inaugurated and named Aqua Pia Antica Marcia. Different in technical conception and layout, it carries to Rome only part of the water from springs collected during Roman times.

The population of Rome gradually increased after 1970 by about 2.5% to 3.5 million, and more springs had to be developed to provide water to meet the increased demand. The mean discharge of springs that supply Marcio aqueduct today is 5.0 m^3/s while Roman aque-

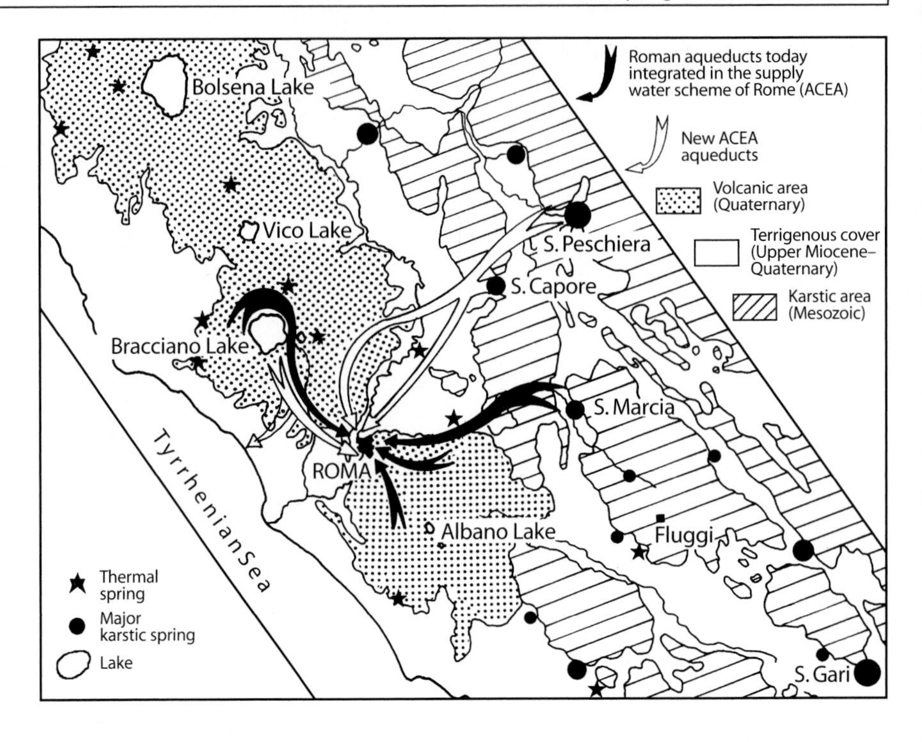

Fig. 8.53. Schematic hydrogeological map of central Italy

ducts produced from the same area about 8.8 m^3/s. These springs are from the karst system of Simbruini Mountains, which has its base level in the Aniene River (321–327 m a.s.l.). The springs occur along the riverbed (linear springs) and along the contact between the karst aquifer and an impervious belt of sandstone (upper Miocene flysch).

The water chemistry is typical of that in the karst aquifer. The mean annual effective infiltration is about 910 mm.

8.18.17 New Vergine Aqueduct

The ancient Aqua Virgo reached Rome at a low elevation and therefore hydraulic constraints did not allow its use extensively for a drinking water supply. In 1901 and 1930 a lifting plant and a new aqueduct 12.7 km long were planned. The springs (25 m a.s.l.) are at the northern slope of the volcanic area of Albano, southeast of Rome. Their mean discharge is 1.1 m^3/s, 0.3 of which is conveyed to the ancient Vergine aqueduct, and 0.8 to New Vergine aqueduct. The groundwater chemistry is not uniform in the catchment area. The water from each spring has a different mineral content. The variability is due to the heterogeneity of the aquifer and to the upwelling of hydrothermal fluids.

8.18.17.1 *Peschiera-Capore Aqueduct*

This aqueduct supplies Rome with about 14 m^3/s. Its construction started in 1935 when the population of Rome was one million. It collects water from two groups of springs: Peschiera (9.0 m^3/s) and Capore (5.0 m^3/s) that belong to different hydrogeological units in the central Apenines range. The aqueduct system is X-shaped and Peschiera and Capore Springs are at two ends. The collecting pipes converge at the Salisano hydroelectric plant, and from there two separate aqueducts run along the two banks of the Tiber River to Rome.

8.19.17.2 *Peschiera Springs*

Located along Velino River with an altitude of 410 m. The mean discharge is 18 m^3/s. The springs are from a homogeneous karst aquifer with a discharge of 26 m^3/s.

Table 8.16. Aqueducts of Rome: mean discharge and chemistry of spring waters[a]

	Peschiera-Capore			Marcio		Nuovo Vergine	Appio-Alessandrino	Paolo-Traiano	
Reference	1, 2	1	2	3	4			5	6
Mean discharge, Q (m^3/s)	13.75	9	5.0	5.0	0.95	0.8	1.2	8 (max)	
Temperature (°C)	11	11	12	10	15	15.5	16	10	15
pH	7.2	7.2	7.3	7.3	6.7	7.3	7.25	8.02	8.0
Conductivity 25 °C (μS/cm)	590	640	580	540	1 460	650	550	540	270
TDS (110 °C)	370	410	375	320	1 030	460	360	354	230
Hardness (°F)	32.5	34.8	29.5	31	74.2	25.5	21.8	11.5	8
Alkalinity ($CaCO_3$)	315	340	270	300	510	275	240	164	112
Organic compounds (O_2)	0.25	0.2	0.25	0.3	0.5	0.35	0.45	0.8	0.5
NH_4									
Ca	110	115	95	87.5	234	78	74	32	15.5
Mg	12.5	15	11.5	21.1	38	18	7.7	8.4	10
Na	4.5	2.5	10	2.51	52.5	31.8	16	43.5	23
K	1.55	2	1.3	0.7	3	36.5	20	39.4	19
NO_2									
NO_3	3.3	2.3	4.3	2	3.5	20.6	14.5	0.7	9
Cl	4.6	3.8	7	5	80	20.5	14.3	45.9	15.9
SO_4	14.5	9.9	27	4.5	215	48	16	40	8.7
P	<0.02	<0.02	<0.02	<0.02	<0.02	<0.02	<0.02	<0.02	<0.02
F	0.1	0.07	0.11	0.1	0.25	1.1	1	1.7	1.9
HCO_3	385	415	165	360	622	335	293	200	138
Si	4.5	3	8.2	4.5	10.5	45.5	48	1.5	50
CO_2	14.5	23	12	20.9	210	22	n.d.	2	3.5
O_2	12	7	13.5	10.9	3.5	7.3	n.d.	9.5	13.7

[a] ACEA data: *1* Peschiera Springs; *2* Capore Springs; *3* Acqua Marcia Springs; *4* Acquoria Springs; *5* Bracciano Lake; *6* Acqua Traiana Springs.

The water chemistry of the springs is uniform. The extension of the recharge area (about 1 000 km^2) contributes to the regularity of the aquifer discharge. The mean annual effective infiltration is about 980 mm.

8.18.17.3 *Capore Springs*

These springs are along the bed of Farfa River at an altitude of 246 m. Their mean discharge is 5 m^3/s with a remarkably steady regime. The recharge area is about 280 km^2, while the mean annual effective infiltration is about 570 mm.

8.18.17.4 *Appio-Alessandrino Aqueduct*

This new aqueduct draws groundwater from Roman catchments (drainage tunnels) connected to the Aqua Appia and Aqua Alexandriana and from recently drilled wells. The catchment areas are along the foothills of the volcano Albano at Pantano Borghese, Finocchi, Torre Angela and have a total discharge of 1 200 l/s. Different piezometric levels in the catchment area are related to a heterogeneous reservoir with groundwater flow in perched aquifers. Structural conditions reflect a wide variability of groundwater chemistries related to several hydrothermal gaseous emissions in the area.

8.18.17.5 *Paolo-Traiano Aqueduct*

In the early 1600s the population of the Trastevere urban area had no option but to use the water of the Tiber. The inhabitants of the areas on the left bank of the river could benefit from good quality water from the Vergine and Felice aqueducts. Pope Paolo V Borghese (1605–1621) therefore ordered the reconstruction of the Aqua Traiana imperial aqueduct that had fallen into neglect. In 1946 Pope Innocenzo X Pamphili promoted the development of the aqueduct with a catchment of more springs in the Anguillara volcanic area and later diverted the outlet of the Bracciano volcanic lake (Arrone River).

At present, ACEA's management has further increased the discharge of Paolo-Traiano aqueduct by pumping water from Lake Bracciano to a yield of about 1 m^3/s. In case of emergency due to the breakdown of Roman aqueducts, a temporary additional discharge of 8 m^3/s is possible by drawing the water from Bracciano Lake to the Bracciano aqueduct. To reduce the fluorine content of the water, Paolo-Traiano aqueduct converges at a mixing plant supplied by karst water from the Peschiera-Capore aqueduct.

8.18.18 Water Wells

The increasing population on the outskirts of Rome, new residential settlements, and more agricultural and industrial water demand resulted in the drilling of many private wells. The wells are up to 200 m deep and top aquifers in volcanic terrains whose base level is represented by Tevere (Tiber) and Aniene Rivers. In this area, the so-called mineral water is a commercial activity with several bottling plants and brands. The pH of the water ranges between 6.7 and 7.2, has a salinity between 0.3 and 0.6 g/l, and a temperature between 14 and 18 °C.

I. Povara

8.19 Thermal Springs in Băile Herculane (Romania)

8.19.1 Introduction

In southwestern Romania, in the neighborhood of Băile Herculane Spa, a major positive geothermal anomaly occurs. The anomaly occupies the southern extremity, a transcrustal intra-Carpathian fault, over 300 km long, that starts in the central part of the southern Carpathians, crosses the Danube and extends southward into the Republic of Yugoslavia. North of Băile Herculane, over a distance of 60 km, the fault splits to form a narrow graben, sunken by some 1 000 m with respect to the adjoining structures. The 600-m-thick carbonate formations and the upper part of the underlying granite, trapped inside the graben, form water reservoirs. Thermo-mineral sources occur along the entire length of the western fault of the graben, the most important of them being situated in the neighborhood of Băile Herculane (see Fig. 8.54).

8.19.2 Băile Herculane – Historical References

1. The first facilities were built by the Romans, after A.D. 105, the Roman "termae" being under the auspices of god Hercules. Many votive tables dating from between A.D. 107 and 287 confirm the therapeutical qualities of the water.
2. In 1736, Count Hamilton, sent by Charles III, rebuilt some of the facilities on the old Roman sites.
3. After 1817 (Franz Joseph), the first modern baths and hotels were built.
4. In 1847, Prince Carol presented the spa of Băile Herculane with the bronze statue of Hercules.
5. In 1884 the first thermal water intake well was drilled (Neptun I, 27 m deep).
6. Between 1968 and 1984 seven new hotels, provided with their own balneotherapy facilities were built, and an integrated water distribution system completed. The current lodging capacity of the spa exceeded 3 000 places.
7. Eleven new wells drilled between 1968 and 1978 provided an additional discharge of 23 l/s. The minimum cumulated discharge currently provided by wells and natural springs is 55 l/s (4 750 m^3/d).

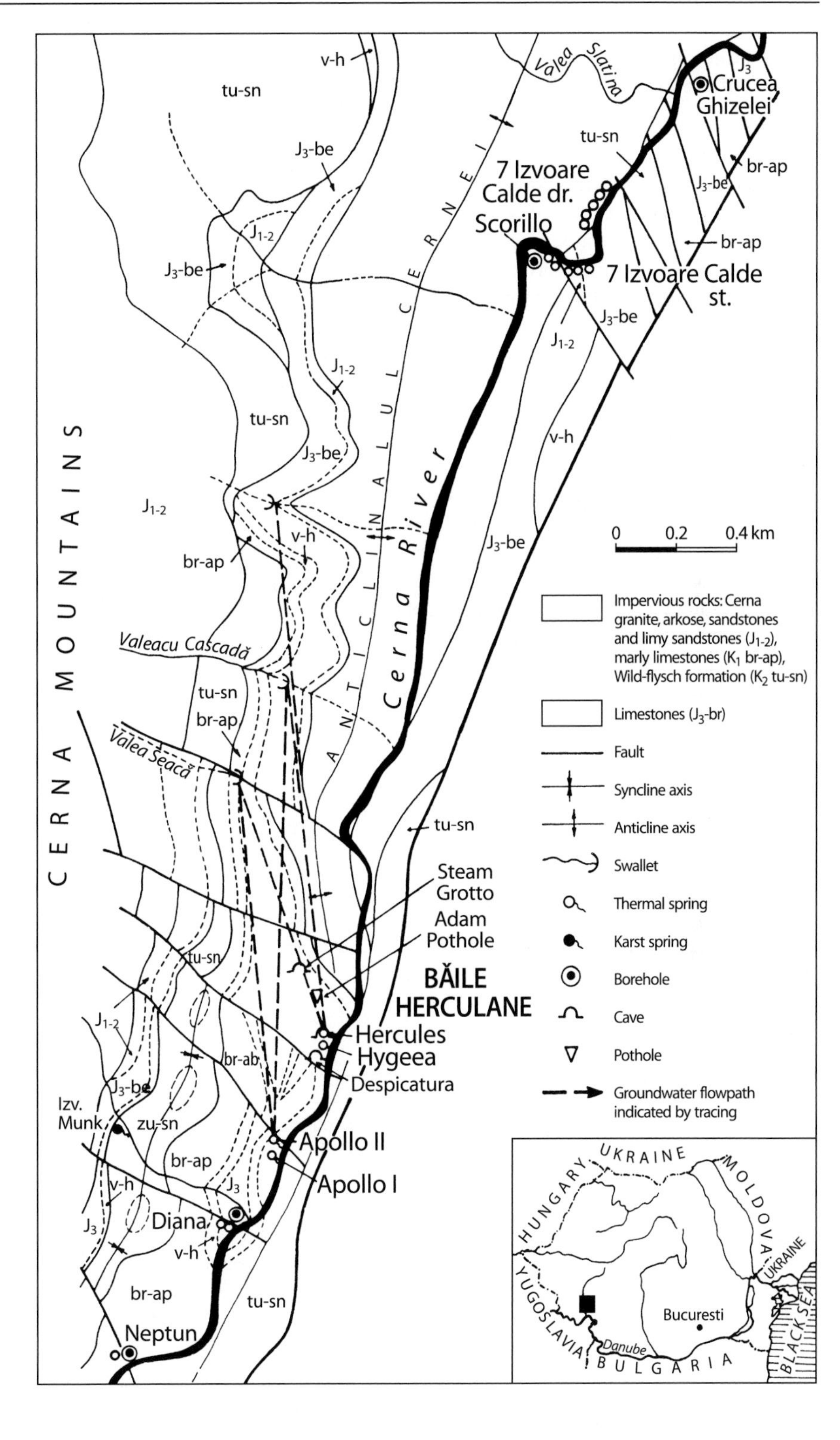

Fig. 8.54. Hydrogeological map of the southern part of the Hercule aquifer (geology modified after Nastaseanu 1982)

8.19.3 Geological and Hydrological Characteristics of the Area

The wells and the natural springs discharge water that originates into two complex thermo-mineral aquifers within the Cerna Graben and within the syncline of Cerna.

The Cerna Graben extends more than 60 km along a NNE-SSW direction. Cerna River closely follows the path of the graben. Deep faults border the graben to the E and the W, the western fault being ascribed to the crustal, or possibly the sub-crustal type (Nastaseanu and Maksimovic 1983). The generalized lithology of the southern section of the graben (about 10 km, located between the railway station of Băile Herculane and the well at Crucea Ghizelei), includes (Fig. 8.54).

1. The basement mainly granitic (the Cerna granite), highly fractured and carrying thermal water; it is dissected by transverse faults, displaying a general trend of southward deepening.
2. The sedimentary formations unconformably cover the basement rocks and include Jurassic and Cretaceous deposits (Nastaseanu 1982).
3. Liassic-Dogger: 30–70 m thick, discontinuous, a sandstone-conglomerate series, spatic limestones.
4. Malm-Berriassian: 80–180 m thick; sandstone limestones, compact limestones, nodules limestones and fine, bedded limestones.
5. Valanginian-Hauterivian: 50–75 m thick; plate limestones, marly limestones.
6. Barremian-Aptian: 200 m; compact or marly limestones.
7. Turonian-Senonian: 300–500 m; wild-flysch facies.

Along the graben, a stack of permeable Late Jurassic-Early Cretaceous carbonate rocks occurs, where important water resources are accumulated (Povara and Lascu 1978; Fig. 8.55a). The recharge of the graben has several origins (Fig. 8.55b):

1. Downflow, mainly of karst origin (stream bed losses of Cerna River occurring 14 km upstream of Băile Herculane, possibly also on the mountain slope on the left side of the river)
2. Heated upflow of deep origin
3. Connate water of high salinity

It should be noted that possible inflows of gas and soluble ions that could have originated in deeper sections of the Earth's crust have been alternatively considered. Assuming that this is the case, the previous

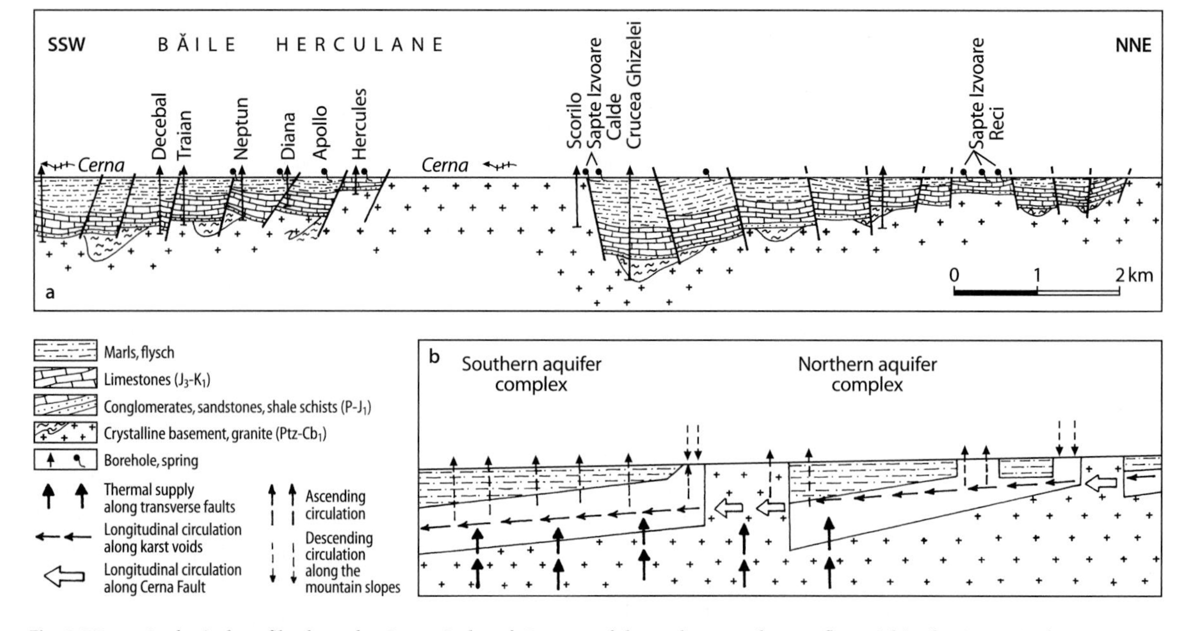

Fig. 8.55. **a** Geological profile along the Cerna Graben. **b** Pattern of the underground water flow within the Cerna Graben (after Povara and Lascu 1978)

(2) and (3) components would be included in a single one, which would transfer to the cold water both the thermal and the mineral characteristics.

The Cerna Syncline extends over 20 km, on the right side of Cerna River, parallel to the valley. It has a disymmetric shape, overturned to the east. Roughly the same, previously described litho-stratigraphic Late Jurassic-Early Cretaceous series, overlies the crystalline-granitic basement. Major interest concerning the thermo-mineral water accumulation is related only to the southern one third of the syncline.

The synclinal bend displays a general southward deepening, being also vertically (and locally horizontally) displaced by transverse faults. In the area of Băile Herculane, the syncline plunges deeply, merging into the Cerna Graben (Simion 1987, unpubl. data)

The recharge of the syncline originates in concentrated sinks into the stream beds, as well as in direct seepage of rainfall. In the southernmost section, there is also a noticeable influence due to the Cerna Graben, whence the thermo-mineral water comes.

The area in the proximity of Roman Hotel acts as a "mixing system" between the karst water coming from the northern part of the syncline and the thermo-mineral water. During rainfall periods, cold seepage water contaminates the thermal water supply. The result is a large variability of the flow rates, temperatures and chemical compositions of the water from the most important springs in the area, namely Hercules and Apollo (Povara 1973; Simion et al. 1985).

Four natural karst cavities are directly connected to the karst aquifer:

1. The caves Hercules and Despicatura, that are paths for discharge of groundwater outside the mountain slope;
2. The Steam Grotto and Adam Pothole (Fig. 8.56), which display permanent emissions of steam 41–54°C (on 16.10. 1972 when rainfall amounted to 108 l/m^2 per 24 h, the steam emission in Adam Pothole ceased for 2 days).

A mining gallery has been excavated into the eastern flank of the syncline, in order to separately tap the two mixing components (i.e. cold and thermo-mineral).

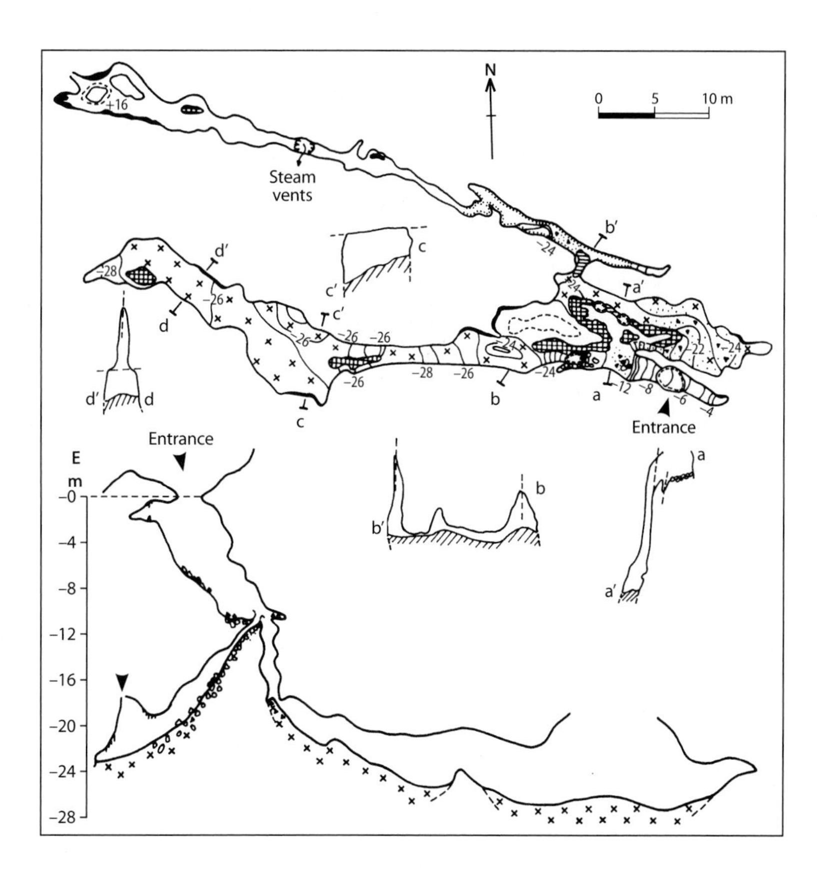

Fig. 8.56. Adam pothole (after Povara et al. 1972)

8.19.4 General Characteristics of the Thermo-Mineral Sources

8.19.4.1 *Flow Rates*

The wells and natural springs display daily fluctuations of small amplitude, and more important seasonal or annual fluctuations. The range of flow rates recorded over a period of about 20 years is shown in Table 8.17. The maximum range of fluctuation displayed by the natural springs Hercules and Apollo is due to the mixing of thermo-mineral water with cold karst water supplied by streamlets and rainfall. Their flow rates are directly correlated to rainfall.

For the wells, filters clogging resulted in progressive reduction of the flow rates.

8.19.4.2 *Temperatures*

The temperatures (Table 8.17) are generally more constant than the flow rates, except for the springs Hercules and Apollo. For the wells, the temperature fluctuations can be ascribed to variations in the flow velocity. As a general trend, temperatures increase southward. This trend closely correlates with the Na-K-Mg geothermometer results (Giggenbach 1988), which suggest that thermo-mineral sources situated to the north undergo a stronger chemical re-equilibration with respect to the deep, 180 °C reservoir of origin, than sources situated to the south (Fig. 8.57). The results of the pumping tests performed in several wells are indicated in Table 8.18.

8.19.4.3 *Chemical Composition*

According to their chemical composition (Table 8.19), the thermo-mineral sources can be divided into:

1. The northern group (Crucea Ghizelei, 7 Izvoare Calde, Scorillo, Hercule):
   a Small TDS, compared with the southern group
   b Small H_2S concentrations
   c The largest sulphate and bicarbonate ions concentrations, in reverse correlation with H_2S concentrations
2. The southern group:
   a Higher TDS, with an obvious southward increasing trend
   b High H_2S concentrations and small sulphate concentrations
   c Chloride is the prevailing anion (up to 50% of the total), closely correlated with the TDS

The pH decreases across the thermal water accumulation, from 7–7.8 in its northern part, to 4.5–5.5 to the south.

8.19.4.4 *Natural Radioactivity of the Water*

The natural radioactivity of the water is maximum in the northern part and decreases to the south (22 mCu to 2.1 mCu). It is assumed that the radioactivity of the water originates in the granitic rocks of the basement.

The springs, as well as the wells, display discontinuous gas eruptions. The composition of a gas sample collected

Table 8.17. Flow rates and temperature fluctuation ranges over the interval 1975–1994

Spring/Borehole	Flow rate (l/s)			Temperature (°C)		
	Maximum	Minimum	Δq	Maximum	Minimum	Δt
Crucea Ghizelei	11.70	7.24	4.46	36.8	31.1	5.7
Sapte Izvoare Calde	1.50	0.95	0.55	56.5	53.1	3.4
Scorillo	0.77	1.04	0.27	51.2	48.4	2.8
Apollo II	7.50	3.64	3.86	50.2	45.5	4.7
Hercules	105.00	10.20	94.80	53.5	17.0	36.5
Diana	0.053	0.033	0.02	54.4	51.8	2.6
Neptun I+IV	4.42	3.40	0.98	53.2	49.8	3.4
Traian	6.79	6.00	0.79	53.2	50.2	3.0

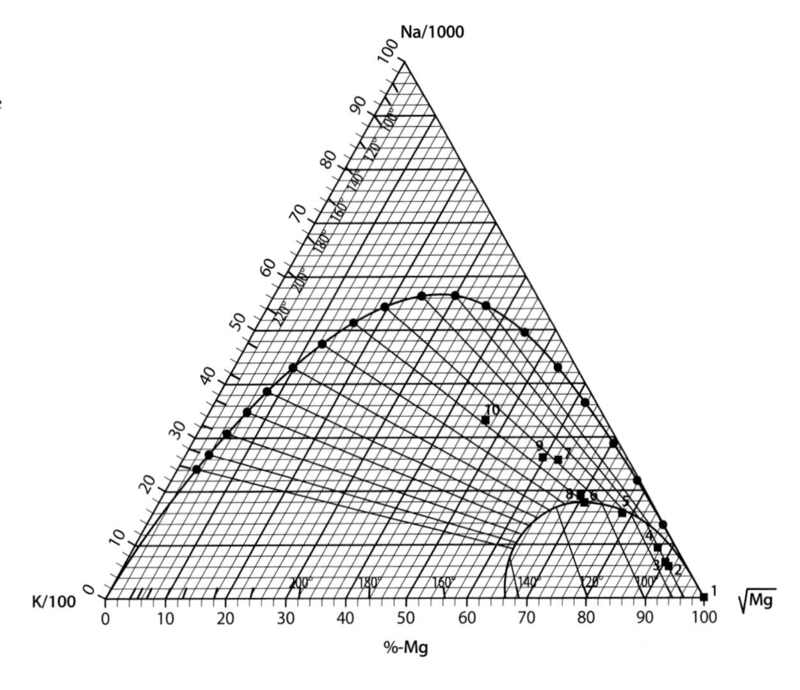

Fig. 8.57.
Triangular diagram for evaluation of Na-K and K-Mg equilibration temperatures: *1* Crucea Ghizelei; *2–4* Sapte Izvoare Calde dr; *5* Sapte Izvoare Calde st; *6* Apollo II; *7* Diana; *8–9* Neptun; *10* Traian

in the well Traian (completed interval 298–456 m) was: He = 0.25; H_2 = 0.085; N_2 = 19.5; CH_4 = 79.4; CH_6 = 0.2; C_3H_5 = 0.018; iC_4H_{10} = 0.0047; nC_4H_{10} = 0.003; iC_5H_{12} = 0.001; nC_5H_{12} = 0.0006.

8.19.5 Geothermal Investigations

Measurements performed in 30–50 m deep wells have indicated five geothermal anomalous areas; two of them are superimposed on the previously described groups of springs. In the northern group, a geothermal gradient of 200 °C/km has been outlined (Veliciu 1988). The SiO_2 geothermometer indicates a 46–92 °C maximum temperature for the water at great depth. In the southern group, the shallow geothermal gradient ranges between 110 and 200 °C/km, while the chemical geothermometer SiO_2 indicates a 60–70 °C maximum temperature for the water at great depth.

The temperature increase with depth, recorded in deeper, groundwater wells displays the following trend: Scorillo 100 °C/km; Traian 89.3 °C/km; Decebal 46.9 °C/km; Crucea Ghizelei 20 °C/km.

In horizontal and ascending boreholes drilled from the mining gallery near Hercules spring, thermal gradients range between 10–980 °C/km (Veliciu 1988, Povara 1992). Thermal gradients' variation in pattern has been used for computing the distance to the unexplored thermal water-carrying karst passage (Mitrofan and Povara 1992). The extent of the "blind" karst passage has been subsequently traced by the electrometric "mise a la masse" method, while possible thermal water inflow paths have been delineated by spontaneous polarization (Mitrofan et al. 1995).

Acknowledgements

The author wishes to express particular appreciation to G. Simion who allowed the use of unpublished data and to H. Mitrofan for his contribution regarding the clarity of the manuscript.

References

Diaconu G (1987) La géologie et la tectonique du périmètre d'influence de la source thermominérale "Hercules", Băile Herculane. Theor Appl Karst 3, Bucuresti, pp 109–116

Gaspar E, Simion G (1985) Tracer research on the dynamics of underground waters in the Cerna Valley (Southern Carpathians, Romania). Theor Appl Karst 2, Bucuresti, pp 183–197

Giggenbach WF (1988) Geothermal solute equilibria. Derivation of Na-K-Mg-Ca geoindicators. Cosmochim Acta 52:2749–2765

Table 8.18. Results of pumping tests performed in several wells

Borehole	Depth (m)	Piezometric level (m above well head)	Q (m^3/d)	q (m^3/(d m))	K (m^3/d)	T (°C)
Crucea Ghizelei	17.3– 125	−60	1.2	0.02	0.006	–
	392 – 802	+30	25.9	0.86	0.005	35
	620 –1201	+29	691.2	20.37	0.11	35.5
Scorilo	5 – 148	−0.1	172.8	2.7	0.07	32
	5 – 298	+15	86.4	5.1	0.036	40
	298 – 498	+15	190.08	12.66	0.12	50
	498 – 550	+24	172.8	7.2	0.3	55
	60 – 550	+17	250.5	14.7	0.51	54
Traian	295 – 453	+52	173	3.24	0.06	47
	461 – 577	+50	484	9.66	0.30	60
	296 – 577	+50	691	13.82	0.17	62
Decebal	241 – 396	+40	86.4	2.1	0.03	35.5
	241 – 476	+35	51.8	1.48	0.012	36
	476 – 598	+35	8.64	0.24	0.006	38
	241 – 598	+45	259	5.76	0.013	39.5

Table 8.19. Chemical data on thermo-mineral springs from Baile Herculane (Marin 1984)

Borehole	TDS	Na^+	K^+	Mg^{2+}	Ca^{2+}	Fe^{2+}	NH_4^+	HCO_3^-	HS^-	SO_4	F^-	Cl^-
Crucea Ghizelei	269.6	3.5	3.0	11.5	42.9	0.64	–	172.5	–	9.8	–	4.9
Sapte Izvoare	782	206	9.0	3.1	25.8	0.1	2.3	70.0	6.1	155	1.2	231
Scorilo	878	247	10.4	2.0	23.3	0.1	1.0	47.5	5.1	157	0.9	296
Hercules	356	739	57.5	17.8	460	0.01	0.3	62.8	0.08	151	0.8	1953
Apollo	2923	667	39	13	306	0.3	5.4	156	14.4	109	1.7	1487
Diana	3563	864	46	12.6	362	0.3	0.08	90.5	54.5	0.0	1.6	1935
Neptun I+IV	5346	1043	65.7	47.5	768	0.3	1.8	54.3	57.3	26.7	1.3	3068
Traian	7262	1548	64.7	33.6	1025	0.4	3.9	48.4	79.4	0.0	1.9	4254
Decebal	7654	1622	62.6	65.9	1033	0.08	1.5	1.18	52	79.2	1.9	4428

Marin C (1984) Hydrochemical consideration of the lower Cerna River basin. Theor Appl Karst 1, Bucuresti, pp 173–183

Mitrofan H, Povara I (1992) Delineation of a thermal water carrying karstic conduit by means of thermometric measurements in the Băile Herculane area (Romania). Theor Appl Karst 5, Bucuresti, pp 39–144

Mitrofan H, Mafteiu M, Povara I, Mitrutiu M (1995) Electrometric investigations on the supply channels of Hercules Spring (Romania). Theor Appl Karst 8, Bucuresti, pp 129–136

Nastaseanu S (1982) Géologie des Monts Cerna. Ann Inst Geo. Geoph LIV, Bucuresti, pp 153–280

Nastaseanu S, Maksimovic B (1983) La corrélation des unités structurales alpines de la partie interne des Carpates Méridionales de Roumanie et de Yugoslavie. Ann. IGG, LX Bucuresti

Povara I (1973) Contribution à la connaissance des sources thermo-minérales de Băile Herculane. Trav Inst Spéol, Emile Racovitza XII, Bucuresti, pp 183–195

Povara I (1992) New data of the Hercules thermal aquifer obtained by temperature measurements (Băile Herculane, Romania). Theor Appl Karst 5, pp 127–138

Povara I, Lascu C (1978) Note sur la circulation souterraine de l'eau par le graben de Cerna. Trav Inst Spéol, Emile Racovitza XVII, Bucuresti, pp 193–198

Povara I, Marin C (1984) Hercules thermomineral spring. Hydrogeological and hydrochemical considerations. Theor Appl Karst 1, Bucuresti, pp 183–195

Povara I, Diaconu G, Goran C (1972) Observations préliminaires sur les grottes influencées par les eaux thermo-minérales de la zone Băile Herculane. Trav Inst Spéol Emile Racovitza XI, pp 335–365

Simion G, Ponta G, Gaspar E (1985) The dynamics of underground waters from Băile Herculane, Cerna Valey, Romania. Ann Soc Geol Belg 108

Veliciu S (1988) Contributii privind prospectiunea geotermica a apelor termale cu aplicatii in Romania. St teh econ, Geofizica D 15

K. D. Arakchaa

8.20 Ecology, Geochemistry and Balneology of Natural Medicinal Water Springs Used for Health Treatment (Tuva's Arzhaan Project, Russia)

8.20.1 Introduction

There are many natural water springs in Tuva that are used by native people for medicinal purposes. They are called Arzhaans. According to people's opinions, the treatment at the Arzhaans is very effective. It's an old natural tradition closely connected to the Tuvinian way of life. Every year more than 10 000 people are treated at the Arzhaans without medical control. The therapeutic methods used by the Arzhaans traditionally are drinking the spring water, and taking showers and baths.

This tradition is a part of the way of life of people in different central Asian countries and, perhaps, of other countries of the world. Studying this tradition is important because it is a part of world ethnic and cultural diversity that is one of the elements of sustainable development.

The Tuva's Arzhaans can be divided into two groups: mineral springs and fresh springs. The mineral Arzhaans have chemical and physical characteristics which correspond to present teaching about medicinal mineral waters. The balneological properties and the mechanism of balneofactor action are more or less known.

From a generally accepted point of view, the fresh Arzhaans are considered to be non-effective because they do not contain balneologically active components. However, the native people are treated by using them for different diseases. The medicinal effect of freshwater springs is not explained by well-known balneological concepts.

Ongoing research includes scientific substantiation of natural spring waters balneological factors and the mechanism of its action on humans, using the physiological effects on Tuvan native people treated with fresh Arzhaans.

Specific tasks for research are:

1. Determination of chemical and microbiological compositions of water from fresh Arzhaans during the year. Ecological monitoring of them and creation of the corresponding data bases.
2. Studies on the hydrogeological conditions of formation of the natural medicinal water springs. Finding out possible ways for protecting them.
3. Investigation of the physical properties and, if possible, molecular structure of Arzhaan's waters.
4. Determination of the geophysical field structure at the outlet places of the fresh Arzhaans.
5. Epidemiological research on uncontrolled medicinal treatment by the fresh Arzhaans. Classification of them on the basis of the balneological effects.
6. Determination of the correlation between data bases.
7. Scientific substantiation of the medicinal effect.
8. Ethnographic study of native peoples' traditions of using natural waters for treatment.

Results so far:

1. The National Scientific Laboratory on Natural Medicinal Water Springs of the Republic of Tuva was created in 1993.
2. Hydrochemical research was carried out in 1989–1994. About 80 Arzhaans were investigated and a data base developed.
3. Microbiological, geophysical and radiological studies have been carried out at some of the mineral and fresh Arzhaans.
4. Medical-sociological research was carried out in 1991 during uncontrolled medicinal treatment at 3 Arzhaans.
5. Medical-epidemiological research was carried out on uncontrolled medicinal treatment at 3 Arzhaans in 1993. More than 300 people with various clinical symptoms were interviewed.

The results of these investigations were presented at the Second USA/CIS Joint Conference on Environmental Hydrology and Hydrogeology (Washington DC, USA, 16–21 May 1993), and were published in a local scientific publication (1994). The book íArjaan Saga Tuvaî was published by author of this project (Moscow 1995).

8.20.2 Expected Outcomes

1. Elaboration of a multidisciplinary research methodology for natural medicinal water springs and national traditions of using them for health treatments.
2. Ecological, hydrochemical and microbiological characteristics of Tuva's fresh Arzhaans.
3. Balneological classification of the natural medicinal water springs.
4. Scientific substantiation of medicinal effects of the freshwater springs.
5. Recommendations to the Tuvinian Government on health tourism development of the Natural Medicinal Water Springs.

Á. Lorberer

8.21 Lake Spring Hévíz – The Greatest Thermal Spring of Hungary

The natural thermal lake of Hévíz is the largest thermal spa in Hungary and now belongs to the World Heritage. It is one of the greatest thermal lakes in the world. It is supplied by subaqueous thermal springs. The curative effects of this medicinal thermal water was known by the Romans as some remains found on this site date from the Roman period, however, no bathing cult existed here in the Middle Ages. The therapeutic-balneological development of this spa started at the turn of the 18th and 19th centuries, after the drainage and reclamation of the area, and the present-day building complex of the bath was constructed.

Lake Hévíz, with Lake Balaton, is one of the main base levels and discharge areas of a large karstic water system of the Transdanubian Middle Range (Fig. 8.58a,b). Lake Hévíz is supplied by thermal springs issuing at the bottom of the lake from a cave (Fig. 8.59). The spring cave "Plózer István" has been formed in an Upper Pannonian (Pliocene) quartzose sandstone formation along a neotectonic fault-line striking a nearly N to S direction (Fig. 8.60).

Thermal springs at 17 °C and 41.5 °C issue onto the bottom of the cave, whereas the colder components of Lake Hévíz come directly from the neighboring Keszthely Mts. The source of the warm components are mainly from the Bakony Mts. (from the surroundings of Sümeg and Pápa, 23–66 km to NE from Hévíz) which flow across the subsided and buried Mesozoic basement of the NW-foreground of the Transdanubian Middle Range. Within the neighboring area of Lake Hévíz, a block system of the Triassic basement (50 to 200 m depth below the ground surface) are overlain by Miocene and Pliocene littoral basal conglomerates. Within this block system of Triassic rocks a "warm water lens" can be encountered. Above and below this lens, colder water occurs (Fig. 8.60). The cold and warm karstic waters near the deep part of the lake are believed to be mixing with near-surface waters from the Pannonian waters and with phreatic waters from peat-bog deposits. Only 3% of the total discharge can originate from water in the near-surface formations.

Due to the indirect character of water flow from the thermal component, changes in discharge of the spring, and changes of infiltration from precipitation to the open karstic terrains is small and delayed. A direct impact on the water discharge is caused by changes in the spillway level of the lake and the yield of thermal wells drilled in the vicinity since 1908 that tap the "Pannonian thermal water lense". Since the 1970s, a water-decreasing trend has been observed in response to overdraft from the karstic system of the Transdanubian Middle Range. The decrease in discharge from Lake Hévíz has been caused by water withdrawal from the bauxite mines in the Nyiràd area about 30 km NE from the lake spring. The withdrawal of water from the mines at 300 m^3/min between 1978 and 1981 will affect for the coming decades the pressure and piezometric conditions of the karstic aquifer within the SW part of the Transdanubian Middle Range. At the end of 1990, all of these bauxite mines were closed and the water withdrawn from karstic water was decreased to 50 m^3/min for drinking purposes only (Fig. 8.58b).

The discharge of the spring of Lake Hévíz, based on measurements made at the end of the last century, was about 600 l/s (that are 36 m^3/min). The average discharge in the years between 1951 and 1970 was 540 l/s which decreased during 1988 to below 300 l/s. Since 1991, the discharge has continuously increased and in 1995/1996 exceeded 400 l/s. Simultaneously, the discharge of the thermal water from wells around the lake varied from 30 to 40 l/s. The decrease and increase in discharge are directly proportional to the karstic water level around the lake (Fig. 8.60) as determined by regional measurements of pressure effects related to karstic water withdrawals.

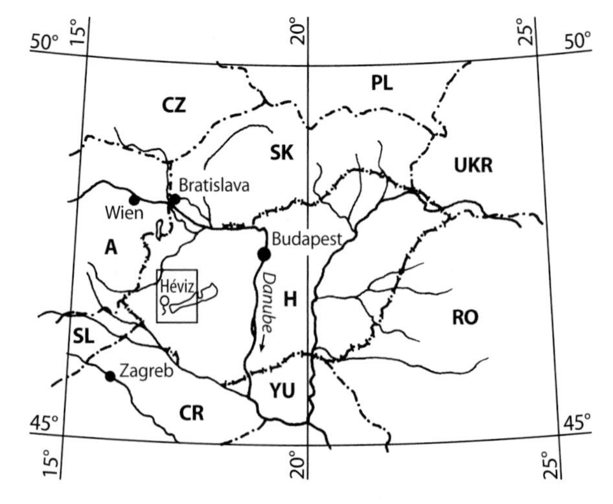

Fig. 8.58a. Location map of lake spring Hévíz

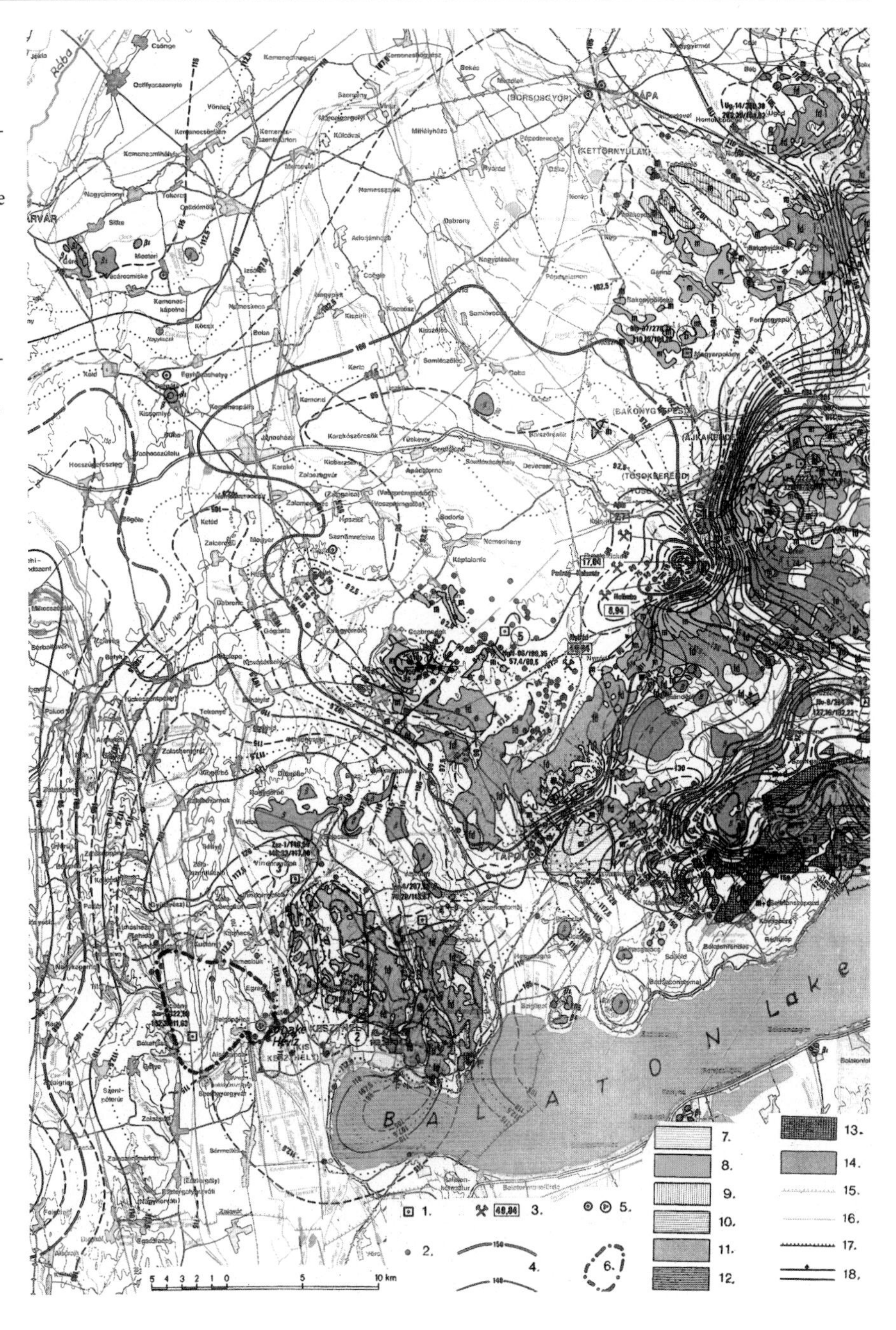

Fig. 8.58b.
Karst hydrogeological map around Lake Hévíz.
1–2 Observation boreholes of the Mesozoic "main karst" aquifer; *3* mean karstic water withdrawal of mine in the year 1992 (m^3/min); *4* piezometric surface of the "main karst" aquifer on 01.01.1993 (m a.s.l.); *5* karstic thermal water producing wells; *6* geothermal inversion due to heat-convection in the Pannonian aquifer of Lake Hévíz; *7–13* outcropping karstic formations (*7* Miocene, *8* Eocene, *9* Upper Cretaceous, *10* Jurassic, *11* Upper Triassic, *12* Middle Triassic, *13* Lower Triassic); *14* neovolcanics; *15* subsurface boundary of the Mesozoic formations of the Transdanubian Middle Range; *16* neotectonic lineaments localized by remote sensing; *17* upthrust-faults; *18* normal faults

The lake water has a chemical character of Ca-Mg bicarbonate with minor Na content along with a total hardness of 17.5 to 19.5. The total dissolved solid content of the water ranges from 515 to 800 mg/l. The ranges of components are as follows: $Na^+ + K^+$ 26.5 to 52.0; Ca^+ 55 to 90; Mg^+ 30 to 45; SO_4^- 15 to 95; Cl^- 15 to 115; HCO_3^- 260 to 440 mg/l. The free CO_2 content is highly variable and ranges from zero to a maximum of 225 mg/l. The sulphide content is relatively low, a maximum of 0.4 mg/l, as it is reduced to sulphur and forms a film on the sur-

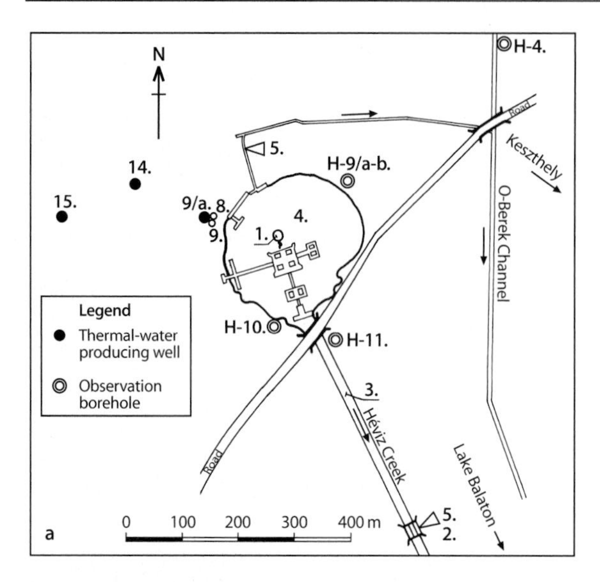

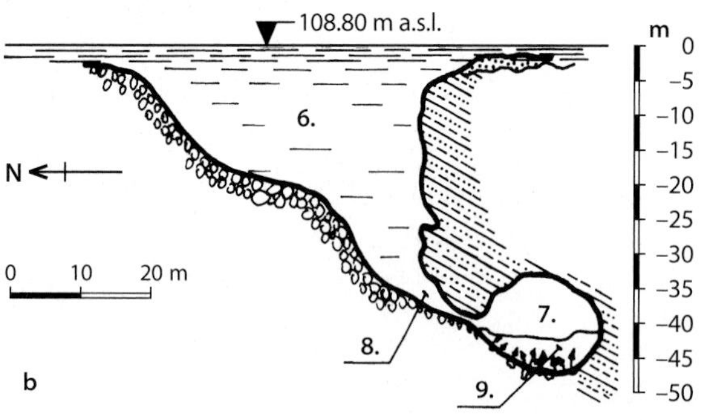

face of the lake. In the deeper part of the lake, pyrite is also formed under the influence of bacterial activity. Objects lost by the bathers many years ago have been enveloped by pyrite. Although direct changes in the chemical composition of the water have not been observed in the past 90 years, the decrease in discharge during the 1980s has caused some hygiene problems during summer months.

The isotope composition of the lake water differs from that of the nearby shallow water from the Pannonian formation. In the warm water component of the lake-spring, the radium content is significant (^{226}Ra activity is of 225.0 mBq/l). In contrast, in the cold karstic waters, the uranium isotope is dominant. Among the dissolved gases, radon, in relation to thoron, is the most significant (^{222}Rn activity: 2.2 mμCi/m^3). The cold spring waters (17 °C) issuing from the cave are determined, according to ^{14}C-dating, to be 5 900 years old, while the warm spring waters (41.5 °C) are 11 100 years old. The thermal waters from wells along the westside of the lake (Fig. 8.59) are between 12 000 and 16 000 years, old whereas cold karstic waters along the eastern side of the lake are 900 to 3 000 years old as determined by ^{14}C-dating.

The bottom of the lake is covered by slightly radioactive peat mud (50% organic material). The smooth silica needles of the "medicinal mud" have a good effect on the blood supply of the human skin and it is applied in the lakehospital for treatment (medicinal mud pack).

Fig. 8.59.
a Map showing Lake Hévíz. **b** A cross section of its spring cave: *1* spring cave "Plózer István"; *2* station for measurement of discharge; *3* drain channel; *4* lake bath; *5* sluice; *6* "crater"; *7* spring orifice; *8* warm springs; **c** aerial photograph showing Lake Hévíz (by Vituki, Argos Studio, 13. Dec. 1982)

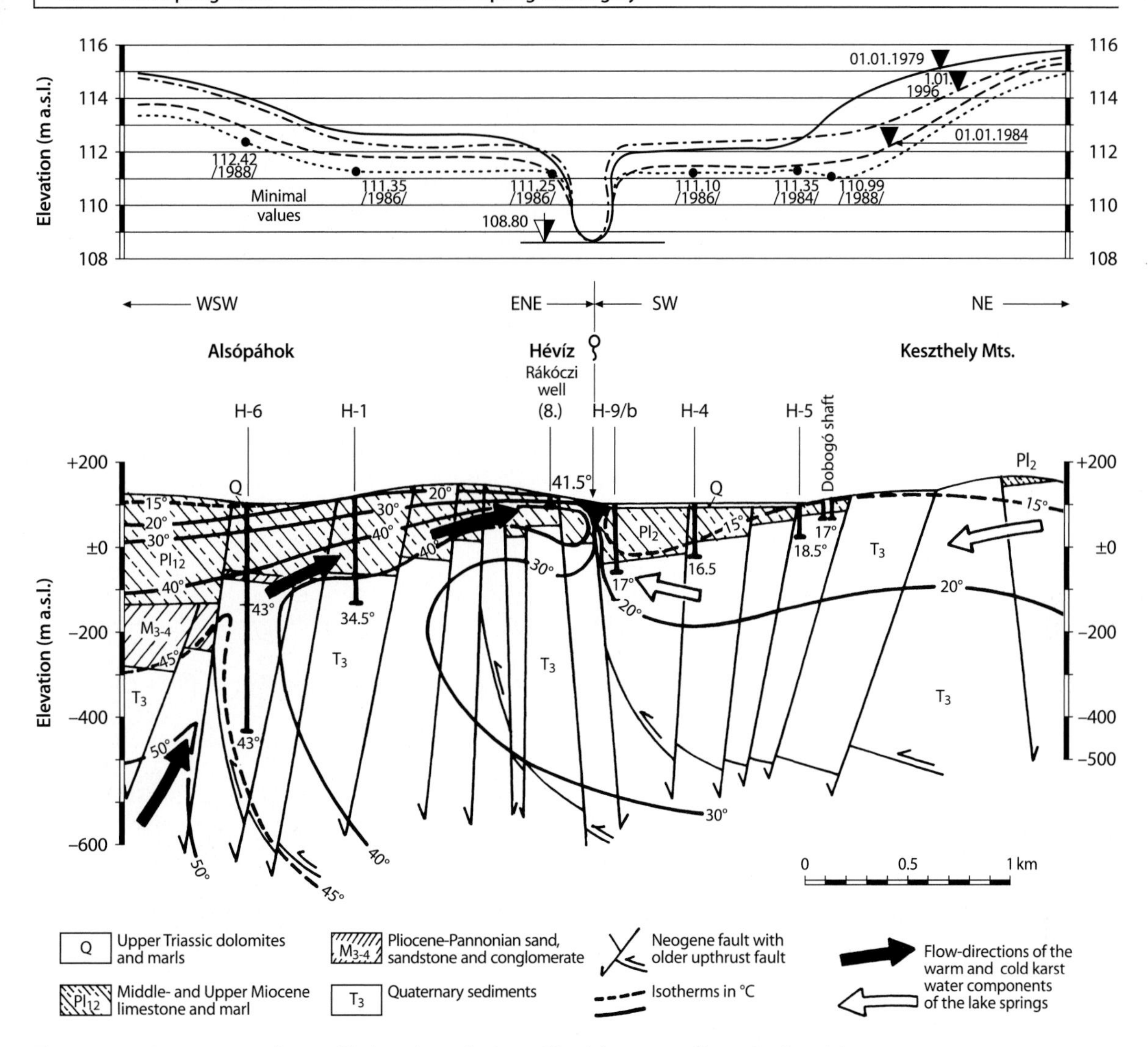

Fig. 8.60. a Piezometric surface profile. b Hydrogeologic profile of the surroundings of Lake Hévíz

The temperature of lake water varies as a function of the air temperature from 22 to 36 °C. The annual mean water temperature in the area is 29 °C. It is worth mentioning that in spite of the great difference between the average temperature of the lake and that of the cave springs, the temperature varies only 1 to 2 °C as measured at many points on the lake surface with the exception of the water in the immediate center of the crater or deeper part of the lake. This is explained by the vertical current from the springs – a cave channel affected by a lateral convectional current with greater speed in the lake. The uniform distribution of water temperature and water quality in the lake is governed by this permanent turbulence. The motion of the peat mud is also governed by this current. Opposite the mouth of the spring cave, a part of the mud is slipping down along the slope of the crater accumulating at the bottom from which it ascends with the vertical warm water flow (Fig. 8.59).

The typical Nymphearoses on Lake Hévíz do not represent endemic plants; they were transplanted from a small thermal water spring lake near Oradea 345 km east of Hévíz.

QuDengNiMa Mineral Water Company

8.22 The Mineral Spring Water of QuDengNiMa, Tibet

QuDengNiMa is a beautiful place in the north of the Himalayas. In the middle of the 8th century A.D. the famous Buddhist Monk – Lian Hua Sheng Master in India was invited to Tibet to propagate Mi Buddhism. Temples and towers were built in the area, as recorded in Tibetan Buddhist Sutra. From then on, QuDengNiMa became one of the famous Buddhist holy lands in Tibet.

QuDengNiMa mineral spring is 5 km south of QuDengNiMa temple, and the springhead is 5 128 m a.s.l. It is said that the Lian Hua Sheng Master distributed the spring water to his believers to cure diseases, and it is believed that the water can cure 360 kinds of acute or chronic diseases and 420 kinds of infectious diseases. The spring has been called "magic water" by Zang Nationas as well as by people in India, Nepal, Sikiim and Bhutan. Every spring and autumn, millions of devout Buddhists arrive at BuDengNiMa to show their respect to Buddha and take the "magic water".

In August 1990, the mineral water was approved by the government of the Xizang Autonomous Region, and in October 1992, it passed certification by the authorized national departments. The QuDengNiMa mineral water meets standard technical demands and contains various kinds of healthful elements, such as zinc, lithium, calcium iodine, and selenium. It is believed that if this mineral water is taken regularly, it will be good for your health. It is reported that this mineral water has a special curable effect on diseases of the stomach, eyes, and the gynecoid organs. The QuDengNiMa mineral water has been bottled and placed on the market, and accepted by a mass of consumers (Fig. 8.61).

Fig. 8.61. Label of the bottled QuDengNiMa mineral water

L. Melioris

8.23 Mineral Waters at Dudince Spa

The mineral and thermal waters at Dudince flow from the inner side of the Western Carpathians arc, the south-western margin of the central Slovak Neovolcanics, along the Levice spring line. The mineral and thermal waters of Dudince, with its physico-chemical composition and content of gases are used extensively for bathing therapy.

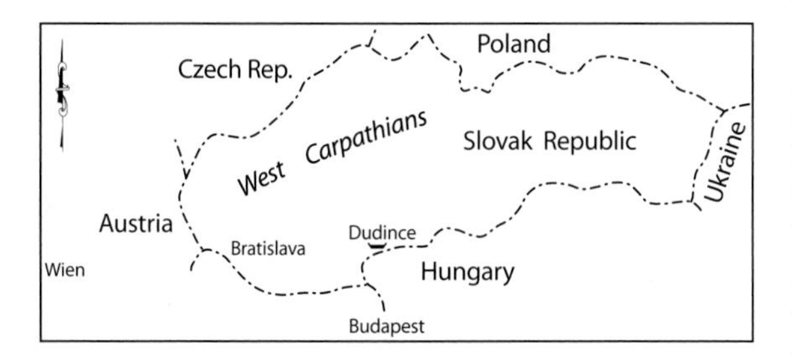

Fig. 8.62. Location of Spa Dudince

The spa at Dudince is in the basin of the river Štiavnica on its left bank at an elevation of 139 m a.s.l. (Fig. 8.62). The average annual precipitation at Dudince is 616 mm, the annual total evapo-transpiration (E) is 464 mm, and the annual air temperature over 10 °C. Beginning in 1953, research on the hydrogeology of Dudince was carried out and after a decade of observation it was determined that the total output of the Dudince Spring was regulated at a consumption of 17 l/s. It was determined that the optimal discharge rate is 6.9 l/s.

The mineral waters of Dudince, with 22 small basins cut into the travertine, are called Roman Baths and were used by Roman legions. In the book *Ungarns Kurorte und Mineralquellen* (1859), D. Wachtel described 9 springs with temperatures of 13–19 °C at Dudince, and described the water as containing iodine, alkaline-saline, hydrogen sulfide acidulous water. According to Tognio (1843), the waters contain Na_2CO_3, $MgCO_3$, NaCl, Na_2SO_4, $CaSO_4$, Si, I, H_2S, and free CO_2. The basin, Merovce Spring, was established in 1990, and according to some information, Dudince mineral water was bottled in white, thick walled bottles, however, the bottling of mineral water at Dudince has stopped.

Table 8.20. Chemical composition of mineral water "Dudince type" (source: Dudince borehole S-3, depth 57.2 m)

Cations	mg/l	(mmol/l) z^+	%	Anions	mg/l	(mmol/l) z^+	%
Li	3.40	0.4899	0.6	F	0.83	0.0436	0.0
Na	810.00	35.2311	47.0	Cl	567.20	15.9986	21.3
K	121.30	3.1021	4.1	Br	1.0	0.0125	0.0
NH_4	2.15	0.1191	0.1	J	<0.10	–	0.0
Ca	497.79	24.8398	33.1	SO_4	509.79	10.6135	14.1
Mg	132.78	10.9237	14.5	NO_2	<0.01	–	0.0
Fe	0.09	0.0032	0.0	NO_3	<0.50	–	0.0
Mn	0.16	0.0058	0.0	PO_4	<0.01	–	0.0
Sr	9.66	0.2204	0.2	HCO_3	2953.27	48.4006	64.4
Ba	0.03			CO_3	0.00	0.0000	0.0
Summary	1577.33		99.6		4032.09		99.8
H_4SiO_4	21.80			Free CO_2	1421.23		
Fe total	0.13			H_2S	5.44		
HBO_2	35.72						
Mineralization (mg/l): 5667.07							

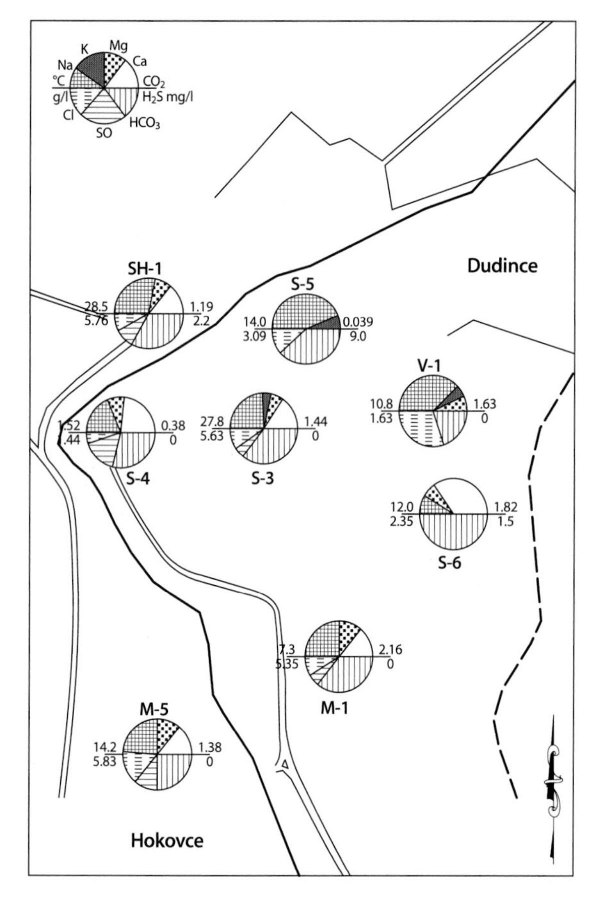

Fig. 8.63. Situation of boreholes at Dudince and the chemical composition of mineral waters

The water at Dudince has a variable representation of ions with mineralization of 5–6 g/l and a temperature of 27–28 °C. The water is of the Na-Ca (Mg)-HCO_3-Cl-(SO_4) type. The content of free CO_2 varies around 1.4 g/l, to an extreme content of 1.8 g/l, and the content of H_2S from 6 mg/l, to an extreme of 10 mg/l. Among trace elements, there are concentrations of B, Br, Li, Rb, Cs, Sr, Ni, As. The chemical composition of the waters of the Dudince type are given in Table 8.20.

The following hydrogeological model of Dudince Spa area is the most probable. The infiltration area of the mineral waters is in the foreland of the central Slovak Neovolcanics, where a complex of volcanic and sedimentary rocks, plus gravels of the basal Badenian formation create favorable accumulation and transport of mineral waters. A proportion of underground waters from Mesozoic carbonates is supposed. Rocks underlying the Neogene at Dudince are believed to be directly connected by fault tectonics and hydraulic connections with the younger rocks. A recent borehole near Hontianske Moravce found mineral water of the "Dudince type" in Mesozoic carbonates (Fig. 8.63).

The mineral spring area at Dudince issues from a marginal block north of Gestenec where there were natural springs in the past. Mineral waters flowing from the northeast dammed by the horst of Permian rocks that enables their significant concentration (Fig. 8.64 and 8.65). The origin of the mineral waters corresponds to their chemical composition. The basic processes of which the result is the presence of the Ca-Mg-HCO_3, Ca-SO_4, and Na-HCO_3 components are

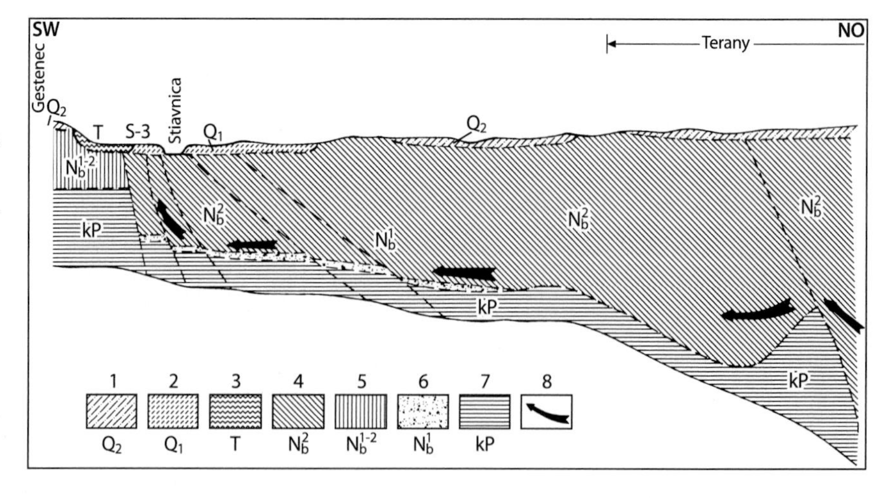

Fig. 8.64. Schematic section through exsurgence area of Dudince mineral waters. *1* Eluvial-deluvial sediments; *2* fluvial sediments, travertines; *3* epiclastic volcanic rocks – tufitic sandstones – weak intergranular-fissure permeability; *4* lower Middle Badenian Plastovc beds – tufitic and epiclastic volcanic claystones, weakly permeable to impermeable; *5* Lower – Middle Badenian; *6* quartzitic gravels and sands with good permeability ($k_i = 2975.10^{-6} - 81616.10^{-1}$); *7* Permian – variegated shales and arcoses; *8* direction of mineral water inflow into the exsurgence area

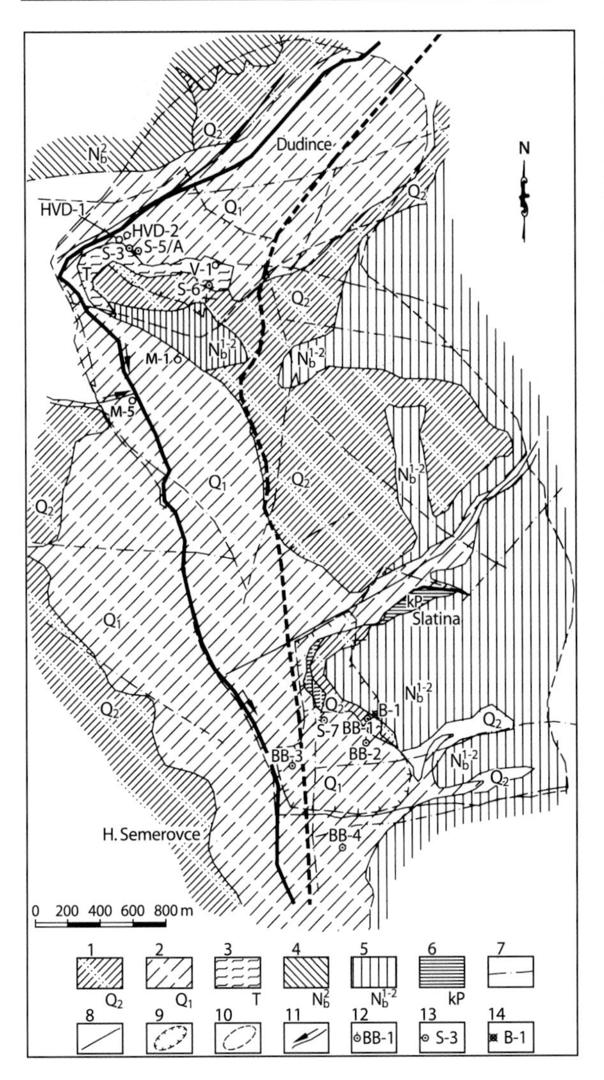

the dissolution of carbonates, sulphates and the hydrolytic breakdown of silicates. As a result of the processes of ion exchange, the components $Na\text{-}SO_4$ (from original $Ca\text{-}SO_4$) and $Na\text{-}HCO_3$ appear. Oxidation-reduction processes also have an essential role in the creation of the chemical composition of the mineral waters, especially in relation to the presence of hydrogen sulphide. Apart from this, the important activities in the mineralization processes are thermodynamic conditions of the circulation of waters and their gasification by carbon dioxide.

Sufficient uncertainty remains at present on the origin of the salt component of the mineral waters, which apart from Na-Cl and $Na\text{-}HCO_3$ also accompany some elements such as Br, B, or others. Their presence in the water indicates the influence of marinogenic mineralization.

Fig. 8.65.
Geological conditions of surroundings and protection area of Spa Dudince. *1* Eluvial-deluvial sediments – loess and loessy loams (Holocene – Pleistocene), low- to impermeable; *2* fluvial sediments – clayey and sandy loams and sands (Pleistocene and Holocne) small thicknesses of creek alluviums (up to 15 m), loamy, low permeable; *3* travertines (Quaternary and Neogene) – occurrence of travertines indicates recent, or old abandoned springs of natrual or mineral waters; *4* epiclastic volcanic rocks – tufitic sandstones, low intergranular-fissure permeability; *5* Lower – Middle Badenian – Plastovce beds – tufitic and epiclastic volcanic claystones, low permeable to impermeable; *6* Permian – variegated shales and arcoses; *7* tectonic lines inferred, or observed (one evidence only); *8* boundaries of geologic wholes; *9* protected areas of natural therapeutic sources of the 1st order; *10* protected areas of natural therapeutic sources of the 2nd order; *11* surficial courses; *12* well and mineral water of Slatina type; *13* well with mineral water of Dudince type; *14* liquidated well

Table 8.21. Content of secondary and trace elements in "Dudince type" mineral water

Element	Average content (mg/l)	Element	Average content (mg/l)	Element	Average content (mg/l)	Element	Average content (mg/l)
As	0.121	Cr	0.002	Pb	0.008	V	0.0001
Ba	0.036	Cs	0.120	Rb	0.170	F^- max.	4.00
Be	0.013	Cu	0.008	Sr	7.800	Br^- max.	7.64
Cd	0.005	Li	3.080	Ti	0.0009	I^- max.	0.93
Co	0.012	Ni	0.010	Zn	0.111	B^- max.	10.60

The analyses were carried out in the laboratory of the Geological Institute of Fac. of Natural Sci., Comenius University, Bratislava.

Among the mineral waters of Slovakia, the Dudince water has elevated concentrations of alkaline metals, as well as strontium, nickel and arsenic. Among the halogenides, elevated contents of bromides are typical. From the therapeutic point of view, the increased concentrations of fluorides, bromides, lithium, strontium, and arsenic are important. Other elements present in the water and their beneficial effects on the human organism are still not clearly proved (Table 8.21).

The relative connection between the representation of the salt component, the representation of chlorides, and the content of boron in the Dudince water is not as clear as in the waters of the neighboring structures. The values of the ratio of Cl/Br is also variable and is in the range of 1 200–1 400. Water from the borehole, M-5, belonging to the Slatina structure, with a significant influence from deep salt waters, has a value which varies around 450.

Characteristic of the "Dudince type" mineral water is the relatively strong gasification with carbon dioxide and hydrogen sulphide (Table 8.22). This is a rare phenomenon in the Western Carpathians. The content of free CO_2 varies around 1.4 g/l, reaching 1.8 g/l, and H_2S around 6 mg/l, with an extreme of 10 mg/l.

The radioactivity of water content of Ra is 36.3 pCi/l. The age of the water has been determined as 28 000 years based on ^{14}C-dating. Values of K N_2/O_2 over 10 indicate a strongly reducing environment, along with the presence of hydrogen sulphide as well. A value of K He/Ar, equal to 0.02, indicates extensive water exchange, which correlates with the results of hydrogeological relations. Tyćler (1975) mentions the presence of the gases, indicating the transition between intensive and limited water exchange.

The mineral water of the "Dudince type" as based on a long period of record (Table 8.23), has a stable chemical composition, with the exception of its gas content which is variable. The balneological characteristic of this mineral water can be described: natural, medium mineralized, Na-Ca, HCO_3-Cl, overgassed with CO_2 and H_2S, lukewarm, hypotonic.

Table 8.22. Gases in mineral water (quantity %) from Dudince S-3(2) (Tyćler 1975)

CO_2	99.74
N_2	1.8×10^{-1}
O_2	4.0×10^{-3}
Ar	1.2×10^{-3}
He	3.0×10^{-4}
H_2	4.0×10^{-4}
CH_4	5.9×10^{-1}
C_2H_4	–
H_2S	6.0×10^{-2}
Total gas content (ml/l)	3 231.75
Temperature (°C)	28
Phase ratio gas/water	25
Pressure (Mpa)	–

Table 8.23. Chemical composition from a long-term point of view

Year	Source	Na^+	K^+	Ca^{2+}	Mg^{2+}	Cl^-	SO_4^{2-}	HCO_3^-	Analyst
1836	Neoznačený	696	22	627	182	404	568	1 614	B. Wehrle
1893	Hlavný pr.	851	129	498	133	564	559	1 965	B. Lengyel
1957	Vrt S-3		1 097	459	137	597	567	3 216	IGHP Žilina
1986	Vrt S-3	819	121	491	131	539	522	1 972	IGHP Žilina
1995	Vrt S-3	810	121	498	135	567	510	2 953	INGEO a.s. Žilina

References

Franko O, Bodiš D (1989) Paleohydrogeology of mineral waters of the Inner Western Carpathians. Zapadne karpaty ser Hydrogeologia 8, Geologicky ustav D. Stura, Bratislava, pp. 145–163

Hyánková K, Melioris L (1993) The unusual chemistry of mineral waters at Dudince. Geologica Carpathica, Bratislava, 44(2): 123–131

Melioris L, Vaas D (1982) Hydrogeological and geologicla relations of the Levice spring line. Zapadne karpaty ser Hydrogeologia 8, Geologicky ustav D. Stura, Bratislava, 4:7–56 (in Slovak)

Tyćler J (1975) Dudince – Slatina – tracing of gas in underground water. Geol Pruzkum Ostrava

PHILIP E. LAMOREAUX

8.24 Les Bouillens, Source Perrier, Nimes, France

8.24.1 Introduction

Les Bouillens or "Source Perrier" is a spring in the Vistrenque plain in the Vergez Township of France, 17.7 km from Nimes, 34.4 km from Montpelier, and 19.3 km north of the Mediterranean Sea (Fig. 8.66). Les Bouillens, bubbling waters, was so named by the French peasants because water from this natural spring bubbled upon reaching the surface as dissolved gas was released to the atmosphere.

Early studies concerned with volume and physical characteristics of the water indicated that the spring naturally discharged at a rate of 22 l/s and had a constant temperature of 16 °C (61 °F), staying warm all year. Studies of origin and chemical characteristics of the water indicated that its high temperature was caused by some water coming in contact with intrusive volcanic rocks at great depths. This is summarized and published in a report of the Academy of Sciences of Paris.

Since 1962, studies of the origin, source, occurrence, and movement of groundwater at Les Bouillens have been carried out, including data collected from about 200 observation wells, hundreds of water samples analyzed, and a piezometric map prepared to delineate the recharge area and determine the direction of movement of groundwater in the Villafranchian geological system which supplies the spring (Fig. 8.67). The most recent hydrogeologic map of the Nimes area was prepared by the French research agency Bureau des Recherches Geologique et Minieres by Bayer and Poul (1975).

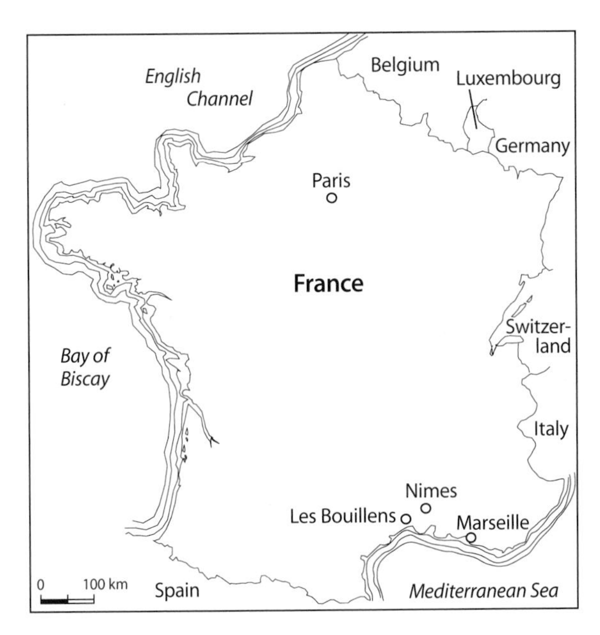

Fig. 8.66. Location of Les Bouillens, France

Les Bouillens produces bottled water that is sold as Perrier Water. The spring occurs in a unique hydrogeologic setting from alluvial beds of the river system underlain by older bedrock of dolomitic limestone and marl which is underlain by volcanic basalt beds. The gas and temperature, and some chemical characteristics of the water at Les Bouillens come from association of CO_2 gas and groundwater deep underground associated with limestone in contact with igneous intrusive rocks that migrate upward to a shallow bed of alluvial sediments. The silica content of the spring water is derived from shallow groundwater in contact with the siliceous sands and gravels in the Villafranchian alluvial aquifer. The gas, which is 99.7% carbon dioxide (CO_2), also contains traces of nitrogen and five rare gases. Isotope studies by Professor Fortes of the University of Orsay, determined that for water samples collected on 31 January 1979, and as stated in his report, the CO_2 was primarily the result of thermo-carbonization which confirmed the earlier geological findings that thermal action of igneous magma occurred associated with limestone at depth. Dr. Fortes states, "The analyses of ^{14}C and ^{18}O indicate that the gases in the samples come from the thermo-decomposition of the deep carbonate rocks decomposed by the intrusion of magmatics known to occur in this region along the fault of the Nimes."

8.24.2 History of Les Bouillens

The original spring, Les Bouillens, bubbled up in a small basin. The carbon dioxide gas bubbles rose to the surface of the spring and escaped to the atmosphere; the water then discharged along natural drainage to the Vistre River. Early man and animals used the spring waters, and it is reported that Hannibal's army used Les Bouillens. The oldest archaeological evidence dates the spring to Roman times. In a report on the water of Vergeze, M. Gobley provided a chemical analysis of the spring water dated 13 January 1863. On 23 June 1863, Napoleon III, through an Act by the Secretary-Minister of Agriculture and Commerce, authorized the production of mineral water from Les Bouillens for medicinal

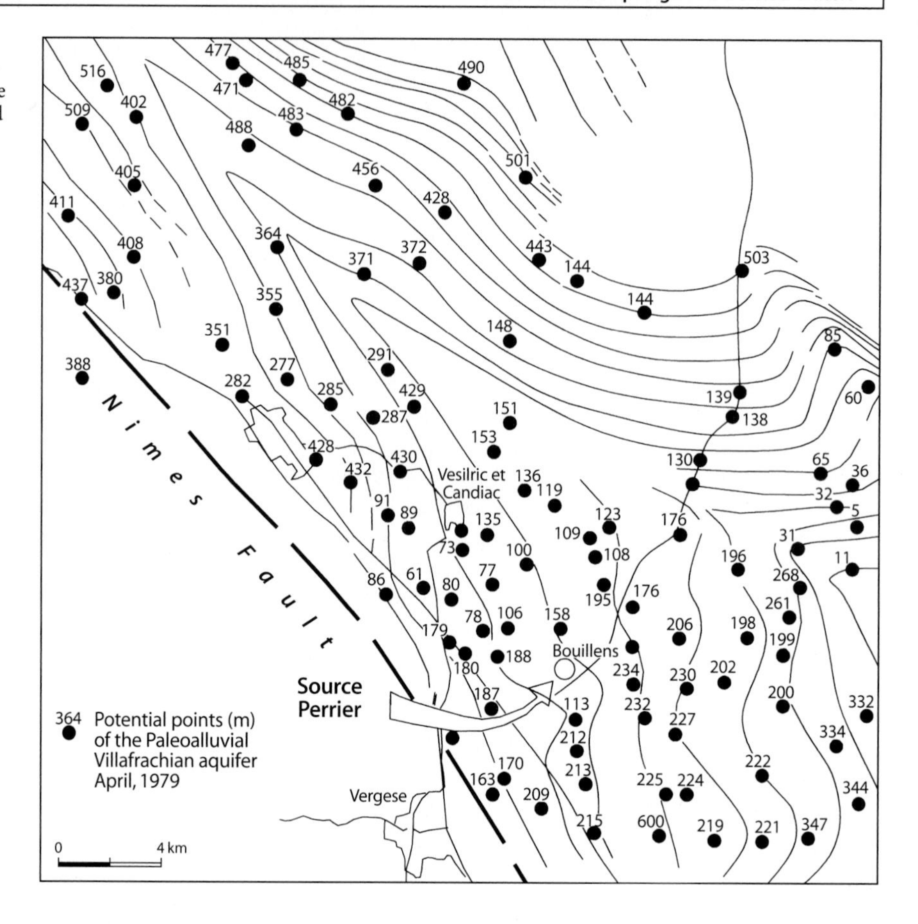

Fig. 8.67. Piezometric map of the Villafranchian aquifer in the Vistre River valley, France (modified from BRGM Bayer 1975)

use. Beginning in 1898, Dr. Perrier studied the use of the thermal-mineral water, and in 1906 the "Compagnie De La Source Perrier" was founded. In 1914, two million bottles were produced and in 1964, 260 million.

Prior to about 1890, the spring water flowed into a basin and was captured along with the gas in a unique system of cone-shaped retorts (Fig. 8.68). With the increased demand for Perrier Water, a well was drilled to the source of the water that bubbled up in a natural spring, to increase production (Fig. 8.69). In 1956 the process of monitoring water and gas through a system of scientifically located and constructed wells brought a high degree of sophistication to the bottling of the natural carbon dioxide gas and water. The present-day Perrier water supply is from wells drilled into the artesian Villafranchian Formation and derives water from a screened interval 10.9 to 22 m in depth. Wells are lined with 300-mm diameter casing and filter screens opposite sand and gravel in the same formation that the original Source Perrier spring issued from. The water produced has a constant temperature of 16 °C (61 °F) throughout the year. In April 1964, Dr. J. Avias found the natural artesian flow of the well to be 6.3 l/s.

The water that discharges from Les Bouillens is a natural product of the earth's hydrological cycle. Water from rainfall on the Cretaceous limestones of the uplands, Garriques, begins its slow movement downward under the force of gravity into the labyrinth of cracks, crevices, and solution cavities in the limestone. Part of this water, charged with calcium bicarbonate from association with the limestone, migrates downward along deep faults associated with the Vistrenque Graben and comes in contact with deep limestone beds near a hot magmatic zone in the earth beneath Les Bouillens. Carbon dioxide gas forms in this thermal zone and then moves upward into the shallow artesian aquifer. The aquifer at Les Bouillens provides water presently bottled and sold to over 100 countries around the world as Perrier Water.

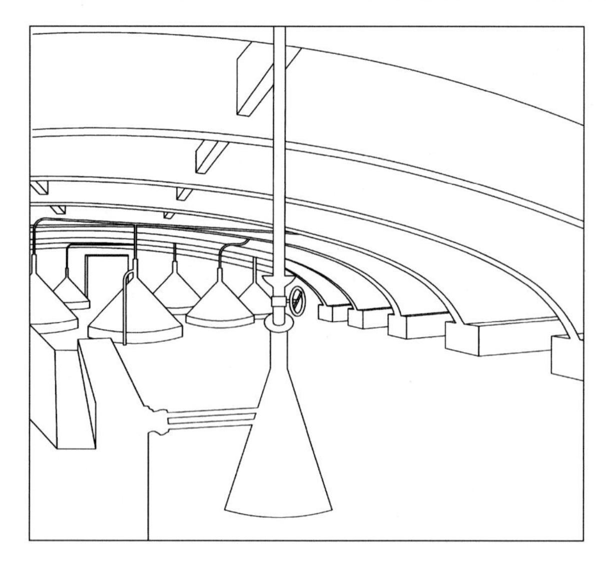

Fig. 8.68. Schematic of retorts originally used to capture carbon dioxide-charged spring water at Les Bouillens, France (schematic about 1890)

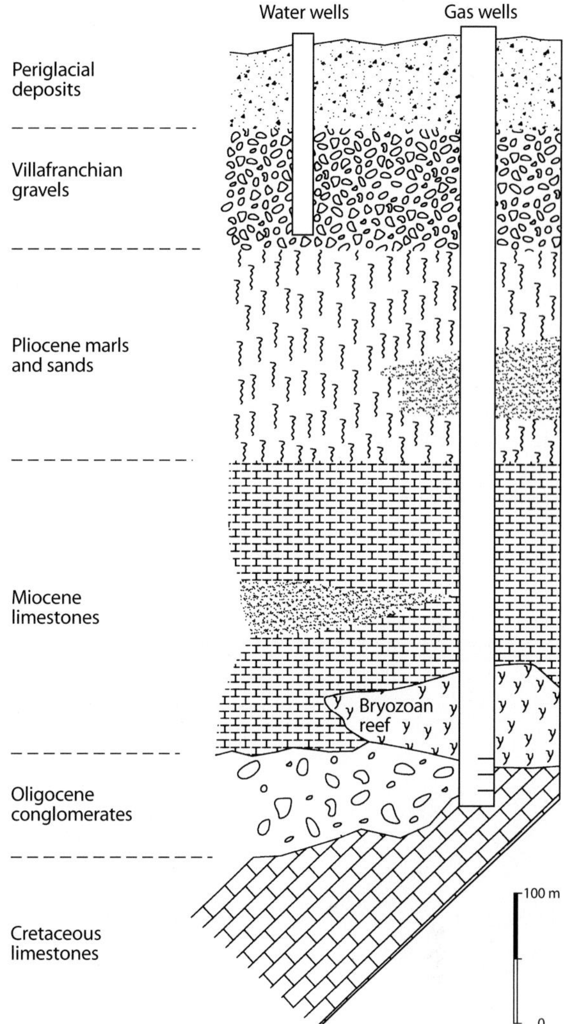

Fig. 8.69. Stratigraphic section penetrated by wells at Source Perrier

References

Avias J (1979) Note sur les charactéristiques géologiques et hydrogéologiques de la source Perrier (Vergéze - France) et des captages actuels d'eau et de gaz aux fins d'embouteillage, 4 p

Avias J, Drogue C, Dijon R, Paloc H, Salvayre H, Verdeil P (1964) Documents sur l'hydrogéologie karstique en territoire periméditerranéen. Mémoires du C.E.R.G.H. Vol. I, 127 p

Bayer F, Poul X (1975) Carte hydrogéologique de la Vistrenque. Bureau de Recherches Géologique et Minières 75 SGN 220 LRO, Annexe 1

Bureau de Recherches Géologiques et Minières (1965) Chronicle of hydrogéology. Bureau de Recherches Géologique et Minières, Paris, No. 7, 148 p

Bureau de Recherches Géologiques et Minières (1971) Carte géologique à 1:50 000 de Sommieres (France), Map XXVIII-42, with explanatory text, 19 p

Bureau de Recherches Géologiques et Minières (1970) Carte géologique à 1:50 000 de Lunel (France). Map XXXVIII-43, with explanatory text, 20 p

Centre d'Êtudes et de Recherches Hydrogéologiques, Institut de Géologie Faculté des Sciences (1966) Rapport de synthese sur les recherches entreprises et les résultats obtenus (géologiques et hydrogéologiques) sur le Périmétre de la Source Perrier. Montpelier, France, Dossier 42

Dole RB (1906) Use of fluorescein in the study of underground waters. US Geol. Survey Water-Supply Paper 160, p. 73–85

Drouge C (1963) Essay on delimitation of the basin of recharge of a spring of the drowned karst of Languedoc. Annales Spéléologiques 18(4):409–414

Drogue C (1963) On the repartition of a karst of waters of a rainfall of constant intensity. Annales Spéléologiques 19(4):631–634

Gobley M (1863) Rapport sur l'eau de Vergéze. Seance du 13 Janvier 1863, p. 264–265

Paloc H (1965) Hydrogeological research in collecting and management in calcareous regions. Information drawn from several recent observations. Bureau de Recherches Géologiques et Minières, Chronique d'Hydrogéologie 7:87–109

Paloc H (1979) Presentation des études hydrogéologiques en cours sur le réseau karstique de la source du Lamalou et le Causse de l'Hortus. Excursion géologique du 12 Novembre 1978, Extrait des Annales de la Société d'Horticulture et d'Histoire Naturelle de l'Hérault 119(1–2–1979), 12 p

Passavy J (1976) Inventaire des sources thermo-minérales de Languedoc-Roussillon et esquisse d'une application géothermique des eaux chaudes: Théses doctorat, Sci de la Terre, Géologie appliquée, Académie de Montpelier, Université des Sciences it Techniques de Languedoc

J. G. Zötl

8.25 Badgastein Spa of Austria

8.25.1 Location

The communities of Badgastein (47°14' N, 13°08' E; 930 to 1 083 m a.s.l., population of 5 600), Bad Hofgastein (869 m a.s.l.), Dorfgastein (830 m a.s.l.) and Böckstein (1 331 m a.s.l.) occupy the area of the Gastein Valley, a political district of Salzburg country (Fig. 8.70 and 8.71).

Being the most important spa in Austria, Badgastein has 7 519 guest beds, among them 2 769 in four- and five-star hotels. Badgastein is easy to reach by car, rail or plane; bus and taxi transfer to the next airport, Salzburg (see Sect. 8.26.7). The importance of Badgastein is attributed to its thermal springs and the Badgastein-Böckstein geothermal heat and radon gallery which are considered to be special geological phenomena.

Fig. 8.70. Gastein Valley; view N-S; background "Hohe Tauern"

8.25.2 History

Gastein thermal springs are mentioned for the first time in a document which reports the healing of the emperor Friedrich III after a hunting accident in 1350. In 1725 Gastein thermal waters were described in medical remarks by Paracelsus. The use of the springs by people of importance began in the 19th century. The high aristocracy included Kaiser Franz Josef I (Austrian-Hungarian monarchy), the German emperors Wilhelm I and Wilhelm II, the Czar of Russia and Bismarck and King Carol of Rumania had villas or apartments at Badgastein. Other well-known visitors include Grillparzer, Schubert, Toscanini, and Thomas Mann.

The most important step in the development of the spa was the discovery of the radon in water from the thermal springs by H. Mache, M. Curie and A. LaBorde during the second half of the 19th century.

In 1936 the Research Institute of Gastein, later Research Institute of the Austrian Academy of Science, was established. Under the long-term director F. Scheminzky (†1979), the institute was very active in radon research. The success of the spa and Badgestein-Böckstein heat gallery in treatment of rheumatic diseases made the spa well known worldwide.

8.25.3 Geology

With the Tertiary era, the elevation of the Alps became a dominant agent. Despite drastic local tectonic events, the general elements were the uplift which began with the Oligocene and is still going on. Its rate is episodic as is the intensity of denudation and erosion. A. Tollmann

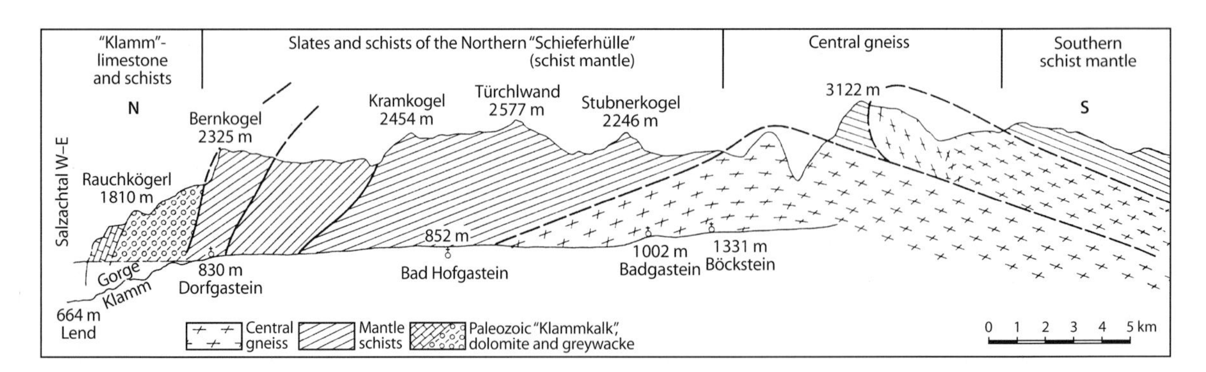

Fig. 8.71. Geological profile of Gastein Valley. Central gneiss window with remnant of mantle schist

(1986) estimates the amount of denudation around the Tauern region west of Gastein as much as 11 000 m since the Upper Oligocene.

The fundamental tectonic rules together with the difference of rock types of nappes and basement resulted in the separation of the Central Alps and the Northern Limestone Alps. In the Central Alps the continuing uplift caused the denudation of the "Schieferhülle" (schist cover). Thus, in wide areas, the "Zentralgneise" (Central gneiss), the oldest rock of the basement, built up the highest peaks of the Eastern Central Alps (Großglockner 3 797 m a.s.l.). But all of these areas are surrounded by rocks of the former schist cover. The geologist therefore gives these areas the term "window" (Fig. 8.71).

Gastein lies within the "Tauern window". The Central gneiss has the quality of a fine crystalline rock with the joint-systems necessary for ore enrichment and (hot) water circulation.

8.25.4 Hydrology

The individual thermal springs of Badgastein discharge from the main joint system or bedding joints of the Central gneiss.

Since the uplift of the Central Alps goes on in the form of vaulting, the main joints are tension joints, and are open to great depths. They provide the avenues for infiltration of rain and melt water, as well as rising gas and hot water and the formation of zones of mineralization (ancient gold mines).

All the thermal water of Badgastein is meteoric water from its infiltration around the "Reed See" (Fig. 8.72) that outflows at Gastein thermal springs. The age of the water is 3 600 to 3 800 years as calculated (Job and Zötl 1969).

The temperature of the thermal water does not depend on the thermal gradient of the region (1 °C/49 m) but on the ascending heat, hot water and gases from the earths mantle, as shown by traces of mantle helium. This indicates open joint systems down to a depth of 30 km.

The total discharge from the developed Gastein thermal spring (No I–XIX), is 4 500 m^3/d (52 l/s), captured directly at the joints of the bedrock.

The water from the springs is a mixture of hot water from great depths and cold water. The mixing occurs in a joint zone at a certain depth. This is shown by a lower tritium content than that of precipitation and shallow groundwater. Table 8.24 shows the percent of cold water in selected thermal springs.

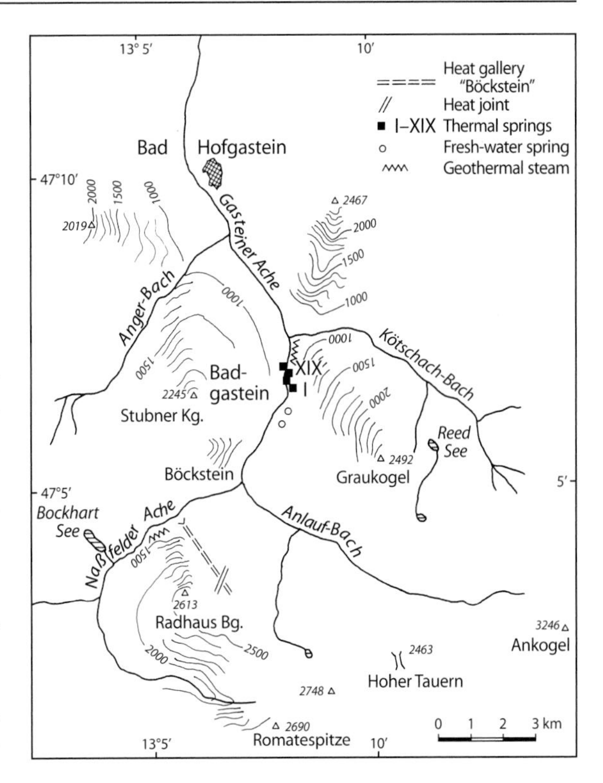

Fig. 8.72. Gastein region: location of Badgastein with thermal springs. Reed See infiltration area, Böckstein heat gallery

The high temperature of Gastein Springs is one of the special features of the spa. Another unusual characteristic is the radon content of the thermal water. Radon (^{222}Rn) is a daughter of radium (^{226}Ra) from the uran-thorium-radium succession row of radioactive elements born in the core of the earth. Being a gas of 3.8 days half-life it emanates from radium into water and air. Consequently, radon is introduced into the human body mainly by respiration and a minor amount through the skin. The question with regard to being helpful or toxic depends on the quantity. Radioactivity of Gastein thermal springs lies between 4.8×10^{-9}Ci/l (= 4.8 nCi/l) and 124.0 nCi/l (spring No. X, Table 8.24). Distinctly influenced by altitude are tritium content, mineralization and temperature. The Rn-content goes its own way (depending on the different Ra traces in the joints). Spring No. XIX discharges at the lowest position on the Gastein cascade from rock fall material, the great part of the water comes from the Gastein cascade.

Table 8.24. Chemical and physical characteristics of various springs at Badgastein Spa

Spring No.	Altitude (mNN)	Freshwater percentage (%) T^a	Na	Cl	SO_4	Average	Thermal water temp. (°C)	Spring discharge (m^3/d)	Radon content (*n*Ci/1)	Remarks
I	1 034	5.4	6.2	5.7	6.5	6.5	45.6	195	4.8	Not used
VIIa	1 005	6.5	–	–	–	–	41.4	18	53.5	3 Outflows
IX/1,2,3	995	13.3	11.3	12.1	10.5	11.8	46.3	68 – 1 887	21.0 – 51.0	5 Outflows
X	983	21.7	21.7	23.3	19.3	21.15	33.3 – 37.0	14	60.6 – 124.0	
XII	975	24.0	26.0	24.9	23.8	24.7	39.5	378	52.8	
XIV	968	23.8	25.8	27.6	22.8	25.0	36.8	106	39.6	
XVIII	954	38.7	39.5	37.1	36.1	37.8	23.0	?	3.5	Not used
XIX	937	75.8	72.8	62.8	68.0	70.0	16.1	261	0.19	Not used

[a] Calculated from the tritium content (1967; 608 TU = 100% in freshwater).

8.25.5 The Gastein-Böckstein Thermal Gallery

In 1940 a last effort was started to reactivate the gold mining in the Central gneiss sequences of the Radhausberg SW of Böckstein (Fig. 8.73). The entrance to the gallery is at an altitude of 1 280 on the NW slope of the Radhausberg. The total length of the gallery is 2 425 m, where the mining was terminated.

At that time, only old miners were available to work on the project. A remarkable number of them had the unexpected experience of being cured of rheumatism and arthritis, painful illnesses which had bothered them for decades.

In the 1940s people of the research institute in Gastein began systematic ivestigations. Measurements of the air temperature showed the highest value (ca. 45 °C) at that point where the gallery crosses the main joint system, gallery meter 1 999 (Fig. 8.74).

The zone of the highest air temperature is around that of a joint set between the gallery meters 1 600 and 2 000. The rock temperature of the heat joints also reaches 45 °C. Towards the inner part of the gallery, the air and rock temperatures decrease, though those in the overlying rock increase. The radon content of the air in the gallery has an average of 4.99 nCi/l air. The gallery became the most important place for treatment of rheumatic diseases and polyarthritis (Fig. 8.74, station IV in the gallery).

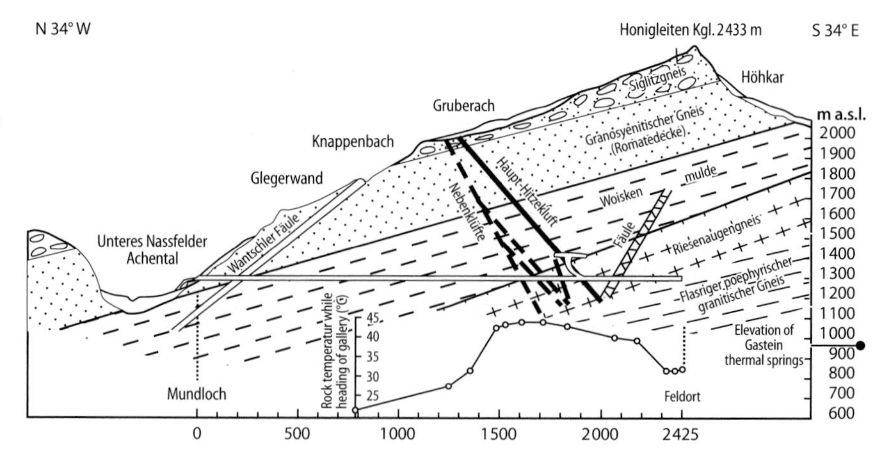

Fig. 8.73. Geological profile of the Radhausberg with Böckstein heat gallery. Pay attention to the main joint system and the scale of temperature in the gallery

Fig. 8.74. Treatment station No. IV in the Böckstein heat gallery. Distance of entrance 2 238 m, air temperature 41.5 °C, humidity 99%

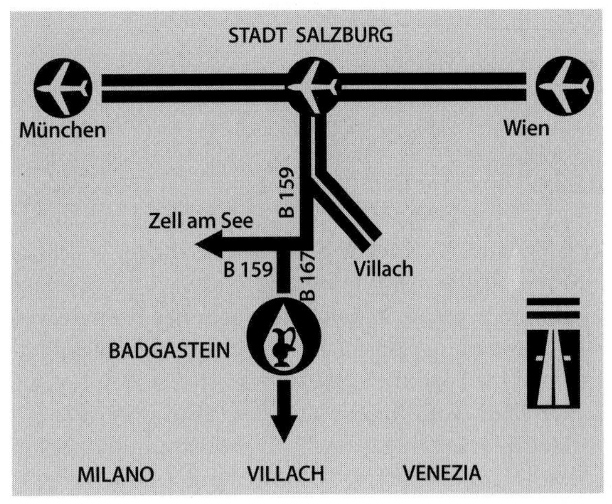

Fig. 8.75. Traffic service

8.25.6 Indications and Medical Treatment: Bottled Water

The knowledge and therapeutic use of radon have developed rapidly. At the spa Badgastein together with Gastein-Böckstein, radon is successfully used in all kinds of chronic inflammatory and degenerative rheumatic diseases including rheumatic nerve and muscle diseases and polyarthritis, collagen diseases like scleroderma and priasis, sequelas after traumatic and infectious paralysis, chronic inflammatory diseases of the respiratory tract, chronic inflammatory adnexa diseases, geriatric symptoms, premenopausal and menopausal disorders, endocrine and vegetative disorders and last but not least, healing of wounds and ulcers. The Böckstein gallery has four treatment stations at different distances from the gallery entrance (Fig. 8.74).

The thermal water of Badgastein has a larger treatment hall as well as thermal water pipes to some of the hotels for individual treatment. Bottled water: registered under "Gasteiner Tafelquellwasser" the former Gasteiner thermal water, cool and purified from the high fluoride content is sold as refreshing drinking water; 1992 capacity of sale 30 million l.

8.25.7 Conventions and Meetings – Recreation and Sports

Badgastein has a well equipped congress center (capacity 1 220 people) with a simultaneous interpreter's system for six channels. For summer relaxation, Badgastein offers 39 tennis courts, a 9-hole golf course, riding excursions and parcours, paragliding with instructions, and more.

Highlights of the wintersports are international ski races (World Cup and Company championships), 250 km of ski runs with 50 lifts and cable ways ranging from 800–2 800 m a.s.l. are available (30% of ski runs are easy ones). Three ski schools offer their assistance (Fig. 8.75).

References

Job C, Zötl J (1969) Zur Frage der Herkunft des Gasteiner Thermalwassers. Steir Beitr z Hydrogeologie 21, Graz

Scheminzky F (Hrsg; 1965) Der Thermalstollen von Badgastein-Böckstein. Innsbruck

Tollmann A (1986) Die Entwicklung des Reliefs der Ostalpen. Mitt Österr Geogr Ges 128, Wien

Zötl JG (1997) The spa Deutsch-Altenburg and the hydrogeology of the Vienna Basin (Austria). Environ Geol 29 (3/4):176–187

Zötl J, Goldbrunner JE (1993) Die Mineral- und Heilwässer Österreichs, geologische Grundlagen und Spurenelemente. Springer-Verlag, Wien-New York, 321 S

B. Blavoux · J. Mudry · J.-M. Puig

8.26 The Karst System of the Fontaine de Vaucluse (Southeastern France)

8.26.1 Introduction

The Fontaine of Vaucluse is the largest spring in France and one of the most important in the world (average discharge 20 m^3/s). Its catchment area has been delineated based on geological, hydrological, and speleological data: the Fontaine is the only discharge point of the Lower Cretaceous limestone series (1500 m thick).

Mediterranean Europe is a mountainous region whose intricate structure is due to the alpine orogenesis. In these mountains, the thick limestone series have been karstified since their emergence: they show typical karst forms along the Mediterranean (Andalusia, Pyrenees, southeastern France, Italy, Dalmacy, Greece).

In southeastern France, the Rhône River (coming from the Swiss Alps, the lake of Geneva) separates the Massif Central basal complex (westward) from the Subalpine ranges (eastward).

About 30 km east of Avignon, a limestone mountain range overlooks the Rhône Valley. To the east and in the south, the Durance Valley, originating from the internal Alps, bounds it. The east-west range formed by Mount-Ventoux, Albion Mountain, and Lure Mountain constitutes the northern part of this mountain range. It overhangs the Jabron and Toulourenc Valleys that separatesit from the Baronnies Mountains. In this defined area, perennial surface run-off is rare and only occurs in the areas where impervious or semipermeable strata overly the limestone (e.g., in the Sault rift, Apt, and Forcalquier Basins).

At the edge of the limestone, several springs, whose average discharge does not generally exceed 50 l/s, combined, drain less than 1 m^3/s. Therefore, almost all drainage of surface water runoff is through only one outlet – the Fontaine de Vaucluse.

8.26.2 Geography and Geology

Based on geography and topography, this area can be divided into four parts from north to south (Fig. 8.76–8.78):

1. The north facing range, with the Mount-Ventoux at an altitude of 1909 m and Lure Mountain at an altitude of 1826 m, the highest relief

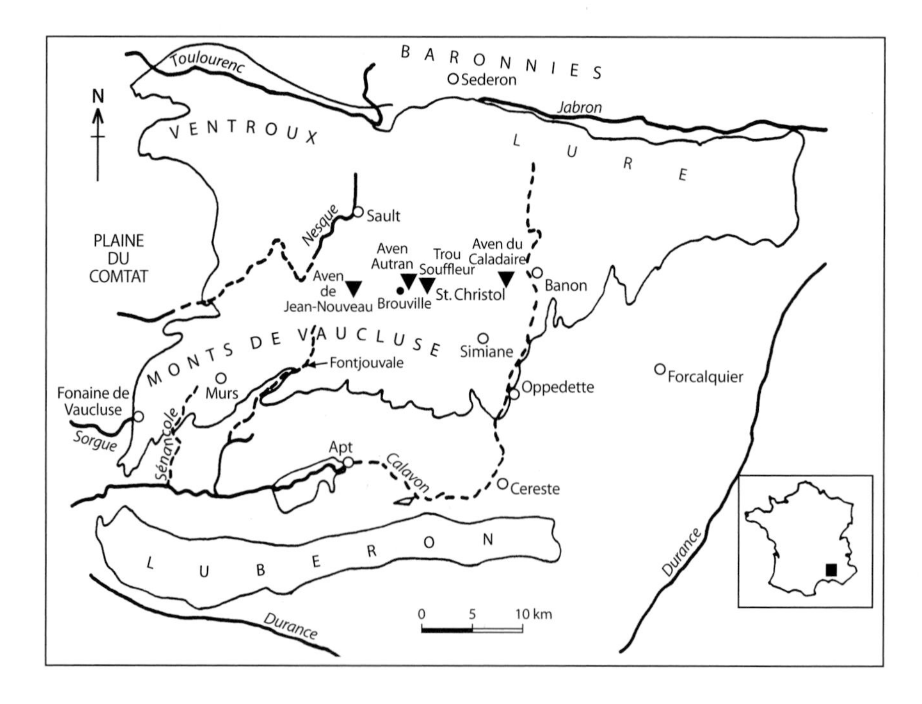

Fig. 8.76.
Geographic sketch map

2. Extending from this range at an average altitude of 600–1000 m are the Vaucluse Mounts and St. Christol Plateau
3. A low zone: the Apt-Cereste Basin
4. A southeastern range: Petit Luberon at an altitude of 719 m and Grand Luberon at an altitude of 1125 m

8.26.3 Climate

The climate is typical of the Mediterranean area, being subtropical with moderate temperatures, mild and rainy winters, hot and dry summers. The distribution of the rainfall, ranges in temperature, and amount of evapotranspiration and seepage are controlled by the altitude.

In this area, the graded altitudes (100–1900 m) are responsible for about +55 mm per 100 m rain gradient, and −0.5 °C per 100 m thermal gradient. These gradients clearly influence the computed effective rainfall (Turc formula with a 20 mm RFU) which is 120 mm on the lowest belt of the intake area and increases to 1380 mm on the upper section (Fig. 8.77).

Thus, the part of the intake area that is situated above the mean altitude provides about three quarters of the yearly effective rainfall.

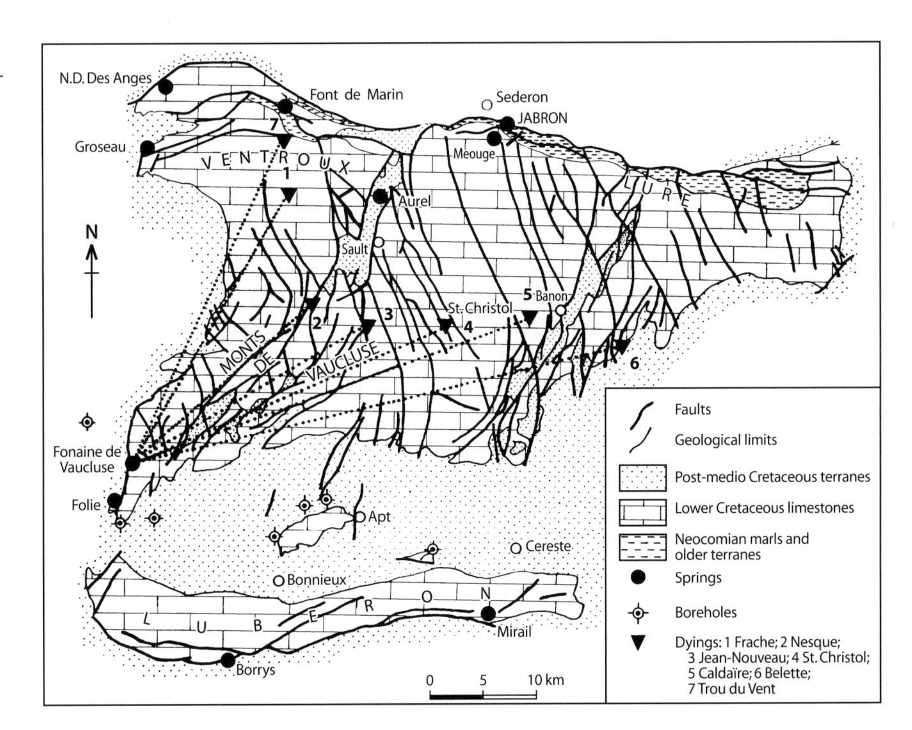

Fig. 8.77. Geological context and hydrogeology

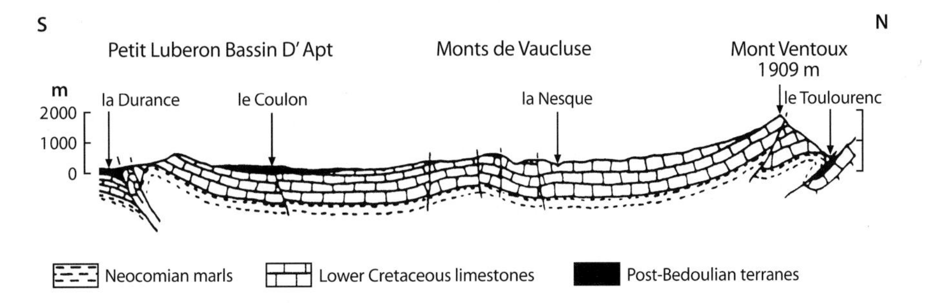

Fig. 8.78. North–south geological cross section

8.26.4 Stratigraphy

The limestone karst series is of Lower Cretaceous age. The substratum of the limestones consist of the Valanginian marls and the Lower Hauterivian clayey limestones. It is overlain conformably by the upper Aptian marls (Gargasian) and unconformably by the sandy and marly Upper Cretaceous and the Tertiary soils.

The upper part of the limestone series is the Upper Hauterivian and the basal part is the Lower Aptian. In the southwestern part of the mountain range, limestone outcrops with the Urgonian facies: limestones with rudistids and perireef bioclastic limestones, whereas in the northeastern part and in the Great Luberon, the lithologic characteristics of the Urgonian has less detritic facies.

The total thickness of the series is as much as 1 000 m in the Lure Mountain and probably exceeds that in the Monts de Vaucluse.

8.26.5 Structure

This area has been affected by alpine tectonics. In the southern subalpine ranges, several major tectonic phases are responsible for the structure. The first major phase, associated with the Pyrenees structuration is of Mid-Cretaceous age. It produced such east-west folds as the Ventoux-Lure and Luberon ranges. The thickness and the competence of these limestones account for this tectonic behavior. This "Haut Provencal block" has transmitted with a minimum of flexible deformations, submeridian constraints, which are responsible for the northward overthrustings of the Ventoux-Lure range, and for the one southward of the Luberon.

Between the south of Mount Ventoux and the south of Luberon, the limestone series only shapes a panel that is a monoclinal with a slight dip to the south on the southern side of Mount Ventoux and Lure Mountain, tabular on the St. Christol Plateau, synclinal under the Apt Basin, and anticlinal on the Luberon. Two groups of fractures (N 30 and N 145) crisscross the plateau. They were active at several times, though generally their vertical and horizontal throw are weak. Along the N 30 family, the Oligocene distension has opened several grabens on the plateau (Sault, Murs, Banon, Simiane). This is also responsible for the normal displacement of the western bounding faults of the Monts de Vaucluse. In the grabens, the limestones are covered with the Upper Cretaceous and Tertiary impermeable formations.

M. Maggiore · F. Santaloia · F. Vurro

8.27 Frido Springs – San Severino Lucano Territory (Southern Italy)

The San Severino region (Fig. 8.79) lies on the meridional Lucanian Apennine in southern Italy. Several springs occur in the San Severino Lucano area. The recharge areas of this springs differ based on the lithology (Fig. 8.79) and other hydrogeological features (Table 8.25).

The groundwater system of the Fosso Arcangelo, Timpa della Gatta and Caramola Springs consist of metaophiolites. The Timpa della Gatta Spring drains continental rocks (amphibolites) with the associated serpentinized peridotites. The Acquafredda and Frido Springs are fed by a water circulation through the Mesozoic calcareous-dolomitic sequence (Alburno-Cervati Unit).

The metamorphosed pelitic-quartz and arenites (Frido Unit) are an aquiclude and represent the impermeable unit for all the springs studied. The various relationships between the permeable and impermeable units distinguish the different types of spring sources (Fig. 8.80).

The Frido Springs are important for the volume of water discharged, maximum discharge is 520 l/s, and for their water quality. Some of the water has been channeled off by the Apulian Aqueduct and supplies a part of the Lucanian region. The remaining amount of water supplies the town of San Severino Lucano. Their index of variability proposed by Meinzer (1923) is equal to 0.3, therefore these springs are considered as having a subvariable regime.

They consist of 10 single water sources at the foot of a steep fault (oriented in an approximate north-south direction). The contact between the permeable Alburno-Cervati Unit and the impermeable metamorphic sequence of the Frido Unit follows this structural feature. The springs overflow this tectonic barrage. They are typical springs in the southern Apennine area where the groundwater system occurs in a limestone-dolomite environment where the water flows are obstructed by the presence of overthrusted impermeable units.

The Frido hydrogeological basin extends to the top of Mt. Pollino (Fig. 8.79) and the catchment area is calculated to be 25.9 km^2. The average altitude is 1 674 m a.s.l.

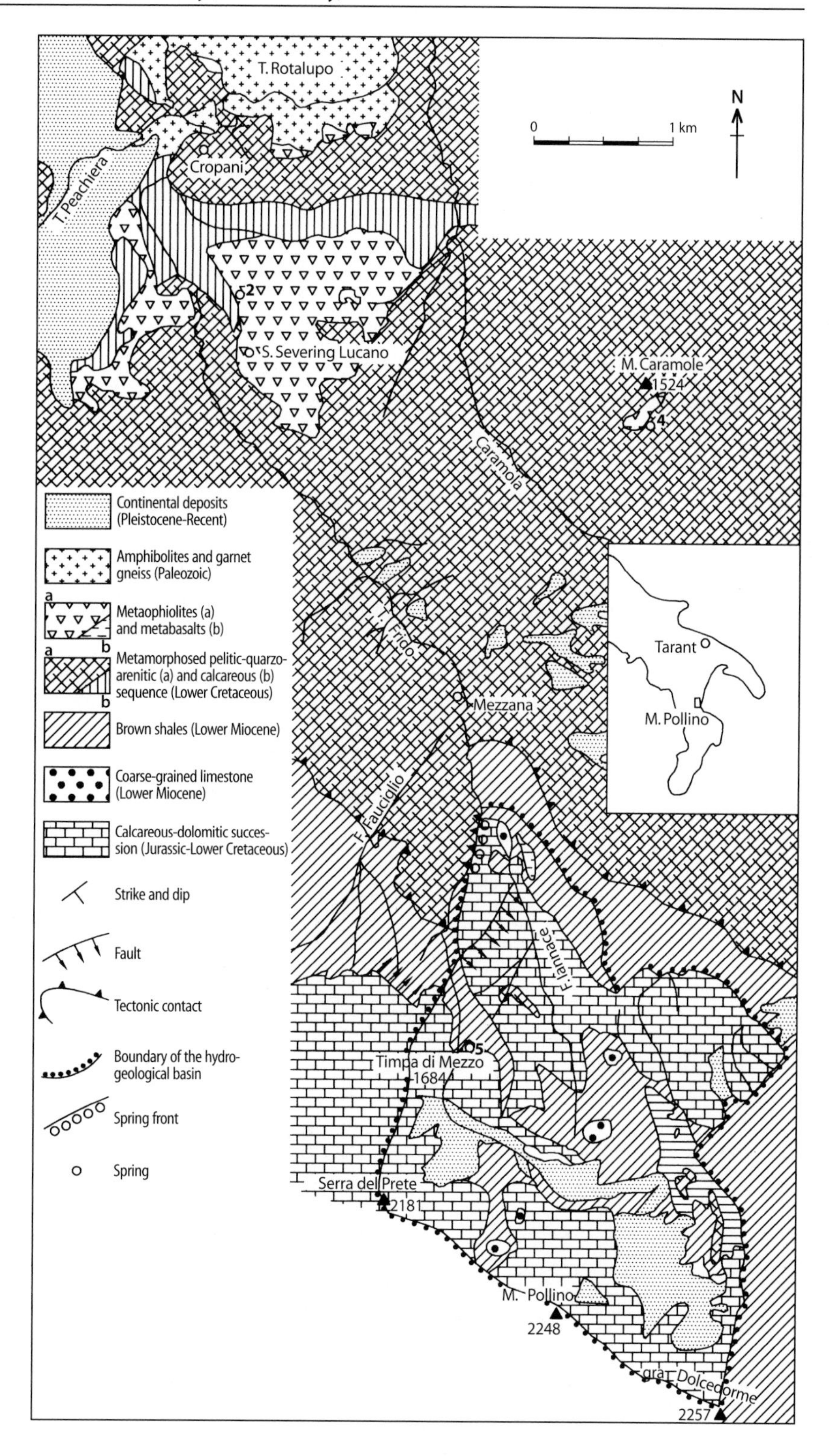

Fig. 8.79.
Map of the study area showing locations of the springs studied

Table 8.25. General features of springs studied

No.	Spring	Altitude (m)	Intake work	Spring type	Discharge Q (l/s)	Temp. (°C)		Aquifer lithology	Boundary rocks lithology
						Air	Water		
1	Timpa della Gatta	725	Chamber	Contact	1	10	12	Amphibolites	Pelitic schists
2	Fosso Arcangelo	780	Trench	Contact	1	14	12	Serpentinites	Metacalcareous layers
3	Timpa della Guardia	1025	Chamber	Contact	1	–	–	Serpentinites	Pelitic schists
4	Carambola	1400	Chamber	Contact	1	7	6	Serpentinites	Pelitic schists
5	Acquafredda	1280	Trincea	Barrier spring	3	12	7	Dolomitic limestone	Shale
6A	Frido	1035	Chamber	Barrier spring	3	5	6	Dolomitic limestone	Pelitic schists
6B	Frido	1028	Chamber	Barrier spring	3	5	6	Dolomitic limestone	Pelitic schists
6C	Frido	1031	Chamber	Barrier spring	3	5	6	Dolomitic limestone	Pelitic schists
6D	Frido	1026	Tunnel	Barrier spring		5	6	Dolomitic limestone	Pelitic schists
6E	Frido	1024	Tunnel	Barrier spring	494	5	6	Dolomitic limestone	Pelitic schists
6G-H-I	Frido	1026	Tunnel	Barrier spring		5	6	Dolomitic limestone	Pelitic schists

Note: The discharge and the temperature are expressed by average values.

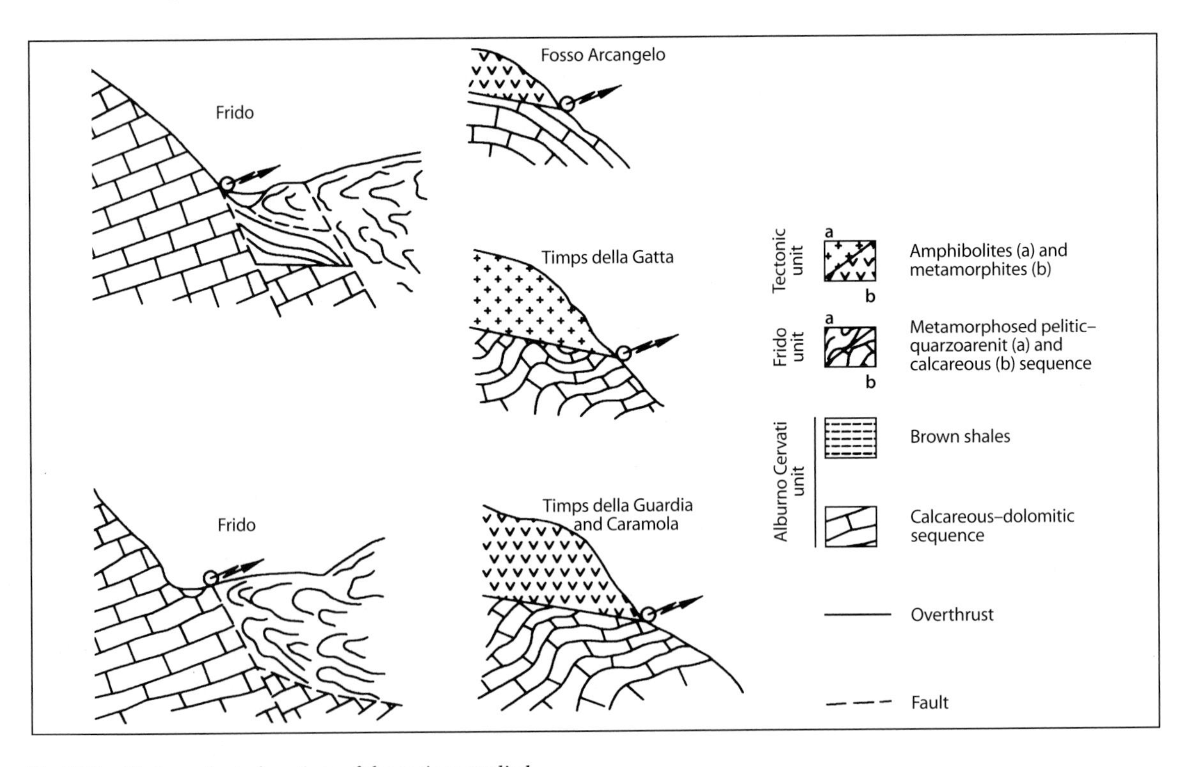

Fig. 8.80. Hydrogeological sections of the springs studied

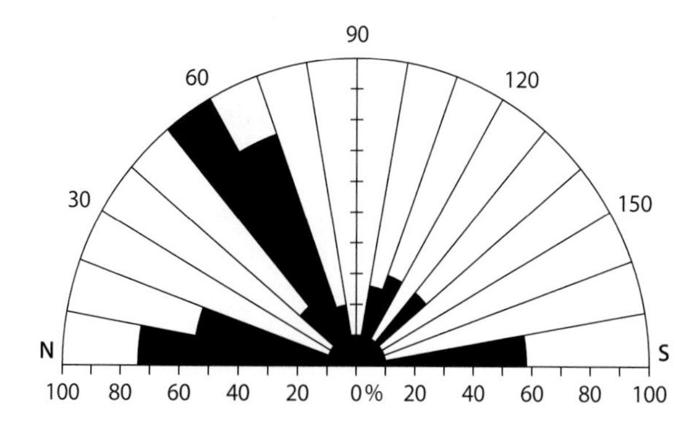

Fig. 8.81. Frequency diagram for the fracturing state of the Frido Springs aquiufer

The permeability of the aquifer feeding the springs is controlled by secondary porosity from dense fracturing. The direction of the structural openings is approximately N 60, N 120, and N 180 and is shown by the frequency diagram (Fig. 8.81). Major permeability occurs primarily in a N-NE direction. Assuming that these openings continue to a substantial depth, they become the preferential flow zone of least resistance for rainwater to seep into the rock, thus recharging the Frido groundwater system.

Other springs in the area have minor discharge values (Table 8.25) and their hydrogeological basins are not large. These springs are used for local water requirements. The hydraulic conductivity of the aquifers has an average value of about 2×10^{-3} cm/s, which is primarily of secondary origin, however, it must also be related to the extreme weathering of the rocks. The recharge to the hydrogeological basins of the Timpa della Gatta, Caramola and Acquafredda Springs is influenced by the continuous and dense forest covering their recharge areas.

This area is representative of a Mediterranean climate, with rainy winters and dry summers. July and August show the lowest values of rainfall and the highest temperatures. Maximum precipitation occurs during the period from October to December, the coldest months.

The average annual rainfall during the period of 1978–1989 for the San Severino Lucamo region was 1 538 mm (Table 8.26). During this period, annual rainfall values ranged from a maximum of 2 353 mm in 1980 to a minimum of 1 077 mm in 1982. A fairly long period of extreme dry weather occurred from 1987 to 1989. This

Year	*P* (mm)
1978	1 431
1979	1 948
1980	2 353
1981	1 324
1982	1 077
1983	1 509
1984	1 906
1985	1 527
1986	1 478
1987	1 298
1988	1 305
1989	1 301
Average value	1 538

Table 8.26. Average annual rainfall (1978–1989) for the San Severino area

pluviometric anomaly influenced the groundwater system because the aquifers are never fully replenished. The lowest discharge of, 225 l/s, for the Frido Springs was recorded in December 1990 (Table 8.27).

The hydrograph in Fig. 8.82 illustrates the close correlation between rainfall and the outflows of the Frido group. The springs highest discharges occur after the rainy season with a lag time of about 5 months. The highest discharge of 477 l/s, recorded in March and April of 1986, are related to the rain that fell on the recharge basin during December of the previous year. This recorded lag time can be attribued to the existence of deep water circulation.

Figure 8.83 shows three recession lines determined for the Frido Springs. Their recession coefficient is 0.0012 l/d. The period of May 1987–January 1988 has been considered in calculating the value of the effective infiltration of the basin. The result was shown to be 442 mm or 32% of the rain (1 402 mm) that fell during the previous recharge period. The hydraulic balance of the Frido Springs drainage area is based on a long observation period 1938–1958 (Table 8.28).

The infiltration has been calculated indirectly considering the value of the runoff from the examination of some physiographic features of the land in question.

Table 8.27. Average monthly discharge (l/s) of Frido Springs intercepted by the Apulian Aqueduct

	1982	1983	1984	1985	1986	1987	1988	1989	1990	1991
January	440	350	387	449	413	348	321	332	280	285
February	425	342	383	451	419	400	336	324	276	285
March	415	347	384	453	477	397	360	361	273	310
April	440	374	398	456	477	398	393	379	281	400
May	435	376	448	460	476	416	408	392	285	440
June	442	374	400	464	468	414	401	380	285	450
July	430	362	478	494	443	398	388	364	285	520
August	415	337	471	488	403	381	368	346	271	495
September	382	322	456	471	403	362	351	327	256	450
October	375	309	433	472	379	345	333	314	245	430
November	355	324	405	462	365	327	332	296	230	410
December	378	349	440	430	349	327	349	293	225	410
Annual average	411	347	424	463	423	376	362	342	266	407

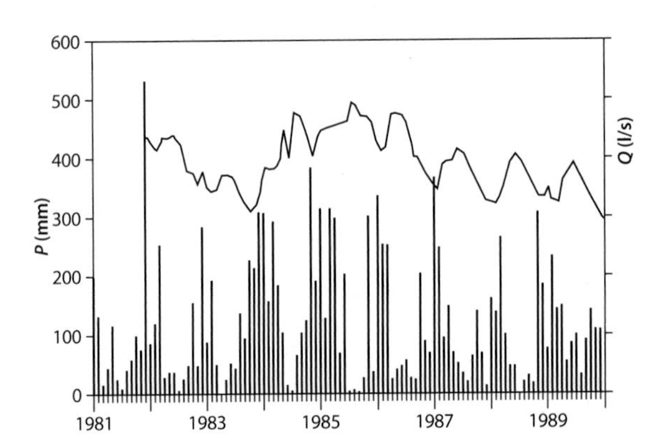

Fig. 8.82. Hydrogram for the Frido Springs: the rainfall *P* is represented by the histogram; the average discharge *Q* by the curve

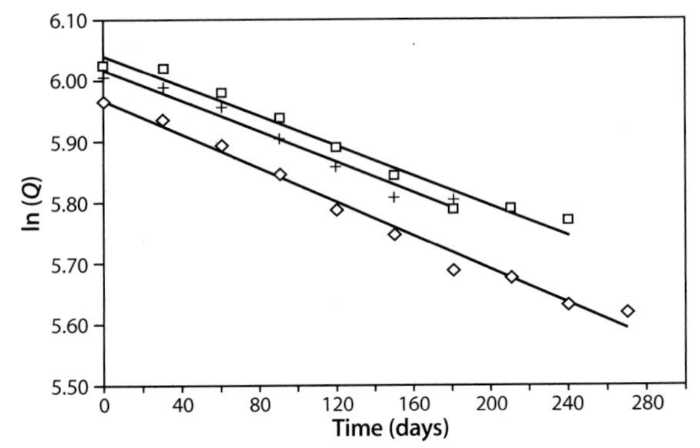

Fig. 8.83. Recession lines for the Frido Springs

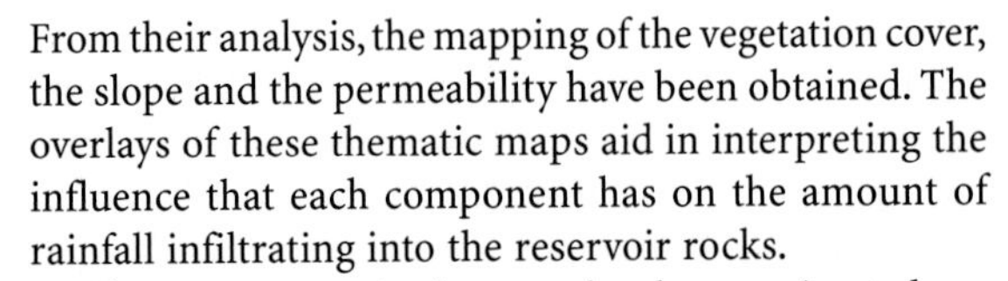

From their analysis, the mapping of the vegetation cover, the slope and the permeability have been obtained. The overlays of these thematic maps aid in interpreting the influence that each component has on the amount of rainfall infiltrating into the reservoir rocks.

The evapotranspiration rate has been estimated using Thornthwaite's method with an aridity index equal to 22.2 and a humidity index of 146.8. The value of the runoff has been calculated using the difference from the

Table 8.28. Hydrogeologic balance for hydrogeological basin of Frido Springs (observation period 1938–1958)

Parameter	Average annual value
Temperature (°C)	8.7
Rainfall (mm)	1315
Evapotraspiration (mm)	585
Infiltration (mm)	279
Runoff (mm)	451

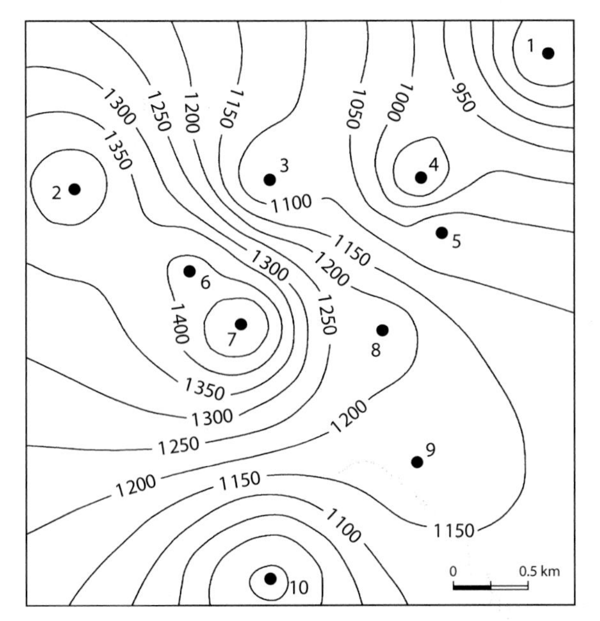

Fig. 8.84. Annual isohyet for the period 1938–1958. The pluviometric stations are: *1* Valsinni; *2* Agromonte; *3* Francavilla sul Sinni; *4* Noepli; *5* Cresosimo; *6* San Severino Lucano;*7* Mezzana di Lucania; *8* Terranova del Pollino; *9* San Lorenzo Bellizzi; *10* Castrovillari

other rates. At this point, the annual rainfall results are 1315 mm (Fig. 8.84), of which evapotranspiration represents 54.8% (585 mm), runoff 21.2% (279 mm), and infiltration to groundwater 34% of rainfall. The values of the infiltration obtained in this study are consistent with those calculated for similar spring basins in carbonatic environment of the central and meridional Apennine.

Discharge response for the Timpa della Gatta, Caramola, Timpa della Guardia and Fosso Arcangelo Springs occurs during and almost immediately (a few days) after intense periods of rainfall. This hydraulic behavior is related to a shallow groundwater flow path. Although meteoric recharge is the same for all of these springs, they have different chemical compositions and the reservoir rock is the only discriminating factor for the water chemistry of each spring (Table 8.29; Meinzer 1923).

References

Meinzer OE (1923) The occurrence of groundwater in the United States, with a discussion of principles, Water Supply Paper 489, 321 p

Table 8.29. Chemical data for the spring waters analysed

Spring No.	Date of sampling	Temp. (°C)	pH	Na^+ (meq/l)	K^+ (meq/l)	Ca^{2+} (meq/l)	Mg^{2+} (meq/l)	Sr^{2+} (meq/l)	HCO_3^+ (meq/l)	SO_4^{2-} (meq/l)	Cl^- (meq/l)	NO_3^- (meq/l)	TDS (mg/l)	P_{CO2} (atm)
1	Jan. 1989	11	7.4	0.29	0.02	1.60	2.14	0.01	3.73	0.04	0.51	0.02	280	0.010
	May 1989	11	7.3	0.29	0.01	1.65	2.22	0.01	3.83	0.11	0.32	0.02	278	0.011
	Jun. 1991	12	7.4	0.30	0.02	1.79	2.59	0.01	4.12	0.11	0.28	0.02	275	0.010
2	Jan. 1989	12	8.2	0.15	0.02	0.92	2.76	0.01	2.49	0.93	0.17	0.02	240	0.001
	May 1989	12	8.1	0.14	0.02	0.80	2.43	0.01	2.17	1.04	0.20	0.03	245	0.001
	Jun. 1991	13	8.2	0.15	0.02	0.94	2.40	0.01	2.61	0.88	0.18	0.03	248	0.001
3	May 1989	12	7.8	0.14	0.01	0.77	1.97	0.01	2.51	0.20	0.21	0.02	190	0.002
4	May 1989	6	8.2	0.30	0.03	1.15	2.04	0.01	2.99	0.20	0.36	0.01	260	0.001
5	Oct. 1989	7	7.8	0.16	0.01	2.70	0.15	0.01	2.54	0.12	0.17	0.02	176	0.003
6A	Jan. 1989	6	7.8	0.11	0.01	2.68	0.61	0.06	3.16	0.02	0.16	0.02	203	0.003
	May 1989	6	7.5	0.10	0.01	2.63	0.58	0.06	3.14	0.02	0.16	0.02	205	0.006
6B	Jan. 1989	6	7.5	0.08	0.01	2.33	1.00	0.10	3.05	0.05	0.11	0.02	215	0.006
6C	Jan. 1989	6	7.4	0.09	0.01	2.29	1.00	0.09	3.04	0.05	0.11	0.02	210	0.008
6D	Jan. 1989	6	7.3	0.09	0.01	2.34	0.97	0.07	3.04	0.05	0.10	0.02	195	0.009
	Jun. 1989	6	7.5	0.10	0.02	2.24	0.98	0.05	2.99	0.20	0.12	0.02	216	0.006
6E	Jan. 1989	6	7.5	0.08	0.01	2.24	0.99	0.11	3.03	0.06	0.11	0.02	220	0.006
6G-H-I	Jan. 1989	6	7.3	0.08	0.01	2.27	1.05	0.10	3.07	0.05	0.12	0.02	215	0.009

G. A. Kellaway

8.28 Environmental Factors and the Development of Bath Spa, England

8.28.1 Introduction

Bath is situated in the valley of the Bristol Avon about 32 km upstream of the confluence of the Avon and the Severn Estuary (Fig. 8.85 and 8.86). In the course of its long and fascinating history, it has suffered many vicissitudes, but in 1978 the spa was closed down owing to contamination of the thermal water by a pathogenic ameba. As a result of this disaster, a series of medical and scientific investigations were carried out during the next decade in order to restore the thermal water to use. The results of this work have recently been described in a book published by the Bath City Council entitled, *Hot Springs of Bath* (Kellaway 1991a). This volume contains work by a number of authors. The account that follows gives a brief summary of some of the results and discusses other environmental and historical factors that have a bearing on the life and development of the city and its spa from Roman times onwards and the conservation of the thermal water sources.

Mesolithic implements provide evidence for the earliest known human occupation of Bath hot springs about 7 000 years B.P. (Cunliffe 1969). Evidence for Iron Age use of the springs is also substantial, although no large structures were built around the springs until an early date in the Roman occupation, possibly about A.D. 60. This marks the first appearance of Bath as a spa, and it survived in this form until the 5th century. There appears to have been no formal settlement during the Dark Ages, when the buildings decayed, but the hot springs were still issuing on the surface. They were restored in some form in the 7th century and continued in use from this time until 1978 when the spa was shut down.

Modern Bath has expanded far beyond its Roman and medieval confines (Kellaway and Taylor 1968). Most of

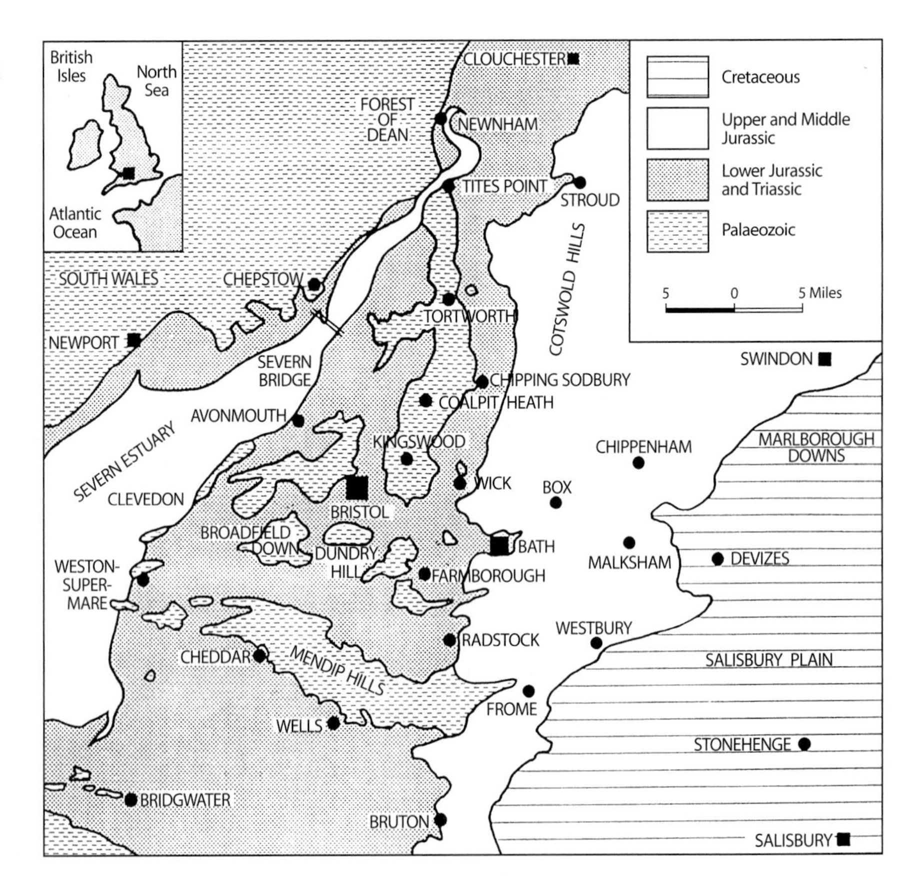

Fig. 8.85.
Geology of the region around Bristol and Bath

this growth was initiated in the 18th and early 19th centuries as a result of the development of the spa.

On the north side of the Avon Valley the Lansdown plateau, formed of Middle Jurassic limestone, is still largely open country. It rises to 238 m above Ordnance Datum (OD = mean sea level at Newlyn) and forms the highest point of the south Cotswold plateau (Fig. 8.85). From its western extremity, the view embraces Bristol and Dundry Hill in the west with the Severn Valley and the hills of Wales beyond. The steep slopes below the plateau edge of Lansdown are composed of slipped and broken Lower Jurassic rocks extending to the valley floor. However, Roman Aquae Sulis and medieval Bath were situated on a gently sloping stable area in the valley floor and were not affected by landslipping. The low ground was, however, liable to flooding by the River Avon in Roman times as it is at the present day.

8.28.2 The Bath at Aquae Sulis

Aquae Sulis was linked with other Roman settlements by a series of Roman roads of which the Fosse Way was one of the most important. This ran from Lincoln in the northeast to the English Channel coast near the mouth of the River Axe in east Devon. South of Bath, roughly midway between Bath and Ilchester, the road crosses the Mendip Hills. These limestone hills extend from Frome in the east to the Bristol Channel coast at Uphill near Weston-super-Mare in the west and attain a maximum height of 323 m OD at Blackdown (Fig. 8.86). They are of great geological and scenic interest, being formed of folded and thrust faulted Carboniferous Limestone (Mississippian) with Devonian and Silurian rocks in the anticlinal cores. Thin Mesozoic strata of Triassic and

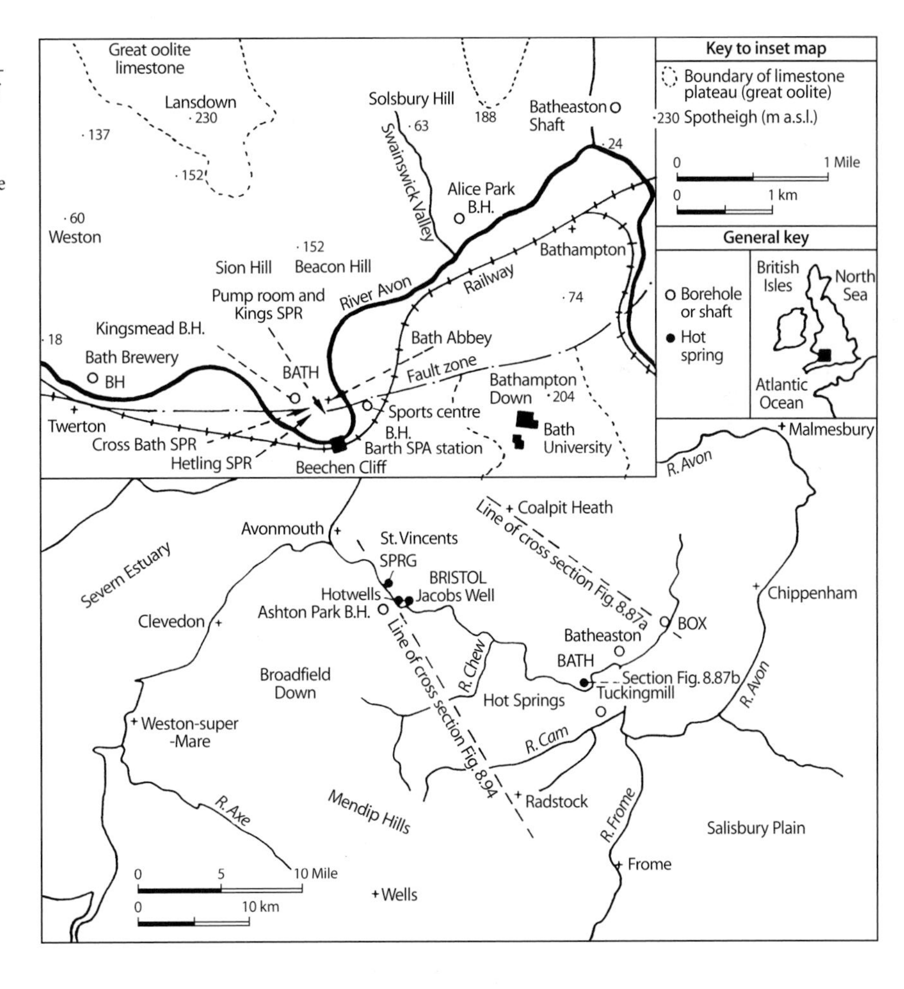

Fig. 8.86.
The Bristol Avon and its tributaries showing the position of the principal hot springs and boreholes. The inset map of Bath shows the site of the hot springs and boreholes in more detail

Jurassic age rest unconformably on the surface of the Paleozoic rocks in the central and eastern Mendips (Fig. 8.85). Both the Mesozoic and the Paleozoic rocks are quite strongly mineralized, although most of the ore bodies only extend to shallow depths (10–20 m) beneath the Mesozoic unconformity. The Mendips were worked for lead in Iron Age times, and the Romans, who were well aware of the mineral wealth of Britain, made haste to develop these and other resources as soon as their military campaign was sufficiently advanced.

One architectural feature of the Roman baths deserves mention here. This is the Roman drain (Cunliffe 1969), which is of importance as it still carries most of the water issuing from the Kings Spring. The system of thermal water monitoring devised after the closure of the spa in 1978 involved the construction of a V-notch weir in the Roman drain. This is still in use, although other methods of measurement have now been implemented.

Bath hot springs were first placed in the trusteeship of the civic authorities by Queen Elizabeth I in 1591. By the 17th and 18th centuries, Bath had become a unique center of medical and scientific activity (Williams and Stoddart 1978). Unlike many continental spas, it was isolated from war and serious civil insurrection and was therefore able to develop in peace from the mid-17th century onwards, although the town suffered serious damage from air raids in World War II. It continued in the tenor of its ways until the winter of 1978 when the spa was closed without warning owing to the discovery of a pathogenic ameba (*Naegleria fowleri*) in the thermal

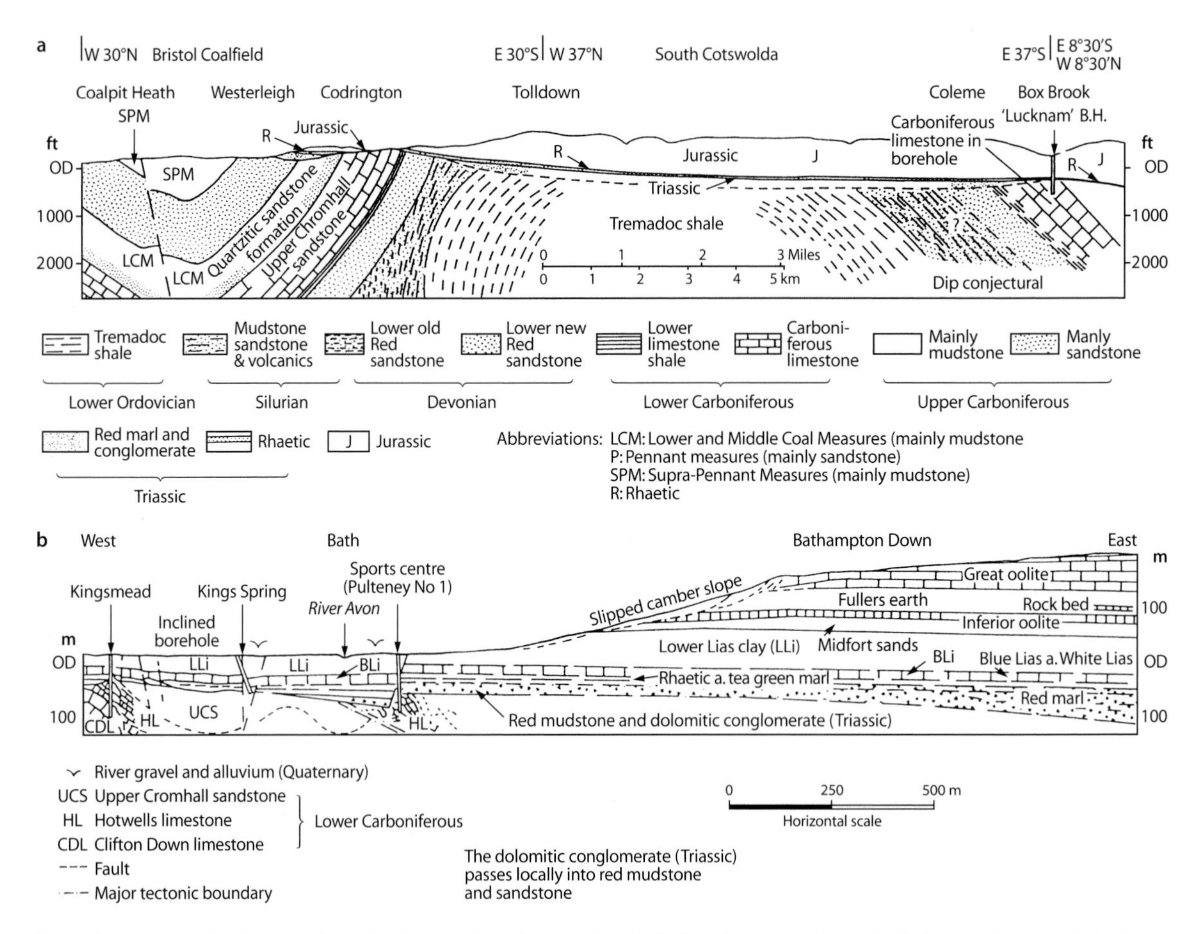

Fig. 8.87. **a** Generalized section from Coalpit Heath to Box showing the inferred structure beneath the Mesozoic cover of the South Cotswolds. **b** Section showing the geological structure of the Avon Valley between Bath and Bathampton Down. The length of section is about 2.7 km

water installations. This dangerous organism is though to have been responsible for the tragic death of a little girl who contracted amebic meningitis after swimming in the thermal water (Kilvington et al. 1991).

Examination of water, mud, and clay samples taken at the surface and at shallow depth from the Kings Spring, and from boreholes, have shown conclusively that *Naegleria* is absent at all depths below the surface of the spring basins wherever the rising thermal water has not made contact with the air. *Naegleria* is an aerobic organism, and the absence of oxygen in the unoxidized thermal water is inimical to its survival. The method of ensuring the exclusion of *Naegleria* is based on the recovery of unoxidized thermal water through lined boreholes connected to a continuous pipe conveying the water directly to the place where it is to be used – thus avoiding oxidation.

This result was achieved at the Kings Spring by sinking an inclined borehole that taps the rising thermal water in a big fissure at about 50 m below OD (Fig. 8.87). Thence it travels up the lined borehole under hydrostatic pressure and a continuous stream of thermal water issues at the Pump Room fountain (Fig. 8.88 and 8.89). In the Cross Bath Spring a vertical tube well (Kellaway 1991c) now performs the same function. The Hetling Spring is not at present in use and still rises in the stone cylinder in which it was confined in the 18th century.

Fig. 8.88. Front of the Grand Pump Room, Bath

Fig. 8.89. Fountain in the Pump Room fed by thermal water from the Kings Spring (*photo:* Bath City Council)

8.28.3 Thermal Water Springs of the Avon Valley

The mean annual (sea level) temperature of the region around Bristol and Bath is 10 °C, and springs having a constant temperature of more than 10 °C are likely to be of deep-seated origin. The highest surface groundwater temperatures in the United Kingdom are found at Bath where the springs have constant temperatures of 41–47 °C at ground level (17 m OD).

About 16 km west of Bath, a group of thermal springs is present in the Avon Valley at Bristol (Fig. 8.86). These rise from fissured Carboniferous Limestone. The temperature of the water is lower than at Bath, and the chemistry suggests that this is due to dilution by cold groundwater (Andrews et al. 1982). However, investigation of these springs is very difficult. The largest and most famous of them is at Hotwells at the southern entrance to the Avon Gorge. Now unused, it rises from fissured Carboniferous Limestone, its exit being below high water mark in the bed of the muddy river, which has a tidal range of 13 m. Nevertheless, the spring is not affected at depth by the saline water, although it is contaminated at its outlet owing to the fissured condition of the bedrock. The full extent of the hydrostatic pressure has not been determined, but it is clearly sufficient to exclude the muddy seawater from the fractured bedrock.

Hotwells thermal spring water has a temperature of 24 °C. It has a pleasant taste and was used primarily for drinking water rather than immersion (Hawkins and Kellaway 1991). Following the closure of Hotwells Spa in the 19th century, the spring was damaged by excavations designed to improve navigation on the river. In the 17th and 18th centuries, however, the port of Bristol supported a major seaborne export trade in bottled Hotwells water. This lasted until the decline of Hotwells Spa at the end of the 18th century and did much to foster the growth of the local glass industry.

Another warm spring supplied Jacobs Well (Fig. 8.86) situated in the floor of a deep valley between Clifton and Brandon Hill, Bristol. The precise site was long forgotten but has been recently discovered (Vaughan and Martelett 1987; Hawkins and Kellaway 1991). The spring has a winter temperature of 13.8 °C and was used for a Jewish Mikveh, or purification bath, in which there was total immersion. It is thought to have been used between 1097 when the Jews arrived in England and 1290 when they were expelled by Edward I. The Jews returned to England in the 17th century, but there is no evidence to show that they resumed their use of the well.

The thermal waters of the three Bath hot springs issue in the center of the city west of the abbey (Fig. 8.86). They have chemical compositions that are basically similar, with a fairly high concentration of sodium, calcium, chloride, and sulfate ions. The chemical and isotopic composition of the Kings Spring water has been recorded and discussed in detail by Edmunds and Miles (1991). Hot wells and the adjacent springs in the Avon Valley at Bristol are geochemically related to those of Bath, notably in the dominance of Ca^{2+}, HCO_3^-, and SO_4^{2-} ions. This affects the hypothetical reconstruction of the route or routes taken by the water that now discharges at Hotwells. Thus, for structural reasons it is difficult to envisage the passage of water either from the Mendips or the Cotswold uplands to the Avon Gorge at Clifton. Yet, the water has had an appreciable residence time in calcareous rocks and has much in common with that of Bath hot springs.

Rising from tectonically formed fissures in the Carboniferous Limestone and its unconformable cover of Mesozoic strata (Fig. 8.90, the thermal water eventually flows into the River Avon. The Kings Spring is drained partly through the Great Roman Bath (Fig. 8.91) and partly from the reservoir surrounding the Kings Spring, the combined flow being diverted into the Roman drain where it is measured. The Cross Bath Spring and Hetling Spring are drained independently to the river.

Owing to the hydrostatic pressure, the thermal water is well able to penetrate the Mesozoic cover rocks in the valley floor by way of open faults and fissures. In some boreholes and shafts, it has been found that water is rising vertically from the Carboniferous rocks. In others it is migrating through fissured Triassic sandstone and conglomerate at depths at or below sea level. In the Batheaston shaft, strong flows were also noted in the

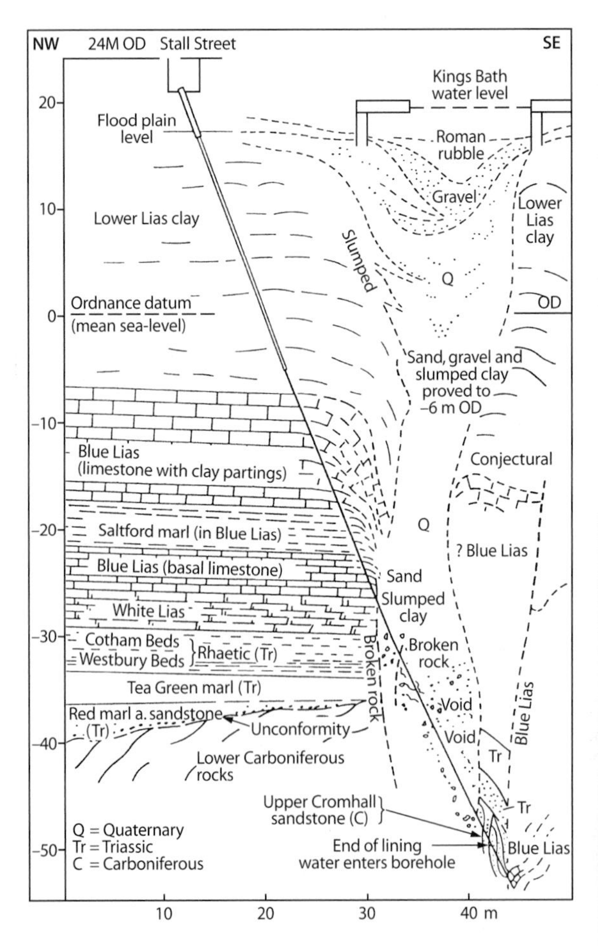

Fig. 8.90. Section along the line of Stall Street inclined borehole, Bath. The Triassic–Jurassic boundary is taken at the base of the White Lias

Fig. 8.91. Great Roman Bath (*photo:* Bath City Council)

White and Blue Lias (Jurassic) limestones. The same is true of the thermal water issuing from faulted Triassic and Jurassic strata proved in the Kingsmead borehole (Kellaway 1991a). In general, there is a gradual fall in temperature away from the central area around the Bath hot springs.

The recharge mound of the hot springs Bath extends beneath the valley floor for about 1 km in a westerly and about 1 km in a northeasterly direction. There is unfortunately no evidence as to the conditions beneath the steep valley sides or the high-level plateaux to the north and south of Bath. Nor is the fault and fissure pattern known in sufficient detail to determine its full extent. Many trial borings made during engineering investigations in the areas immediately west and southwest of the hot springs have proved warm or tepid water in the White Lias and Blue Lias limestones that underlie the Lower Lias Clay. In most cases there is also evidence of faulting and fissuring, but it is seldom possible to determine the relationship of the small faults proved on adjacent sites. It is clear, however, that in addition to major Mesozoic or Tertiary faults, such as the Pennyquick and Newton faults (Kellaway 1991a), there are numerous smaller dislocations, including some structures that resemble cambers with dip-and-fault structure and trough gulls. These may be of Quaternary age. There can be no doubt that the fracturing of the rocks in the valley floor, and the partial blocking of the exit of the spring pipe by river gravel has encouraged the thermal water to spread underground, particularly in the harder Mesozoic sandstones and limestones. Highly impermeable clays and shales of Rhaetic and Lower Jurassic age have performed as effective aquicludes where the intersecting faults have been insufficient to cause complete severance and loss of continuity in individual formations. Under these circumstances lateral movement of water is dominant and most of the water in the recharge mound (RCM) of the hot springs at Bath may be held in the Mesozoic cover.

In the case of major thermal water flows originating in deeply penetrating fissures cutting both the Carboniferous and Mesozoic formations (e.g., the Kings Spring), the thermal water has found its way to the surface, rising through fissures in the hard rocks and forming cylindrical channels (or spring pipes) through the overlying Mesozoic clays and shales (Fig. 8.87). It is possible that there may be some thermal water rising in small springs at Bathwick east of the river and between the Hetling spring and Bath Spa Railway Station (Fig. 8.86), but some of the thermal water recorded near Bath Station may be leaking from damaged installations at the main centers of emission. Further investigation is required, but in ancient occupation sites work of this kind can be very expensive. In the central area of Bath, the ground above the ruins of the Roman city averages 6 m in thickness and access to the natural surface of the ground requires large excavations.

8.28.4 Recharge Mound at Bath

Approximately 1.25×10^6 l of water with a temperature of 41–47 °C reaches the surface daily at Bath. Rising from deep fissures in the Carboniferous Limestone, it penetrates an unconformable cover of gently dipping Mesozoic strata before flowing into the River Avon (Fig. 8.90b). Its hydrostatic head is about 30 m OD but may have been somewhat greater before the spring pipe was drilled for investigation purposes. The RCM extends from Bath Brewery, 2.5 km west of the Kings Spring to Batheaston Shaft, 4 km to the northeast of Kings Spring (Fig. 8.85). In the center of Bath the thermal water zone in the fissured Mesozoic rocks is up to 1 km in width, although the thermal water appears to be very uneven in its distribution, being mainly concentrated in fault and fissure belts. Stanton (1991) has calculated that the RCM contains more than 100 million l.

The value of the recharge mound is that it acts as a buffer in the event of short-term over abstraction of thermal water at the main springs. It also controls the flow of cold groundwater, which is mainly derived from the crop of the Triassic and Jurassic rocks west of Bath. This migrates downdip beneath the impervious Lower Lias Clay, which is penetrated at Bath by spring pipes from which the thermal water issues under hydrostatic pressure (Fig. 8.87).

Much of the information relating to the RCM at Bath has been obtained from tests carried out during the progress of the investigations from 1978 onwards. These show that the heat output of the springs is more stable than their flow. The total heat output is estimated by Stanton (1991) at 55 008 mcal/d.

8.28.5 Statutory Control of Thermal Water at Bath

From the beginning of the 19th century onwards, there have been a number of instances where the sinking of shafts and boreholes in the floor of the Avon Valley at

Bath has affected the flow of the hot springs. One of these, the Batheaston Coal Shaft (Fig. 8.85) was constructed in 1805–1813 under the supervision of William Smith, who recorded a slight diminution of flow of the hot springs when an influx of water in the shaft was being pumped away. Subsequently, in 1836, an attempt to deepen Pinch's Well at Kingsmead Square (Fig. 8.85) caused temporary but serious losses in the flow of all three hot springs at Bath. The affair of Pinch's Well greatly alarmed the corporation and from this time onwards a close watch was kept on any developments that were likely to cause loss of pressure in the thermal springs. It was not until 1925, however, that legislation was passed to enable the civic authorities to protect the springs from damage other than legal action in open court. The Bath Act (1925) laid down a number of conditions that had to be fulfilled by persons wishing to make boreholes, shafts, tunnels, or other deep excavations. Under the terms of the act, application had to be made to the council for permission to construct such holes within the limits of the city boundaries.

When, in 1982, the County of Avon Act was passed through Parliament, it was drawn up to include Batheaston and was formulated with precisely defined depth limits for proposals to drill, excavate, or tunnel. With regard to areas lying outside the area defined in the County of Avon Act, the City Council is a license holder of the National Rivers Authority (NRA), which is responsible for dealing with the conservation of all groundwater, including thermal water resources. Cooperation is maintained between Bath City Council and the NRA as well as with planning authorities. This is considered to provide adequate environmental protection for the thermal water springs at the present time.

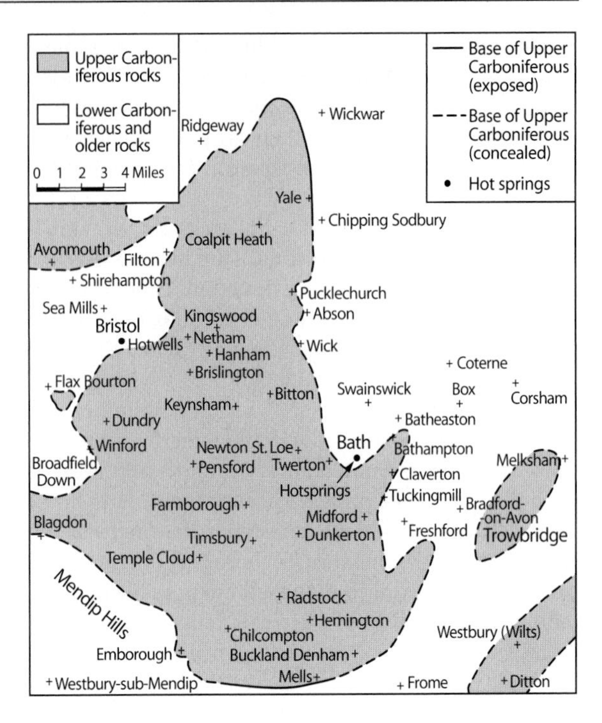

Fig. 8.92. Distribution of the Upper Carboniferous rocks of the Bristol and Somerset Coalfield and the position of the hot springs. The outline of the coal basin has been generalized and Variscan faults omitted

8.28.6 Structural Elements in Thermal and Groundwater Conservation

Reconstruction of the details of the structure of the Variscan elements in the Bristol and Somerset coalfield is difficult, as a large part of the basin is concealed by Mesozoic rocks. Nevertheless, it can be seen that Bath and Bristol are situated at the concealed margin of the coal basin (Fig. 8.92) where the Coal Measures and Millstone Grit (Westphalian and Namurian) rest on Carboniferous Limestone (Dinantian) rocks. This suggests that the movement of the groundwater is restrained by the cover of Upper Carboniferous rocks, primarily by the thick mudstones of the Coal Measures.

The Carboniferous Limestone (Mississippian) is the most important aquifer. Totaling about 760 m in thickness, it consists almost entirely of limestone and dolomite in the Mendips, with an important aquiclude (Lower Limestone Shale) at the base (Kellaway and Welch 1993; Fig. 8.93). Large quantities of water issue from the Carboniferous Limestone of the Mendips, the water being collected in three major reservoirs, from which public supplies are pumped to the region as a whole. A generalized section (omitting the thin Mesozoic rocks) extending from the Mendips in the southeast to Bristol in the northwest shows the complexity of the faults and contortions affecting the Paleozoic formations (Fig. 8.94). The near-vertical faults in the central part of the section form part of the southern continuation of the Malvern Fault Zone (Fig. 8.95). The position of the Farmborough Fault and its relationship with the Wick Thrust near Bath has recently been reconsidered in the light of seismic and borehole data (Kellaway 1993).

The writer now considers it to be most unlikely that water is traveling at depth in the Carboniferous Limestone by a direct route from the eastern Mendips to Bath. This does not, however, rule out groundwater movement northwards from the central Mendips (Fig. 8.93) towards the Pensford Basin and thence eastward to Bath by way of reactivated Mesozoic structures such as the Newton and Pennyquick faults (Kellaway 1991a).

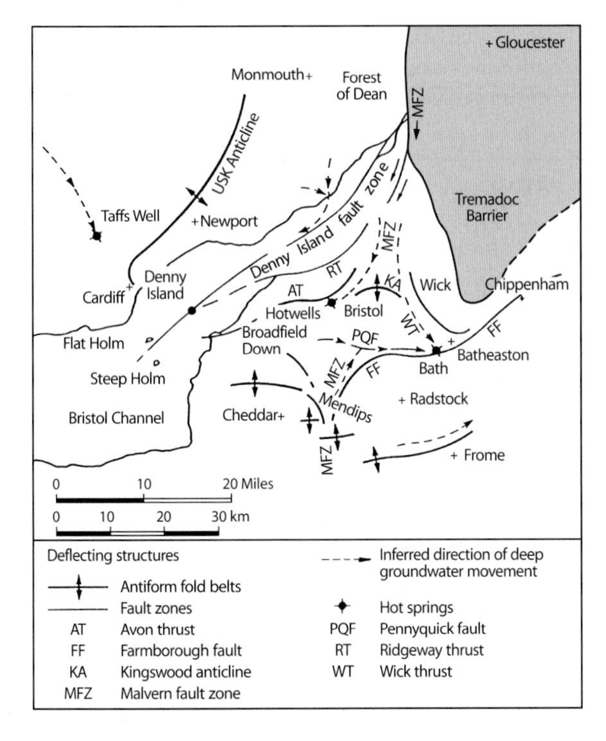

Fig. 8.93. Inferred deep groundwater movements in mid-Quaternary to Recent times in relation to deflecting Variscan structures and the Tremadoc barrier

While the question of the precise source of the Bath thermal water may not appear to have any immediate environmental impact, it is a matter of great concern in the long term. Groundwater levels in the Carboniferous Limestone of the central Mendips are relatively high, rising to about 213 m OD, and this has encouraged a widespread belief that this region is the principal, possibly the sole source of the resurgence at Bath (Andrews et al. 1982). In order to establish the validity of this assumed hydrological connection, however, it is necessary to show that the geological structure is uniquely favorable to groundwater movement from the Mendips towards Bath. Ideally, it should also be possible to demonstrate that any substantial lowering of the Mendip water levels due to quarrying can be correlated with diminished flow at Bath Springs.

Large-scale quarrying of limestone in the Mendips is active and in the eastern Mendips the groundwater supply has been adversely affected. Locally, as at Whatley near Frome, the water table has been lowered by as much as 50 m. Despite the fact that systematic monitoring of the Bath hot springs has been maintained since 1978, there is as yet no evidence of changes due to external causes.

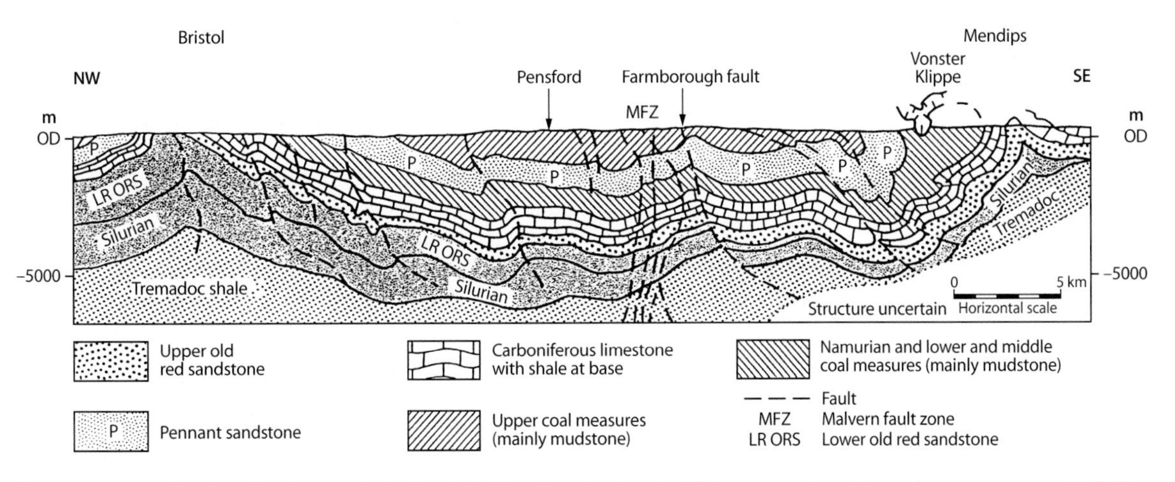

Fig. 8.94. Generalized section between Bristol and the Mendips showing the Variscan structure. Mesozoic cover rocks omitted. The section is approximately 32 km in length

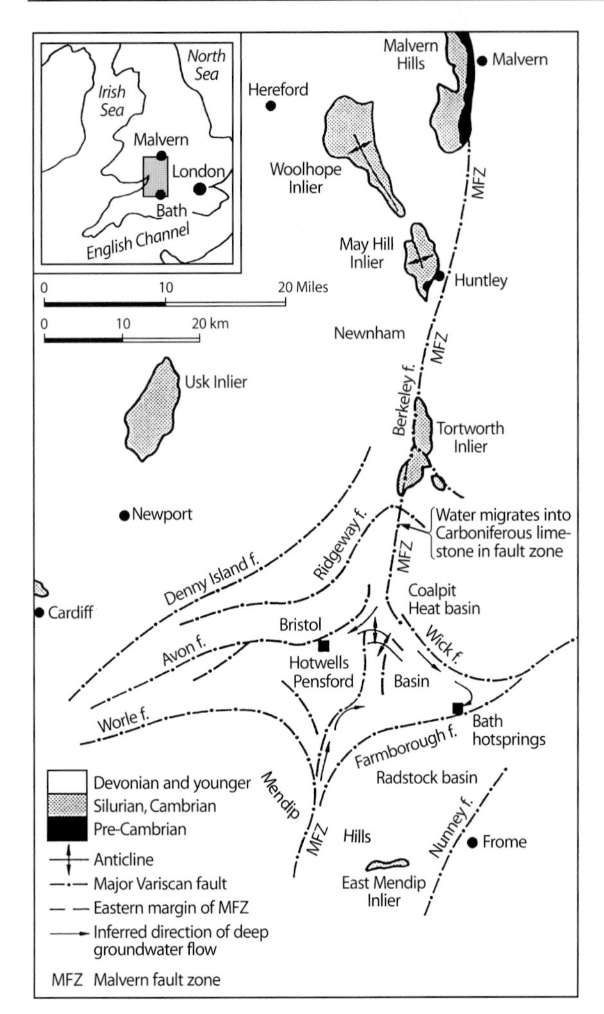

Fig. 8.95. The Malvern Fault Zone (MFZ) showing the possible route of northerly and southerly migrating groundwaters in relation to Bath and Hotwells

8.28.7 Geothermal Gradients

There is little evidence as to regional geothermal gradients in the Bristol and Somerset coalfield and the surrounding areas. The strong vertical thermal water flows at Bath show no significant increase in temperature down to depths of at least 80 m. All indications are that vertical movement of the water may originate at a depth of several hundred meters, with the water moving freely up very steeply inclined vertical features. Estimates of the regional geothermal gradient, based on a number of different lines of evidence, range from 20 to 24 °C/km with the possibility that gradients are lower in the southern part of the coalfield than they are in the north. Temperature gradients within the zones of fracturing (e.g., in the Avon Gorge at Clifton) may differ greatly from those in the adjacent blocks, where descending cold groundwater is dominant, as at Ashton Park south of the Avon Gorge (Kellaway 1967).

8.28.8 Origin of Bath Thermal Water

In the absence of any evidence for the presence of a slowly cooling pluton, all the current explanations of the origin of the thermal water springs of the Avon Valley depend on deep circulation (i.e., to depths of at least 1.66 km) of groundwater. It is generally accepted that the water has had significant residence time in marine Carboniferous Limestone. The distribution of this formation (which is partly exposed and partly concealed) is shown in Fig. 8.96. In dealing with the origin of thermal water at Bath it is not possible to exclude some consideration of the spring at Hotwells, Bristol, since there are geochemical similarities that indicate that the water at Hotwells is of the same general type as that of Bath. The differences are due mainly to the greater dilution of Hotwells water by cold groundwater.

Three theories have been put forward to explain the origin of the thermal water springs at Bristol and Bath: (1) the water is derived from the Mendips, having traveled through the Carboniferous Limestone to a resurgence in the Avon Valley at Bath (Andrews et al. 1982); (2) a modification of (1) proposed by Burgess et al. (1991), which considers the possibility that the Carboniferous Limestone is recharged by vertical leakage through the Jurassic and Triassic cover rocks as well as the Coal Measures (including the Pennant Sandstone); and (3) that the thermal springs are fed primarily by water circulating in the Malvern Fault Zone and its southerly continuation in the Bristol and Somerset Coalfield and the Mendips (Kellaway 1991a).

1. The Mendip theory (Andrews et al. 1982) has already been mentioned. It is dependent on assumed hydrological continuity of the exposed Carboniferous Limestone beneath the Avon Valley at Bath. On structural grounds, a direct connection between the eastern Mendips and Bath is unlikely (Kellaway 1993).

   In the case of Hotwells, the structure and hydrology of the intervening Ashton Park-Broadfield Down

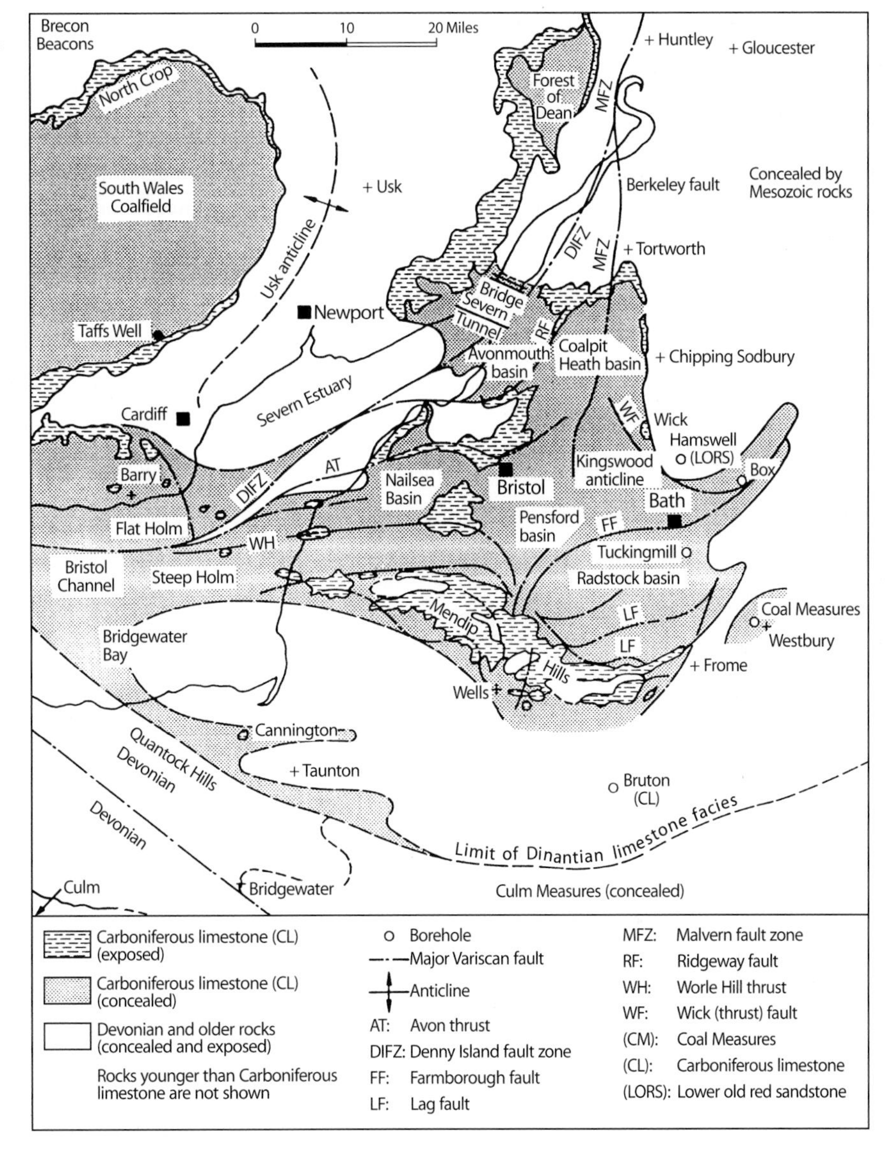

Fig. 8.96. Distribution of the Carboniferous Limestone in the Bristol and Somerset coalfield and surrounding regions. South of its structurally defined southern boundary the Dinantian is represented by an aquitard Culm facies. South Wales and the Forest of Dean are structurally isolated from the region east of the Severn Estuary. Variscan structures of the Radstock Basin are described by Chadwick et al. (1983)

area rules out a direct connection with the Mendips. However, both Hotwells and Bath could be connected with the central north-south fault zone of the coalfield by way of transverse post-Jurassic (E-W) fault and fissure belts. Nevertheless, the water issuing at Hotwells is likely to be derived from areas north of the Kingswood Anticline (Fig. 8.96).

Systematic monitoring of the flow of the Bath hot springs since 1978 has not yielded any evidence indicating interference with the flow of the hot springs by lowering the water table in the Carboniferous Limestone of the Mendips.

2. The modified Mendip theory proposed by Burgess et al. (1991) evaluates the effect of piezometric head due to the presence of a substantial Mesozoic cover, which forms the uplands of the South Cotswolds. The longest and most continuous plateau-like mass of Mesozoic strata is bounded by the Cotswold escarpment on the west and the Avon Valley on the south. The Mesozoic strata extend west of the main escarp-

ment, which is formed of Middle Jurassic rocks (Fig. 8.85) to the margin of the Coalpit Heath Basin. However, the lowest Jurassic formations are aquitards or aquicludes and, being mainly argillaceous, have a very low permeability. The crop of the Carboniferous Limestone is exposed between Tortworth and Chipping Sodbury (Fig. 8.96), but south of Chipping Sodbury it is concealed, except for some small inliers at Codrington and Wick (Fig. 8.96). However, the area of Carboniferous Limestone concealed by the Mesozoic strata in this part of the south Cotswold is very small.

Groundwater descending from the Mesozoic cover of the south Cotswolds would encounter rocks older than the Carboniferous Limestone beneath the greater part of the Middle Jurassic uplands. This water could only penetrate the Carboniferous Limestone of the Coalpit Heath Basin by traveling laterally through the barrier formed by the Lower Limestone Shale (Table 8.30, Fig. 8.90a).

Table 8.30. Pre-Cambrian and Paleozoic rocks of the central part of the Bristol and Somerset Coalfield, Tortworth inlier, and the Malvern Hills

Permian rocks are absent; the main Variscan movements took place between 290 and 260 Ma (Rotliegendes)			
Upper Carboniferous (333–290 Ma)	Coal Measures (2 500 m)[a]		
	Millstone Grit = Quartzitic Sandstone Group (ca. 300 m)		
Lower Carboniferous (363–333 Ma)	Carboniferous Limestone (800 m)[a]	Hotwells Formation (350 m) Clifton Down Formation (250 m) Rock Formation (150 m) Lower Limestone Shale (50 m)	Upper Cromhall Sandstone Hotwells Limestone Black
Devonian (409–363 Ma)	Upper Old Red Sandstone	Tintern Sandstone (125 m) Quartz Conglomerate (30 m)	Portishead Beds of Bristol and the Mendips
	Unconformity		
	Lower Old Red Sandstone[b]	Thornbury Beds (600 m) Downton Castle Sandstone (8 m)	
Silurian (439–409 Ma)		Ludlow and Wenlock (ca. 300 m)[c] Llandovery, basal beds not represented (ca. 500 m)	
Unconformity			
Ordovician (Upper Cambrian) (510–493 Ma)	Tremadoc (or Shineton Shale)[d]	Micklewood Beds and Breadstone Shales (1 500–2 000 m)	
Cambrian		Whiteleaved Oak Shales (150 m)[f]	
(570–510 Ma)[e]		*Unconformity*	
		Hollybush Sandstone (270 m)[f] Malvern Quartzite (90 m)[f]	
Pre-Cambrian	(Before 570 Ma)	Longmyndian Grit and Shales (Huntley)[f] Uriconian (Warren House Group) Volcanics[f]	
	(Before 1200 Ma)	Malvernian (Gneiss and Schists)[f]	

[a] The Carboniferous Limestone and Millstone Grit of the British Isles = Mississippian of USA; Coal Measures = Pennsylvanian. For detailed stratigraphy see Kellaway and Welch (1993).
[b] The Lower Old Red Sandstone is missing in the central area of the Malvern Fault Zone, having been removed by erosion prior to the deposition of the Upper Old Red Sandstone.
[c] Represented by volcanic rocks in the eastern Mendips.
[d] 300 m in Malvern area.
[e] Dates for Cambrian-Carboniferous are based on Harland and others (1989).
[f] Malvern sequence.

A fairly substantial, although deeply dissected, cover of Triassic and Jurassic rocks extends from Bath to the eastern Mendips (Fig. 8.86), but the valley sides are strongly cambered, causing rapid loss of water by excessive drainage of the influves. It is most unlikely that calculations based on theoretical sections such as those examined by Black and Barker (1981) or Burgess et al. (1991) would have any application to big cambered hill masses such as Dundry Hill near Bristol (Fig. 45 in Kellaway and Welch 1993) or Bathampton Down near Bath.

With regard to the possible penetration of water draining vertically downwards into the Coal Measures from the Mesozoic cover, there is critical evidence from the central and eastern part of the Radstock Basin where the Upper Coal Measures are overlain by Triassic and Jurassic strata (Richardson 1925). Two groups of coals were formerly worked in the highest measures. The Upper or Radstock Group is separated from the lower or Farrington Group by barren mudstones and sandstones, in which strongly saline water is present (MacMurtrie 1886). The Farrington seam workings were dry or carried freshwater, and it is not possible for descending groundwater to penetrate the Barren Red Group without affecting the salinity of the Farrington Group below.

Downward migration into the Carboniferous Limestone is therefore out of the question in the central and eastern part of the Radstock Basin. Had the Mesozoic rocks of the Cotswolds formed a continuous cover over a larger area of the concealed drop of the Carboniferous Limestone, the concept advanced by Burgess et al. (1991) could not be lightly dismissed. However, the Mesozoic cover at Bath and Radstock is deeply dissected, and the rocks include a high proportion of shales, clay, and silt that would be described as aquitard or aquiclude. Under more favorable conditions, the piezometric head due to the Mesozoic rocks could have made a significant contribution to the maintenance of the flow of the hot springs of Bath.

3. The Malvern Fault Zone hypothesis (Kellaway 1991b) is based primarily on observation of the way in which thermal-water movement is concentrated in zones of intense fracturing and fissuring commonly associated with reactivation of Variscan and post-Variscan structures. The Malvern Fault Zone is unquestionably the major tectonic feature in the region between the West Midlands and the Mendips. Many of the large thrust faults of Variscan Age, which have attracted attention in the past, were, in the opinion of the writer, formed contemporaneously as a result of strong lateral movement along the major structure.

A general account of the stratigraphy and structure of the Malvern Fault Zone in the Abberley Hills, the Malverns, and the Newnham area southwest of Glouchester has been given by Earp and Hains (1971). Farther south in the Tortworth Inlier (Fig. 8.95), the fault zone, which is partly of mid-Devonian Age, has been described by Cave (1971). Mid-Devonian earth movements led to the destruction of the Lower Old Red Sandstone and its nonmarine cornstones over much of the Malvern Fault Zone (Kellaway 1991b). The calcium carbonate dissolved in the thermal water is of marine origin. Had the Lower Old Red Sandstone remained intact, it would have been difficult to establish movement of the thermal water through the faulted rocks of the Malvern Fault Zone. Some of the large north-south faults in the northern part of the Bristol coalfield are an extension of mid-Devonian structures that have been reactivated by the Variscan movements. A number of these exhibit evidence of Mesozoic and post-Mesozoic activity.

The Malvern hypothesis assumes that the Bath thermal water and the thermal water content of the Hotwells Springs are derived, at least partly, from a northerly source (or sources) by way of the Malvern Fault Zone (Fig. 8.95 and 8.96). This could involve the underground migration southward of a proportion of the water over distances of up to 64 km, as opposed to 32 km from Mendips to Bath. The Malvern source rocks would be Pre-Cambrian and Lower Paleozoic, including those of the Malvern and Abberley Hills, which attain a maximum altitude of over 400 m OD. The Pre-Cambrian rocks of the Malvern Hills are noted for the softness and purity of their ground-water, but in the course of its journey southward along the fault zone, contact would be made with faulted and folded limestones and calcareous sandstones of Silurian (Llandovery and Wenlock) age. Other elevated masses of Silurian rocks on the western side of the fault zone, such as the Woolhope Anticline and May Hill, may contribute groundwater to the fault zone, which continues southward by way of Tortworth to the Coalpit Heath Basin (Fig. 8.90a). The resurgent water at Bath may therefore contain only a small percentage of water from the Pre-Cambrian rocks of the Malverns.

The composition of Hotwells water is given by Edmunds and Miles (1991) as Bath-type thermal water and shallow Carboniferous Limestone water in the ratio of 1:2:3. They believe that "both the thermal water sources" (i.e., Hotwells and Bath) "have a similar structural control over their flow". This can best be explained by derivation primarily from a northerly source.

It is interesting to consider the effect of the Malvern hypothesis on some of the problematical features of the other explanations applied to Bath. One of these is the difficulty of accounting for the uniform volume and temperature of thermal water emitted at Bath. The effect of bypassing the gravel and sand plugging in the spring pipes at Bath is to produce a dramatic increase in flow. The buffering effect of the impeded flow from the spring pipes is consonant with a source that contains a very large amount of water moving very slowly yet constantly towards its resurgence. While there are undoubtedly karstic cavities in the calcareous Triassic conglomerate in the recharge mound at Bath, the presence of karstic conditions at great depth in the Carboniferous Limestone is not supported by deep borings and tunnels made in that formation. The karstic phenomena seen in the Mendips and other upland limestone regions around Bristol undoubtedly extend below sea level, possibly in extreme cases for up to 100 m, but they are notable by their absence in borings of over 300 m, where the water in the Carboniferous Limestone is held in tectonically generated fissures differing in no significant respect from those found in hard sandstone or relatively insoluble quartzite.

To obtain a steady flow of water at a uniform temperature over a length of time that is almost certainly measured in millennia demands a continuous zone of open fissures of great length and depth of penetration. The Malvern Fault Zone is the only regional structural element which, in the author's opinion, is large enough to satisfy these conditions.

In terms of equilibrium in SiO_2, Edmunds and Miles (1991) suggest temperature limits for Bath thermal water of 64 °C for chalcedony and 96 °C for quartz. Taking the regional geothermal gradient as 20 °C/km, the appropriate maximum circulation depths would be 2700 or 4300 m. Both these limits could be applied in terms of the deep structure of the Malvern Fault Zone, but it is interesting to note that if equilibrium with quartz is taken as a guide, then the thermal water could be circulating in Pre-Cambrian and Lower Cambrian rocks below the Tremadoc aquiclude as far south as the margin of the Coalpit Heath Basin. If the smaller depth figure is accepted, then the water has probably migrated into faulted Cambrian and Silurian rocks between May Hill and Tortworth. This would imply that it could have already commenced taking up $CaCO_3$ of marine origin from the calcareous sandstone and limestones of Silurian (Llandovery and Wenlock) age and could be penetrating the Upper Old Red Sandstone and Carboniferous Limestone beneath a cover of Namurian and Westphalian rocks near the northern end of the Coalpit Heath Basin. The distance traveled in the Carboniferous Limestone en route to Bath would probably be about 25–32 km, much the same as that inferred by the Mendip for the route from the Central Mendips in Bath.

There is also a possibility that Mendip water traveling northwards in the Carboniferous Limestone and then eastward to Bath (Fig. 8.93) may be providing a proportion of the thermal water that is mixing with a smaller but constant amount of much older, hotter water derived from northern sources.

8.28.9 Age of Thermal Water Springs

Structural and geochemical evidence suggests that the origin of the westward draining Bristol Avon may lie somewhere between 200 000 and 700 000 years B.P. The evidence from the spring pipes at Bath points to the date of formation of the hot spring as being ca. 150 000 years B.P. It has to be borne in mind, however, that the formation of the spring pipes may postdate the formation of the Twerton Terrace and could be as late as early Devensian (Kellaway 1991a, pp. 223, 229–231). The hydrological problems are more easily solved if a younger rather than an older date is attributed to the appearance of the hot springs, since the development of a fairly high relief assists in providing the necessary hydrostatic pressure required to maintain flow at the point of resurgence. Nor can the tectonic movements responsible for the generation of the fissure belts be regarded as confined to one short period of activity. They may have been episodic and particularly intense in mid-Quaternary times, but the date of commencement is indeterminate, and activity has not necessarily ceased.

8.28.10 Restoration of the Spa

By late 1982, sufficient evidence had been accumulated to show how the principal springs at Bath were fed by thermal water rising from structurally disturbed Paleozoic rocks and then invading a cover of unconformable Mesozoic strata that form the recharge mound. Trial borings in the basin of the Kings Spring showed that the thermal water issues from a funnel-shaped opening penetrating the Lower Lias Clay and infilled with deposits of fluviatile sand and gravel similar to those of the river terraces on the lower slopes and floor of the Avon Valley. Cylindrical and conical spring pipes have been formed in Lower Lias Clay at Bath by the penetration of rising thermal water driven by hydrostatic pressure and probably aided more than once by seismic pumping. The White and Blue Lias limestones, which underlie the Lower Lias Clay, are cut by fracture zones and fissures; some of these are located on reactivated post-Variscan faults (Fig. 8.90b).

The Stall Street inclined borehole sunk in 1983 now bypasses the gravel-filled spring pipe of the Kings Spring (Fig. 8.87) and is capable of supplying sufficient clean thermal water to permit redevelopment of the spa. At the present time only the Fountain in the Pump Room is supplied with clean unoxidized thermal water from the Stall Street borehole. At the Cross Bath, a vertical tube well now taps the thermal water rising up the principal fissure. The yield is adequate to supply the Cross Bath, although an additional supply could be obtained from the Cross Bath inclined borehole; this is tapped water in fissured Carboniferous Limestone beneath the Mesozoic cover. However, the small size of the Cross Bath limits its practical value, although it is of great historical interest and is an integral part of the spa complex, occupying an important site at the western end of Bath Street.

Work has also been carried out on the top of the stone cylinder in which the Hetling Spring was enclosed in the 18th century. This is situated in front of John Wood's Hot Bath (1775–1778, Fig. 8.97), which was supplied by water from the Hetling Spring cylinder. The Hetling Spring is the scene of William Smith's restorative work (1808–1809) where he dealt successfully with the problem of leakage (Kellaway 1991b). At the present time, the Beau Street baths, which lie at the rear of the Hot Bath, are unused. Restoration of facilities for immersion or bathing in the hot water at Bath still lies in the future.

The thermal water establishment was the raison d'etre of Roman Aquae Sulis and, nearly 2 000 years later, gave rise to the development of 18th-century Bath, in which the classical orders reasserted their influence on architectural development. Georgian Bath provides remarkable illustrations of 18th century town planning but

Fig. 8.97. Hot Bath (*right*, build by John Wood after 1773) and Iron Bath (*center*)

its center, like that of Aquae Sulis and medieval and Tudor Bath, lies at the Kings Spring. The Cross Bath and the Hot Bath, situated a short distance to the west of the Pump Room, are linked to the main center by colonnaded Bath Street. Restoration of the use of the hot springs and the principal elements of Bath Spa is essential if continuity is to be maintained and the environmental and historical character of Bath is to be preserved.

Acknowledgements

The author is indebted to Bath City Council for permission to base Fig. 8.87, 8.89–8.93 and 8.96 on published illustrations of the Hot Springs of Bath (1991). He is indebted to his wife for her unswerving help in typing and editorial work whithout which this would not have appeared.

References

Andrews JN (1991) Radioactivity and dissolved gases in the thermal water of Bath. In: Kellaway GA (ed) The hot springs of Bath. Investigations of the thermal waters of the Avon Valley. Bath City Council, Bath, pp. 157–170

Andrews JN, Burgess WG, Edmunds WM, Kay RLF, Lee DJ (1982) The thermal springs of Bath. Nature 298: 1–4

Armstrong W (1838) An account of the tapping and closing of the Springs of hot water at Mr. Pinch's Brewery. Bath, Bristol, United Kingdom

Black JH, Barker JA (1981) Report of the Institute of Geological Sciences. Her Majesty's Stationery Office, London, pp. 81–83

Burgess WG, Black JH, Cook AJ (1991) Regional hydrodynamic influences on the Bath-Bristol springs. In: Kellaway GA (ed) The hot springs of Bath. Investigations of the thermal waters of the Avon Valley. Bath City Council, Bath, pp 171–177
Cave R (1971) Geology of the Malmesbury District. Memoirs of the Geological Survey of Great Britain. Her Majesty's Stationery Office, London
Chadwick RA, Kenolty N, Whittaker A (1983) Crustal structure beneath southern England from deep seismic reflection profiles. J Geol Soc London 140(6):893–911
Cunliffe B (1969) Roman Bath. Rep Res Comm Soc Antiquaries London 24:1–224, plates I–LXXXIV
Earp JR, Hains BA (1971) The Welsh Borderland. British Regional Geology, Institute of Geological Sciences. Her Majesty's Stationery Office, London
Edmunds WM, Miles DL (1991) The geochemistry of Bath thermal waters. In: Kellaway GA (ed) The hot springs of Bath. Investigations of the thermal waters of the Avon Valley. Bath City Council, Bath, pp. 143–156
Harland WB, Armstrong RL, Cox AV, Craig LE, Smith AG, Smith DG (1989) A geologic time scale 1989. Cambridge University Press, Cambridge
Hawkins AB, Kellaway GA (1991) The hot springs of the Avon Gorge, Bristol, England. In: Kellaway GA (ed) The hot springs of Bath. Investigations of the thermal waters of the Avon Valley. Bath City Council, Bath, pp. 179–204
Kellaway GA (1967) The Geological Survey Ashton Park Borehole and it bearing on the geology of the Bristol District. Bull Geol Surv GB 27:49–153
Kellaway GA (ed; 1991a) The hot springs of Bath. Investigations of the thermal waters of the Avon Valley. Bath City Council, Bath
Kellaway GA (1991b) The work of William Smith at Bath (1799–1813). In: Kellaway GA (ed) The hot springs of Bath. Investigations of the thermal waters of the Avon Valley. Bath City Council, Bath, pp 25–54
Kellaway GA (1991c) Investigation of the Bath hot springs (1977–1987). In: Kellaway GA (ed) The hot springs of Bath. Investigations of the thermal waters of the Avon Valley. Bath City Council, Bath, pp 97–126
Kellaway GA (1993) The hot springs of Bristol and Bath. In: Williams BJ (ed) Proceedings of the Ussher Society. R.W. Publications, Newmarket, Suffolk, pp 83–88
Kellaway GA, Taylor JH (1968) The influence of landslipping on the development of the city of Bath, England. Report of the 23rd International Geological Congress, Vol 12. Academia, Prague, pp 65–76
Kellaway GA, Welch FBA (1993) Geology of the Bristol district. Memoirs of the British Geological Survey. Her Majesty's Stationery Office, London
Kilvington S, Mann PG, Wadhurst DC (1991) Pathogenic *Naegleria* amoebae in the water of Bath: a fatality and its consequences. In: Kellaway GA (ed) The hot springs of Bath. Investigations of the thermal waters of the Avon Valley. Bath City Council, Bath, pp 89–96
MacMurtrie J (1886) Notes on the occurrence of salt springs in the coal measures at Radstock. Proc Bath Nat Hist Antiquarian Field Club, Bath, 6:84–94
Richardson L (1925) Section of No. 1 Pit, Dunkerton Collieries, Dunkerton, Somerset. Proc Somerset Nat Hist and Archaeol Soc Taunton 70:117–121
Vaughan R, Martelett J (1987) Jacob's Well rediscovered. Temple Local History Group, Bristol, 3-1987:7–15
Williams WJ, Stoddart DM (1978) Bath: some encounters with science. Kingsmead Press, Bath

A. Afrasiabian · S. B. Siadati

8.29 Sasan and Taq-e-Bostan Karstic Springs, Iran

8.29.1 Sasan Karstic Spring

Sasan Karstic Spring is in the southwest of the Tang-e-Chogan Valley in the northwest of Kazerun in southwest Iran. It is a spring that discharges from the Asmari limestone of Oligocene/Miocene Age. It is the source of water that creates a magnificent oasis in the desert of Iran, and its use by man dates to antiquity. Its specific location is lat. 29°48'27" and long. 51°37'50" with an altitude of 480 m a.s.l.

The spring has been a famous source of water near Nishabour City, Iran and was first recorded in 466 B.C. It is close to the famous Shahpour Cave that is correlated with the Sasanian Dynasty of Iran. The Sasan has been used as a source of drinking water and irrigation since prehistory. Shahpour Cave is in the Asmari limestone and a very prominent karstic feature (Fig. 8.98). The spring discharges to the Shahpour River (Fig. 8.99).

The average yearly spring discharge is 116 million m^3. Its average discharge is 3 693 l/s. Like most water typical

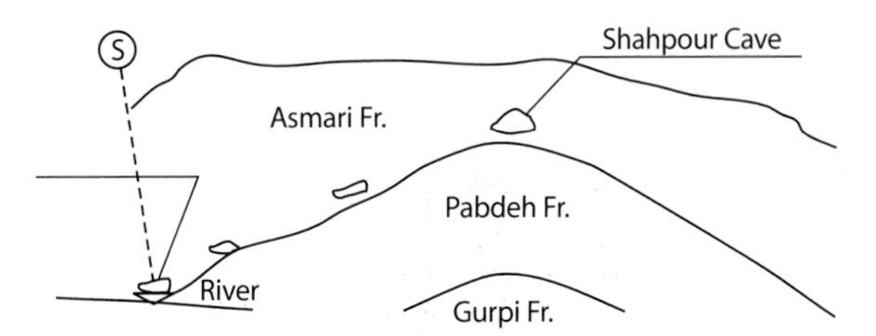

Fig. 8.98. Shahpour Cave is in the Asmari limestone

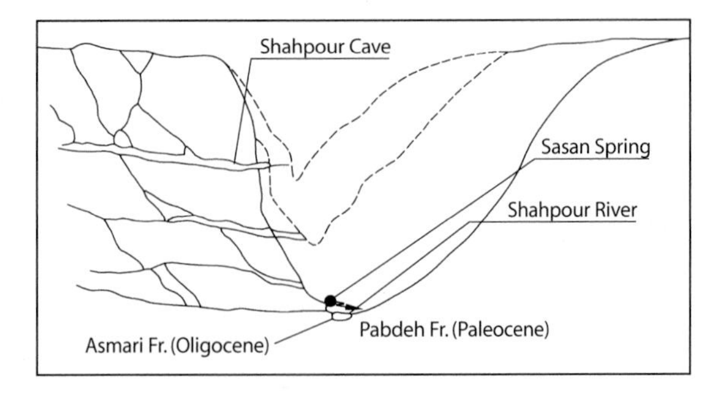

Fig. 8.99. Sasan Spring discharges to the Shahpour River

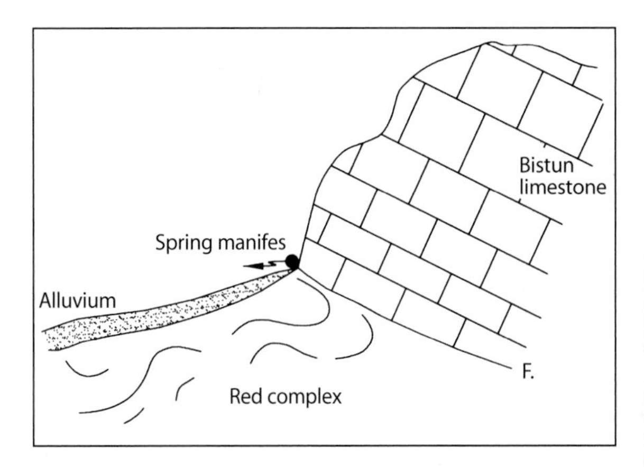

Fig. 8.100. Spring descharges to alluvium

of that associated with limestone it is of a calcium bicarbonate type with an electrical conductivity of 531 μS/s.

8.29.2 Taq-e-Bostan Karstic Spring

A second major karstic spring in Iran is Taq-e-Bostan, in the northern Kermanshah northwestern part of the country and discharges near the base of the Mioleh Mountain. Its specific location is lat. 34°23'18" and long. 47°08'00" with an altitude of 1360 m a.s.l. The average discharge is 1493 l/s with an average yearly discharge of 47.2 million m^3.

This spring has been recorded since the Sasanian Dynasty (A.D. 226–652), and has been used for drinking water and irrigation since ancient times.

Taq-e-Bostan Spring discharges as a calcium bicarbonate water with an average conductivity of 275 μS/s.

Fig. 8.101. Taq-e-Bostan Spring

The spring discharge point is near the base of the Bistun limestone at a fault contact with the underlying "Red Complex". The spring discharges to the alluvium at an elevation of 1360 m a.s.l. (Fig. 8.100 and 8.101).

E. Rosenthal

8.30 The Roman Springs of Israel

8.30.1 Introduction

The Romans and later the Byzantines who ruled the Eastern Mediterranean lands between 37 B.C. and 640 C.E., developed thermal water sources for curative and recreation purposes. In Israel, two springs – En Noit and Hammei Tveria – were always known as "The Roman springs of …". Both springs are geographically located in the Jordan-Dead Sea Rift Valley. The water of En Noit is a mixture of small amounts (1%) of Ca-chloride brine occurring in the Rift and paleowater flowing from Sinai towards the Dead Sea in the Kurnub Group aquifer. The water of Hammei Tveria is a mixture of fresh groundwater draining the eastern Galilee Cenomanian-Turonian and Eocene calcareous aquifers with the deep-seated brines of the rift.

8.30.2 General Setting

Based on local conditions in Israel, thermo-mineral waters were defined as "… groundwater with a minimal TDS-content of 1000 mg/l and a constant temperature exceeding 25 °C in the temperate highlands and 33 °C in the Rift Valley …" (Rosenthal 1994). This classification covers all known springs and wells characterized by thermal anomalies as well as subaqueous hot springs outflowing in and along the shores of the Dead Sea and of Lake Tiberias. Following this classification, three major hydrothermal provinces are outlined in Fig. 8.102:

1. The Sinai-Negev Nubian Sandstone province
2. The Jordan-Dead Sea Rift Valley province
3. The Hammat Gader province

The Romans and later the Byzantines who ruled the Eastern Mediterranean lands paid particular attention to thermal water sources and developed them for curative and recreational purposes. Evidence of their engineering ingenuity can be recognized among the remnants of ancient waterworks around thermo-mineral springs: En ("the spring of") Noit in the Dead Sea area and the Hot Springs of Tiberias (Hammei Tveria or Hammath). Both springs are characteristic of two different hydrogeothermal provinces as defined by Rosenthal (1994).

The Sinai-Negev Nubian Sandstone province includes central Sinai in Egypt and most of the Negev in Israel. The province spreads fan-wise from the Gulf of Suez around the northern and eastern margins of the igneous massif of Sinai and reaches the Dead Sea and to the Gulf of Eilat within the Rift Valley. The extent of this province coincides with the areal occurrence of the reginal Lower Cretaceous Nubian Sandstone aquifer (locally known as the Kurnub Group). It comprises ferruginous terrigenous sandstone, arkose and sandy shales interbedded with marine carbonates and shales. In central and eastern Negev the Kurnub Group attains 350–400 m.

Groundwater in the Kurnub Group is of immense dimensions, estimated at billions of m^3 (Issar et al. 1972).

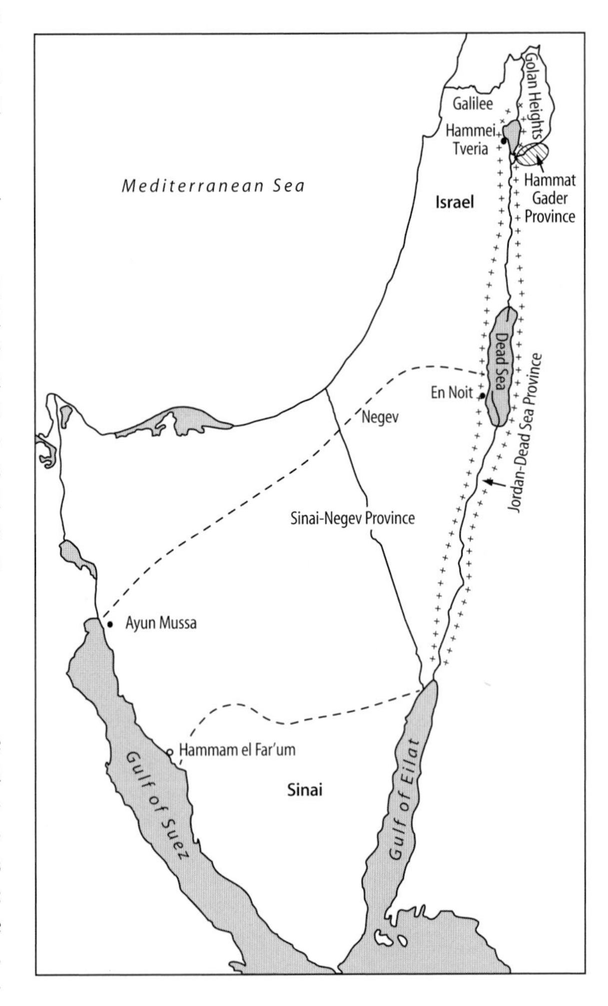

Fig. 8.102. Location map

It contains ancient water (up to 30 000 years) (Issar et al. 1972; Galai 1983) which flows from a piezometric level of +200 m in the Sinai highlands towards the Dead Sea, the Gulf of Eilat and to the Gulf of Suez. Throughout large parts of Sinai and most of the Negev, this aquifer is confined beneath a 500–900 m sequence of mostly marine sediments. Due to aquifer depth and deep confinement throughout all of this province, there are very few natural aquifer outlets, springs that occur only along the faulted margins of this province. Issar (1971) identified two of these springs along the eastern rims of the Suez Graben: Ayun Mussa (water temperature 27 °C) emerges from tension joints in Tertiary beds overlying uncomformably the Nubian Sandstone. Hammam el Far'un (water temperature 72 °C) issues from faulted Tertiary carbonates. The high temperature, high iron content, and the mineralogical assemblage of surrounding spring deposits indicate that the emerging water originate in the deeply buried Nubian Sandstone aquifer.

En Noit or the Roman spring at Boqeq, near the Dead Sea emerges on the western fault-escarpment overlooking the Dead Sea at an elevation of about 150 m above the level of the sea. Though nowadays it is a small seepage of about 1 m^3/h, the rich green vegetation around the spring surrounded by the rust-colored ferruginous sinters, stands clearly out against the beige-gray background of calcareous strata of the escarpment (Fig. 8.103). The location of the spring coincides with the intersection of two longitudinal faults which create a wedge-shaped tectonic block built of Mastrichtian-Paleocene bituminous shales (Eckstein and Rosenthal 1965; Fig. 8.104). Prior to the mid 1960s, the outflow of the spring was 7–8 m^3/h. During the last two decades the outflow decreased to less than 1 m^3/h. However, water temperature (39 °C) and the overall chemical composition has remained unchanged throughout the last 45 years.

The water of En Noit is brackish to slightly saline (TDS = 5 910 mg/l) (Table 8.31). It stands out by its relatively low pH (6.4), high Fe-content (3.0 mg/l) and high sulfate concentration (1 687 mg/l) which causes the high SO_4/Cl equivalent ratio (0.62). The water is of the Na-Ca-Cl-SO_4 type with the following typical ionic ratios: Na/Cl = 0.93; Mg/Ca = 0.35; SO_4/Cl = 0.62; Ca(SO_4 + HCO_3) = 0.8. The Cl/Br weight-ration is 412. The water is oversaturated with calcite, aragonite and siderite and undersaturated with gypsum and anhydrite. The calculated assemblage of dissolved minerals is halite = 52.7 mmol/l and anhydrite = 17.7 mmol/l. The high Fe concentration causing oversaturation with siderite explains the wide distribution of calcareous-ferruginous sinters forming a prominent morphological terrace around the spring and on the slope leading

Fig. 8.103. a En Noit surrounded by ferruginous sinters. b Remains of aqueduct leading spring water to the nearby Roman fortress (*photos:* E. Rosenthal)

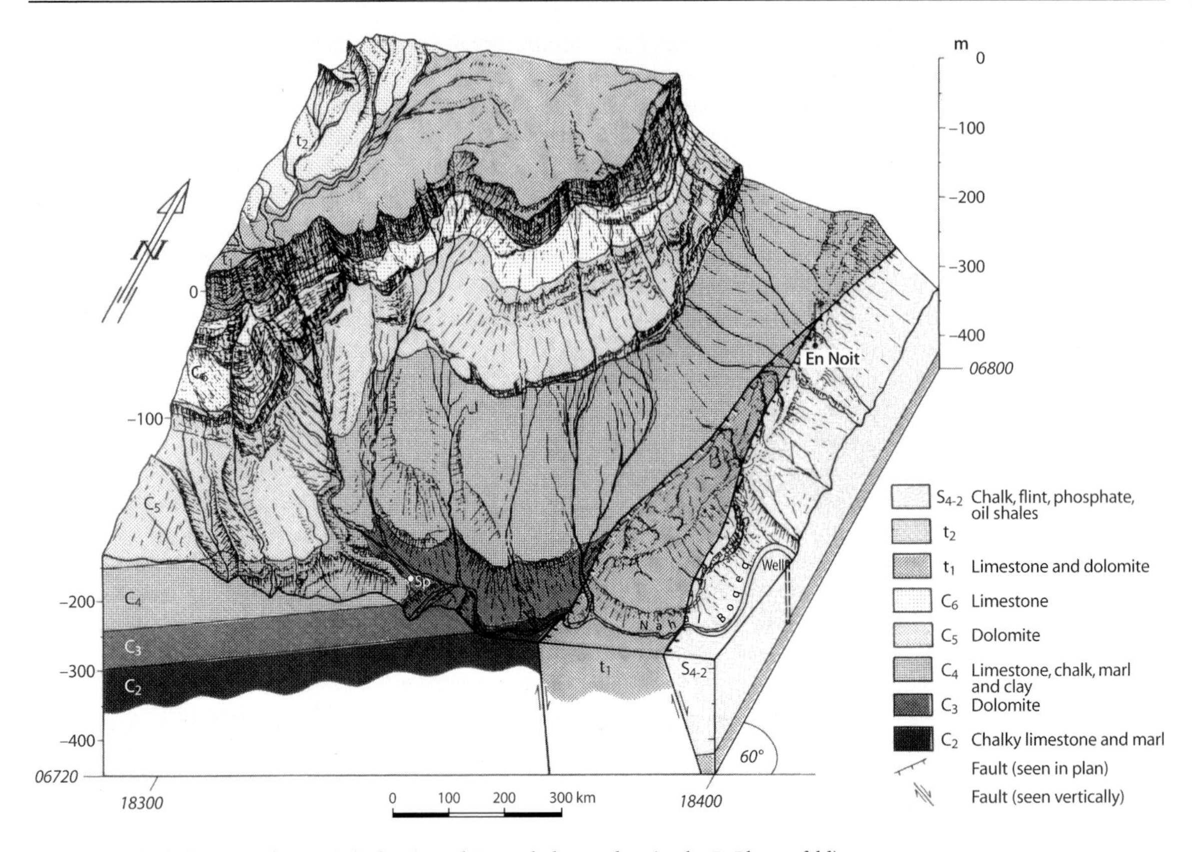

Fig. 8.104. Block diagram of En Noit (Eckstein and Rosenthal 1965; drawing by D. Blumenfeld)

to the Dead Sea. Illani et al. (1985) studied the mineralogy of the sinter and reported the occurrence of Fe, Si and Cu (>1%) and of Zn, Ni and Co (<1%). The authors suggested that Fe could have been leached from sedimentary rocks by water circulating through the tectonic zones and deposited in fissures formed in host rocks. However, they did not exclude contribution of hydrothermal solutions which rose along fracture zones. Signs of tectonic and volcanic activity associated with the rift system are plentiful in the region (Illani et al. 1985).

Though there are no precise records, it appears that in the 1950s, the water of En Noit was characterized by a low H_2S-content of 1–2 mg/l which diminished in time to nothing (Rosenthal 1986).

There is ample archaeological evidence indicating that the water of En Noit was diverted by an aqueduct to the nearby fortified settlement of Boqeq which was erected as early as the Hellenistic II-Hasmonean period (152–37 B.C.). Boqeq became famous because of the production of scents extracted from desert plants grown around the settlement. At a later stage, because of above-human-body temperature of the spring water, excessive salinity, sulphurous odor and iron content, the Romans developed En Noit as a spa which was used by the garnison of Boqeq. It became a military outpost, later to become a part of the *limes* i.e., an intricate system of strategic military fortifications protecting the eastern borders of the Roman empire. Eckstein (1968 pers. comm.) excavated the vicinity of the seepage and discovered (under a 1.5-m-thick cover of debris cemented by sinter) the remains of the Roman spa. It included two stone-protected outlets of the spring, and the remains of a Roman bath house. Of special interest is an aqueduct chiseled out in the sinter leading from the bath house down-slope to the fortress and to the soldier's settlements. The Boqeq fortified settlement and its wa-

Table 8.31.
Results of chemical analyses of water emerging from the Roman springs and comparison with relevant hydrochemical and endmembers

Parameters	Roman spring En Noit Dead Sea area	Wildcat LOT 2A Dead Sea area	Roman spring Hammei Tveria Lake Tiberias area	Wildcat Devora 2A Rift related area
Ca (mg/l)	606.00	474.00	3467.00	34820.00
Mg	130.00	121.00	614.70	1950.00
Na	1212.00	940.00	6622.00	40200.00
K	59.00	47.00	372.40	400.00
Cl	2018.00	1573.00	17715.00	129393.00
SO_4	1687.00	1140.00	770.20	178.00
HCO_3	190.00	460.00	150.80	171.00
TDS (mg/l)	1574.00	1963.00	29712.10	207112.00
Density	1.00	1.00	1.02	1.14
React. error	0.03	0.06	0.31	0.03
Ionic strength	0.13	0.10	0.64	4.61
*r*Ca	30.24	23.65	173.00	1737.52
*r*Mg	10.69	9.95	50.55	160.36
*r*Na	52.74	40.91	288.16	1749.35
*r*K	1.51	1.20	9.53	10.24
*r*Cl	56.91	44.36	499.58	3648.98
rSO_4	35.09	23.72	16.02	3.70
$rHCO_3$	3.12	7.54	2.47	2.80
Total cations	*95.18*	*75.71*	*521.25*	*3657.47*
Total anions	*95.12*	*75.62*	*518.07*	*3655.49*
Cl/Br	411.84	326.00	77.02	74.36
*r*Na/Cl	0.93	0.92	0.58	0.48
*r*Q	0.79	0.76	9.35	267.00
*r*Mg/Ca	0.35	0.42	0.29	0.09
*r*Na/K	34.92	33.99	30.22	170.82
*r*Ca/Na	0.57	0.58	0.60	0.99
rNa/SO_4	1.50	1.72	17.98	472.42
rCl/SO_4	1.58	1.87	31.18	985.43
rCl/HCO_3	18.26	5.88	201.98	1301.04
rSO_4/HCO_3	11.26	3.14	6.48	1.32
Saturation indices				
Anhydrite	−0.291	−0.485	−0.773	1.135
Aragonite	0.113	0.330	0.832	0.050
Calcite	0.247	0.465	0.954	0.140
Dolomite	−0.102	0.435	0.794	−1.357
Gypsum	−0.143	−0.330	−0.781	0.266
Magnesite	−0.503	−0.214	0.298	−1.180
Siderite	0.027	−0.008	−7.192	−6.052

Table 8.31.
Continued

Parameters	Roman spring En Noit Dead Sea area	Wildcat LOT 2A Dead Sea area	Roman spring Hammei Tveria Lake Tiberias area	Wildcat Devora 2A Rift related area
Dissolved minerals				
NaCl	53.060	52.827	56.601	49.357
KCl	1.799	1.931	2.387	0.368
$MgCl_2$	2.280	2.465	7.897	3.657
$CaCl_2$			20.038	46.432
Total chlorides	*57.139*	*57.223*	*86.923*	*99.814*
$CaSO_4$	35.354	35.593	3.891	0.121
K_2SO_4				
Na_2SO_4				
$MgSO_4$	4.954	0.092		
Total sulfates	*40.308*	*35.685*	*3.891*	*0.121*
$MgCO_3$	2.261	7.031	0.176	0.029
$CaCO_3$			0.209	0.034
Total carbonates	*2.261*	*7.031*	*0.385*	*0.062*

terworks persisted until the Arab conquest in the 7th century. During the following 1 200 years, bedouins from all over the Negev, Sinai, and Transjordan made the strenuous journey through the desert and the exhaustive climb of the escarpment to enjoy the curative properties of the spring.

The water emerging from En Noit is hydrochemically identical to sulfate- and iron-rich Kurnub Group paleowater flowing in the northwestern parts of the Sinai-Negev province. This is confirmed by direct hydrochemical evidence from wildcat Lot 2A drilled approximately 7 km southwest of the spring which penetrated the whole section of the Kurnub Group (Table 8.31). Rosenthal et al. (1996) showed that within the highly faulted zone close to the fault-escarpment which outlines the western margins of the Dead Sea area, the paleowater of the Kurnub Group flowing through the confined aquifer from Sinai Highlands towards the Dead Sea change their chemical composition owing to mixing with the typical Ca-chloride water of the rift. Netpath modeling (Plummer et al. 1994) revealed that the mixing process (with a very small percentage of brines) also involved the dissolution of dolomite, gypsum, halite, and CO_2 gas and the precipitation of calcite and K-smectite. The dissolving halite and gypsum are most likely residuals from the higher levels of the proto Dead Sea. Radon emanation typical of En Noit waters is explained in the following way: in the immediate vicinity of the spring, the confined, artesian groundwater of the Kurnub Group rises along the longitudinal faults leading to the spring and leach (along the flow path) U-bearing bituminous shales of Campanian age (Rosenthal 1986).

Balneologists suggested that En Noit waters could be therapeutically exploited both by bathing (to treat various skin ailments) and by ingestion (to treat chronic constipation) (Atlas 1962).

The Jordan-Dead Sea River Valley Province forms part of the Syrian-Red Sea-East African Rift System. The groundwater of this vast province are thermal, radioactive, and sulphurous. The dominant hydrochemical feature of groundwater in this province is the occurrence of Mg- and Ca-chloride brines mixing with fresh groundwater draining from replenishment areas adjacent to the rift (Rosenthal 1994).

The mountainous area of eastern Lower Galilee slopes down to the Tiberias Basin in a series of step-faulted blocks built of Upper Cretaceous to Tertiary calcareous rock formations covered by Neogene fluvial sediments and multiple Neogene to Pliocene basalt flows. The Golan Heights in the east are built essentially of a similar section. A host of warm and hot springs are scattered along the shores of Lake Tiberias and on its floor. It seems that their occurrence is tectonically controlled as in other parts of the rift. The most important springs are those emerging along the western shore of the lake coinciding with the western fault of the rift (Fig. 8.105).

Fig. 8.105.
Geological section of Tiberias Hot Springs (Hammei Tveria) (Golani 1962)

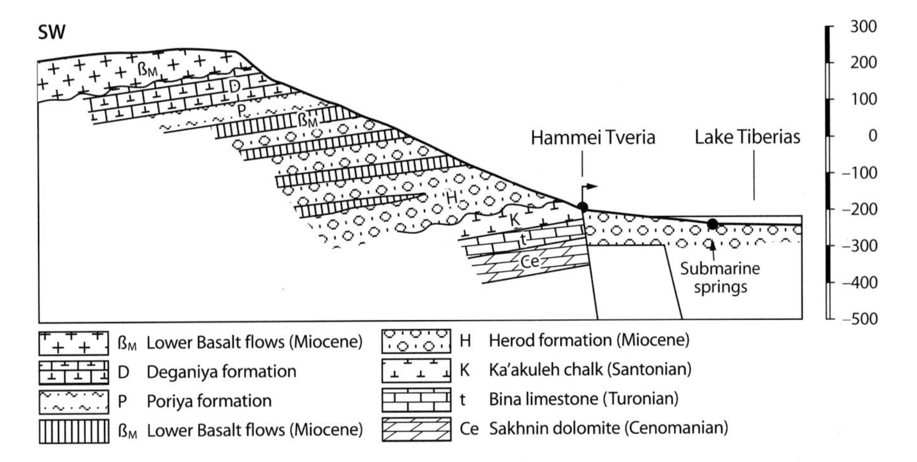

Hammei Tveria or Hammath (Tiberias Hot Springs) stand out due to their high salinities (up to 17 800 mg/l Cl) and elevated temperatures (59–62 °C). The three main springs forming this group emerge close to a major fault of the rift system (Golani 1962). Among the three springs, the Roman Spring contains the highest concentration of Cl (17 714 mg/l) and the temperature of outflowing water is 59 °C (Eckstein 1975; Rosenthal 1994). The outflow of the springs is 100 m^3/h. Eastward, at a distance of 100 m, a host of submarine springs of similar salinities and temperatures emerge at the bottom of the lake. Only 10% of the water outflowing from the spring is exploited for baleneological purposes and the remaining amount is diverted.

The water of the Roman Spring is saline, very clsoe to that of seawater and has a pH of 6.6 (Table 8.31). The water is of the Na-Ca-Cl type and is characterized by the following ionic equivalent ratios: Na/Cl = 0.58; Mg/Ca = 0.29; SO_4/Cl = 0.03; Ca/(SO_4 + HCO_3) = 9.4. The Cl/Br weight-ratio is 77. The water is oversaturated with all common carbonate minerals (calcite, aragonite, dolomite and magnesite) and undersaturated with sulfates. The calculated assemblage of dissolved minerals (in mmol/l) is halite = 288; calcite = 53; dolomite = 25 and anhydrite = 8. The water contains 136 mg/l CO_2 and 1.9 mg/l H_2S.

Hammath or Hammei Tveria has been known since prebiblical times. It was mentioned as an area of recreation in the 13th century B.C. Egyptian Anastasy A papyrus. In the Old Testament it was mentioned twice as a fortified city (Book of Joshua, 19 and 21). The curative properties of its water were so important that Jewish rabbinical law allowed bathing in the spring even on Saturdays. The Talmud recommends bathing in the spring to cure skin diseases. Pliny, the Roman historian praised the curative properties of the water and described the luxurious baths and spa visited by sick people from all over the Roman empire. A coin minted during the reign of Traianus Caesar (1st century C.E.) depicts Hygea the goddess of health, residing by the hot spring of Tiberias. Excavations nearby revealed a sequence of archaeological remains covering the time-span between the Hellenistic period (3rd century B.C.) and the late Jewish settlements in the 5th century C.E. (Fig. 8.106). Friedman (1913), renown for his pioneering studies of the Dead Sea, made a detailed historical survey of this spa. It goes back as far as the Roman conquest and includes descriptions of travelers during the Talmudic and Arab periods to the survey of the first analyst of the waters of the thermae, the Swedish naturalist Hasselquist (1761; Eckstein 1975). The modern medical evaluation of the thermo-mineral water of Hammei Tveria was made by Lachmann (1933, 1934), by Buchman (1928, 1943, 1971), and by Kurland (1971). Nowadays, the water of the Roman spring is utilized for the treatment of rheumatic diseases. This is the balneotherapeutic center in Israel.

The source of salts in the thermo-mineral water of Hammei Tveria was investigated by numerous authors who suggested different and often contradictory mechanisms and processes (Vengosh and Rosenthal 1994). Considering that the maximal Cl-content in Hammei Tveria water is almost identical to that of seawater, Mazor and Mero (1969), Kafri and Arad (1979), and Kafri (1996) suggested that the salinity is of marine origin and is derived from penetration of Mediterranean Sea water inland into the northern Rift Valley. On the other hand, Starinsky (1974) showed that the previously mentioned

Fig. 8.106. **a, b** Outlet of Roman Spring near Tiberias and the nearby excavated remains of ancient bathhouse and aqueduct (*photos:* E. Rosenthal)

ionic ratios indicate influence of Ca-chloride brines which were formed in the Dead Sea area and migrated northward within the Rift. Starinsky (1974) also suggested that the source brine might be similar to the water encountered in wildcat Rosh Pina 1 (i.e. 105 230 mg/l Cl) which is of typical Ca-chloride composition. Simon and Mero (1992) reported the occurrence of thermal Ca-chloride brine (approximately 180 000 mg/l Cl and 130 °C) in a thick sequence of Neogene evaporites in wildcat well Zemah 1 located in the Rift Valley, less than 10 km south of Hammei Tveria. This brine might have been related to a saline lake which existed there during the Neogene (Starinsky 1974; Rosenthal 1988). According to Starinsky, the low Mg/Ca ratio values in the hot spring and in other saline sources could be the result of dilution of evaporated seawater which was also involved in secondary diagenetic processes such as dolomitization. Gat et al. (1969) argued that the oxygen isotope composition of the saline water indicates dilution of evaporated seawater by freshwater. The deviation from Ca-chlroide brines were also confirmed by the occurrence of high $^{11}B/^{10}B$ ratio ($\delta^{11}B = 44\%$) in the water of Hammei Tveria, indicating mixing with a brine component characterized by a high $\delta^{11}B$ value, such as found in the Dead Sea brines (Vengosh et al. 1991).

The origin of the Rift Valley brines and the occurrence of Ca-chloride groundwater bodies has been investigated by several authors (Bentor 1969; Starinsky 1974; Rosenthal 1988). The latter described the Devora type of deeply confined, thermal and hypersaline Ca-chloride brines. The end-member of this brine-type was identified in drill-stem tests in wildcats Devora 2A, Rosh Pina 1 and Sarid 1 which were drilled in areas adjoining the rift and hydraulically connected to it. The chemical characteristics of these brines are as follows (Table 8.31) (r = meq/l; w = mg/l):

- $rC1 > rHCO_3 > rSO_4$
- $rCa > rNa > rMg$
- pH = 4.7
- TDS = 209 g/l
- rMg/rCa = 0.1
- rNa/rCl = 0.47
- rNa/rK = 37–49
- $rCa/(SO_4 + HCO_3) = 186$
- wCl/wBr = 57

The piezometric heads of these brines were calculated from pressure release tests and were found to be, in all cases, in the 40–105 m below msl (mean sea leve) range, i.e., usually exceeding ground surface elevation.

According to Rosenthal (1988) and Illani et al. (1988), the Devora-type brines were formed in the southern part of the country as a result of seawater evaporation that occurred during the Cambrian-Lower Cretaceous continental time span. Due to subsequent structural events, the initial Mg-rich brines migrated northward under confined conditions and with increasing contact with carbonate rocks. These inferred flow conditions could provide a suitable environment for the chemical diagenesis of brine, i.e. the gradual exchange of Mg with Ca-ions inducing dolomitization of marine limestones interfingering and replacing north- and northeastward continental formations. Another brine-forming mechanism was proposed by Rosenthal (1980). It is based on the fact that these deep-seated brines (up to depths of 5 227 m) have been found to occur together with high concentrations of methane gas. Very low pH values (4.1) are characteristic of these brines. Saline and acid fluids occurring at these depths could react with surrounding limestone formations creating Ca-chloride solutions.

The high pressures characterizing the rift brines could be related to their confinement under a 3 000–4 000 m thick column of sediments. The lithostatic pressures exerted on the brines could have been augmented by the Late Cretaceous to Late Tertiary orogenic folding and by plate and block movements (Freund et al. 1975). On the other hand, the high pressures could also be generated by gases such as methane measured at depths and CO_2, which could be assumed to exist at depths (if one considers the chemical reactions previously suggested) and the measured pH values.

8.30.3 Conclusions

The Jordan-Dead Sea Rift Valley with its fault-conditioned structures, is an ideal area for the present-day manifestation of brines occurring in its deep horizons. Due to deep and dense faulting along the margins of the rift, hydrological contact was established with the deep-seated formations storing the brines, thus faciliating their upflow and intermixing with water from other aquifers. Hence, the water of En Noit is a mixed product of small amounts (1%) of Ca-chloride brine occurring in the rift and paleowater flowing from Sinai towards the Dead Sea in the Kurnub Group aquifer. The water of Hammei Tveria is a mixture between fresh groundwater draining the eastern Galilee Cenomanian-Turonian and Eocene calcareous aquifers (Kafri 1996) and the similar Ca-chloride deep-seated brines. The approximate mixing ratio is 9% brine and 81% fresh groundwater from calcareous aquifers.

References

Atlas M (1962) The newly discovered medicinal drinking spring water in Israel. Unpublished report to the Ministry of Development

Bentor YK (1969) On the evolution of subsurface brines in Israel. Chemical Geology 4:1–2, 83–110

Buchman M (1928) Die Thermen von Tiberias and Tiberias als Winterkurort. Palaestina 8–10, Wien

Buchman M (1943) Tiberias Hot Springs. Harefuah – Journal of the Pales Jewish Medical Assoc 25, 8

Buchman M (1971) Therapeutic results in Tiberias Hot springs. In: Yaacov E (ed) Tiberias Hot Springs. Tiberias Hot Springs, Inc

Eckstein Y (1974) Terrestrial heat flow in Israel. PhD Thesis at the Hebrew University in Jerusalem (in Hebrew, English abstract)

Eckstein Y (1975) The thermomineral springs of Israel. Rep of the Israel Health Resorts Authority, Jerusalem

Eckstein Y, Rosenthal,E (1965) The En Noit Spring. Rep of the Geological Survey of Israel (report to the Ministry of Development – in Hebrew)

Freund R, Goldberg M, Weissbrod T, Druckman Y, Derin B (1975) The Triassic-Jurassic structure of Israel and its relation to the origin of the Eastern Mediterranean. Israel Geol Surv Bull 65:1–26

Friedman A (1913) Beiträge zur chemischen und physikalischen Untersuchung derThermen Palaestinas. 4e Veröff d Gesellsch für Palaestina-Forschung

Galai A (1983) Geology and hydrology of the Arava Valley (based on water-well data). Internal, unpublished report of the Oil Exploration-Israel Co., Ltd., Tel Aviv

Gat JR, Mazor E, Tzur Y (1969) The stable isotope composition of mineral water in the Jordan Rift Valley. J Hydrol 7:334–352

Golani U (1962) The geology of Lake Tiberias region and the geohydrology of the saline springs. Report of Tahal, Water Planning for Israel (in Hebrew, English abstract)

Hasselquist A (1761) Reise nach dem Morgenlande von 1749–1752. Rostock

llani S, Kronfeld J, Flexer A (1985) Iron-rich veins related to structural lineaments and the search for base metals in Israel. J of Geochemical Exploration 24:197–206

llani S, Rosenthal E, Kronfeld J, Flexer A (1988) Epigenetic dolomitization and iron mineralization along faults and their possible relation to the paleohydrology of southern Israel. Appl Geochem 3:487–498

Issar A, Rosenthal E, Eckstein Y, Bogosh R (1971) Formation waters of springs and mineralization phenomena along the eastern shore of the Gulf of Suez. Bull of the IASH 16,3:25–44

Issar A, Bein A, Michaeli A (1972) On the ancient water of the Upper Nubian sandstone aquifer in Sinai and southern Israel. J Hydrol 17:353–374

Kafri U (1996) Main karstic springs of Israel. Environmental Geology 27:80–81

Kafri U, Arad A (1979) Current subsurface intrusion of Mediterranean sea water – a possible source of groundwater salinity in the Rift Valley system, Israel. J Hvdrol 44:267–287

Kurland L (1971) Radioactivity of mineral springs of Israel with special emphasis on Tiberias Hot springs. In: Yaacov E (ed) Tiberias Hot Springs. Tiberias Hot Springs, Inc.

Lachmann S (1934) Les eaux thermales sulphurees de Tiberiade et de EI-Hamme. Folia, Medicinae Internae Orientalia 1:3–4

Mazor E, Mero F (1969) The origin of the Tiberias-Noit mineral water association in the Tiberias-Dead Sea Rift Valley. J Hydrol 7:318–333

Plummer LN, Prestemon EG, Parkhurst DL (1994) An interactive code (NETPATH) for modeling NET geochemical reactions along a flow PATH, Version 2.0. Geological Survey, Water Resources Investigations Report, 94-4169, Reston, Va
Rosenthal E (1980) Hydrogeology and hydrogeochemistry of the Bet Shean, Harod and Mehola Valleys, Israel. PhD Thesis, Hebrew University in Jerusalem, pp 247 (in Hebrew, English abstract)
Rosenthal E (1986) The oasis of En Boqeq. Unpublished report of the Hydrological Service of Israel (in Hebrew)
Rosenthal E (1988) Ca-chloride brines at common outlets of the Bet Shean-Harod multiple aquifer system, Israel. J Hydrol 97:89–106
Rosenthal E (1994) The hydrogeothermal provinces of Israel. IAH-International Contributions to Hydrogeology, vol 15:113–136
Rosenthal E, Jones BF, Weinberger G (1996) The chemical evolution of Kurnub Group paleowater in the Sinai-Negev province, a mass balance approach
Simon E, Mero F (1992) The salinization mechanism of Lake Kinneret. J Hydrol 138:327–343
Starinsky A (1974) Relationship between Ca-chloride brines and sedimentary rocks in Israel. PhD Thesis, Hebrew University in Jerusalem (in Hebrew, English abstract)
Vengosh A, Rosenthal E (1994) Saline groundwater in Israel: its bearing on the water crisis in the country. J Hydrol 156:389–430
Vengosh A, Starinsky A, Kolodny Y, Chiras AR (1991) Boron isotope geochemistry as a tracer for the evolution of brines and associated hot springs from the Dead Sea, Israel. Geochim Cosmochim Acta 55:1689–1695

V. Karise · P. Vingisaar

8.31 Estonian Mineral and Bottled Waters

In Estonia, bottling of recently discovered local mineral water was started in the 1960s. The interest in mineral water grew quickly and the production increased and export began. According to the requirements of the former Soviet Union standards, in Estonia the groundwater is considered a mineral water when its total dissolved solids (TDS) is >2 g/l. The curative properties of Estonian mineral waters have been assessed at the former All-Union Institute of Physiotherapy and Spa Treatment, comparing them with well-known mineral waters of the Soviet Union.

In Estonia, mineral waters occur in the southern part of the country (Fig. 8.107) where deeper groundwater strata (usually >200 m from the surface) are less affected by infiltrating freshwater. Mineral waters are found mainly in Cambrian and Late Proterozoic (Vendian), less often in Devonian sandstones (Fig. 8.108). The aquifers are under artesian pressure, with piezometric levels 10–40 m above the surface and free outflow yields usually vary within 2–4 l/s. Due to the relatively shallow depth of the aquifers as well as the platform tectonic regime, the waters are cold (approximately 10 °C) and do not contain carbon dioxide. Hydrochemically, the waters are of similar composition, predominantly of Cl-Na-type, where heightened contents of Ca and Br often occur. Sulphate waters are found in Devonian rocks. The first occurrence of mineral water in Estonia was discovered in the crystalline basement near Pärnu town. Later, several others were discovered, however, attempts to develop these mineral waters by wells has not been successful.

There are no mineral water springs in Estonia, and to date all mineral water is obtained from wells. Some Estonian mineral waters have been accepted in Estonia and neighboring countries for their natural curative properties. The natural conditions, occurrence, and location are good and they should be described more precisely, especially the mineral water at Värska and Kuressaare.

The wells supplying Värska mineral water are in the southeastern part of Estonia near a creek discharging to Lake Peipsi. The location is ecologically undisturbed and famous for its beautiful nature. During 1968–1975, four different mineral water aquifers were discovered there. It was also discovered that mud on the bottom of Lake Peipsi proved to have curative properties. There-

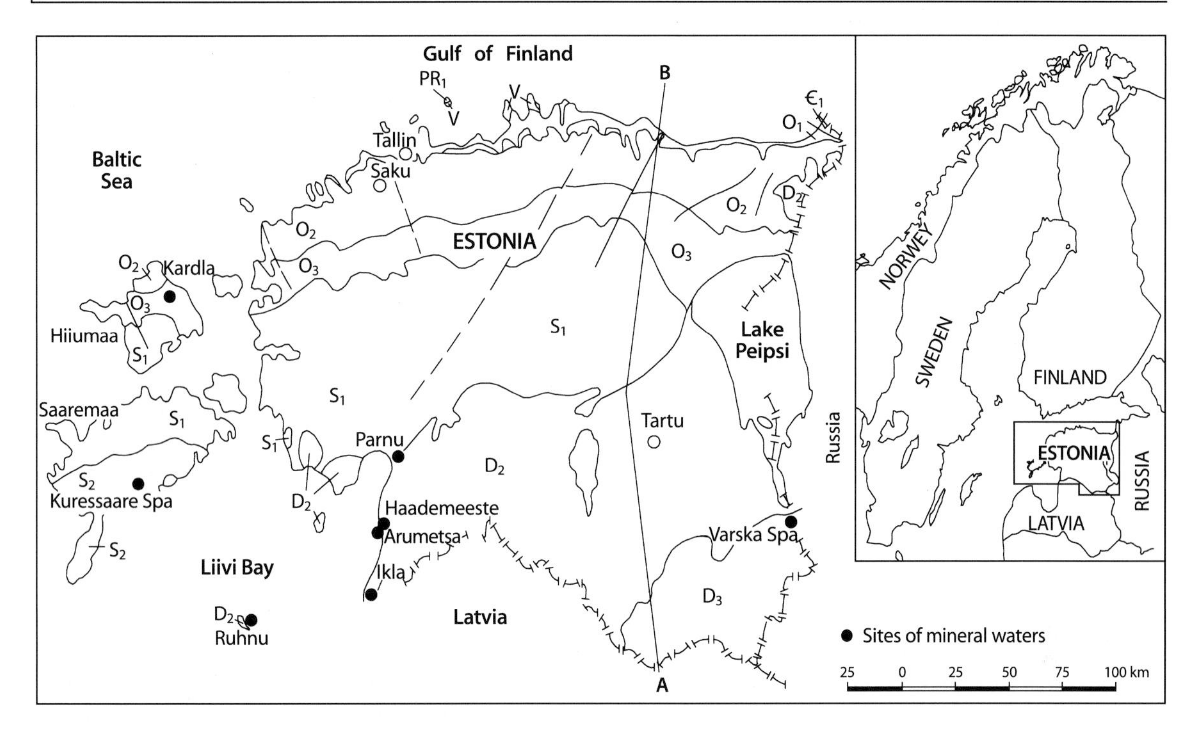

Fig. 8.107. Distribution of mineral waters of Estonia. Mineral waters occur in the southern part of the country

fore, the combination of mineral water and curative mud led to the establishment of Värska Spa where treatment began in 1980.

The mineral water at Värska was found in four aquifers occurring in sandstones and siltstones: Middle-Lower Devonian (D_{2-1}), Ordovician-Cambrian border zone (O-C), and the two Cambrian-Vendian (Neoproterozoic) aquifers separated by a clay layer ($C\text{-}V^2$ and $C\text{-}V^1$). The basic data of these aquifers are presented in Table 8.32.

The bitter-tasting Värska-I mineral water is of SO_4-Cl-Ca-Na-Mg type. After discovery, it was bottled in insignificant amounts for only a couple of years. The bottling of this water was used in some hospitals for the treatment of the digestive tract, bile, and liver diseases. In the mineral waters of lower aquifers, Cl and Na ions predominate. The Värska-II mineral water was the first mineral water in Estonia and has been bottled since 1968. It is and has been the best known and bottled water in Estonia. It has been recommended for drinking to patients suffering from chronic digestive tract diseases (mainly hypoacidity) and some metabolic diseases. To date there are two companies bottling Värska-II mineral water. Descriptions are presented as follows according to their trade denominations of the bottled water.

Värska originaal (Värska original) is produced by Värska Vesi Ltd. (Värska, Estonia). The company started bottling in 1973. Värska-II (Cl-Na-Ca-[Mg] type) mineral water with 2.2 g/l TDS is obtained from a borehole located at Värska. It is bottled naturally, adding carbon dioxide. In the 1980s annual production of this mineral water was 17 million 0.51-l bottles. Värska Vesi Ltd. is the largest producer of mineral water in Estonia.

Värska ravi-lauavesi (Värska curative table water) is produced by Värska Vesi Ltd. is diluted in ratio 1:1 with groundwater of HCO_3-Mg-Ca type and a TDS of 0.4 g/l. Carbon dioxide is added. The groundwater is obtained from a local borehole at a depth of 145 m. TDS of the water bottled is 1.3 g/l. The operating capacity of the plant is about 1.5 million 1-l bottles annually.

Mineraalvesi Värska (Mineral water Värska) is produced by Tartu Brewery Ltd. (Tartu, Estonia). This bottling was begun in 1968 after discovering Värska-II mineral water. The mineral water is mixed with HCO_3-Mg-Na-Ca-type groundwater with a TDS of 0.5 g/l at a ratio of 1:1, with the fresh groundwater being pumped from a

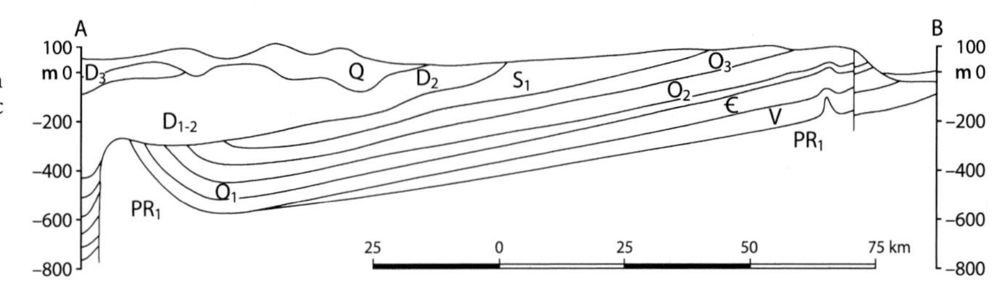

Fig. 8.108.
Geological cross section. Mineral waters are found mainly in Cambrian and Late Proterozoic

Table 8.32.
Main data of mineral water aquifers at Värska, Estonia

Aquifer	Depth of ceiling (m)	Water level over surface (m)	Yield on outflow (l/s)	TDS of water (g/l)	Name of water
D_{2-1}	259	12.8	17.1	4.6	Värska-I
O-C	451	≈20	4.8	2.2	Värska-II
$C\text{-}V^2$	520	20.1	12.5	5.9	Värska-IV
$C\text{-}V^1$	572	30.6	2.2	15.7	Värska-III

210 m deep borehole at the brewery. During bottling, carbon dioxide is added. The TDS of the water bottled is 1.3 g/l. In 1994, annual production was 8.8 million 0.33-l glass bottles.

Mineral water from the deeper aquifer Värska-III is not suitable for drinking because of its high TDS (Br is 42.6 mg/l). However, it is recommended for curative baths for treatment of chronic diseases of the skeletal-articular system, gynecological and peripheral nervous system diseases, and also some heart and blood vessel system diseases. At Värska Spa this mineral water has been used for baths since 1980.

The Ikla mineral waters from the southwestern border of Estonia on the coast of the Liivi (Riga) Bay and Ruhnu (on Ruhnu Island in the Liivi Bay) are almost similar to Värska-III. Ikla and Ruhnu mineral waters occur in the Ordovician-Cambrian border zone, at depths of 645 and 707 m, with TDS of 14.0 and 17.1 g/l and Br content 31 and 48 mg/l. In 1969–1974, Ikla mineral water diluted with freshwater at ratios 1 : 4 and 1 : 1 was bottled by Pärnu Brewery under the trade names Ikla-20 and Ikla-50. Ruhnu mineral water has never been used.

At Kuressaare Spa on Saaremaa Island, two boreholes drilled in 1989 tapped basal layers of the Cambrian, producing two varieties of Kuressaare mineral water; at a depth of 458–502 m, $Cl\text{-}HCO_3\text{-}Na$ type-water with TDS of 2.1 g/l was found, and at a depth of 540–555 m, Cl-Na-Ca type-water with TDS of 3.8–4.0 g/l was found. During 1990–1992, both were bottled with annual production reaching 0.17 million 0.33-l bottles. These mineral waters are used at Kuressaare Spa for curative drinking water and baths. The resort at Kuressaare utilizes mud excavated from the bottom of a nearby lake, mixed with mineral water for a curative effect. Mineral waters bottled at an earlier period in Estonia, 1970 to 1992 (Häädemeeste, Pärnua) included Arumetsa and Häädemeeste by Pärnu Brewery. Kärdla mineral water (Kärdla) on Hiiumaa Island was bottled in 1985–1992.

There are three main freshwater bottled products on Estonia's market today. "Siller" Spring water, Tartu Brewery Ltd. bottles $HCO_3\text{-}Mg\text{-}Na\text{-}Ca$ type-water with TDS of 0.5 g/l and fluorides 1.2–1.5 mg/l, adding carbon dioxide. The water originates from Silurian carbonaceous rocks at a depth of 210 m. This water has been produced since 1994, and in 1996 the production was 1 million 0.33-l glass bottles. Saku Vichy Classique is produced by Saku Brewery Ltd. (Saku, Harju County, Estonia) and is produced from groundwater obtained from wells in the Cambrian-Vendian sandstones at a depth of 220 m. It is a $Cl\text{-}HCO_3\text{-}Na\text{-}Ca\text{-}Mg$ type with TDS of 0.33 g/l. Manganese and iron are separated from the water and magnesium and potassium are added along with carbon dioxide. The TDS of this bottled water is 0.6 g/l. Bottling of this water began in May 1995 and in 1996 the production was 2 million 0.33-l bottles.

Bonaqua bottled water has been produced by AS Coca-Cola Joogid (Tallinn, Estonia) since 1995. Part of this production is exported from Estonia.

Y. Daoxian

8.32 The History of Mineral Water Exploitation in China

8.32.1 Introduction

According to historical documents, Chinese people knew about mineral water, especially hot springs, long ago. A book, *Annotation on water scripture,* written by Li Daoyuan, who lived in Beiwei Dynasty (A.D. 386–543), described 41 hot springs, most of them in North China. In Volume 38 of Geography Dictionary compiled in the time of the Kangxi Emperor (A.D. 1662–1722) of the Qing Dynasty, 78 hot springs were recorded and knowledge of mineral and hot springs grew rapidly in this century. In the book, *Study on hot springs in China* (Chen Yanbing 1939), 584 springs were described, including 103 in Guangdong Province, and 1 in Tibet. In 1956, the book *Major hot springs in China* (Zhang Hongzhao 1956) was printed by the Geological Publishing House and contained records for 972 hot springs in China. In 1973, the report *Methodology for geothermal water survey and exploration* was compiled by the Institute of Hydrogeology and Engineering Geology, and published by the Geological Publishing House. It contains records of more than 2 000 hot springs and exploration wells in China.

The political center of China was originally in North China where the exploration for mineral and hot water was started in the middle reach of Yellow River, the cradle of Chinese Culture. It expanded from the North to the South. For example, Lishan hot spring, 25 km to the east of Xian City, now the provincial capital of Shaanxi province, at an earlier date was the capital of China for about 2 000 years (1134 B.C.–A.D. 907). The city, or Lishan Spring, was used by many monarchies of China for medical treatment by King Zhouyou (781–771 B.C.), the last emperor of Zhou Dynasty. The Anning hot spring, 30 km southwest of Kunming City, the capital of Yunan province in south China, has been known since the time of King Guangwu (A.D. 25–56) of the Eastern Han Dynasty. The famous General Su Wenda visited the spring for treatment of pernicious malaria, a subtropical disease. It is recorded that he was not able to return to the capital (Luoyang, Henan province) with his army, but was eventually cured by daily bathing in the hot spring.

The Xinzi hot spring at the south of Lushan Mt. is a well-known summer resort in Jiangxi province. It is in the middle reach of the Yangtze River and was recorded from the time of the Eastern Jin Dynasty (A.D. 317–420). It has been highly praised by many ancient poets. The hot spring in Huangshan Mt., another famous summer resort of China, is in Anhui province, eastern China. Its use began in the Tang Dynasty (A.D. 618–907), when Xue Yong, the governor of Xixian county was cured from an epidemic sickness by bathing in it.

The first scientific summary of mineral water in China was made by Li Shizhen (A.D. 1518–1593), a well-known ancient Chinese pharmacologist in Ming Dynasty. In his book *Compendium of Materia Medica*, he classified mineral waters in China according to chemical contents, into sulphur, cinnabar, vitriol, realgar, and arsenic springs. On the basis of water taste, he classified mineral waters into sour, bitter, salty, cold, and hot springs. He also described the medical effects of mineral waters, and treated dermatosis, rheumatism, and other diseases.

8.32.2 Uses of Mineral Water in Ancient China

The most popular use of mineral water in ancient China was for medical treatment, but there were also records of employing hot springs for cooking and agriculture.

8.32.3 Medical Use

In the book *Annotaton on water scripture*, published in Beiwei Dynasty (A.D. 386–543), Li Daoyuan wrote: "the Huangnu hot spring on the Shahe River, Lushan, Henan is so hot that rice can be cooked in it. It was reported that a Taoist Priest drank water from the spring three times per day and bathed in it. All his diseases were cured in 40 days."

According to historical records, the King Yingzheng, the first Emperor of Qin Dynasty (221–210 B.C.) cured his dermatosis and King Li Shiming (A.D. 627–649) of Tang Dynasty recovered from rheumatism, both by taking baths in the Lishan hot spring at Xian City.

8.32.4 Agriculture Use

In the book *History of Han Dynasty* written by Sima Biao in the period of the Jin Dynasty (A.D. 265–420), it is recorded that there is a hot spring 3 km to the south of Cunzhou City, Hunan province. Irrigated by the warm water, several hectares of ricefield downstream from the spring can be sowed in December, and cropped in March of the next year, thus the hot spring makes it possible to raise three crops every year. Similar practices are recorded at Dongze hot spring, Xinyang county, Hubei province.

In Tang Dynasty (A.D. 618–907), melon irrigated by Lishan hot spring east of Xian can be harvested in February.

8.32.5 Modern Bottled Mineral Water Industry in China

It is reported that a mineral water spring in the Wutaishan Mt., Shanxi province was used for drinking 1 000 years ago, and the Weina Mineral Water in Inner Mongolia Autonomous Region was exploited 250 years ago. This information however has not been verified. The first commercial bottled mineral water industry in China was Laoshan Mineral Water, produced near the coastal city of Qingdao, Shandong province. It was started in 1931 and stopped producing during the World War II. However, since 1962, the production of Laoshan Mineral Water has been restored and it is sold on the Hong Kong market. Production from several other mineral water enterprises have also been put onto the market since the 1960s, including the Weina Mineral Water of Inner Mongolia Autonomous Region, the Longchuan Mineral Water of Guangdong province, and the Pikou and Tanggangzi Mineral Waters of Liaoning province.

In 1987, the Standard for Natural Drinking Mineral Water of the People's Republic of China (GB-8537-87) was issued by the National Technology Supervision Bureau. In less than 10 years, the known sites of mineral water in China have increased to more than 2 000, with an exploitable resource of 280 000 000 m^3/yr. Among them, more than 500 sites have been authenticated, and about 250 sites exploited, with an annual production of more than 1 million t. It is expected that the annual production of bottled mineral water in China will be about 5 million t in the year 2000.

G. Dörhöfer · G. Goldberg

8.33 The Rhume Spring in the Foothills of the Harz Mountains, Northern Germany

8.33.1 Geological Framework

The Rhume Spring is situated a few kilometers south of the Harz Mountains, a Variscian outcrop of mainly Carboniferous and Devonian rocks, which in the south are overlain by Zechstein deposits. The Harz Mountains reach up to about 1 000 m elevation and are the major orographic element in northern Germany, which otherwise is characterized by glacial Quaternary deposits of the last ice ages and bordering Mesozoic hills of mainly halotectonic origin.

The occurrence of the spring is closely related to the geological framework of the Pöhlde Basin, a horst structure which is developed at a width of about 10 km to the south of the Harz Mountains containing Zechstein rocks of up to 300 m thickness (Fig. 8.109). In the south, the Pöhlde Basin is limited by the Bunter Sandstone complex of the Eichsfeld swell region. Karst features, like sinkholes and caves, can be found throughout the basin due to the solution of the Zechstein anhydrite and gypsum rocks.

The Eichsfeld swell runs in a southwesterly direction. The substratum of the Pöhlde Basin is made up of Hauptanhydrit (main anhydrite) of the Odertal horst between the two Bunter Sandstone blocks of the Rotenberg and Herzberg.

The geological position of the spring is determined by a NW-SE trending fault with a displacement height of about 120 m, where the silty claystone sequence of the Lower Bunter is thrown down against the Zechstein dolomites (Herrmann 1969).

8.33.2 Hydrogeological Situation

The best water conducting units within the Zechstein sequence are karstified gypsum and dolomites. A Quaternary cover developed due to subrosion of the Zechstein sulphates in a topographic depression filled with river deposits of the Oder and Sieber, which originate in the Harz. These gravel and sand deposits constitute an aquifer of less significance.

Because of the existence of alternating layers of low and good permeability, coupled with diverse stages of

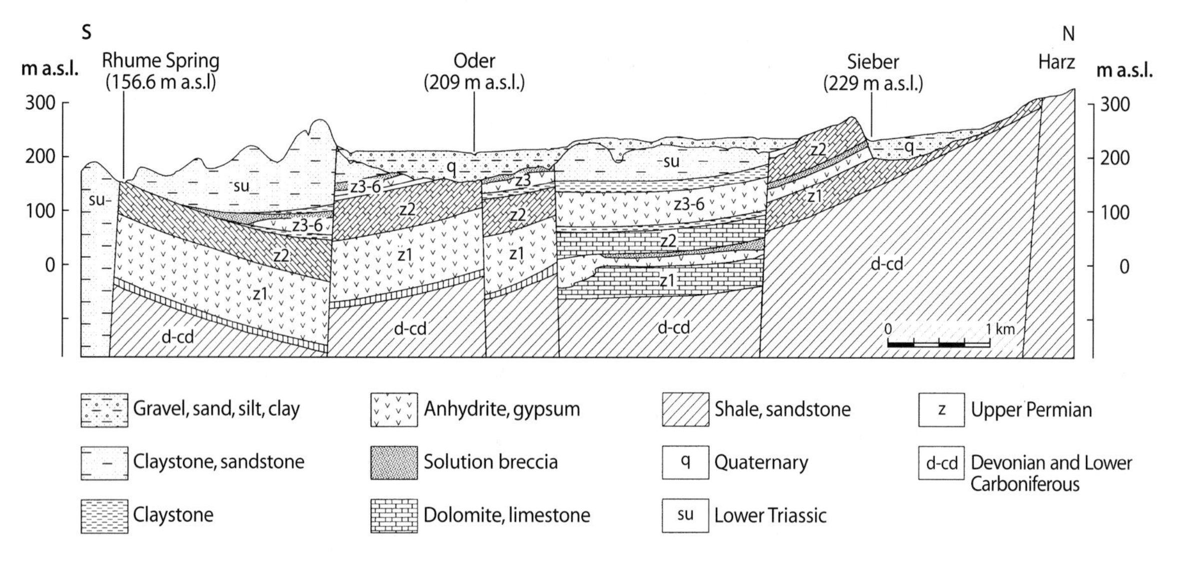

Fig. 8.109. Geological cross section of the catchment area of the Rhume Spring (modified after Grimmelmann 1992)

karstification, several aquifer horizons are developed in the Zechstein formation. In general, in most of the area, the groundwater table of the shallow aquifer in the river deposits lies above the piezometric surfaces of the deeper aquifers in the Zechstein. The karst groundwater table lies at about 50–60 m below land surface and fluctuates by up to 10 m in some places. Hydraulic contacts between all aquifer horizons are obvious. In many places along the river courses, surface water seeps away either directly into the karst aquifers or into the shallow aquifer, which in turn feeds the deeper aquifer via sinkholes.

The Rhume Spring develops along the fault in a lake measuring about 20 m in diameter, containing a number of sources of which the main outlets are located at a depth of 3.6 m. The mechanism of discharge is governed by the pressure conditions of the karst groundwater body within the Pöhlde Basin. Heavy rain or snowmelt causes a rise in the karst water table, which consequently, due to the higher hydraulic gradient, immediately produces higher discharge rates at the spring. The change in storage can be observed after short time intervals when turbidity becomes visible and water conductivity values decrease.

Surface water losses of the Oder and Sieber Rivers are large. They are calculated to reach up to 40 million m^3/yr, representing up to 60% of the annual discharge of the Rhume Spring.

8.33.3 Groundwater Tracer Tests

In 1909/10 Thürnau carried out tracer tests within Uranin to establish groundwater flow paths and velocities. Figure 8.110 is a facsimile reproduction from his work which shows the presumed pathways in the karstified underground between the edge of the Harz Mountains and the Rhume Spring. As can be seen, Thürnau postulated the presence of major conduits concentrated in the northeast of Pöhlde into a large flow system directed towards the spring. In 1980, some of these tracer tests were repeated and further extended into other parts of the Pöhlde Basin. From one of the tracer injection points in a sinkhole near Herzberg, the tracer reached the Rhume Spring within 78 h over a distance of about 7 500 m, which results in a groundwater flow velocity of 96 m/h. Taking into account that a certain retention time might exist for the vertical movement of the tracer down into the main karst groundwater body, a horizontal groundwater flow velocity well over 100 m/h can be assumed.

A large number of water samples were taken from various points in the Pöhlde Basin in order to characterize the spring water as a source for drinking water abstraction. Major ions of concern are Cl^- and SO_4^{2-}, whereas the chloride content in general stays well below 30 mg/l, there is some regional variation in the sulphate content of collected ground and surface water.

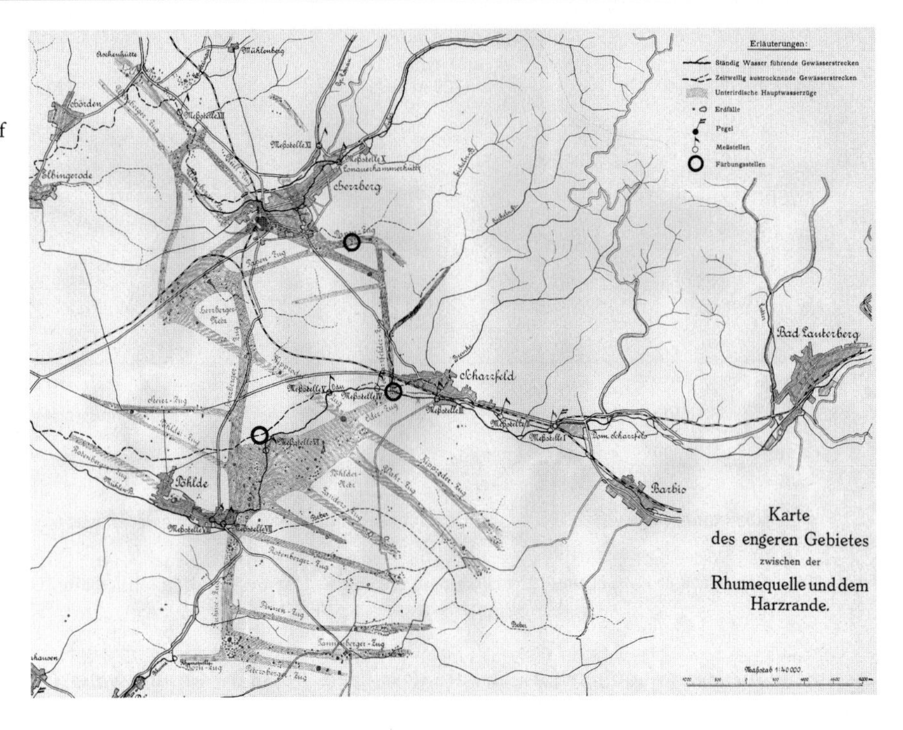

Fig. 8.110. Facsimile of Thürnau's map which shows the presumed pathways in the karstified underground between the edge of the Harz Mountains and the Rhume Spring

The SO_4^{2-}-concentration, in general, increases towards the southwest (BL5, HB1 and Rhume Spring in Table 8.33). This is probably due to the residence time that groundwater is in contact with sulphate rocks. In places where the groundwater velocity is assumed to be low (at borehole BLM for instance), high sulphate concentrations are common. Wells HB4, BL5 and EEW can be regarded as exceptions, because there dolomitic rocks prevail and significant infiltration of surface water is most likely to occur.

The Rhume Spring water hydrochemically represents mixed conditions in between the two extremes cited above. On the one hand the Rhume Spring is the combined outlet of a major karst channel sytstem with high groundwater flow velocities, on the other hand, it is located in an area where the groundwater comes into contact with sulphate rocks.

Table 8.33. Chloride and sulphate concentrations in rivers, boreholes, and wells in the Pöhlde Basin and in Rhume Spring water, 1980–1983

Sampling point	Cl^- concentration (mg/l)	SO_4^{2-} concentration (mg/l)
Oder River	<40	<80
Sieber River	<40	<80
HB 4	<40	<120
BL 5	<30	<120
EEW	<30	<120
BL 3	<30	<120
HB 1	<30	150–270
Johannis Spring	<150	<1 000
BL 4	<30	250–410
BLM	<30	>1 500
Rhume Spring	35–50	170–400

8.33.4 Groundwater Use and Protection

The towns of Bad Lauterberg, Herzberg and a few villages make use of the groundwater of the Pöhlde Basin by means of deep abstraction wells and directly from the Rhume Spring. Before the production wells were sunk, 14 exploratory boreholes were drilled to establish groundwater potential and quality. Some of them intersected greywackes and schists of the underlying Paleozoic bedrock at about 250 to 300 m depth. The investigation program has revealed that, apart from 2.8 million m^3/yr which are presently abstracted, another 4 million m^3/yr of good quality groundwater can be withdrawn safely. Seven wells are producing between

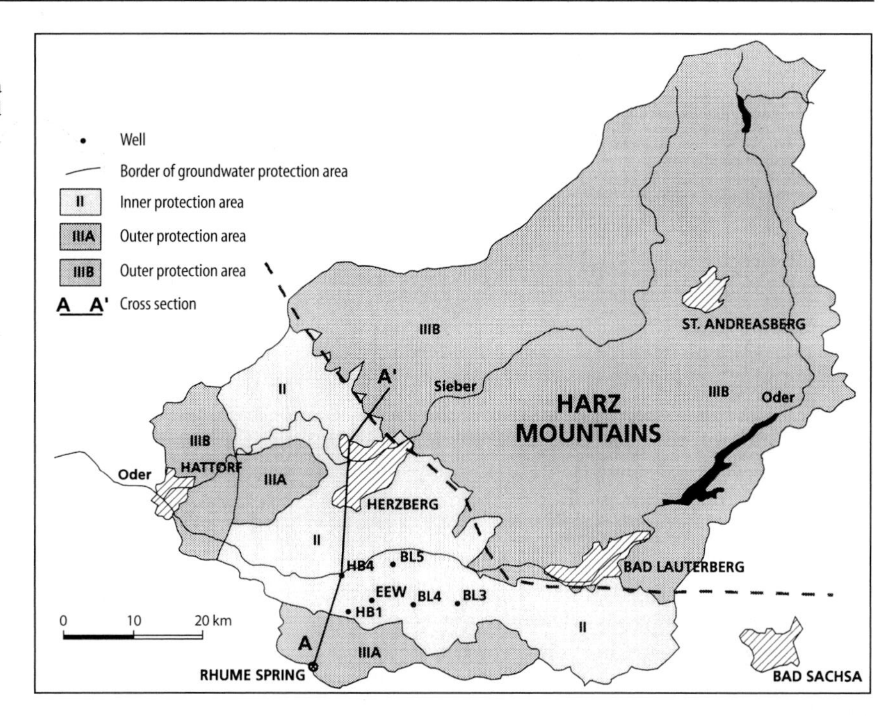

Fig. 8.111. Proposed groundwater protection zone for the catchment area of the Rhume Spring (modified after Grimmelmann 1992)

60 and 200 m^3/h at drawdowns of 4 to 26 m; well BL5 alone at present supplies 150 m^3/h at a reasonable drawdown of about 15 m. These figures may differ slightly from well to well depending upon the subsurface structural features. At the Rhume Spring, water is being tapped and piped to a reverse osmosis unit in a nearby treatment plant for removal of dissolved solids.

The major outstanding task in connection with safeguarding the water supply from the Pöhlde Basin and the Rhume Spring against pollution is to delineate groundwater protection zones. Therefore some 30 shallow holes were drilled down to a maximum depth of 25 m to find out whether low permeability silty claystone layers of the Lower Bunter Sandstone shielding the karst aquifers are present beneath the Quaternary. These findings, coupled with results from geoelectric soundings, assisted in defining zones II and III of the protection area. The delineation of the different zones (Fig. 8.111) of the drinking water protection area is based on the major hydrogeological features discussed above. An inner protection (zone II), which generally is based on the calculated 50-day travel time isochrone, covers large areas around the wells and the spring. The outer protection area (zone III), divided into zone IIIA up to a distance of 2 km away from the abstraction points and zone IIIB beyond that area, covers the outcrop of the Paleozoic rocks in the remote parts of the catchment areas of the rivers Sieber and Oder. So far, this zoning proposal has not been put into formal legal status, because the protection of such a large zone would require major restrictions to land use and regional planning. However, it is clear that effective groundwater protection of this vulnerable catchment area can only be accomplished if major restrictions to potentially hazardous activities are forced upon the land use.

References

Grimmelmann, W (1992) Hydrogeologisches Gutachten zur Bemessung und Gliederung des Trinkwasserschutzgebiets Pöhlder Becken – Report Niedersächsisches Landesamt für Bodenforschung, 29 p

Haase H (1936) Hydrologische Verhältnisse im Versickerungsgebiet des Südharz-Vorlandes. Dissertation, Universität Göttingen

Herrmann A (1969) Die geolgische und hydrogeologische Situation der Rhumequelle am Südharz. Jh Karst- und Höhlenkde 9 (Der Südharz):107–112

Jordan H (1979) Der Zechstein zwischen Osterode und Duderstadt (südliches Harzvorland). Z dt geol Ges 130:145–163

Thürnau K (1913) Der Zusammenhang der Rhumequelle mit der Oder und Sieber. Dissertation, Techn Universität Hannover. Mittler & Sohn, Berlin (25 p, 10 tables)

A. W. Creech · R. D. Dowdy, Jr.

8.34 Camp Holly Springs, Henrico County, Virginia, USA: A Case Study in "Springhead" Protection

8.34.1 Introduction

Groundwater resources that were once believed to be safe from impact by virtue of the filtering properties of sediments and rocks are now known to be susceptible to contamination from the activities of h umans. When these resources are used as a drinking water source (Fig. 8.112), risk management and resource preservation via wellhead protection programs are warranted. Wellhead protection safeguards groundwater supplies through judicious application of scientific evaluation and land use planning, defined simply as "managing a land area around a well to prevent groundwater contamination" (EPA 1993). The Safe Drinking Water Act (SDWA) provides the standard by which the success of a wellhead protection program is judged. A drinking water source is considered contaminated, and wellhead protection breached, if well water quality is contaminated over an SDWA maximum contaminant level (MCL).

Springs used by the water bottling industry present a similar, potentially more challenging, scenario. The standards of resource preservation still apply, but often to a much more stringent degree than is the case with wellhead protection. While the International Bottled Water Association (IBWA) and similar industry groups defer to the SDWA standards, public perception more often defines the market reality. Also, spring water bottlers have little or no access to the contingencies available to the water well users in the event contamination occurs (e.g., filtration, chemical treatment, development of a new source, etc.). The protection of aquifers that sustain springs tapped by water bottlers is therefore an interesting problem from a geological point of view, and often a bureaucratic issue that may require close interaction with local government.

This article will discuss Camp Holly Springs in Henrico County, Virginia, as a case study demonstrating the importance of "springhead" protection, and as an example of effective coordination between geologists, attorneys, the spring owner, and local government officials.

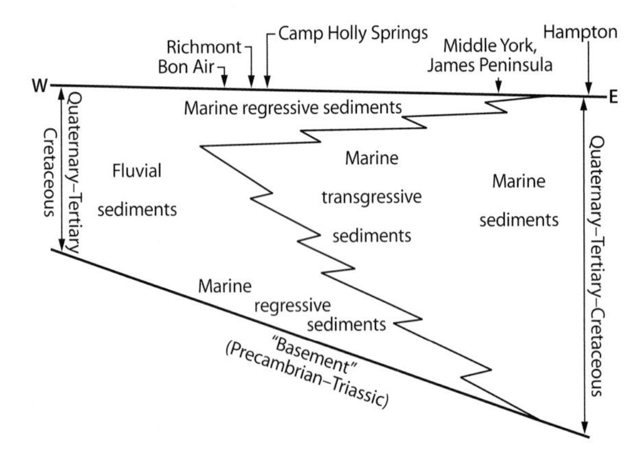

Fig. 8.112. West-East generalized cross section showing the relationship of fluvial and marine transgressive/regressive coastal plain sediments in east central Virginia

8.34.2 History of Camp Holly

The only known military post in Henrico during the War of 1812 was located at Camp Holly; the hill is said to have been fortified by the Cocke family, who lived nearby along the river at Bremo. The place was used as a camping ground by the soldiers of both the War of 1812 and the Civil War. It was ideally located because of its excellent spring and because of its situation at the edge of the river bluffs 1.5 miles north of the James, where the river makes a hairpin turn at the tip of Jones Neck.

This spot is also claimed to have been used during the Revolution as a cavalry campground. In 1923 the Camp Holly Water Company was established; the water was bottled here by the Clark family and sold in the Richmond area (Henrico County 1976).

An old house here dates from the first half of the 19th century. Originally the dwelling was a hall-and-parlor structure of one and a half stories, but in 1893 a room was added to the west gable end, and the entire roof raised to a full two stories. Frame: English basement under original section; center-hall plan; three-bay; exterior end chimneys of American bond.

A chemical analysis by Robb and Moody, Richmond, Virginia, is dated 26 June 1928. It is a part of a label for the "Pure Spring Water" at this location (Fig. 8.113). On 20 February 1991, a sample was taken from the source and submitted to the Commonwealth of Virginia, Bureau of Chemistry for analysis. The results are given in Table 8.34.

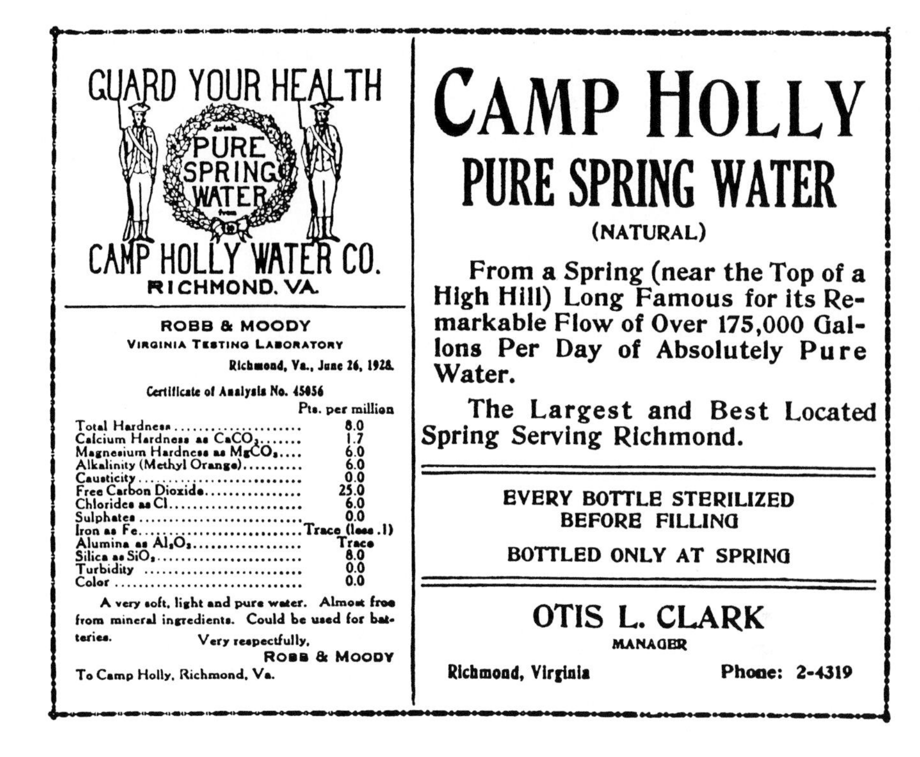

Fig. 8.113. Label – Camp Holly Pure Spring Water

8.34.3 Camp Holly Springs – Location

Camp Holly Springs, Inc., is in the Varina District of Henrico County, Virginia, near Richmond (Fig. 8.114). The company owns two springs (Camp Holly Spring and Diamond Spring) that provide a source of water to Diamond Springs Distributors, Inc., which in turn handles distribution under the trade name "Diamond Springs". Water from the springs is distributed in Virginia and North Carolina under the Diamond Springs label; in Virginia, North Carolina, Maryland, Delaware, West Virginia, and the District of Columbia under the "Richfood" label; and elsewhere in the US, and in Central America and Europe, under other labels. The IBWA News has identified Camp Holly Springs as a "top-rated bottled water plant" (IBWA 1994), and in December 1995, Camp Holly Springs was for the third time recognized for "Excellence in Manufacturing" by the IBWA. It is also certified by the National Sanitation Foundation, and authorized to display the NSF seal on its product.

The springs are not a public drinking water supply source per se; however, they have acted as such following hurricanes and other emergencies that have contaminated public supplies or otherwise created excess demand for potable water.

The near surface geologic regime at Camp Holly Springs is Coastal Plain, consisting of stratified sands, and gravels deposited during Tertiary and Quaternary regression events, then reworked by fluvial processes (Fig. 8.112; Daniels et al. 1974). Average depth for shallow wells in the general area is 39 ft (Wigglesworth et al. 1984).

8.34.4 Spring Water Quality Standards

Regulation of the bottled water industry varies from state to state. The IBWA and similar industry associations require that water quality standards (typically SDWA parameters) be met as conditions of membership. However, spring water quality may need to exceed such standards in order for the bottler to compete successfully in the marketplace. Many customers purchase bottled water because they want something "cleaner" than the water coming from their faucets. As an example, nitrate concentration can be instrumental to a bottler's success. The SDWA MCL for nitrate is 10 mg/l. In Henrico County,

Table 8.34. Commonwealth of Virginia, Bureau of Chemistry analysis of Camp Holly Springs

	Contaminant name	Maximum allowable level	Analysis results
Metal	Arsenic	0.05	<0.010
	Barium	1.0	0.03
	Cadmium	0.010	<0.010
	Chromium	0.05	<0.010
	Lead	0.05	<0.010
	Mercury	0.002	<0.0003
	Selenium	0.01	<0.010
	Silver	0.05	<0.010
	Aluminium		<0.1
	Calcium		0.5
	Iron	0.30	<0.01
	Magnesium		0.5
	Maganese	0.05	<0.01
	Strontium		<0.01
	Zinc	5.0	<0.01
	Copper	1.0	<0.01
	Potassium		1.0
	Sodium	20.0	3.3
	Nickel		<0.01
	Calcium hardness		1.0
	Magnesium hardness		2.0
	Ca and Mg hardness		3.0
	Total hardness		3.0
	Antimony		<0.020
Inorganic	pH	6.5–8.5	6.0[a]
	Alkalinity-total		3.5
	Alkalinity-bicarbonate		3.5
	Alkalinity-carbonate		0.0
	Hardness-EDTA		5.0
	Phenolics, total		<0.0005
	Corrosion index		6.24
	Fluoride	1.8	<0.10
	Chloride	250.0	4.8
	Color (Alpha)	15 CU	3.0
	Turbidity (FTU)	1.0	0.19
	Hydrogen sulfide	0.05	<0.03
	Sulfate	250.0	0.30
	Nitrogen-nitrate	10.0	0.14
	Nitrogen-ammonia		0.02
	Nitrogen-Nitrite		<0.01
	Phosphate-ortho A and P		<0.01
	Specific conductance μ mhos/cm		32.0
	Total dissolved solids	185 °C	27.0
	Volatile	550 °C	10.0
	Fixed	550 °C	17.0
	Silica		7.99
	Langlier index	15 °C	–5.74
	Ca hardness		0.5
	Bromide		<0.01
Radiological	Gross alpha	15 pCi/L	0.3
	Gross beta	50 pCi/L	1.4
Trihalo-methane			none detected
Volatile organics			none detected
Pesticide			none detected

[a] Acceptable range is 5.0–9.0 for bottled water, 6.8–8.5 is the standard for piped systems.

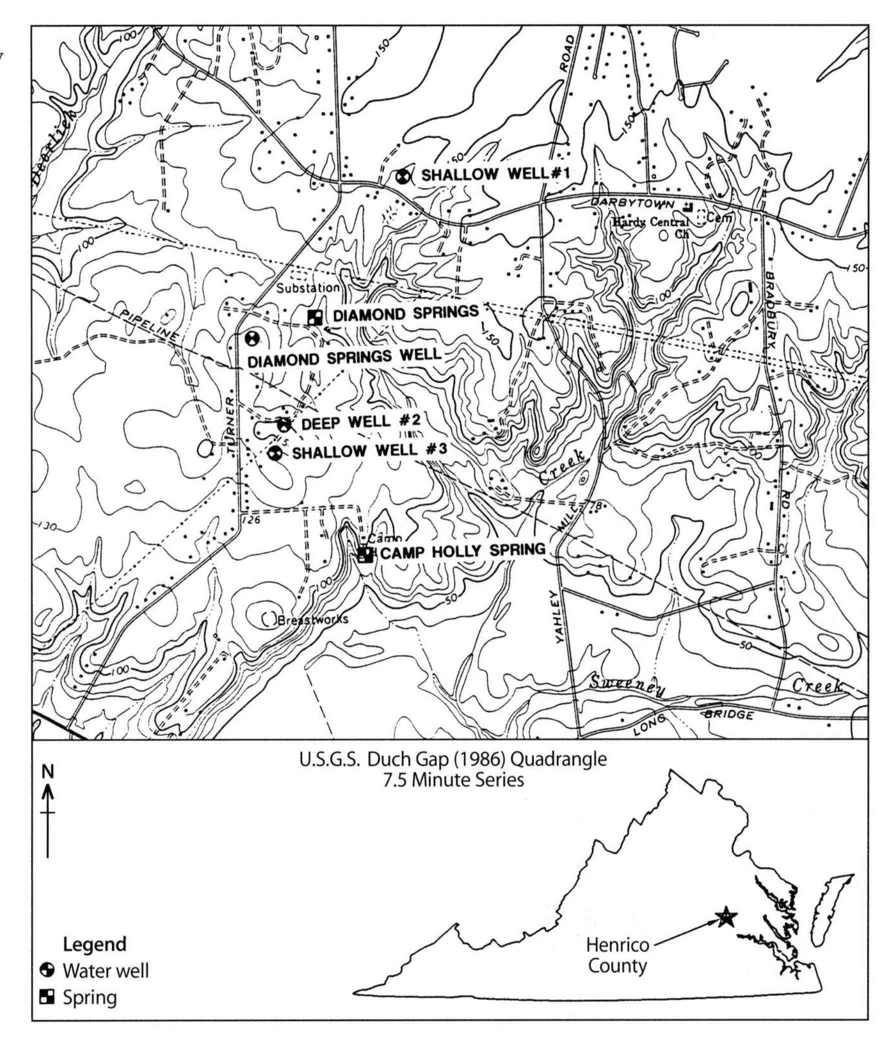

Fig. 8.114. Area location map, Camp Holly Springs and Diamond Springs

Virginia, however, where Camp Holly Springs is located, typical nitrate concentration in unimpacted groundwater is less than 3 mg/l (Wigglesworth et al. 1984). To some customers, this "background" concentration becomes a de facto water quality standard because an increase over background, even one considerably below the SDWA MCL, could be construed as implying that the spring water is less than "natural".

The market realities of spring water quality extend beyond parts per million chemical concentrations and deal also with "truth in labelling" issues. Water from the same aquifer collected from a well that is unassociated with a spring can be comparable in quality in every measurable way, but might not be allowed to be labeled as "spring" water (IBWA 1995). Similarly, the IBWA Model Bottled Water Regulations and many state regulations do not allow water that has been filtered or chemically treated (other than for sterilization purposes) to be labeled or marketed as "natural," "pure," or "spring" water. The value of the spring is based solely on its natural quality or the market perception that this water is superior. There is little allowance for the corrective measures afforded to water utilities.

In 1990, Camp Holly Springs (at that time Diamond Springs Water Company, Inc.) recognized the need to consider protection of the aquifer serving the springs. To achieve this it was necessary to identify the subsurface source of the springs (i.e., a deep or a shallow aquifer). The evaluation of susceptibility was based on the assumption that a deep source would be protected from potentially contaminating surface activities (e.g., septic tanks), whereas a shallower source would be at higher risk.

Water samples were collected from the springs supplying the bottling plant, other springs in the area, and from shallow and deep water wells in the general vicinity. They submitted the water samples to a laboratory for quantitative identification of various isotopic relationships in the waters. For this study, the waters were analyzed for specific hydrogen and oxygen isotope content. Because hydrogen and oxygen are intrinsically associated in the water molecule, the isotropic ratios of these elements are usually discussed together. To characterize a flow system, the ratio of deuterium (D) to hydrogen (H) is plotted against an oxygen isotope ($\delta^{18}O$). When these data are plotted, waters from similar flow systems show a linear relationship. The study indicated that both Diamond Spring and Camp Holly Spring have a shallow aquifer as a source – either as separate shallow sources or distinct horizons within the same source, with Diamond Spring being the upper aquifer or horizon (Fig. 8.115).

Groundwater dating was done by measuring the radioactive isotope of hydrogen, called tritium. In this study, tritium dating indicated that the springs receive recharge from relatively modern (post-1950s) precipitation. There was no indication that deep groundwater influences the springs.

Based on these conclusions, and with the sandy/gravelly nature of the aquifer in mind, it was concluded that the springs are susceptible to impact from surface contamination. Accordingly, the study included a recommendation for shallow groundwater recharge protection

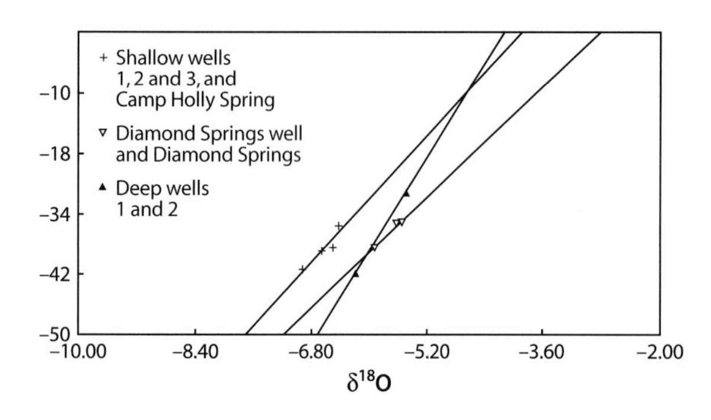

Fig. 8.115. Isotopic relationship of sampled waters

areas (developed using wellhead protection protocols) to safeguard the springs. Of the potential sources of contamination, septic tanks presented the greatest concern. This is because nitrate released from septic tanks is typically not physically or chemically attenuated by soil, but by dilution in groundwater (Kaplan 1987). And, as noted above, spring water bottlers can be vulnerable to the effects of market perception: nitrate concentrations that exceed regional "background" can be detrimental to this perception.

8.34.5 Groundwater Resource (Spring) Protection

These efforts have allowed Camp Holly Springs to implement a long-term resource protection plan. The company has purchased land within the shallow groundwater recharge protection area as it has become available. In addition, it has undertaken several capital improvements at company facilities for the sole purpose of shallow aquifer protection (e.g., improvements to vehicle parking areas, relocation of activities downgradient of the springs). The company even advises the truck drivers emptying dumpsters not to compress loads on site, to prevent discharge of residual liquids to the ground.

Perhaps more significantly, the company and its attorney and consultant have worked closely with the Henrico County Government to develop resource protection guidelines for preservation of the springs and the recharge area. For example, the 30 April 1993, *Final Report: Henrico County Wellhead Protection Pilot Project*, (Wigglesworth 1993) noted that it was "necessary to include Diamond Springs and Camp Holly Springs in this project." More recently, the County Planning Department's March 1996 *Henrico Comprehensive 2010 Land Use Plan* identifies the springs as important environmental resources. In the section addressing the protection of potable water, the Plan indicates that it shall be a County policy to "protect the quality of the Camp Holly Springs and Diamond Spring recharge area to the extent reasonably practicable" (Henrico County 1996).

8.34.6 Summary and Conclusions

Springs are unique and outstanding natural features worthy of preservation. Such efforts can be modelled after wellhead protection programs, but may need conservative application to account for the elevated quality standards associated with the bottled water industry. In the case of Camp Holly Springs, an evaluation of isotopic relationships in the springs and in shallow and deep wells provided the necessary hydrogeologic understanding needed to develop shallow aquifer recharge protection areas. Subsequent coordination with local government officials allowed "springhead" protection to become a reality for Camp Holly Springs.

Acknowledgements

The authors are indebted to John F. Deal, Esq., of Deals & Lacheney, P.C., Richmond, and to James K. Richard, C.G., Senior Hydrogeologist with Jacques Whitford, Inc., Freeport, Maine, for assistance in the preparation of this article.

References

Daniels PA Jr, Onuschak E Jr (1974) Geology of the Studley, Yellow Tavern, Richmond, and Seven Pines Quadrangles, Virginia. Virginia Division of Mineral Resources, Report of Investigations No. 38
EPA (1993) Wellhead protection: a guide for small communities. EPA/625/R-93/002
Henrico County (1996) Henrico 2010 comprehensive land use plan. County of Henrico Planning Department
Henrico County (1976) Inventory of early architecture and historic and archeological sites
IBWA (1994) Rep Bliley Tours Camp Holly Springs, IBWA News Vol. 12, No. 2
IBWA (1995) Model bottled water regulation
Kaplan OB (1987) Septic systems handbook. Lewis Publishers
Wigglesworth HA (1993) Final report: Henrico County wellhead protection pilot project. County of Henrico, Virginia
Wigglesworth HA, Perry TW, Ellison RP III (1984) Groundwater resources of Henrico County, Virginia. State Water Control Board, Planning Bulletin No. 328

O. Franko

8.35 Piestany Thermal H_2S Water and Mud

8.35.1 Introduction

Slovakia belongs to a region with a very rich source of mineral and thermal water (1093 sources). The flow of thermal waters from Piestany placed it among the highest flow and warmest water in the region. These are Ca, Mg, SO_4, HCO_3, Cl types with TDS of 1.4 g/l, 10 mg/l of H_2S, and a temperature of 68 °C. The flow is 40 l/s. The waters seep on the spa island, which lies between the river Vah and its tributary. The characteristics of this water and mud predestine the existence of this spa as the greatest in Slovakia. The Piestany thermal water is widely used in the treatment of rheumatic and neurological diseases.

8.35.2 Location

Piestany is one of the world-famous spas for treatment of rheumatic and neurological diseases. It owes it fame to its thermal waters with curative effects and its unique sulphur-containing mud. Piestany is the greatest therapeutical center for diseases of the locomotion system in the Slovak Republic. It is in the western part of Slovakia in the Vah Valley, 80 km from Bratislava and 150 km from Vienna (Fig. 8.116). Piestany is considered to be one of the warmest locations in Slovakia and the climate is typical of the Slovakian lowlands which is slightly dry. The average annual air temperature is 9.4 °C. The monthly

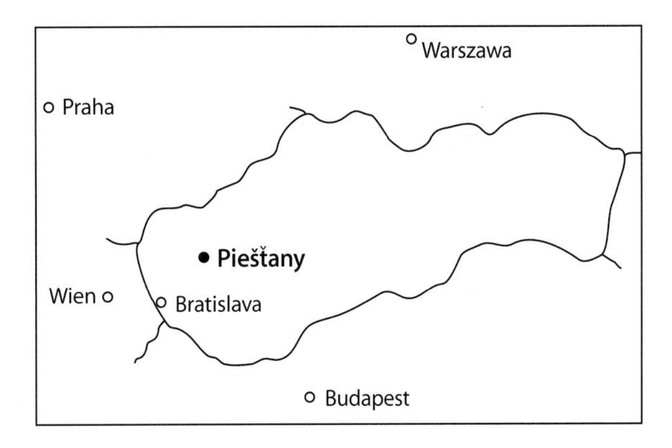

Fig. 8.116. Piestany location

average daily temperature in May at 2 p.m. is 19.3 °C, in July 24.5 °C, and in September 20.5 °C. The annual average of sunshine is 2 085 h, with an average of 9 h per day in summer. Only 75 days each year are without sunshine. The average annual precipitation is 611 mm at the spa which has an altitude of 162 m a.s.l.

8.35.3 History

The name Piestany was first recorded in writing in the year 1113 in the Zobor Monastery document from Ungar king Koloman I. During later centuries, when the inhabitants learned about the influence of these waters, the settlement Teplice (derived from hot water) was established. Later the settlement merged with Piestany. The first time these thermal waters were recorded was in the publication by Juraj Werher in the year 1549, called *De admirandis Hungariae aquis hypomnemation*. J. Werher wrote, "The bank of the river Vah over the Hlohovec ... in that time was very famous for the hot springs and their admirable origin. There is not one stable spring but several along the river Vah. When the stream increases or decreases its flow, the springs consequently appear or disappear." This was reported in the year 1571 by Andrej Baccius Elpidianus, personal doctor of Pope Sixtus V. The memory of Piestany thermal water, spa and life around it is recorded by Adam Trajan in a celebrated poem in the year 1642. The work *Schediasma de thermis Postheniensibus*, written by physicist Justus Jan Torkos from Bratislava (1745) states, "I do not find any other reason why the springs of thermal water are still replaced, only that the riverbed of river Vah is also still replaced from one place to other ... These waters are still flowing up from the depth, water from river press them and retain them in the permeable sands and gravel, until the thermal water will have greater pressure than water from river and than flow out on the bank of the river as the spring." According to the origin, he mentioned, "We have no doubts that our thermal waters have their origin in these mountains." In the second half of the 19th century the opinion that the thermal water has a volcanic origin prevailed. At the beginning of the 20th century, J. Knett (1920) decided that the occurrence of these waters belongs to the Povazie fault, which provides H_2S thermal waters not only to Piestany but also to Belusske Slatiny, Trencianske Teplice (NE Piestany) and Svaty Jur near Bratislave. Current views concerning the origin and classification of Slovak thermal waters including Piestany, are mentioned in the works of M. Mahel (1952), O. Hynie (1963), G. Rebro (1966), O. Franko et al. (1975), and O. Franko and D. Bodis (1989).

8.35.4 Discharge Rate

Thermal water gushes out of the natural springs from the river Vah Quaternary alluvials, which are composed of sandy gravel, and is tapped by the well Trajan at a depth of 11.2 m (Fig. 8.117). Hot water discharge has a temperature of 60.5 °C. A free flow of 22 l/s is recorded from the wells V-1 (50.5 m deep at 67.6 °C), V-4a (56.2 m

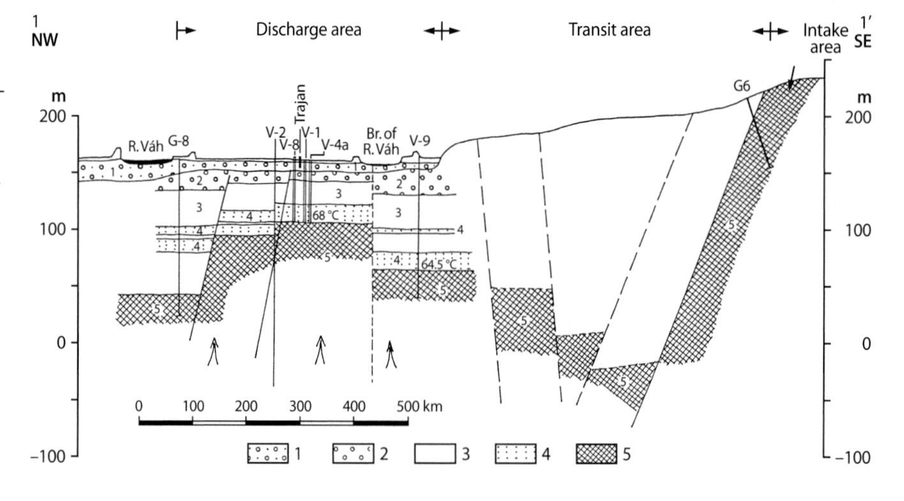

Fig. 8.117. Transverse section across the hydrogeological structure of Piestany thermal waters: *1* Quaternary river sediments, gravels and sands; *2* Neogene conglomerates; *3* Neogene clays; *4* Neogene sandstones; *5* Mesozoic sediments with Triassic carbonates and evaporites

deep at 67.7 °C), and V-8 (54 m deep at 68.2 °C), which tap the thermal water in the Neogene sandstones. It is assumed that total flow of this formation reaches 100 l/s. Maximum temperature of water measured at the bottom of well V-9 at the depth of 79.5 m was 71.5 °C.

8.35.5 Use

The curing source is sulphurous mud. It is the sediment of stream alluvials (loess and loess loams) in the branch of river Vah (Obtokove ramcno) that are physically and chemically under the influence of thermal H_2S water. The yearly consumption of mud is ca. 500 m^3.

Thermal waters are used for curing. J. Torkos (1745) wrote, "I found that these thermal waters have a big influence for the coxarthrosis, paralysis, and paresis. This water also cures these diseases." The modern medical point of view concerning these thermal waters was written in the work of a doctor famous at Piestany; balneologist and surgeon Frantisek Ferdinand Ernest Sherer, 1837, *Die Quellen und Bäder zu Posteny (Pjestjen) in Ungarn.*

The real growth of spa Piestany is dated to the year 1821 when the Graf Jozef Erdody started refurbishing and adding more buildings after the 1813 flood destruction of the spa to accommodate many Austrian soldiers who were injured by Napoleons army. One of the cure houses is named Napoleon Spa. The historical part of the spa today (Thermia Palace, spa Irma etc.) was constructed after 1889 at a time that Alexander Winter and Sons Co. rented the spa from Graf Erdody for 30 years. The construction of the historical part of the spa was continued up to 1945. The new part of the spa (Balnea Grand-Splendid, Balnea Esplanade-Palace, The Arts Centre) was built after 1945. At present, the spa is operated by a shareholders' company which takes care of 35 000 patients yearly, using 2 500 employees, including 75 doctors. The spa is considered international as it treats 12 000 expatriates from all over the world.

The physical and chemical properties of the thermal waters and sulphurous mud make Piestany ideal for the treatment of chronic rheumatic diseases. Successful treatment has been achieved for non-inflammatory diseases of the joints like coxarthrosis, gonarthrosis, spondylosis, and spodylarthrosis, diseases of the inter vertebral discs, rheumatoid arthritis and extraarticular forms of rheumatism. Additional applications are post-traumatic lesions of the joints and bones, particularly following fractures and orthopedic operations. The Piestany treatment is recommended for those suffering from organic diseases of the nerves, including painful afflictions of the peripheral nerves (plexitis, polyneuritis, and radiculitis) and vertebrogenous painful syndromes (lumbosciatic and cervici-brachial syndrome).

8.35.6 Hydrogeology

The source of all thermal waters of natural springs in Slovakia is the Triassic dolomites and limestones of the Inner West Carpathians. Due to an alpinotype folded nappe structure of the West Carpathians, thermal waters seep from Triassic carbonates of Envelope units of crystalline rocks; krizna (lower), choc (middle), and upper nappe. Infiltration areas of thermal waters are in mountains, transit (transi-accumulation) in the inner mountain depressions and bays and discharge areas are mainly on the margin of depressions, sometimes in the middle of depressions (high blocks), or in mountain valleys (elevation of folds).

Water seepage takes place at the crossings of longitudinal (marginal) old faults with the transverse young faults. The thermal waters belong to open hydrogeological structures (Franko et al. 1975). When the carbonates are dissolved, then the waters are of Ca-Mg-HCO_3 type, when carbonates and evaporites (gypsum, anhy-drite) are dissolved, the water is of the Ca-Mg-SO_4-HCO_3 type. In the case of presence of marinogenetic mineralization, the water is of a mixed type with presence of NaCl. Total dissolved solids (TDS) of thermal carbonate waters reach 1 g/l, TDS of carbonate-sulphatic waters reach 5 g/l and mixed waters reach up to 7 g/l.

The thermal water aquifers of Piestany are in the Triassic limestones and dolomite of the Manin-Inovec subunit of the Envelope unit exposed in the Povazsky Inovec Mountains, at the left bank of the river Vah (Mahel 1952). At the northern slope of the Povazsky Inovec Mountains, these rocks plunge below the valley plain of the river Vah under the Krizna nappe. The karst water flow descends in them towards the Vah Graben (Trnava bay) changing into thermal water that rises along the boundary fault to the surface. In the upper layers the thermal water flow enters the faults of the local longitu-

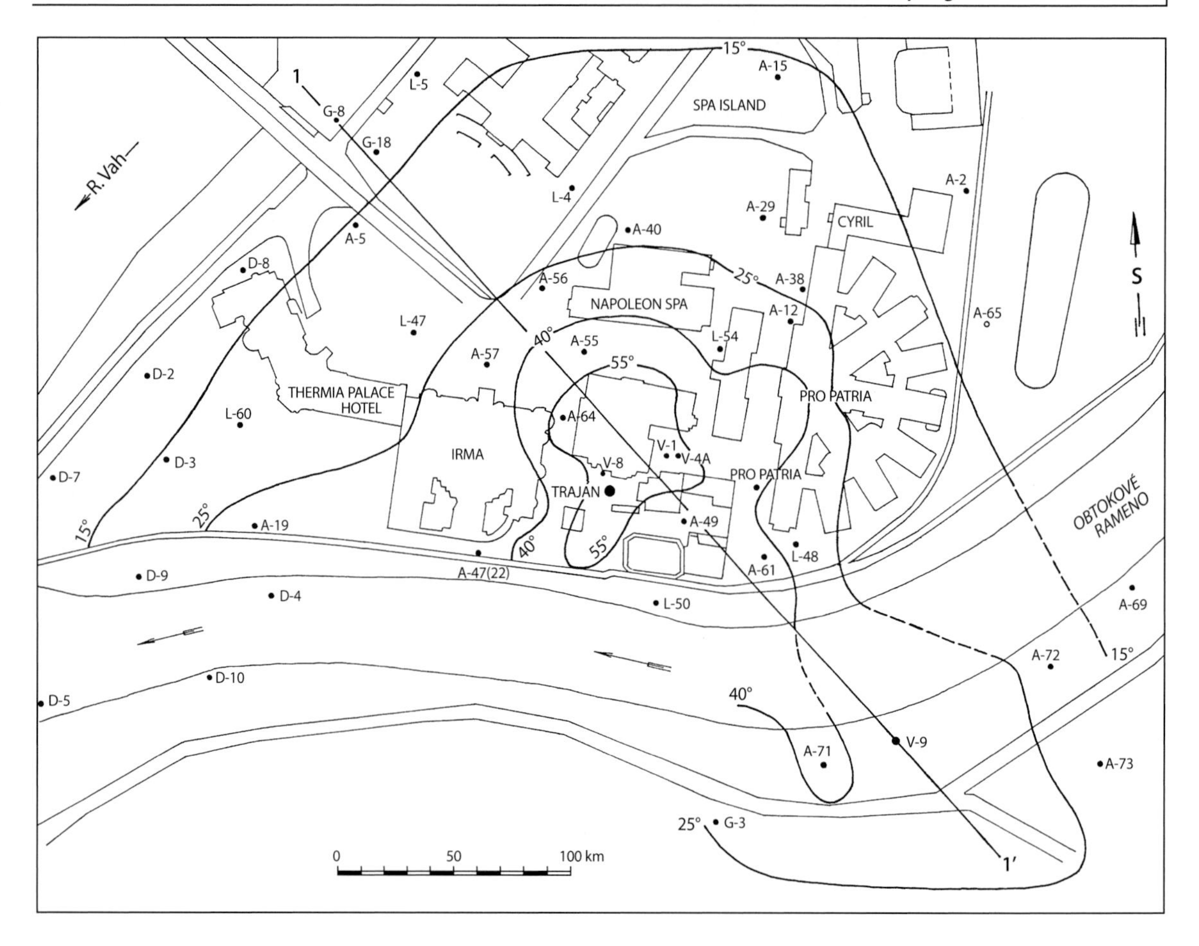

Fig. 8.118. Thermoisohypses in the discharge area of Piestany thermal waters, dated 17.03.1983

dinal (NNE-SSW) horst in Late Tertiary (Panonian-Pontian) filling of the Mesozoic Vah Basin.

The thermal water circulation is along the longitudinal faults linked through a system of transverse faults (WNW-ESE). The center of the thermal water ascent is on the crossing of the faults in the most elevated part of the horst (Fig. 8.117). During the passage through the tectonic paths of ascent in Late Tertiary rocks, thermal water is accumulated for the first time, forming an aquifer in the basal permeable Late Tertiary layer. For the second time, the thermal water is accumulated after its entrance from the faults and fissures of the Tertiary substratum into Pleistocene gravel alluvium of the Vah Valley. Groundwater flows in a thermal water discharge into the alluvium at the margin (Fig. 8.118). Therefore, the thermal water of the gravel alluvial deposits is in places in direct hydrological connection with the thermal water in the permeable upper Tertiary layer. Initially, the thermal water was utilized only by means of shallow wells sunk in the thermal water body with a free water table in the gravel deposits, having a variable circulation influenced by the natural water fluctuation of the river Vah (well Trajan, Fig. 8.119). In 1953, a thermal water artesian aquifer was found in the basal Tertiary layer at a depth of 50 m (Hynie 1963). This aquifer was successfully developed by a system of borings. The value of the water tapped in this confined accumulation is greater as its composition is standard and independent from the water level fluctuation in the river Vah. The age of the thermal water according to ^{14}C reaches 26 000–28 000 years.

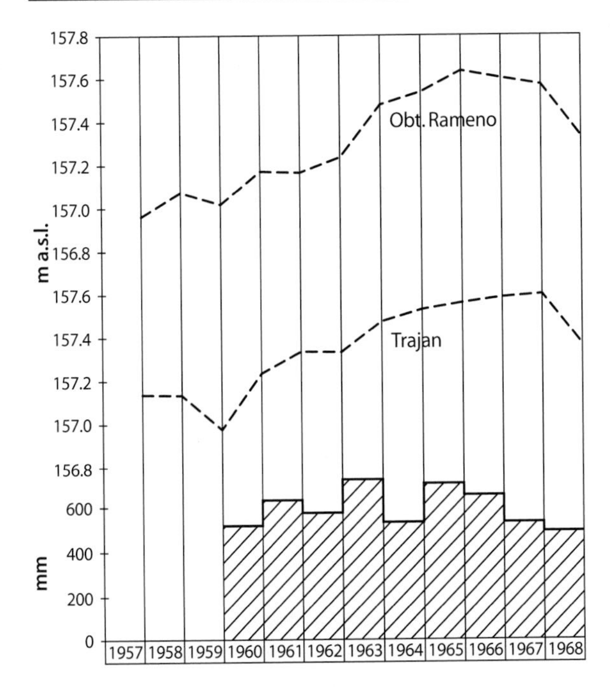

Fig. 8.119. Relation of Piestany thermal water in the well Trajan and surface water in the branch (btokove rameno) of the river Vah (Franko 1995)

Table 8.35. Major ions and other characteristics in Piestany thermal water (well V-1)

Parameter	Value	Parameter	Value
T_{vodv}	67.2 °C	Fe	0.21 mg/l
pH	6.7	Al	0.07 mg/l
CO_2	151.0 mg/l	Cl	118.75 mg/l
H_2S	12.0 mg/l	Br	0.14 mg/l
TDS	1 407.0 mg/l	J	0.21 mg/l
Li	0.87 mg/l	F	1.66 mg/l
Na	90.90 mg/l	NO_2	0.00 mg/l
K	15.90 mg/l	NO_3	1.20 mg/l
NH_4	0.08 mg/l	SO_4	558.40 mg/l
Mg	41.34 mg/l	HPO_4	2.97 mg/l
Ca	228.46 mg/l	HCO_3	340.10 mg/l
Sr	5.90 mg/l	CO_3	0.00 mg/l
Ba	<0.10 mg/l	OH	0.00 mg/l
Mn	0.00 mg/l		

8.35.7 Quality of Water

According to rMg/rCa (0.30), the aquifers are mainly limestone. According to values of $8^{34}S$ (26.7%), waters dissolved gypsum of the upper Triassic. The chemical analysis of water from 1968 is shown in Table 8.35.

References

Franko O, Gazda S, Michalicek M (1975) Genesis and classification of mineral water in West Carpathians (Tvorba a klasifikacia mineralnych vod Zapadynch Karpart). Geologicky ustav D. Stura, Bratislava, 230 p

Franko O, Bodis D (1989) Paleohydrogeology of mineral waters of the Inner West Carpathians. Zapadne karpaty ser Hydrogeologia 8, Geologicky ustav D. Stura, Bratislava, pp. 145–163

Hynie O (1963) Hydrogeologie CSSR – Mineralni vody, CSAV. Praha, p. 797

Mahel M (1952) Etudes des rapports entre les sources minérales et la structure géologique en Slovaquie (Mineralne pramene Slovenska so zretelom na geologicku stavbu). Prace SGU 27, Bratislava, p. 84

Mahel M (1996) Unique and admired waters of Slovakia (Vzacne a obdivovane vody Slovenska). Balneologicke muzeum Piestany, p. 182

S. O'Neill

8.36 Ballygowan Natural Mineral Water – History of the Source

8.36.1 Introduction

Ballygowan is a recognized spring source of natural mineral water with an annual output of some 80 million l of water. It is abstracted by wells from a limestone aquifer with a naturally Ca-Mg rich water. The area is essentially rural with moderate to high rainfall. The sources are located deep within the limestone and are protected by both the natural characteristics of the aquifer and a very extensive aquifer protection policy and management strategy.

The original source is considered to be 800 years old. It was discovered by the Knights Templar, upon their return from the Crusades and became known as St. David's Holy Well with reputed therapeutic qualities. It has been used since the 1940s as a source of both mineral waters and soft drinks. The well was rediscovered as a source of natural mineral water in the early 1980s. St. David's Well was used initially but as the volume of production increased, it was considered more efficient to abstract the water from strategically located boreholes upgradient of the source. The Ballygowan company has continued to grow throughout the 1980s and 1990s. A state-of-the-art bottling facility was completed in 1987 and effectively doubled in size again in 1991. Since then, Ballygowan has continued to grow both in Ireland and more especially the UK (see labels in Fig. 8.120). Other markets include France, Germany, the New England States, Japan and the Middle East.

Ballygowan was recognized as a source of Natural Mineral Water by the EC in 1988 under Directive 80/777/EC. It was the first certified as complying with the Bottled Water Standard of Ireland, I.S. 432. It was the first company to obtain ISO9 002, the international manufacturing management system in 1993, and the first to obtain ISO14 000, the international environmental management system in 1997.

Originally wholly owned by its founder Geoff Read, 51% was bought by Budweiser in 1987. It was subsequently sold back to Geoff Read in 1989. 51% was bought by Allied Domecq Group and 49% by Guinness Group in 1993.

8.36.2 Location, Topography, and Drainage

The well field is in the SW corner of Ireland in West Limerick. It is over 30 miles from the nearest city. The area has a mean elevation of 62 m OD with average annual precipitation of 1054 mm and an annual average temperature of 10 °C. The land around the well field is unimproved pasture being used principally for dry stock and dairy farming. The soils range from being well-drained loam and clay loam of the Brown Podzolic and the Grey-Brown Podzolic Groups to poorly drained soils of the Gley and Podzol Groups. The elevation varies from 62 m OD in the area of the well field to over 300 m OD in the hills on the western perimeter of the catchment. The overall topography is that of hills on the Western fringes of the catchment, forming a lip to an arcuate basin which gives way to rolling hills and knolls to form a plain in the area of the source. The catchment area is some 18.2 km^2 and is drained by the Arra River and the Dooally River.

8.36.3 Geology

The Dinantian and early Namurian of the south-western province is the general stratigraphic setting for the area. The Lower Limestone Shales and the overlying Ballysteen Limestone Formation give way to the massive carbonate build ups of the Waulsortian Mudbank Limestone Formation. The well field lies on the boundary between the Waulsortian Mudbank Limestone and the bedded Middle Limestones of late Dinantian age.

The Newcastle West region was tectonically deformed during the Variscan Orogeny and is about 35 km north of the Variscan Thrust Front and has undergone significant deformation in the hinge region of a large open anticline called the Newcastle West Anticlinorium.

The area is described as being glacio-karstified. It is a covered barrier karst area and as such tends to concentrate recharge into preferentially weathered fracture zones. The process of karstification has been stopped since the end of the last glaciation and the limestone is fully saturated throughout the year. Karstification extends down to between 10 to 15 m below ground level, and contains up to 60% of the storage of the whole aquifer. This layer is very important for water storage. The overburden has a buffering effect on recharge, permitting it to percolate more slowly.

8.36.4 Hydrogeology

The hydrogeology of the source is complicated. Even though the aquifer is a single geological unit, it can be

Fig. 8.120. Labels of bottled Ballygowan mineral water

divided into four hydrogeological units based on the fracture intensities. The epikarst, confined to the top 15 to 20 m is considered to be the first unit and is directly beneath the clays. The fractures in this layer are open and interconnected. Recharge entering this unit moves quickly. The permeability of this layer is high and the overall groundwater velocity is high with transmissivities from 50 up to 400 m^2/d. It can have a specific yield of up to 5%.

Unit 2 extends from below the epikarst down to 150 m below ground level, the maximum depth of karstification and is a massive, largely unfractured, limestone. It permits the slow movement of water from the fractured unit 1 to the fractures in unit 2. It acts as a buffer zone between the more permeable unit 1 and the fractures present in the lower units. The lower fracture intensity results in transmissivities of between 0.1 m^2/d to 10 m^2/d. The storativity is between 10^{-3} to 10^{-4}.

Unit 3 is the remainder of the Waulsortian Reef limestone below 150 m. The permeability is 10^{-4} to 10^{-2} m^2/d. The storativity is 10^{-4} or lower. Unit 4 is the fault zone. These can be spatially diffuse but where they do occur they are linearly extensive areas of very high transmissivity and storativity. They tend to be preferentially dolomitised and karstified. The high permeability zones extend down to between 250 m and 350 m below ground level. They have transmissivity values from 100 to 1 500 m^2/d in the most dolomitised and karstified areas. Storativity ranges from 10^{-4} to 5^{-2} (5%).

The spring/well source exploits fracture zones deep within the aquifer where permeabilities are high enough to provide a yield of between 400 and 850 m^3/d. It is a compromise between obtaining an economic yield and adequately protecting the source.

The aquifer responds in a leaky confined manner and exhibits dual porosity responses in the fault zones and porous media responses at a large scale in the other units. When abstraction occurs at rates below 600 m^3/d, the majority of the water is derived from units 3 and 4 and so the chemical characteristics of those waters predominate. Leakage occurs from the unit 1 via unit 2 to unit 3 and 4 under pumping conditions, so that there is an input of water that has the chemical characteristics of unit 1. Unit 1 acts like a header tank, supplementing the abstraction requirement from units 2, 3 and 4. The fractures in unit 2 act as a series of micro conduits slowly percolating water downwards. The higher the pumping rate from a well, then the higher is the proportion of unit 1 water in the overall blend of water being abstracted from the units 3 and 4.

8.36.5 Hydrochemistry

The water is a calcium (magnesium) bicarbonate rich water. It has total dissolved solids of around 450 ppm. The pH is around 6.9 and the temperature of the water is between 12 and 13.5 °C depending on how deep the water is abstracted. Piper diagrams demonstrate that the water is $CaCO_3$ water with a predominance of Ca^{2+} and Mg^{2+}. The Na^+ and K^+ and SO_4^{2-} and Cl^- make small contributions to the overall chemical characteristics of the water. It is classified as a low sodium water, with nitrates at around 9 ppm. Radiocarbon dating indicates an age of between 800 and 1 200 years for water sampled from a depth of 48 m.

As the recharging water enters unit 1 and moves downgradient, there is degassing of CO_2 due to changes in temperature and partial pressures. This leads to a decrease in HCO_3^- and hardness as the effects of the recharge decline and the water moves further and further away from the recharge area. Ions such as Cl^-, NO_3, and SO_4^{2-} are also taken up by recharging waters entering this unit.

The hydraulics of the aquifer system under natural conditions leads to a stratification of the waters in the aquifer. The recharge areas and unit 1 are high in HCO_3^- and hardness but low in Mg^{2+} while the waters of units 2 and 3 are lower in HCO_3^- and hardness and higher in Mg^{2+}. This is reflected in changes in conductivity values after a recharge event and after periods of low or absent rainfall.

The system is complex with positive increases in water quality parameters in one part of the system being balanced by decreases in the same parameters in other parts of the system due to recharge events and changes in abstraction rates. There is natural stratification of the water chemistry in the aquifer, which is disturbed by pumping. There is the blend of water being abstracted which is dependent on the pumping rate. There are recharge events driving the whole system and finally there is a single set of chemical data taken at a single point in time whose analyses depends not only on the vagaries of nature but also on the problems of sampling and analytical protocols.

8.36.6 Well Field Management

The production wells all tap the units 3 and 4. Cement grout seals off the annulus of the original open hole in units 1 and 2. This prevents water from these units percolating down via the original open hole annulus of the well to the lower units. The yield of unit 3 is 350 m^3/d to 430 m^3/d in fractures and considerably higher in sucrosic dolomite zones. Once this discharge rate is exceeded, water from unit 2 and ultimately unit 1 must be drawn down. When these wells are pumped at discharge rates of less than 430 m^3/d, the majority of the water is derived from the lower units. Water abstracted from the lower units is replaced by water moving slowly from the recharge areas to the north and by percolation from unit 1 via unit 2. Hence some of the abstracted water would have been derived from the less chemically stable unit 1. If the abstraction rate exceeds 430 m^3/d to 520 m^3/d then the percentage of water from unit 1 and unit 2 will increase in proportion to the increase in discharge and there will be a component of vertical flow.

8.36.7 Aquifer Protection

A very extensive aquifer policy was implemented in late 1992. A series of aquifer vulnerability maps were drawn up, and based on these and the use of MODFLOW, a groundwater flow model, a series of defensible aquifer protection zones were further generated. A series of codes of practice covering agricultural, industrial and amenity activities in the catchment were drawn up. These were implemented in co-operation with the local regulatory authorities, planing authorities, farming bodies and resident associations. A further model was developed to model discrete fracture flow as an aide to further protecting and managing the aquifer.

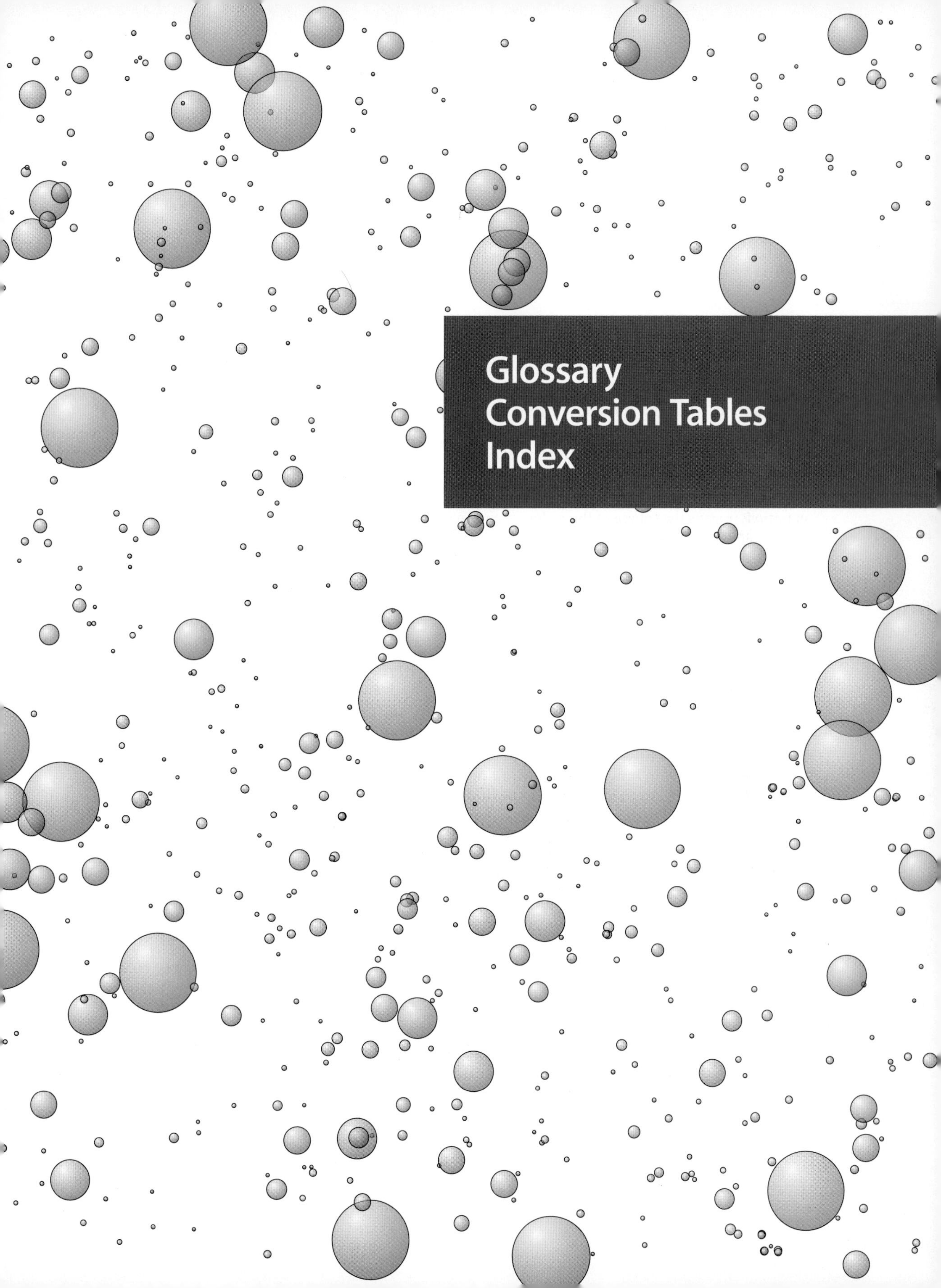
Glossary
Conversion Tables
Index

Glossary

Alluvial/slope spring. Boundary spring (AGI 1997).

Ambient air. Any unconfined portion of the atmosphere: open air, surrounding air (ERA 1994).

Anticlinal spring. A contact spring occurring along the outcrop of an anticline, from a pervious stratum overlying one that is less pervious (ASCE 1962; AGI 1997).

Approved source. When used in reference to a bottled water plant's product water, or water used in the plant's operations, means the source of the water. The source could be from a spring, artesian well, drilled well, public or community water system. Or the water could be from any other source that has been inspected and the water sampled, analyzed, and found of a safe and sanitary quality with or without treatment.

Aquifer. A body of rock that contains sufficient saturated permeable material to conduct groundwater and to yield significant quantities of water to wells and springs (Lohman 1972; AGI 1997).

Aquifer test. A test involving the withdrawal of measured quantities of water from, or addition of water to, a well and the measurement of resulting changes in *head* in the aquifer both during and after the period of discharge or addition. The results are used to estimate hydraulic properties of the aquifer (AGI 1997).

Area source. Any small source of non-natural air pollution that is released over a relatively small area but which cannot be classified as a point source. Such sources may include vehicles and other small engines, small businesses and household activities (ERA 1994).

Artesian. An adjective referring to groundwater confined under hydrostatic pressure. Etymol: French *artésian,* "of Artois", a region in northern France (AGI 1997).

Artesian flow. Movement of water from a well or spring under conditions in which the artesian head is sufficient for the water to flow above the land surface (AGI 1997).

Artesian spring. A spring from which the water flows under artesian pressure, usually through a fissure or other opening in the *confining bed* above the aquifer (AGI 1997).

Artesian water. Confined groundwater (AGI 1997).

Attenuation. The process by which a compound is reduced in concentration over time, through absorption, adsorption, degradation, dilution, and/or transformation (ERA 1994; AGI 1997).

Bacterium. A single-celled microorganism that lacks chlorophyll and an evident nucleus. Most bacteria are capable of breaking down extraneous matter; some are pathogens. Range, Pre-Cambrian to the present (AGI 1997).

Barrier spring. A spring resulting from the diversion of a flow of groundwater over or underneath an impermeable barrier in the floor of a valley (Schieferdecker 1959; AGI 1997).

Black smoker. A submarine hot spring that occurs on the deep sea floor near spreading centers. It ejects hot water, hydrogen sulfide, and other gases under great pressure and provides nutrients for local communities of unusual organisms (AGI 1997).

Boiling spring. (a) A spring, the water from which is agitated by the action of heat. (b) A spring that flows so rapidly that strong vertical eddies develop (AGI 1997).

Boiling spring. A Jamaican term for a fountaining resurgence (AGI 1997).

Border spring. Boundary spring (AGI 1997).

Bottle spring. A freshwater spring that issues through the floor of a saline lake or pool. The name is derived from the fact that fresh water can be obtained by submerging a stoppered bottle directly over the spring and then removing the stopper (AGI 1997).

Bottled water. Water that is placed in a sealed container or package, and is offered for sale for human consumption or other consumer uses. Bottled water may be with or without natural or added carbonation, and may be prepared with added flavors, extracts, and/or essences derived from a spice or fruit and comprising less than one percent by weight of the final product. Said products shall contain no sweeteners, acidulants, or additives other than said flavors, extracts, or essences.

Boundary spring. A type of gravity spring whose water issues from the lower slope of an alluvial cone (AGI 1997).

Brackish water. An indefinite term for water with a salinity intermediate between that of normal seawater (35) and that of normal fresh water (0) (AGI 1997).

British Thermal Unit (BTU). The quantity of heat required to raise the temperature of a standard pound of water from 63° to 64°F.

Brook. (a) A small stream or rivulet, commonly swiftly flowing in rugged terrain, of lesser length and volume than a creek; especially a stream that issues directly from the ground, as from a spring or seep, or that is produced by heavy rainfall or melting snow. Also, one of the smallest branches or ultimate ramifications of a drainage system. (b) A term used in England and New England for any tributary to a small river or to a larger stream. (c) A general literary term for a creek (AGI 1997).

Carbonated spring. A spring whose water contains carbon dioxide gas. This type of spring is especially common in volcanic areas (Comstock 1878; AGI 1997).

Carbonated water or sparkling water. Bottled water containing carbon dioxide (AGI 1997).

Chalybeate. An adjective applied to water strongly flavored with iron salts or to a spring yielding such water. Etymol: Greek, an ancient tribe of ironworkers in Asia Minor (AGI 1997).

Channel spring. A type of depression spring issuing from the bank of a stream that has cut its channel below the water table (AGI 1997).

Chlorination. The application of chlorine to drinking water, sewage, or industrial waste to disinfect or to oxidize undesirable compounds (ERA 1994).

Clear well. A reservoir for storing filtered water of sufficient quantity to prevent the need to vary the filtration rate with variations in demand. Also used to provide chlorine contact time for disinfection (ERA 1994).

Cold spring. A spring whose water has a temperature appreciably below the mean annual atmospheric temperature in the area; also, a nonpreferred usage for any nonthermal spring in an area having thermal springs (Meinzer 1923; AGI 1997).

Conduction. The transport of heat in static groundwater, controlled by the thermal conductivity of the geologic formation and the contained groundwater and described by a linear law relating heat flux to temperature gradient (AGI 1997).

Conduit. A passage that is filled with water under hydrostatic pressure (AGI 1997).

Confined groundwater. Groundwater under pressure significantly greater than that of the atmosphere. Its upper surface is the bottom of a confining bed (AGI 1997).

Contact spring. A type of gravity spring whose water flows to the land surface from permeable strata over less permeable or impermeable strata that prevent or retard the downward percolation of the water (Meinzer 1923; AGI 1997).

Contaminant. (a) Any physical, chemical, biological, or radiological substance or matter that has an adverse effect on air, water, or rocks (ERA 1994). (b) An undesirable substance in water, air, or rocks that is either not normally present or is an unusually high concentration of a naturally occurring substance (AGI 1997).

Contamination. The addition to water of any substance or property that changes the physical and/or chemical characteristics of the water and prevents the use or reduces its usability for ordinary purposes such as drinking, preparing food, bathing, washing, recreation, and cooling. Sometimes arbitrarily defined differently from *pollution,* but generally considered synonymous (AGI 1997).

Convection. The transport of heat by flowing groundwater (AGI 1997).

Corrosion. (a) A process of erosion whereby rocks and soil are removed or worn away by natural chemical processes, especially by the solvent action of running water, but also by other reactions such as hydrolysis, hydration, carbonation, and oxidation. (b) A term formerly used interchangeably with *corrosion* for the erosion ("gnawing away") of land or rock, including both mechanical and chemical processes. The mechanical part is now properly restricted to "corrasion" and the chemical to "corrosion" (AGI 1997).

Cubic Feet Per Minute (CFM). A measure of the volume of a substance flowing through air within a fixed period of time. With regard to indoor air, refers to the amount of air, in cubic feet, that is exchanged with indoor air in a minute's time, i.e., the air exchange rate. Also applies to liquid flows (ERA 1994).

Debouchure. (a) Debouchment. (b) The place where an underground stream reaches the surface; the opening from which a spring issues. (c) The point in a cave where a tubular passage connects with a larger passage or chamber (AGI 1997).

Decontamination. A variety of processes used to clean equipment that has contacted formation material, groundwater, or surface water that is known to be or suspected of being contaminated (AGI 1997).

Deep-well injection. Deposition of raw or treated, filtered hazardous waste by pumping it into deep wells, where it is contained in the pores of permeable subsurface rock (ERA 1994).

Depression spring. A type of gravity spring, with its water flowing onto the land surface from permeable material as a result of the land surface sloping down to the water table (Meinzer 1923; AGI 1997).

Dike spring. A spring issuing from the contact between a dike composed of an impermeable rock, such as basalt or dolerite, and a permeable rock into which the dike was intruded (AGI 1997).

Dimictic. Said of a lake with two yearly overturns or periods of circulation, such as a deep freshwater lake in a temperate climate, with overturns in the spring and fall (AGI 1997).

Dimple spring. Depression spring (AGI 1997).

Discharge. The rate of flow of surface water or groundwater at a given moment, expressed as volume per unit of time (AGI 1997).

Discharge area. An area in which subsurface water, including both groundwater and vadose water, is discharged to the land surface, to bodies of surface water, or to the atmosphere (AGI 1997).

Disposal. Final placement or destruction of toxic, radioactive, or other wastes; surplus or banned pesticides or other chemicals; polluted soils; and drums containing hazardous materials from removal actions or accidental releases. Disposal may be accomplished through use of approved secure landfills, surface impoundments, land farming, deep well injection, ocean dumping, or incineration (ERA 1994).

Dissolved solids. Disintegrated organic and inorganic material in water. Excessive amounts make water unfit to drink or use in industrial processes (ERA 1994).

Distilled water. Water which has been produced by a process of distillation and meets the definition of purified water in the most recent edition of the *United States Pharmacopeia* and bottled.

Downgradient. The direction that groundwater flows; similar to "downstream" for surface water (ERA 1994).

Drainage basin. The area of land that drains water, sediment, and dissolved materials to a common outlet at some point along a stream channel (ERA 1994).

Drinking water. Water obtained from an approved source that has at minimum, undergone treatment consisting of filtration (activated carbon or particulate) and ozonation, or an equivalent disinfection process and bottled.

Dump. A site used to dispose of solid wastes without environmental controls (ERA 1994).

Ebbing-and-flowing spring. Periodic spring (AGI 1997).

Ecology. The study of the relationships between organisms and their environment, including the study of communities, patterns of life, natural cycles, relationships of organisms to each other, biogeography, and population changes (AGI 1997).

Economic yield. The maximum estimated rate at which water may be withdrawn from an aquifer without creating a deficiency or affecting the quality of the supply (AGI 1997).

Ecosystem. The interacting system of a biological community and its nonliving environmental surroundings (ERA 1994).

Effluent. (a) A surface stream that flows out of a lake (e.g. an outlet), or a stream or branch that flows out of a larger stream (e.g. a distributary). (b) A liquid discharged as waste, such as contaminated water from a factory or the outflow from a sewage works; water discharged from a storm sewer or from land after irrigation (AGI 1997).

Enthalpy. A thermodynamic variable (typically symbolized as H) that is defined as the sum of the internal energy of a body plus the product of its volume multiplied by the pressure, measured in joules (J) (AGI 1997).

Environment. The sum of all external conditions affecting the life, development and survival of an organism (ERA 1994).

Eye. The opening from which the water of a spring flows out onto the land surface (AGI 1997).

Fault-dam spring. Fault spring (AGI 1997).

Fault spring. A spring flowing onto the land surface from a fault that brings a permeable bed into contact with a less permeable bed (AGI 1997).

Filtration spring. A spring whose water percolates from numerous small openings in permeable material. It may have either a small or a large discharge (AGI 1997).

Fissure spring. A spring issuing from a crack or joint. Several springs of this type may flow out along the same fissure line (AGI 1997).

Fluoridation water. Water containing fluoride and bottled. The label shall specify whether the fluoride is naturally occurring or added. Any water which meets the definition of this paragraph shall contain not less than 0.8 mg/l fluoride ion and otherwise comply with the Food and Drug Administration (FDA) quality standards in Section 103.35(d)(2) of Title 21 of the Code of Federal Regulations (CFR).

Fount. A fountain or spring of water (AGI 1997).

Fountain. (a) A spring of water issuing from the Earth. (b) The source or head of a stream (AGI 1997).

Fountainhead. The fountain or spring that is the source of a stream (AGI 1997).

Fracture spring. A spring whose water flows from joints or other fractures, in contrast to the numerous small openings from which a filtration spring flows (AGI 1997).

Fresh water. (a) Water containing less than 1 000 mg/l of dissolved solids; generally, water with more than 500 mg/l is undesirable for drinking and for many industrial uses (Solley et al. 1983). (b) In general usage, the water of streams and lakes unaffected by salt water or salt-bearing rocks (AGI 1997).

Gamma radiation. Electromagnetic radiation from an atomic nucleus, often accompanying emission of alpha particles and beta particles (AGI 1997).

Geochemical exploration. The search for economic mineral deposits or petroleum by detection of abnormal concentrations of elements or hydrocarbons in surficial materials or organisms, usually accomplished by instrumental, spot-test, or "quickie" techniques that may be applied in the field (AGI 1997).

Geochemistry. As defined by Goldschmidt (1954), the study of the distribution and amounts of the chemical elements in minerals, ores, rocks, soils, water, and the atmosphere, and the study of the circulation of the elements in nature, on the basis of the properties of their atoms and ions; also, the study of the distribution and abundance of isotopes, including problems of nuclear frequency and stability in the universe. A major concern of geochemistry is the synoptic evaluation of the abundances of the elements in the Earth's crust and in major classes of rocks and minerals (AGI 1997).

Geophysical exploration. The use of geophysical techniques, e.g. electric, gravity, magnetic, seismic, or thermal, in the search for economically valuable hydrocarbons, mineral deposits, or water supplies, or to gather information for engineering projects (AGI 1997).

Geothermal gradient. The rate of change of temperature in the Earth with depth measured in °C/m or °C/km. The gradient differs from place to place depending on the heat flow in the region and the thermal conductivity of the rocks. The average geothermal gradient in the Earth's crust approximates 25 °C/km of depth (AGI 1997).

Geothermal heat flow. The amount of heat energy leaving the Earth per cm^2/s, measured in calories/(cm^2 s), The mean heat flow for the Earth is about 1.5 ±0.15 microcalories/(cm^2 s), or about 1.5 heat-flow units. Heat-flow measurements in igneous rocks have shown a linear correlation between heat production in rocks and surface heat flow. The heat production is due to the presence of uranium, potassium, and thorium.

Geyser. A type of hot spring that intermittently erupts jets of hot water and steam, the result of groundwater coming into contact with rock or steam hot enough to create steam under conditions preventing free circulation; a type of intermittent spring (AGI 1997).

Geyserite. A synonym of *siliceous sinter*, used especially for the compact, loose, concretionary, scaly, or filamentous incrustation of opaline silica deposited by precipitation from the waters of a geyser (AGI 1997).

Geyser basin. A valley that contains numerous springs, geysers, and steaming fissures fed by the same groundwater flow (Schieferdecker 1959; AGI 1997).

Geyser cone. A low hill or mound built up of siliceous sinter around the orifice of a geyser. The term is sometimes mistakenly applied to an algal growth on objects (such as wooden snags) occurring along the shores of some Tertiary lakes (AGI 1997).

Geyser crater. The bowl- or funnel-shaped openings of the geyser pipe, which often contains a *geyser pool* (AGI 1997).

Geyser pipe. The narrow tube or well of a geyser extending downward from the *geyser pool* (AGI 1997).

Geyser pool. The comparatively shallow pool of heated water ordinarily contained in a *geyser crater* at the top of a *geyser pipe* (AGI 1997).

Gravity spring. A spring issuing from the point where the water table and the land surface intersect; an outcrop of the water table (AGI 1997).

Groundwater. (a) That part of the subsurface water that is in the *saturated zone*, including underground streams. (b) Loosely, all *subsurface water* as distinct from surface water (AGI 1997).

Groundwater discharge. (a) Release of water from the *saturated zone*. (b) The water or quantity of water released (AGI 1997).

Groundwater level. (a) A synonym of *water table*. (b) The elevation of the water table or another potentiometric surface at a particular place or in a particular area, as represented by the level of water in wells or other natural or artificial openings or depressions communicating with the *saturation zone* (AGI 1997).

Groundwater recession curve. The part of a stream hydrograph supposedly representing the inflow of groundwater at a decreasing rate after surface runoff of the channel has ceased. Because the base runoff to the stream may include some water that had been stored in lakes and swamps rather than in the ground, the lower recession curve cannot be assumed to represent groundwater only (AGI 1997).

Groundwater runoff. The portion of the runoff that has recharged the groundwater system and has later been discharged into a stream channel or other surface-water body. It is the principal source of base flow or dry-weather flow of streams unregulated by surface storage. Such flow is sometimes called *groundwaterflow* (Rogers 1981; AGI 1997).

Groundwater withdrawal. The process of withdrawing groundwater from a source; also, the quantity of water withdrawn.

Hardpan spring. A contact spring occurring above a layer of hardpan and flowing from a perched water table (AGI 1997).

Hard water. Water that does not lather readily when used with soap, and that forms a scale in containers in which it has been allowed to evaporate; water with more than 60 mg/l of hardness-forming constituents, expressed as $CaCO_3$ equivalent (AGI 1997).

Hazardous substance. (a) Any material that poses a threat to human health and/or the environment. Typical hazardous substances are toxic, corrosive, ignitable, explosive or chemically reactive. (b) Any substance designated by ERA to be reported if a designated quantity of the substance is spilled in the waters of the United States or if otherwise emitted to the environment (ERA 1994).

Hillside spring. *Contact spring* (AGI 1997).

Hot spring. A *thermal spring* whose temperature is above that of the human body (AGI 1997).

Hydrogeological cycle. The natural process recycling water from the atmosphere down to (and through) the earth and back to the atmosphere (ERA 1994).

Hydrogeology. (a) The science that deals with subsurface waters and with related geologic aspects of surface waters. Also used in the more restricted sense of *groundwater geology* only. The term was defined by Mead (1919) as the study of the laws of the occurrence and movement of subterranean waters. More recently it has been used interchangeably with *geohydrology*. (b) The study of the laws governing (1) the movement of groundwater; (2) the mechanical, chemical, and thermal interaction of this water with the porous medium; and (3) the transport of energy and chemical constituents by the flow of groundwater (Domenico and Schwartz 1990; AGI 1997).

Hydrograph. A graph showing stage, flow, velocity, or other characteristics of water with respect to time (Langbein and Iseri 1960). A stream hydrograph commonly shows rate of flow; a groundwater hydrograph, water level or head (AGI 1997).

Hydrologic budget. An accounting of the inflow to, outflow from, and storage in a hydrologic unit such as a drainage basin, aquifer, soil zone, lake, or reservoir (Langbein and Iseri 1960); the relationship between evaporation, precipitation, runoff, and the change in water storage, expressed by the hydrologic equation (AGI 1997).

Hydrologic cycle. The constant circulation of water from the sea, through the atmosphere, to the land, and its eventual return to the atmosphere by way of transpiration and evaporation from the sea and the land surfaces (AGI 1997).

Hydrologic properties. Those properties of a rock that govern the entrance of water and the capacity to hold, transmit, and deliver water, e.g. porosity, effective porosity, specific retention, permeability, and direction of maximum and minimum permeability (AGI 1997).

Hydrologic system. A complex of related parts – physical, conceptual, or both – forming an orderly working body of hydrologic units and their man-related aspects such as the use, treatment and reuse, and disposal of water and the costs and benefits thereof, and the interaction of hydrologic factors with those of sociology, economics, and ecology (AGI 1997).

Hydrology. (a) The science that deals with global water (both liquid and solid), its properties, circulation, and distribution, on and under the Earth's surface and in the atmosphere, from the moment of its precipitation until it is returned to the atmosphere through evapotranspiration or is discharged into the ocean. In recent years the scope of hydrology has been expanded to include environmental and economic aspects. At one time there was a tendency in the US (as well as in Germany) to restrict the term "hydrology" to the study of subsurface waters (DeWiest 1965). (b) The sum of the factors studied in hydrology; the hydrology of an area or district (AGI 1997).

Hydrothermal. Of or pertaining to hot water, to the action of hot water, or to the products of this action, such as a mineral deposit precipitated from a hot aqueous solution, with or without demonstrable association with igneous processes; also, said of the solution itself. "Hydrothermal" is generally used for any hot water but has been restricted by some to water of magmatic origin (AGI 1997).

Ice dam. A river obstruction formed of floating blocks of ice that may cause ponding and widespread flooding during spring and early summer (AGI 1997).

Infiltration. (a) The flow of a fluid into a solid substance through pores or small openings; specifical: the movement of water into soil or porous rock. (b) The process of falling rain or melting snow entering a soil or rock across its interface with the atmosphere, or water from a stream entering its streambed across the stream-streambed interface. Infiltration connotes flow into a material, in contrast to *percolation,* which connotes flow through a material (AGI 1997).

Intermittent spring. A spring that discharges only periodically. A *geyser* is a special type of intermittent spring (Meinzer 1923; AGI 1997).

Intermitting spring. *Intermittent spring* (AGI 1997).

Isotope. One of two or more species of the same chemical element, i.e. having the same number of protons in the nucleus, but differing from one another by having a different number of neutrons. The isotopes of an element have slightly different physical and chemical properties, owing to their mass differences, by which they can be separated (AGI 1997).

Karst. A type of topography that is formed on limestone, gypsum, and other rocks, primarily by dissolution, and that is characterized by sinkholes, caves, and underground drainage. Etymol: German, from the Yugoslavian territory Krš; type locality, a limestone plateau in the Dinaric Alps of northwestern Yugoslavia and northeastern Italy. First published on a topographic map, *Ducatus Carnioliae*, in 1774 (AGI 1997).

Karst spring. A spring emerging from karstified limestone (AGI 1997).

Lagoon. (a) The basin of a hot spring; also, the pool formed by a hot spring in such a basin. Etymol: Italian *lagone*, "large lake". (b) A perennial brine pool near the margin of an alkaline lake; e.g., near Lake Magadi in southern Kenya; (c) Any shallow artificial pond or other water-filled excavation for the natural oxidation of sewage or disposal of farm manure, or for some decorative or aesthetic purpose (AGI 1997).

Landfill. (a) Sanitary landfills are land disposal sites for non-hazardous solid wastes at which the waste is spread in layers, compacted to the smallest practical volume, and cover material applied at the end of each operating day. (2) Secure chemical landfills are disposal sites for hazardous waste. They are selected and designed to minimize the chance of release of hazardous substances into the environment (ERA 1994).

Leachate. (a) A solution obtained by leaching; e.g. water that has percolated through soil containing soluble substances and that contains certain amounts of these substances in solution. (b) Groundwater containing pollutants leached from buried solids, especially from waste in landfills (AGI 1997).

Leaching. (a) The separation, selective removal, or dissolving-out of soluble constituents from a rock or orebody by the natural action of percolating water. (b) The removal in solution of nutritive or harmful constituents (such as mineral salts and organic matter) from an upper to a lower soil horizon by the action of percolating water, either naturally (by rainwater) or artificially (by irrigation). (c) The extraction of soluble metals or salts from an ore by means of slowly percolating solutions; e.g. the separation of gold by treatment with a cyanide solution. (d) The chemical process of the removal or extraction of soluble compounds from solid wastes that are exposed to weathering; leached fluids can contaminate groundwater and surface-water systems (AGI 1997).

Log. A continuous record as a function of depth, usually graphic and plotted to scale on a narrow paper strip, of observations made on the rocks and fluids of the geologic section exposed in a well bore (AGI 1997).

Maximum Contaminant Level (MCL). The maximum permissible level of a contaminant in water delivered to any user of a public water system. MCLs are enforceable standards (ERA 1994).

Medicinal spring. A spring of reputed therapeutic value due to the substances contained in its water (AGI 1997).

Microorganism. Living organisms so small that individually they can usually only be seen through a microscope.

Mineral spring. A spring whose water contains enough mineral matter to give it a definite taste, in comparison to ordinary drinking water, especially if the taste is unpleasant or if the water is regarded as having therapeutic value. This type of spring is often described in terms of its principal characteristic constituent; e.g. *salt spring* (AGI 1997).

Mineral water. Water that contains naturally or artificially supplied mineral salts or gases (e.g. carbon dioxide) (AGI 1997).

Mitigation. Measures taken to reduce adverse impacts on the environment (ERA 1994).

Modeling. An investigative technique using a mathematical or physical representation of a system or theory that accounts for all, or some of its known properties. Models are often used to test the effect of changes of system components on the overall performance of the system.

Monitoring. Periodic or continuous surveillance or testing to determine the level of compliance with statutory requirements and/or pollutant levels in various media or in humans, plants, and animals (ERA 1994).

Monitoring well. (a) A well drilled to obtain water quality samples or measure groundwater levels. (b) A Well drilled at a hazardous waste management facility or Superfund site to collect groundwater samples for the purpose of physical, chemical, or biological analysis to determine the amounts, types, and distribution of contaminants in the groundwater beneath the site (ERA 1994).

Mound spring. A spring characterized by a mound at the place where it flows onto the land surface. According to Meinzer (1923), "mound springs may be produced, wholly or in part, by the precipitation of mineral matter from the spring water; or by vegetation and sediments blown in by the wind – a method of growth common in arid regions" (AGI 1997).

National Interim Primary Drinking Water Regulations. Commonly referred to as NIPDWRs (EPA 1994).

National Secondary Drinking Water Regulations. Commonly referred to as NSDWRs (ERA 1994).

Natural water. Means spring, mineral, artesian, or well water which is derived from an underground formation, and is not derived from a municipal system or public water supply and bottled.

Nitrate. A mineral compound characterized by a fundamental anionic structure of NO_3. Soda niter, $NaNO_3$, and niter, KNO_3, are nitrates (AGI 1997).

Nonthermal spring. A spring in which the temperature of the water is not appreciably above the mean atmospheric temperature in the vicinity. A spring whose temperature approximates the mean annual temperature, or a *cold spring,* is considered a nonthermal spring (Meinzer 1923; AGI 1997).

Outcrop spring. *Contact spring* (AGI 1997).

Overflow spring. A type of *contact spring* that develops where a permeable deposit dips beneath an impermeable mantle. Groundwater overflows onto the land surface at the edge of the impermeable stratum (AGI 1997).

Perched spring. A spring whose source of water is a body of perched groundwater (AGI 1997).

Perennial spring. A spring that flows continuously, as opposed to an *intermittent spring* or a *periodic spring* (AGI 1997).

Periodic spring. A spring that ebbs and flows, owing to natural siphon action. Such springs issue mainly from carbonate rocks, in which solution channels form the natural siphons. It is distinguished from a geyser by its temperature – that of ordinary groundwater – and general lack of gas emission (AGI 1997).

Phreatic water. A term that was originally applied only to water that occurs in the upper part of the saturated zone under watertable conditions, but has come to be applied to all water in the saturated zone, thus making it an exact synonym of *groundwater* (Meinzer 1923; AGI 1997).

Pocket spring. A spring at the lower edge of a deposit of pervious material filling an irregular depression of the bedrock (AGI 1997).

Pocket valley. A valley whose head is enclosed by steep walls at the base of which underground water emerges as a spring (AGI 1997).

Pollutant. Generally, any substance introduced into the environment that adversely affects the usefulness of a resource (ERA 1994).

Pollution. Generally, the presence of matter or energy whose nature, location or quantity produces undesired environmental effects. Under the Clean Water Act, for example, the term is defined as the man-made or man-induced alteration of the physical, biological, and radiological integrity of water (ERA 1994).

Pool spring. A spring fed from a deep source, sometimes related to a fault, and forming a pool. A pool spring may develop the shape of a jug, because a peripheral platform is developed over the water by vegetation and sediments blown in by the wind (Meinzer 1923; AGI 1997).

Potable water. Water that is safe and palatable for human use; *fresh water* in which any concentrations of pathogenic organisms and dissolved toxic constituents have been reduced to safe levels, and which is, or has been treated so as to be, tolerably low in objectionable taste, odor, color, or turbidity and of a temperature suitable for the intended use (AGI 1997).

Precipitation. Water that falls to the surface from the atmosphere as rain, snow, hail, or sleet. It is measured as a liquid-water equivalent regardless of the form in which it fell (AGI 1997).

Pulsating water. *Geyser* (AGI 1997).

Purified water. Means water produced by distillation, deionization, reverse osmosis, or other suitable process and bottled and that meets the definition of purified water in the most recent edition of the United State Pharmacopeia. Only water which meets the definition of this paragraph and is vaporized, then condensed, may be labeled "distilled water."

Radioactive spring. A spring whose water has a high and readily detectable radioactivity (AGI 1997).

Radon. A colorless naturally occurring, radioactive, inert gas formed by radioactive decay of radium atoms in soil or rocks.

Recharge. The processes involved in the addition of water to the saturated zone, naturally by precipitation or runoff, or artificially by spreading or injection; also, the amount of water added (AGI 1997).

Recharge area. An area beneath which water reaches the saturated zone following *infiltration* and *percolation*. Beneath it, downward components of hydraulic head exist and groundwater moves downward into deeper parts of the aquifer (Fetter 1994; AGI 1997).

Recommended Maximum Contaminant Level (RMCL). The maximum level of a contaminant in drinking water at which no known or anticipated adverse affect on human health would occur, and that includes an adequate margin of safety. Recommended levels are nonenforceable health goals (ERA 1994).

Safe yield. A synonym of *economic yield* that is also applied to surface-water supplies. The volume of water that can be withdrawn annually from an aquifer (or groundwater basin or system) without (1) exceeding average annual recharge; (2) violating water rights; (3) creating uneconomic conditions for water use; or (4) creating undesirable side effects, such as subsidence or saline-water intrusion. Use of the term is discouraged because the feasible rate of withdrawal depends on the location of wells in relation to aquifer boundaries and can rarely be estimated in advance of development (AGI 1997).

Salt spring. A *mineral spring* whose water contains a large quantity of common salt; a spring of salt water (AGI 1997).

Sand boil. (a) A spring that bubbles through a river levee, with an ejection of sand and water, as a result of water in the river being forced through permeable sands and silts below the levee during flood stage. (b) A spring in an excavation, caused by unbalanced hydrostatic pressure (Rogers 1981; AGI 1997).

Saturated zone. A subsurface zone in which all the interstices are filled with water under pressure greater than that of the atmosphere. Although the zone may contain gas-filled interstices or interstices filled with fluids other than water, it is still considered saturated. This zone is separated from the *unsaturated zone* (above) by the *water table* (AGI 1997).

Scarp-foot spring. A spring that flows onto the land surface at or near the foot of an escarpment (AGI 1997).

Seepage spring. This term may be used as a synonym of *filtration spring*, but is often limited to springs of small discharge (Meinzer 1923; AGI 1997).

Seepage. (a) The act or process involving the slow movement of water or other fluid through a porous material such as soil. (b) The amount of fluid that has been involved in seepage (AGI 1997).

Siliceous sinter. The lightweight porous opaline variety of silica, white or nearly white, deposited as an incrustation by precipitation from the waters of geysers and hot springs. The term has been applied loosely to any deposit made by a geyser or hot spring (AGI 1997).

Sinkhole spring. A spring that flows from a sinkhole in karst terrain (Fetter 1994; AGI 1997).

Siphon. A water conduit in the shape of an inverted U, in which the water is in hydrostatic equilibrium (AGI 1997).

Spa. (a) *Medicinal spring*. (b) A place where such springs occur, often a resort area or hotel. – The name is derived from that of a town in eastern Belgium where medicinal springs occur (AGI 1997).

Spring. A place where groundwater flows naturally from a rock or the soil onto the land surface or into a body of surface water. Its occurrence depends on the nature and relationship of rocks, especially permeable and impermeable strata, on the position of the water table, and on the topography (AGI 1997).

Spring-fed lake. *Spring lake* (AGI 1997).

Spring lake. (a) A lake, usually of small size, that is created by the emergence of a spring or springs, especially one having visibly flowing springs on its shore or springs rising from its bottom. (b) A lake that receives all or part of its waters directly from a spring (AGI 1997).

Springline. A line of springs marking the intersection of the water table with the land surface, as at the foot of an escarpment or along the base of a permeable bed at its contact with a *confining bed* (AGI 1997).

Spring mound. A roughly circular mound of sand and silt, 5–6 m high and 10–12 m across, formed by a spring rising to the surface and depositing its load (AGI 1997).

Spring water. Water derived from a spring (Rogers 1981; AGI 1997).

Submarine spring. A large offshore emergence of fresh water, usually associated with a coastal karst area but sometimes with lava tubes (AGI 1997).

Subsurface water. Water in the lithosphere in solid, liquid, or gaseous form. It includes all water beneath the land surface and beneath bodies of water (AGI 1997).

Sulphur spring. A spring containing sulphur water (AGI 1997).

Surface water. (a) All waters on the surface of the Earth, including fresh and salt water, ice, and snow. (b) A water mass of varying salinity and temperature, occurring at the ocean surface and having a thickness of 300 m or less (AGI 1997).

Talus spring. A spring occurring at the base of a *talus slope* and originating from water falling upon or seeping into the slope (AGI 1997).

Terrace spring. A spring that has built up at its mouth a series of terraces or basins through deposition of material from the flowing water (Rogers 1981; AGI 1997).

Thermal gradient. The rate of change of temperature with distance. When applied to the Earth, the term *geothermal gradient* may be used.

Thermal spring. A spring whose water temperature is appreciably higher than the local mean annual atmospheric temperature. A thermal spring may be a *hot spring* or a *warm spring* (Meinzer 1923; AGI 1997).

Thermal water. Water, generally of a spring or geyser, whose temperature is appreciably above the local mean annual air temperature (AGI 1997).

Tinaja. (a) A term used in southwest USA for a *water pocket* developed below a waterfall, especially when partly filled with water. (b) A term used loosely in New Mexico for a temporary pool, and for a spring too feeble to form a stream (AGI 1997).

Total Dissolved Solids (TDS). All material that passes the standard glass river filter; now called total filterable residue. Term is used to reflect salinity (ERA 1994).

Total Suspended Solids (TSS). A measure of the suspended solids in wastewater, effluent, or water bodies, determined by tests for "total suspended non-filterable solids" (ERA 1994).

Toxic pollutants. Materials that cause death, disease, or birth defects in organisms that ingest or absorb them. The quantities and exposures necessary to cause these effects can vary widely (ERA 1994).

Tracer. Any substance that is used in a process to trace its course, specifcally radioactive material introduced into a chemical, biological, or physical reaction (AGI 1997).

Tubular spring. A gravity spring or artesian spring whose water issues from rounded openings, such as lava tubes or solution channels (AGI 1997).

Transpiration. The process by which water absorbed by plants, usually through the roots, is evaporated into the atmosphere from the plant surface (AGI 1997).

Unconfined aquifer. An aquifer having a *water table*; an aquifer containing *unconfined groundwater* (AGI 1997).

Unconfined groundwater. Groundwater that has a water table, i.e. water not confined under pressure beneath a confining bed (AGI 1997).

Unsaturated zone. A subsurface zone containing water under pressure less than that of the atmosphere, including water held by capillarity; and containing air or gases generally under atmospheric pressure. This zone is limited above by the land surface and below by the surface of the *saturated zone*, i.e., the *water table* (AGI 1997).

Uranium. A radioactive heavy metal element used in nuclear reactors and the production of nuclear weapons. Term refers usually to ^{238}U, the most abundant radium isotope, although a small percentage of naturally occurring uranium is ^{235}U.

Valley spring. A type of *depression spring* issuing from the side of a valley at the outcrop of the water table (AGI 1997).

Vauclusian spring. A fountaining spring in a karst region, generally of exceptionally large discharge. Etymol: from Fontain de Vaucluse in southern France (AGI 1997).

Volcanic spring. A spring with water derived from considerable depths and brought to the surface by volcanic forces (Rogers 1981; AGI 1997).

Warm spring. A *thermal spring* whose temperature is appreciably above the local mean annual atmospheric temperature, but below that of the human body (Meinzer 1923; AGI 1997).

Waste. (a) Loose material resulting from weathering by mechanical and chemical means, and moved down sloping surfaces or carried short distances by streams to the sea; especially *rock waste*. (b) Any solid or liquid material generated by human activity that has no economic value, usually the result of the manufacture, mining, or processing of a material to produce an economic product (AGI 1997).

Waterhole. (a) A natural hole, hollow, or small depression that contains water; especially in an arid or semiarid region. (b) A spring in the desert. (c) A natural or artificial pool, pond, or small lake (AGI 1997).

Water pollution. The presence in water of enough harmful or objectionable material to damage the water's quality (ERA 1994).

Water Quality Standards. State-adopted and EPA-approved ambient standards for water bodies. The standards cover the use of the water body and the quality criteria which must be met to protect the designated use or uses (EPA 1994).

Water well. (a) A well that extracts water from the saturated zone or that yields useful supplies of water. (b) A well that obtains groundwater information or that replenishes groundwater. (c) A well drilled for oil but yielding only water (AGI 1997).

Weeping spring. A spring of small yield (AGI 1997).

Well. (a) An artificial excavation (pit, hole, tunnel), generally cylindrical in form and often walled in, sunk (drilled, dug, driven, bored, or jetted) into the ground to such a depth as to penetrate water-yielding rock or soil and to allow the water to flow or to be pumped to the surface; a water well. (b) A term originally applied to a natural spring or to a pool formed or fed from a spring; especially a mineral spring. (c) A term used chiefly in the plural form for the name of a place where mineral springs are located or of a health resort featuring marine or freshwater activities; a spa (AGI 1997).

Wellhead. The source from which a stream flows; the place in the ground where a spring emerges (AGI 1997).

Wellstrand. A Scottish term for a stream flowing from a spring (AGI 1997).

Well water. Water obtained from a well; water from the saturated zone or from a perched aquifer; *phreatic water* (AGI 1997).

Wetlands. A general term for a group of wet habitats, in common use by specialists in wildlife management. It includes areas that are permanently wet and/or intermittently water-covered, especially coastal marshes, tidal swamps and flats, and associated pools, sloughs, and bayous (AGI 1997).

Yellow-Boy. Iron oxide flocculent (clumps of solids in waste or water); usually observed as orange-yellow deposits in surface streams with excess iron content (EPA 1994).

References

ASCE (American Society of Civil Engineers. Hydraulics Division. Committee on Hydraulic Structures) (1962) Nomenclature for hydraulics. Blaisdell FW et al. (eds), American Society of Civil Engineers, New York, 501 p (Manuals and reports on engineering practice, no. 43)

Comstock TB (1878) An outline of general geology, with copious references. University Press, Ithaca

DeWiest RJM (1965) Geohydrology. John Wiley & Sons., New York

Domenico, Patrick A, Schwartz, Franklin W (1990) Physical and chemical hydrogeology. John Wiley & Sons., New York

Environmental Protection Agency (1994) Terms of environment, glossary, abbreviations, and acronyms. EPA 175-B-94-015, April 1994, Communications, Education, and Public Affairs (1704), U.S. Environmental Protection Agency

Fetter CW (1994) Applied hydrogeology. MacMillan College Publishing Co.

Goldschmidt VM (1954) Geochemistry. Clarendon Press, Oxford

Jackson JA (1997) Glossary of geology, 4th edition. America Geological Institute, Alexandria

Langbein WB, Iseri KT (1960) General introduction and hydrologic definitions. U.S. Geological Survey, Water-supply Paper 1541-A, 29 p. Manual of hydrology:pt. 1

Lohman SW et al. (1972) Definitions of selected ground-water terms – revisions and conceptual refinements. U.S. Geological Survey, Water-Supply Paper 1988

Meinzer OE (1923) Outline of ground-water hydrology, with definitions. U.S. Geological Survey, Water-supply Paper 494

Rogers BC et al. (1981) Glossary – water and wastewater control engineering. American Public Health Association, American Society of Civil Engineers, American Water Works Association, and Water Pollution Control Federation

Schieferdecker AAG (1959) Geological nomenclature. Gorinchem: Royal Geological and Mining Society of the Netherlands

Conversion Tables

Table 8.36. English–SI conversion table

Length	1 inch	= 2.54 cm
	1 ft	= 0.3048 m
	1 mi	= 1.609 km
Area	1 $inch^2$	= 6.4516 cm^2
	1 ft^2	= 0.0929 m^2
	1 acre	= 0.4047 ha
		= 0.4047×10^4 m^2
	1 mi^2	= 2.590 km^2
Volume	1 US ft oz	= 29.54 cm^3
	1 ft^3	= 2.832×10^{-2} m^3
		= 28.32 liter
	1 US gal	= 3.785×10^{-3} m^3
		= 3.785 liter
	1 UK gal	= 4.546×10^{-3} m^3
		= 4.546 liter
	1 US bushel	= 3.524×10^{-2} m^3
		= 35.24 liter
	1 oil barrel	= 0.156 m^3
		= 156 liter
Fluid flux	1 cubic ft/s	= 2.832×10^{-2} m^3/s
		= 28.32 liter/s
	1 US gal/min	= 6.309×10^{-5} m^3/s
		= 6.309×10^{-2} liter/s
	1 UK gal/min	= 7.576×10^{-5} m^3/s
		= 7.576×10^{-2} liter/s

Mass	1 oz	= 28.35 g
	1 lb_m	= 0.4536 kg
	1 s. ton	= 907 kg
	1 l. ton	= 1016 kg
	1 lb_m/ft^3	= 16.02 kg/m^3
Force	1 lb_f	= 4.448 N
Temperature	x °F	= (9/5) x °C + 32
Stress and pressure	1 $lb_f/foot^2$	= 47.88 Pa
	1 psi	= 6.895×10^3 Pa
	1 atm	= 1.013×10^5 Pa
	1 bar	= 10^5 Pa
		= 0.1 MPa
Work or energy	1 ft lb_f	= 1.356 J
	1 calorie	= 4.185 J
	1 BTU	= 1.055×10^3 J
Hydraulic conductivity	1 ft/s	= 0.3048 m/s
	1 US gal/(day ft^2)	= 4.720×10^{-7} m/s
Transmissivity	1 ft^2/s	= 9.290×10^{-2} m^2/s
	1 US gal/(day ft)	= 1.438×10^{-7} m^2/s
Intrinsic permeability	1 ft^2	= 9.290×10^{-2} m^2
		= 9.412×10^{10} darcy
	1 darcy	= 0.987×10^{-12} m^2

Table 8.37. Prefixes for multiplying and dividing of SI units

10^{-1}	tenth	deci	d	10^{1}	ten	deca	da
10^{-2}	hundredth	centi	c (%)	10^{2}	hundred	hecto	h
10^{-3}	thousandth	milli	m ‰	10^{3}	thousand	kilo	k
10^{-6}	millionth	micro	μ (ppm)	10^{6}	million	mega	M
10^{-9}	billionth	nano	n	10^{9}	billion	giga	G
10^{-12}	trillionth	pico	p (ppb)	10^{12}	trillion	tera	T

ppm (parts per million) – particles per million particles

ppb (parts per billion) – particles per billion particles

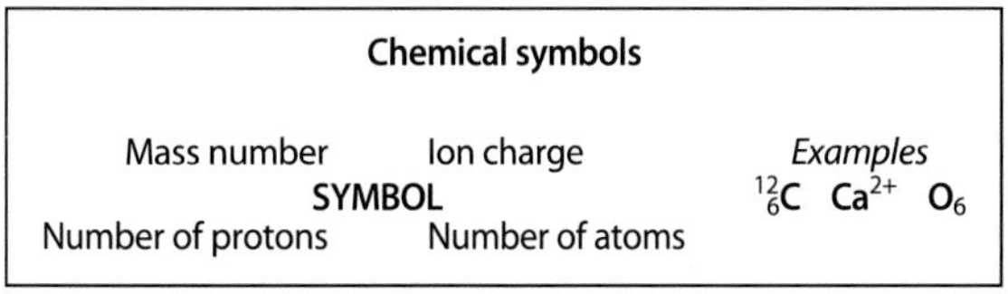

Chemical symbols

Mass number Ion charge

SYMBOL

Number of protons Number of atoms

Examples

$^{12}_{6}C$ Ca^{2+} O_6

Fig. 8.121. Chemical symbols

Index

A

B

C

F

G

H

I

J

K

L

M

N

O

P

Q

R

S

X

Y

Z

Printing: Mercedes-Druck, Berlin
Binding: Buchbinderei Lüderitz & Bauer, Berlin